REACTIONS
OF SULFUR
WITH ORGANIC
COMPOUNDS

REACTIONS OF SULFUR WITH ORGANIC COMPOUNDS

M. G. Voronkov
N. S. Vyazankin
E. N. Deryagina
A. S. Nakhmanovich
V. A. Usov

Institute of Organic Chemistry
Siberian Division of the
Academy of Sciences of the USSR
Irkutsk, USSR

Edited by

J. S. Pizey

Formerly of University of Aston
Birmingham, England

Consultants Bureau • New York and London

Library of Congress Cataloging in Publication Data

Main entry under title:

Reactions of sulfur with organic compounds.

Includes bibliographical references and index.
1. Organosulphur compounds. I. Voronkov, M. G. (Mikhail Grigorévich), 1921–
II. Pizey, J. S.
QD412.S1R37   1984   547′.06   84-4989
ISBN-13: 978-1-4684-0681-8   e-ISBN-13: 978-1-4684-0679-5
DOI: 10.1007/978-1-4684-0679-5

This book is published under an agreement with the
Copyright Agency of the USSR (VAAP).

© 1987 Consultants Bureau, New York
Softcover reprint of the hardcover 1st edition 1987
A Division of Plenum Publishing Corporation
233 Spring Street, New York, N.Y. 10013

Foreword

This book, written by the eminent Russian organic chemist
Professor M. G. Voronkov, exhaustively deals with the reactions of
sulfur with organic compounds. It is the only book on the subject
which covers both work readily available to Western scientists and
research carried out by Russian workers. Since much of the work
discussed in this volume is not readily accessible to Western scientists, it will enable them to find hitherto "unpublished" research.
In addition, Professor Voronkov has dealt with many reactions and
processes of interest to the industrial chemist; e.g., there is an
interesting section on sulfur dyes and related compounds, and the
chemistry of vulcanization is lucidly discussed. Hence this book
should be of value to both academic and industrial chemists inter-
ested in the chemistry and reactions of sulfur, and also to those
synthetic chemists studying new pathways to organic compounds. The
text has recently been updated and revised, and there are numerous
references to publications up to and including 1980. An exhaustive
index is provided to enable the reader to refer rapidly to any of
the compounds, reactions, or processes discussed in the text.

J. S. Pizey

Preface

This monograph is an attempt to present, classify, and critically discuss our knowledge of the reactions of elemental sulfur with organic compounds which has accumulated during the past two centuries. It also summarizes the results of the numerous studies of one of the authors (M. G. Voronkov) carried out at the Leningrad State University, Leningrad (1944-1950), and at the Institute of Organic Synthesis, Academy of Sciences of the Latvian SSR, Riga (1962-1970). These studies have led to the discovery of many new reactions of sulfur with organic compounds and to new types of sulfur-containing organic compounds.

This review is divided into eight chapters covering the basic physical (Chapter 1) and chemical properties (Chapter 2) of elemental sulfur, its reactions with hydrocarbons (Chapter 3), organic halides (Chapter 4), organic sulfur-containing compounds (Chapter 5), oxygen-containing compounds (Chapter 6), nitrogen-containing compounds (Chapter 7), and organic compounds of other elements (Chapter 8). Each chapter contains an exhaustive bibliography. An effort has been made to include all the literature published up to the beginning of 1981.

The authors have attempted to discuss the wide-ranging possibilities of the utilization of sulfur in organic chemistry and chemical technology and to turn the reader's attention to one of the very interesting, fascinating, and promising groups of organic reactions.

It is the authors' hope that this review will be favorably received and that their efforts will thus be adequately rewarded.

Irkutsk

M. G. Voronkov
N. S. Vyazankin
É. N. Deryagina
A. S. Nakhmanovich
V. A. Usov

PREFACE

Contents

4 ORGANIC HALIDES

5 ORGANIC SULFUR COMPOUNDS

6   OXYGEN-CONTAINING COMPOUNDS

7   NITROGEN-CONTAINING COMPOUNDS

8   REACTIONS OF SULFUR WITH ORGANOMETALLIC COMPOUNDS

Introduction

The present review is the first comprehensive presentation of the above-mentioned topic, although a number of reviews [1-21] and chapters in several monographs [22-28] devoted to the reactions of sulfur with various groups of organic compounds have been published during the past years.

However, all these publications are considerably out of date and, furthermore, they do not cover all the available material on this topic. Because of this, they do not provide the reader with complete information about the reactions of various types of organic compounds, especially heterocycles. Also, the information concerning the mechanisms and the practical use of these reactions available in these reviews is necessarily limited.

The oldest source reporting the action of sulfur on organic substance seems to be the Holy Writ. In the book of Genesis (chapter 19, verses 24 and 25) it is told: "Then the Lord rained upon Sodom and upon Gomorrah brimstone (sulfur) and fire from the Lord out of heaven; and he overthrew those cities, and all the plain, and all the inhabitants of the cities, and that which grew upon the ground." The catastrophe caused by burning melted sulfur may be compared, in this case, with an A-bomb explosion.

The ability of elemental sulfur to react with organic compounds with the formation of products useful for practical purposes has also been known for a very long time. However, these products did not initially play a beneficial role. Thus, for example, the excellent Byzantine "napalm" ("the Greek fire") was obtained by the sulfuration of petroleum, vegetable oils, and other organic substances [29]. In ancient Rome a product obtained by heating sulfur with olive oil and rosin was used as a pesticide in vineyards. The products obtained by the sulfuration of unsaturated vegetable oils, i.e., the so-called "sulfurous balsams", were used for medical purposes by the noted 4th-century Greek physician Aegineta [30]. These products continued to be widely used for a long time.

The reactions of sulfur with terpenes were studied for the first time in the 18th century by Homberg [31] and by Boerhaave [32]. In the same century, Scheele [33] proposed one of the first methods for the laboratory preparation of hydrogen sulfide based on heating sulfur with olive oil. Carbon disulfide was obtained for the first time by Lampadius [34] by the interaction of sulfur vapor with charcoal. In the first half of the 19th century, many chemists started to study systematically the reactions of sulfur with unsaturated oils and with oleic acid [35-38]. Thus oleic acid was the first individual organic compound whose reaction with sulfur was studied. Within the framework of these investigations, the conditions for the reaction of sulfur with unsaturated oils and the properties of the sulfurous balsams formed were studied in detail. Attempts to provide a theoretical explanation of the observed processes also appeared at that time.

During the same century, the interaction of sulfur with organic compounds became one of the processes used on an industrial scale, namely the hot vulcanization* of rubber and the preparation of factices, sulfur dyes, and goldplating agents.

The most important impetus for further investigations in the field of reactions of sulfur with organic compounds was the discovery of rubber vulcanization in 1839 (cf. [39]). The interest in this process was initially purely practical and technological. The scientific aspects of the vulcanization process remained a mystery until the present century when detailed and meticulous systematic studies of the structure of vulcanized rubber and of the vulcanization mechanism were initiated which are still being pursued at the present time.

Unfortunately the first studies of the reaction between sulfur and organic compounds were carried out on naturally occurring unsaturated macromolecules, such as rubber, terpenes, higher fatty acids, and glycerides. Their complex structure and the level of chemistry at that time made it impossible to arrive at any reliable conclusions concerning the structures of the products and the direction of sulfuration of organic compounds. The situation improved markedly at the end of the 19th century when the researchers started

---

* The term "vulcanization" was introduced by Brockdon (cf. [39]) and was derived from the word "volcano". As a rule, a volcano is thought to be associated with sulfur and heat.

to study the reactions of sulfur with simple organic compounds, such as ethylene, acetylene, stilbene, styrene, and amines. Nevertheless, until the 1940's the mechanistic knowledge of sulfuration remained underdeveloped compared to the general progress of organic chemistry. Only in the last three decades did we obtain sufficient experimental evidence to establish a satisfactory theoretical basis for the prediction of the direction of the reaction of sulfur with various groups of organic compounds and to establish possible reaction mechanisms.

In spite of the considerable interest in the reactions of sulfur with organic compounds, both scientific and industrial, these processes represent one of the insufficiently studied important groups of organic reactions. The reasons for this are manifold: the complex structure of sulfur itself and its ability to react in several different directions at the same time; evolution of hydrogen sulfide and probably also the formation of polysulfanes which accompany most of the sulfurations and which lead to various side reactions (addition, hydrogenation, condensation, polymerization, etc.); the instability of various intermediates which are often converted into tar-like materials under the reaction conditions; difficulties connected with the isolation of the final reaction products; etc. Thus, it is clear why the mechanism of the interaction of sulfur with organic compounds remains vague in many cases.

Reactions of sulfur with organic compounds are now widely used in practice. The hot vulcanization of rubber with sulfur is carried out in the presence of special accelerators and has not lost its importance [12,40]. The reaction of sulfur with organic compounds is used for the preparation of sulfur and other dyes, plastics, lubricating additives, pharmaceuticals, pesticides, fertilizers, corrosion inhibitors, and other valuable products. Dehydrogenation with sulfur is a frequently used laboratory technique for structural determinations in the chemistry of hydroaromatic compounds. Also, this technique has been patented as a method for the industrial preparation of butadiene from butane and styrene from ethylbenzene. Reactions of coal, paraffinic and olefinic hydrocarbons with sulfur form the basis for the industrial preparation of hydrogen sulfide, carbon disulfide, thiophene, and its homologs.

The reaction of sulfur with petroleum hydrocarbons is especially interesting, both from a theoretical and a practical point of view. This reaction takes place under normal conditions in nature or during the distillation and cracking of petroleum and leads to the formation of various sulfur-containing organic compounds and hydrogen sulfide. These compounds are present in crude sulfur-containing petroleum and in the corresponding distillates and residues. This reaction plays an important role in the formation of asphalt as well. A detailed investigation of the mechanism of sulfuration and dehydrogenation of organic compounds made it possible to explain the origin of sulfur-containing compounds in fuels.

Thus, the words of a well-known Russian chemist, B. V. Byzov, have not lost their meaning today, in spite of the fact that he made this statement more than fifty years ago [1]: "Perhaps there is no other research field in which Aristotle's advice, 'collect the facts first and only then relate them to each other in your mind', could be more useful."

The continuously expanding sources of cheap, varied, and some-times even surplus raw materials, i.e., sulfur-containing substances and especially elemental sulfur available from the chemical industry (the world output of elemental sulfur is about 25 million tons per year [41]) will undoubtedly stimulate further investigations in the field of the reactions of sulfur with organic compounds. One can assume that the joint efforts of an army of organic sulfur chemists, using the modern tools of organic chemistry, will prove fruitful and will lead to the synthesis of new types of sulfur-containing compounds, new chemical reactions, and new industrial processes.

REFERENCES

1. B. V. Byzov, Zh. Russ. Fiz.-Khim. Ova., 53, 1 (1921).
2. L. Szperl, Rocz. Chem., 36, 291 (1945).
3. W. Friedmann, Erdöl Teer, 6, 225, 301, 342, 359 (1930).
4. L. Ružička, Fortschr. Chem. Phys., Phys. Chem., 19, 335 (1928).
5. L. Ružička, Angew. Chem., 51, 5 (1938).
6. J. Cocata, Chem. Rev. (Jpn.), 7, 58 (1941).
7. P. Plattner, Die Chem., 55, 131 (1942).
8. W. N. Jones, Chem. Rev., 36, 291 (1945).
9. H. E. Westlake, Chem. Rev., 39, 219 (1946).
10. J. van Alphen, Angew. Chem., 66, 193 (1954).
11. F. Asinger and M. Theil, Angew. Chem., 70, 667 (1958).
12. B. A. Dogadkin and V. A. Shershnev, Usp. Khim., 30, 1013 (1961).
13. F. Asinger, W. Shäfer, K. Halcor, A. Saus, and H. Thiem, Angew. Chem., 75, 1050 (1963).
14. H. Schumann and M. Schmidt, Angew. Chem., 77, 1049 (1965); Angew. Chem., Int. Ed. Engl., 4, 1007 (1965).
15. R. Mayer and K. Gewald, Angew. Chem., 79, 298 (1967).
16. R. Mayer, Z. Chem., 16, 260 (1976).
17. M. Schmidt, Angew. Chem., 85, 474 (1973).
18. H. Juraszyk, Chem. Ztg., 98, 126 (1974).
19. P. Neumann and F. Vögtle, Chem. Ztg., 98, 138 (1974).
20. R. Mayer, Z. Chem., 16, 260 (1976).
21. S. Bleisch and R. Mayer, Wiss. Z. Tech. Univ., Dresden, 29(1), 89 (1980).
22. O. Lange, Die Schwefelfarbstoffe, ihre Herstellung und Verwendung 2. Aufl., Spamer, Leipzig (1925).
23. J. Scheiber and K. Sändig, Die Kunstlichen Harze, Stuttgart, Wissenschaftliche Verlagsgesellschaft M. B. H. (1929).

24. C. Ellis, The Chemistry of Petroleum Derivatives, Vol. 1, The Chemical Catalog Co., New York (1934).
25. C. Ellis, The Chemistry of Synthetic Resins, Vol. 2, Reinhold, New York (1935).
26. W. A. Pryor, Mechanisms of Sulfur Reactions, McGraw-Hill, New York—San Francisco—Toronto—London (1962).
27. A. Senning (ed.), Sulfur in Organic and Inorganic Chemistry, Marcel Dekker, New York, Vol. 1 (1971); Vols. 2, 3 (1972).
28. D. H. Reid (ed.), Organic Compounds of Sulfur, Selenium, and Tellurium, London Chemical Society, Vol. 1 (1970); Vol. 2 (1973); Vol. 3 (1975).
29. Gmelins Handbuch der Anorganischen Chemie, 8. Aufl., Verlag Chemie, Weinheim/Bergstr., Systemnummer 9, Schwefel, Teil A, S.1 ff. (1953).
30. P. Aegineta, Synopsis of Medicine, transl. by F. Adams, London (1844).
31. W. Homberg, Anat. Botan. Chem. Abh. Konigl. Acad. Wiss. (Paris), B.II, 345–355 (1731).
32. H. Boerhaave, Elementa Chemiae, Paris (1732).
33. C. W. Scheele, Chemische Abhandlungen von der Luft und den Feuer, Berlin, 237 (1793).
34. W. A. Lampadius, Ann. Chim. Phys., 49(1), 243 (1804); Philos. Mag., 20(1), 131 (1805).
35. W. Radig, H. Harff, G. Ulex, and F. Schoy, Pharm. Ztg., 308 (1833); Pharm. Zbl., 315 (1833).
36. W. Radig, H. Harff, and G. Ulex, Arch. Pharm., 11, 15 (1835); Pharm. Zbl., 607, 623 (1835).
37. H. Reimsch, J. Pract. Chem., 13, 36 (1838).
38. T. Anderson, Ann. Chem., 63, 370 (1847).
39. T. Hancock, Personal Narrative of the Origin and Progress of the Caoutchouc or India Rubber Manufacture in England, 2nd edn., London (1920), p. 175.
40. G. A. Blokh, Organic Accelerators for Rubber Vulcanization, Khimiya, Leningrad (1972).
41. Sulphur Institute Journal, Sulphur Institute, Vol. 9, Nos. 2–4 (1973); Vol. 10, Nos. 1, 2 (1974).

1
Structure and Physical Properties of Elemental Sulfur

1.1. ATOMIC AND MOLECULAR STRUCTURE

The sulfur atom possesses six valence electrons. Of these six electrons, the two unpaired 3p electrons are utilized to form covalent bonds with neighboring atoms:

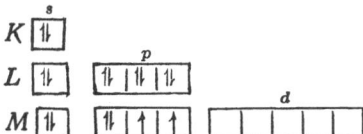

The sulfur atom can also make use of its vacant 3d orbitals and thus form hybrid dσ- and dπ-bonds which are responsible for the high stability of rings and chains consisting of sulfur atoms [1-4].

We are not going to consider in detail the form and properties of the s and p orbitals of the sulfur atom [5,6] but will study the five 3d orbitals [7]. Of these, two orbitals, d_z2 and d_{x2-y2}, take part in the formation of the σ-bonds. When the two orbitals simultaneously participate in bond formation, the central sulfur atom having twelve valence electrons becomes a hexavalent atom. Stable sulfur compounds of type (SF_6) are known [8].

Examination of atomic wave functions [9], radii functions from self-consistent field calculations, and orbital energies (the d-orbital energies are only about 10% of the p orbital energy [10,11]) fails to confirm the suggestion that the d orbitals contribute much [12,13] to the S-S bonds. This is in agreement with the data of authors [14-21]. However, the d orbital contribution to the formation of the S—S bonds greatly increases the magnetic field effect [8].

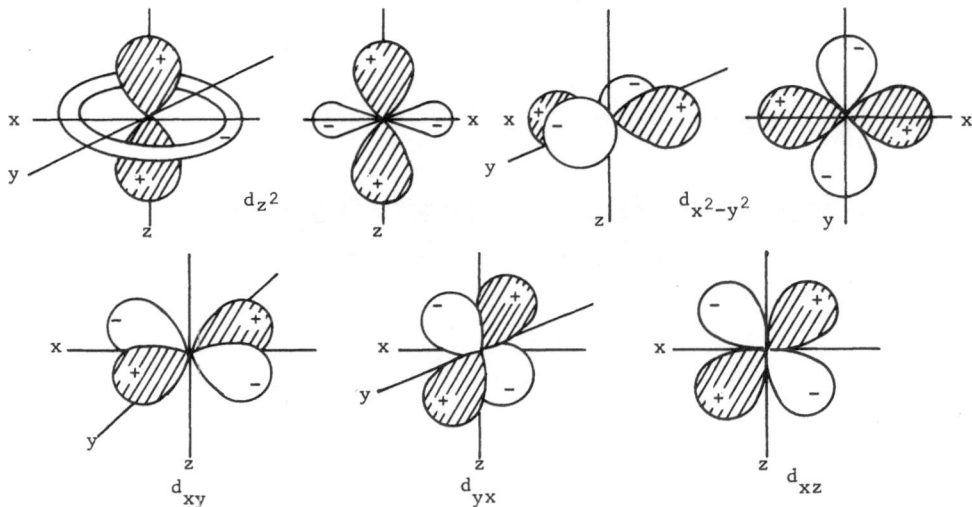

Fig. 1. 3d orbitals.

Wave mechanistic calculations of different electronic configurations of the sulfur atom have been carried out (Table 1.1) [10,11].

The number of different molecular forms of sulfur existing at various temperatures is incredibly large. The number of atoms in sulfur molecules can be anywhere from 2 to 10^6 [1,22,23]. Polyatomic sulfur molecules can possess cyclic structures or they can form polymeric chains.

Under normal conditions, the thermodynamically stable molecular form of sulfur is cyclooctasulfur [24-27]. All other molecular forms of sulfur rearrange into this form at temperatures below 95°C. The eight-membered ring of S_8 is especially stable in the crown form [4,23,28,29]:

Table 1.1. Orbital Ionization Potential (I_v), Electron Affinity (E_v), and Mulliken's Electronegativity (X) of Atomic Sulfur

Configuration	Orbital	I_v	E_v	X
s^2p^2pp	p	12.4	2.4	7.4
$(sp^3)^2(sp^3)^2sp^3sp^3$	sp^3	15.5	4.8	10.1
$(sp^2)^2(sp^2)^2sp^2\pi$	sp^2	16.3	5.4	10.9
	π	12.7	2.8	7.7

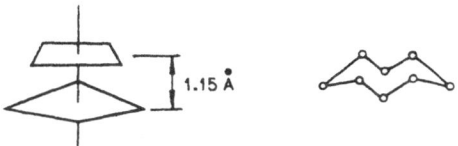

    The four upper and four lower sulfur atoms occupy apexes of two
squares lying in two parallel planes. The distance between the
planes is 1.15 Å. The squares are concentrically rotated with
respect to each other by 45°. The S-S bond lengths in S_8 are 2.06 Å
[30]. It is likely that the cyclooctasulfur molecule, S_8, can also
exist in other, thermodynamically less favorable, "twisted" forms [1,
31, 32]:

    When heated above its melting point, cyclooctasulfur polymerizes
with the formation of polymeric zig-zag-like chains in which the S-S
bond length is 2.04 Å [33]:

$$\cdot\overline{\underline{S}}\diagup{}^{\overline{\underline{S}}}\diagdown\underline{S}\diagup{}^{\overline{\underline{S}}}\diagdown\underline{S}\diagup{}^{\overline{\underline{S}}}\diagdown\underline{S}\cdot$$

1

$$|\overline{\underline{S}}\diagup{}^{\overline{S}}\diagdown\underline{\underline{S}}\diagdown{}^{\overline{\underline{S}}}\diagdown\underline{S}\diagup{}^{\overline{\underline{S}}}\diagdown\underline{S}| \qquad\qquad {}_{\ominus}|\overline{\underline{S}}\diagdown{}^{\overline{\underline{S}}}\diagdown\underline{S}\diagup{}^{\overline{\underline{S}}}\diagdown\underline{S}\diagup{}^{\overline{\underline{S}}}\diagdown\underline{S}_{\oplus}$$

2   3

    Long chains have a biradicaloid structure (1) [34-36]. Short
chains can be resonance stabilized (2) [37,38]. A contribution of
the ionic structure (3) to the structure of polymeric sulfur chains
is unlikely [39]; however, the possibility of the formation of linear
dipoles has been suggested [40].

    Cyclooctasulfur is known to undergo photochemical polymerization
[4]. In the dark, the formed "photosulfur" undergoes rapid depolymerization with regeneration of the original cyclooctasulfur. It is
likely that the formation of the stable cyclic structure S_8 is due
to the thermally induced interaction of the sulfur atom at the end
of the chain preferentially with the eighth sulfur atom:

$$
\begin{array}{ccc}
S\diagup^{\textstyle S-S}\diagdown S & S\diagup^{\textstyle S-S}\diagdown S & \\
\;\;\;\;| \rightarrow \;\; | \;\;\;\;\;\;\;\; | & + -S-S-S-S- & (1.1)\\
S-S-S-S-S-S-S\diagup^{\textstyle S}\diagdown_{S-S}\diagup^{\textstyle S} & &
\end{array}
$$

S_x S_8 $S_{(x-8)}$

Finally, "twisting" of the chains into eight-membered rings converts all the sulfur into the stable cyclooctasulfur [41].

The electronic structure of the S_8 molecule has been calculated from the photoelectronic and x-ray electronic spectra [19,42-44]. The energies of symmetrically resolved optical transitions in the 140-350 nm range have been estimated [43]. The equilibrium S_8 geometry has been determined, and R_s and α have been found to be 2.07 Å and 105°C respectively [14]. From the absorption spectrum, the S_8 energy in the excited singlet state in the liquid phase is found to be 83 ± 3 kcal/mole [45,46].

The smaller unstable sulfur molecules S_2, S_3, and S_4 are characterized by a high reactivity which is clearly due to the fact that these species exist as biradicals and dipoles. The dipoles can be resonance stabilized [37,38]:

$$\cdot\bar{\underline{S}}-\bar{\underline{S}}\cdot \rightarrow \bar{\underline{S}}=\bar{\underline{S}} \leftrightarrow \overset{\oplus}{|\underline{S}}-\bar{\underline{S}}|^{\ominus} \qquad \boxed{S_2} \qquad\qquad (1.2)$$

$$\cdot\bar{S}\diagup^{\displaystyle\bar{S}}\diagdown\bar{\underline{S}}\cdot \rightarrow \overset{\oplus}{|S}\diagup^{\displaystyle\bar{S}}\diagdown\bar{\underline{S}}|^{\ominus} \leftrightarrow |\underline{S}\diagup^{\displaystyle S^{\oplus}}\diagdown\bar{\underline{S}}|^{\ominus} \qquad \boxed{S_3} \qquad (1.3)$$

$$\cdot\bar{\underline{S}}\diagup^{\displaystyle\bar{S}}\diagdown_{\displaystyle\bar{\underline{S}}}\diagup^{\displaystyle\bar{S}\cdot} \rightarrow \overset{\oplus}{\bar{\underline{S}}}\diagup^{\displaystyle\bar{S}}\diagdown_{\displaystyle\bar{\underline{S}}}\diagup^{\displaystyle\bar{S}|^{\ominus}} \boxed{S_4} \qquad\qquad (1.4)$$

$$\begin{array}{ccccc} |\underset{|}{S}|^{\oplus} && S && |\bar{S}|^{\ominus} \\ | && \| && | \\ \overset{\ominus}{|\bar{S}}\diagup^{\displaystyle S}\diagdown_{\displaystyle S} & \leftrightarrow & \overset{\ominus}{|\bar{S}}\diagup^{\displaystyle S}\diagdown_{\displaystyle\bar{S}^{\oplus}} & \leftrightarrow & \overset{S}{}\diagdown_{\displaystyle S|^{\oplus}} \end{array} \qquad (1.5)$$

The constant value of spin-spin coupling for molecules of S_2 has been calculated for interatomic distances of 3.3-3.8 a.u. [47-49]. The emission spectra of S_2 molecules have been measured in the 6983-7760 Å [50] and 7440-8085 Å region [51]. The effective vibrational and rotational constants and Morse function parameters for the S_2 molecule have been determined [52]. The photoelectron spectra of the S_2 molecule have been studied [53-55]. A pure rotation spectrum of combination scattering of S_2 at 900°K has been obtained using an Ar^+ laser (λ 488 nm) of 4 W [56,57].

The dependence of the S-S valence force constant on the S-S interatomic distance has been investigated. Some equations for the dependence of the S-S stretching vibrations on the S-S interatomic distance for the S_2 molecule are given [58]. The electronic structure of the S_2 molecule has been calculated using the method of pseudo-potential [59,60]. The equilibrium compositions of the flame-forming gas within the temperature range of 293-6000°K and at 0.1-30 bars have been calculated [61]. The absorption of S_2 on tungsten mono-crystal faces has been examined by emission electron spectroscopy. At 1850°K desorption of sulfur takes place [62]. The possibility of the formation of a van der Waals complex of sulfur with xenon of the type S_2-Xe has been reported recently [63].

The fluorescence lifetime of S_2 molecules in the presence of quenching gases such as (He, Ar, Xe, N_2, CF_4, C_2F_6) has been measured [64,65].

In the case of the S_3 molecule, the structure S-S-S is more likely that of a biradical; however, a cyclic trimer is even more stable [37]:

The diffuse reflection and scattering spectra for S_3 and S_4 molecules at 1.0-5.4 eV have been examined. Unstable under normal conditions, these molecules stabilize in the cavity of zeolite with pores approximately 10 Å in diameter [66]. The electronic structure of S_3 molecules has been calculated for both the open (C_{2v}) and cyclic (D_{3h}) forms. The C_{2v} form of S_3 has been found to possess total energy 12.6 kcal/mole lower than the D_{3h} form [67].

In the case of the S_4 molecule one cannot exclude the existence of strained cyclic structures (4,5) [32]. However, the structure with a three-membered ring is quite unlikely. Normal geometry of the S-S bond impedes the formation of an unstrained S_4 ring. For this molecule, however, other linear structures are possible. The calculations based on spectroscopic atomic parameters suggest the trans configuration of S (8) and the branched D_{3h} form (10) to be most stable [18,68,69].

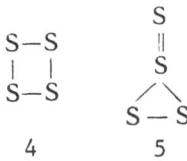

Cyclopentasulfane (S_5), an unstable modification of sulfur, is a liquid which rapidly polymerizes when exposed to sunlight [70]. The S_5 ionization energy is 8.60 eV [71]. The calculation of three configurations of cyclic S_5, the planar (A), envelope (B) and half-chair (C) shows that the energies of configurations (B) and (C) are almost similar, being 18 kcal/mole lower than that of conformer (A) [72].

The orange-red cyclic modification of sulfur corresponding to S_6 cyclohexasulfur (S_6) has been known for a long time [23,73,74]. The molecules of S_6 have the chair form and the S-S interatomic distance is ∿ 2.06 Å:

It seems that cyclohexasulfur has a strained ring and already at 50-60° is converted into polymeric chains. When exposed to sunlight S_6 molecules are converted to S_8 and S_{12} [17].

The IR and Raman spectra of S_6 have been studied [75-77]. An x-ray diffraction study of S_6 at 90° has been undertaken. The crystals are rhombic and the S-S bond length is 2.068 Å. The photoionization energy of S_6 has been found to be 10.2 eV [55].

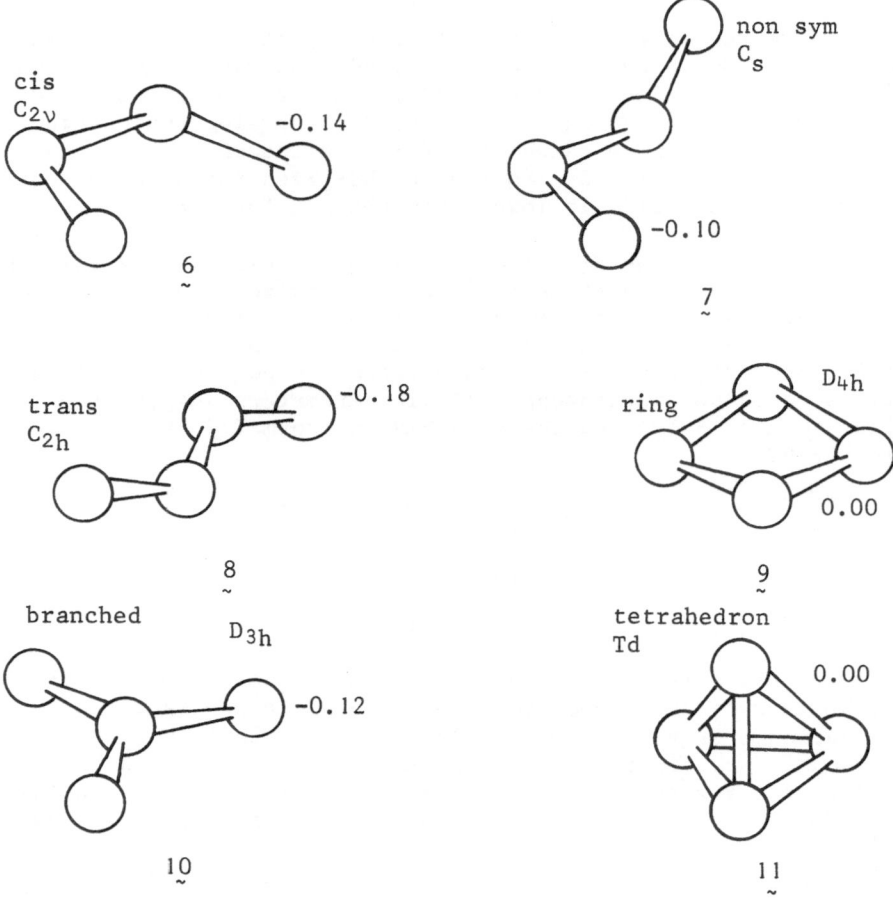

Fig. 2.   Six isomers of S_4. The numbers on the terminal atoms
          indicate the electronic charge.

Equilibrium conformations of cyclic molecules of sulfur have been calculated [78]. The planar form is found to be most stable for the S_4 cycle, the half-chair and envelope for the S_5, and the chair for the S_6 cycle. Several stable conformers have been predicted for S_n molecules (n = 7-12).

The potential curve of the chair-boat conformation transition for the cyclohexasulfur molecule (S_6) has been calculated [79]. The two forms are stable; however, the chair form is more favorable by 3.67 kcal/mole. The boat-chair transition barrier has been estimated to be 22.86 kcal/mole.

In a number of recently performed studies it was possible to establish kinetic control [32,70,81] in the formation of the cyclic molecules S_6 [1,23,73,74], S_7 [74], S_9 [75], S_{10} [74], S_{11} [4,70], S_{12} [80-82], and S_{18} [4,70].

The molecules of cycloheptasulfur (S_7, yellow needles with a melting point of 39°C) are chair-like with four sulfur atoms in the same plane [83]. The crystal structure of S_7 has been examined [84]. The photoionization energy of S_7 is 8.67 eV [85]. The IR and Raman spectra of S_7 have been studied [86]:

The ring system of cyclononasulfur (S_9) is about as stable as that of S_6 [4,87]. Cyclic S_9 molecules occur in sulfur vapor [88].

Cyclodecasulfur (S_{10}) forms yellow crystals, unstable to light and melting at 60°C [4,74,89].

The structure of cyclodecasulfur (S_{10}) at −110°C has been studied by X-ray diffraction. Cyclodecasulfur forms monoclinic crystals. The S_{10} molecules display C_2 symmetry and six sulfur atoms lie in one plane, the other atoms being 1.23-1.24 Å above and below this plane. The average S-S distance is 2.056 Å [90]. The IR and Raman spectra of S_{10} and adduct $S_{10} \cdot S_6$ have been studied [91].

Cycloundecasulfur (S_{11}) is stable only in the dark and at low temperatures [4,92].

Cyclododecasulfur (S_{12}) is one of the stable modifications of sulfur [4,27,93,94]. It forms pale yellow needles and melts with decomposition at 148°C. The twelve-membered ring of S_{12} is highly symmetrical [93,94]. The sulfur atoms lie in three planes, i.e., four atoms are in the middle plane, three atoms above and three atoms below this plane [82]. The S-S bond lengths are 2.05 and 2.06 Å.

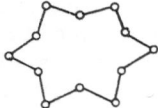

Cyclododecasulfur is considerably less soluble in carbon disulfide
than cyclooctasulfur S_8 [76]. This difference may be used for its
separation from excess cyclooctasulfur.

The atomic force constant K for sulfur atoms in the S_2 molecule
and the cyclic S_6, S_8, and S_{12} molecules has been determined. The K
values are similar in spite of differences in structure, bond lengths
and coordination [94,97,98].

Calculations of S_6, S_7, S_8, and S_{12} normal vibrations with the
force field has been carried out and the dependence of some normal
vibrations on the ring size found [99]. The IR and mass spectra of
S_{12} have been studied [95,96].

Cyclooctadecasulfur (S_{18}, 12) [92,100,104] and cycloeicosasulfur
(S_{20}, 13) [92,100] are stable modifications of molecular sulfur and
possess the following structures:

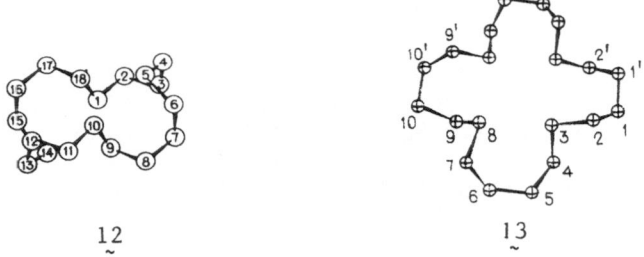

12   13

Cyclooctadecasulfur (S_{18}) has a melting point of 126°C (with
decomposition). The S—S bond lengths are ∿ 2.06 Å [29,100]. The
IR and Raman spectra of cyclic S_{18} have been examined [95]. The
S_{18} crystalline structure has been studied by x-ray diffraction
[100,104].

Cycloeicosasulfur (S_{20}) has a melting point of 121°C (with
decomposition). The S—S bond lengths are ∿ 2.04 Å [100].

From an analysis of 14 crystalline structures made up of molecules possessing from 3 to 27 sulfur atoms the mean statistical van
der Waals radius value has been found for the sulfur atom [101].

It is assumed that cyclic sulfur molecules from S_{23} to S_{33} can
exist in carbon disulfide [102].

A three-dimensional electronic gaseous model enabling adequate
bond energies for sulfur rings of different sizes to be obtained has
been developed [103].

The charge distribution for approximately 100 different molecules and ions of elemental sulfur has been calculated from a Hückel
model [94].

1.2. ALLOTROPIC MODIFICATIONS

Sulfur can exist in various allotropic modifications due to the
formation of molecules with different numbers of atoms and different
crystal structures. At room temperature the stable modification is
rhombic sulfur (S_α) [1,22,27,103]. It is yellow, has low electrical
and heat conductivity, is readily soluble in carbon disulfide and
poorly soluble in organic solvents (cf. Section 1.4). The melting point
of S_α is 112.8°C. However, when slowly heated above 95.6°C, it is
transformed into another stable modification, monoclinic sulfur (S_β)
[105]. Upon rapid heating of S_α, the melt consists of a mixture of
S_λ and S_μ. At 112.8°C, rhombic sulfur is in equilibrium with S_λ only.
Crystallization of rhombic sulfur from an equilibrium melt of S_λ and
S_μ takes place at 110.2°C. Crystals of rhombic sulfur consist of
nonplanar eight-membered rings of S_8 [22,103,106]. (For physical
properties of S_α, see Section 1.3.1.)

Monoclinic sulfur (S_β) is formed when rhombic sulfur is slowly
heated above 95.6°C [1,27,106-108]. The "ideal" melting point of
S_β is 119.3°C. The actual melting point of S_β is 114.6°C, i.e., the
so-called "normal" melting point [109]. At normal temperatures, the
S_β modification changes into the S_α modification [103,110-112]. The
transition point for the transition $S_\beta \rightleftharpoons S_\gamma$ is 88.5°C [113].

Similarly to the case of S_α, the crystals of S_β consist of nonplanar eight-membered cyclic molecules. The physical properties of
S_β have been studied in detail [103,106,114-118].

S_λ is an amorphous yellow modification of sulfur, soluble in
carbon disulfide. An equilibrium $S_\lambda \rightleftharpoons S_\mu + S_\pi$ is established in the
melt. At 160°C, molten sulfur contains 89.2% S_λ, 4.1% S_μ, and 6.7%
S_π. The equilibrium $S_\lambda \rightleftharpoons S_\pi$ is light-sensitive. At higher temperatures, the above equilibrium is shifted towards the formation of S_π.
S_λ consists of nonplanar eight-membered rings. The physical properties of S_λ have been studied [120-127].

The kinetics of the liquid-phase reaction of S_λ and S_π allotrope
intertransition in the equilibrium state region has been studied.
This is almost a first-order reaction [119]. At room temperature,
a slow transition $S_\lambda \rightarrow S_\alpha$ takes place.

S_μ is a red amorphous modification of sulfur, insoluble in
carbon disulfide. It consists of irregularly arranged zig-zag-like
chains [22,128-138]. However, there seems to be some indication
that S_μ can actually consist of S_6 molecules [129]. Ultraviolet

irradiation or irradiation with α-, β-, or γ-rays accelerates the transition of S_μ into S_α by breaking the long sulfur chains [139]. The physical properties of the S_μ modification have been investigated [140-144].

The S_π allotropic modification of sulfur is present in small amounts in liquid sulfur [145-148]. This modification can be isolated by slowly cooling molten sulfur from 119.3°C to 114.5°C. It is less soluble in carbon disulfide than is S_λ. On standing, S_π is fully converted into S_μ and subsequently into S_α. The data concerning the structure of S_π are controversial. These data indicate that S_π consists of chains of eight-membered molecules [147,149], or that it contains molecules of cyclohexasulfur [67], or a mixture of cyclic sulfur molecules consisting of more than eight atoms [151]. Some physical properties of S_π have been studied [149-152].

S_γ is a metastable modification and is formed upon rapid cooling of a saturated solution of sulfur in benzene or ethanol [153-158] and consists of eight-membered sulfur molecules. S_γ forms shiny plate-like crystals with a melting point of 106.8°C. Above 75°C S_γ is rapidly converted into S_β. The kinetic behavior of the system $S_\alpha \rightleftharpoons S_\beta \rightleftharpoons S_\gamma$ has been studied [111,159-162].

The S_ζ modification of sulfur was isolated from a solution of rubber in carbon disulfide or chloroform in the presence of selenium [1]. It forms reddish-brown monoclinic crystals [163-165] containing eight-membered sulfur molecules. S_ζ is less stable than S_γ, but is more stable than S_δ [22].

The S_η modification was also obtained from solutions of rubber in carbon disulfide or chloroform in the presence of selenium [166]. The lifetime of this modification is $\sim$ 10 minutes. However, so far the existence of S_ζ and S_η has not been convincingly proven.

The S_δ modification was obtained, together with S_α and S_γ, on oxidation of an ethanolic solution of ammonium polysulfide at $\sim$ 15°C in the presence of air [128,165,167-170]. Crystals of S_δ consist of eight-membered sulfur molecules, are rather unstable, and undergo transformation into S_γ, S_β, or S_α. No final proof of the structure is available.

The S_θ modification [166,171] was isolated together with S_γ from dilute solutions of sulfur in α-pinene or nitrobenzene, as well as from a solution of raw rubber in carbon disulfide. The unstable S_θ modification is rapidly converted into S_γ, S_β, or S_α under normal conditions.

The S_O modification [22] was discovered on evaporation of solutions of S_δ. It is very unstable and is converted into S_γ within a few seconds [128].

The S_ξ modification of sulfur [171] forms monoclinic helical crystals. It is an extremely unstable modification and its existence is doubtful.

The S_τ modification crystallizes out from a hot solution of sulfur in o-xylene. S_τ is an unstable allotropic modification of sulfur, with a lifetime of several hours under normal conditions [22].

The S_ϕ allotropic modification was obtained from a solution of sulfur in α-pinene at 8°C [164,172,173]. It is assumed that S_ϕ consists of long helical chains. Many authors assume [173-175] that the S_ϕ modification is actually a mixture of the S_ψ and S_γ allotropic modifications of sulfur.

The S_ψ allotropic modification of sulfur was precipitated from a solution of sulfur in α-pinene [166]. It has been suggested that S_ψ forms long helix-like crystals [39,115,176] and is a component of S_ϕ.

The S_κ allotropic modification is monoclinic and was obtained when boiling o-xylene was saturated with sulfur [166,168]. It has been discovered that S_κ is present in molten sulfur [176]. Under normal conditions S_κ is more stable than S_γ [22].

The S_ι allotropic modification was isolated chromatographically from the S_π modification [163]. It is insoluble in carbon disulfide.

The S_ω modification was obtained by the sublimation of flowers of sulfur and by the hydrolysis of disulfur dichloride with water [177,178]. S_ω is the portion of sublimed sulfur which is not soluble in carbon disulfide [179]. The S_ω modification is stable under normal conditions and it can last for thirty-six hours at ∿ 90°C [180,181]. Ammonia and other bases catalyze the conversion of S_ω into S_α [130]. It is assumed that S_ω consists of eight-membered sulfur molecules [182]. It is not clear yet whether S_ω is a pure allotropic modification or whether it also contains other modifications of liquid sulfur [22].

S_ρ is an allotropic modification which was obtained when concentrated hydrochloric acid was treated with a concentrated solution of sodium thiosulfate at 10°C [83]. S_ρ can be extracted, together with S_λ, into toluene or benzene. S_ρ consists of bonded cyclohexasulfur molecules. It is thermodynamically unstable, but can survive for a long time in solution. The physical properties of S_ρ have been investigated [123,184,185].

The S_ϵ allotropic modification is a crystalline form of S_ρ, and consists of cyclohexasulfur molecules [183,186,187]. Under normal conditions S_ϵ crystals are converted into S_α. Carefully purified

S_ϵ is stable for a considerable length of time at 0°C. The crystal structure and physical properties of S_ϵ have been studied [188-190].

The S_ν modification is an insoluble yellow precipitate which is formed during the preparation of red sulfur by the interaction of ammonium polysulfide with disulfur dichloride [163].

The S_m modification was isolated from the products obtained from the distillation of S_ω or by the chromatography of red sulfur [191]. This modification can be sublimed under dry nitrogen.

Yellow sulfur has been obtained by the condensation of sulfur vapor at reduced pressure onto a cold surface at −78°C. It is possible that yellow sulfur is a mixture of allotropic modifications [192,193]. At −100°C all low-temperature forms are converted into this allotropic modification.

Black sulfur is observed when diatomic sulfur vapor is condensed [130,194,195]. No final proof of the existence of this allotropic modification of sulfur is available.

Purple sulfur is formed when diatomic sulfur vapor is condensed on a surface cooled with liquid nitrogen [196-198]. There are indications that purple sulfur is a polymerization product of S_2 molecules [199]. At room temperature purple sulfur is converted into yellow sulfur within several seconds. At -80°C, this transition requires several hours [197]. The physical properties of purple sulfur have been studied [200,201].

Red sulfur has been obtained using several different techniques. This allotropic modification is formed on irradiation of S_α with γ-rays [202], upon condensation of hot sulfur vapor on a surface with a temperature below 20°C, and also upon interaction of ammonium polysulfide with disulfur dichloride at pH 8 [22]. The physical properties of red sulfur have been investigated [203-205].

Green sulfur is formed upon the condensation of hot sulfur vapor, containing diatomic molecules, on a surface cooled to -104°C to -78°C. It was also obtained by the condensation of S_λ on a liquid-air cooled surface. Green sulfur is stable at liquid nitrogen temperature, and at -100°C it is slowly converted into yellow sulfur. At room temperature, this transition requires only several seconds [185].

Violet sulfur was obtained by the condensation of sulfur vapor consisting of diatomic molecules on a surface cooled to a temperature below −180°C. At −160°C this allotropic modification is converted into green sulfur. The structure and physical properties of violet sulfur have been studied [192,193,206].

Orange sulfur was obtained when ammonium polysulfide was treated with disulfur dichloride [207]. Under normal conditions it can exist for about thirty minutes. It is soluble in carbon disulfide.

The colored modifications of sulfur seem to be complex mixtures of allotropic modifications [208] and the elucidation of their structure is still in progress.

1.3. PHYSICAL PROPERTIES

1.3.1. The Solid State

Under normal conditions, the stable solid modification of sulfur is rhombic sulfur (S_α). Other solid allotropic modifications of sulfur are known (cf. Section 1.2); however, these modifications exhibit a tendency to change into α-sulfur. Rhombic sulfur consists of nonplanar cyclooctasulfur molecules [24-26]. The crystal structure of α-sulfur has been studied in detail [209-224]. The parameters of the crystal lattice of α-sulfur are as follows [225,226]: a = 10.44349 Å; b = 12.84009 Å; c = 24.43665 Å. At room temperature, the a:b:c ratio is 0.8134:1:1.9032. At $-72°C$, the ratio becomes a:b:c = 0.8139:1:1.9120, and at $-175°C$ it is a:b:c = 0.8151:1:1.9208 [227]. The growth of α-sulfur crystals has been studied [166,180, 181,218,228,230], as well as their cleavability [212,229]. Rhombic sulfur has D_{2h}^{24} - Fddd symmetry [218,229]. The unit cell of a crystal of α-sulfur equals 3324.1 Å^3, and the atomic cell is 26 Å^3 [225]. The bond energy of the S-S bonds within the eight-membered ring of S_8 is 63.8 kcal/mole if one considers these bonds to be single covalent bonds [231]. It follows from other thermochemical data that the bond energy is only 53.9 kcal/mole [232,235].

The spectral characterisitcs of rhombic sulfur which have been studied include the UV [234,235], IR [236-238], Raman [239-242], and x-ray spectra [243-246]. The mass spectra of elemental sulfur have been examined [247]. The photoelectronic spectra of orthorhombic sulfur with the exciting quantum energy $h\nu \leqslant 21.2$ eV have been studied [248].

The density of α-sulfur, recrystallized from carbon disulfide, has been determined over a wide temperature range [249,250]. At 20°C the density of α-sulfur is 2.0370 [251] or 2.0454 g/cm^3 [252], according to different sources.

The electrical conductivity of α-sulfur is very low [253,254] and the same is true of its thermal conductivity [255,256]. The magnetic susceptibility of α-sulfur at room temperature is 0.4581-0.487 [257-259]. Rhombic sulfur is readily soluble in carbon disulfide and has a limited solubility in various organic solvents (cf. Section 1.4) and is insoluble in water.

The effect of high pressures (up to 31 kbar) upon the polymerization of α-sulfur [260] and its melting point has been studied
[261]. Phase transitions of rhombic sulfur at 25-83 kbar and 25-
400°C have been investigated [262]. Elastooptic coefficients of
rhombic sulfur have been calculated [263]. The electrochemical and
chemical properties of sulfur over a wide temperature range have
been presented [264,265].

Theoretical and experimental studies of the structure and
dynamics of rhombic sulfur have been summarized [266-268].

1.3.2. The Liquid State

Rhombic sulfur melts at 112.8°C and becomes a mobile straw-
colored liquid. The low viscosity of molten sulfur is due to the
fact that the cyclooctasulfur molecules in the crown form can be
easily displaced with respect to each other [220,229,269]. The
viscosity of liquid sulfur decreases slowly with increasing temperature starting from the melting point and ending at 159°C (10.07
poise). Upon further heating, a sharp increase in viscosity is
observed and reaches a maximum at 187°C (937 poise). Further increase of the temperature up to the boiling point (444.6°C) leads
to a sharp decrease of viscosity down to 0.83 poise [270-273].
This anomalous behavior of molten sulfur can be explained by changes
in its molecular structure and by transitions among the various
modifications of sulfur which take place upon heating [175,274-281].

At 120-159°C, molten sulfur predominantly consists of stable
cyclooctasulfur molecules (the S_λ modification) and its viscosity
decreases with increasing temperature, as observed with normal
liquids. Above 159°C, the eight-membered sulfur rings undergo rapid
cleavage with the formation of $\cdot S_8 \cdot$ biradicals. These biradicals
can recombine, or they can attack the ring structures still present
and form polymeric chains [282-284] which reach their maximum length
($\sim 10^6$ sulfur atoms) at the point of highest viscosity [274]:

$$S_8 \rightleftarrows \cdot S\!-\!S_6\!-\!S\cdot \quad \cdot S\!-\!S_6\!-\!S\cdot + S_8 \rightleftarrows \cdot S\!-\!S_6\!-\!S\!-\!S_8\cdot \text{ etc.} \tag{1.6}$$

$$\cdot S\!-\!S_6\!-\!S\cdot + \cdot S\!-\!S_6\!-\!S\cdot \rightleftarrows \cdot S\!-\!S_6\!-\!S\!-\!S\!-\!S_6\!-\!S\cdot \xrightleftharpoons[\quad]{+n\cdot S\!-\!S_6\!-\!S\cdot} \cdot S\!-\!S_n\!-\!S\cdot$$

There are indications that the melt may contain even other
rings, less stable than S_8, at 159-189°C. These rings can contain
more or fewer than eight sulfur atoms [29,285].

Thermodynamic calculations have shown [151] that above 120°C
molten sulfur contains 8.2 wt.% of large rings with an average
number of 13.8 sulfur atoms; at 150°C it contains 20 wt.% of rings

with an average number of 17.6 sulfur atoms, and at 160°C it contains 30 wt.% of large rings.

An ESR study of sulfur [1,34,36] has confirmed that, at higher
temperatures, the melt contains free radicals whose concentration
reaches $6 \cdot 10^{-3}$ M at 300°C. Thus, sulfur chains are indeed biradicals, $\cdot S-S_n-S\cdot$, which seem to be in equilibrium with large rings
which can become intercalated.

It is possible to calculate from the temperature dependence of
the concentration of free radicals that the S-S bond dissociation
energy in this case is quite small, i.e., -33.4 kcal/mole. This
value is in good agreement with a value obtained from specific
thermal conductivity data [276]. In addition to the random cleavage
of S-S bonds and recombination of -S· radicals, the ESR spectra also
indicate a very rapid reaction between radicals and chains:

$$-S\cdot + \;\; -S-S_x-S- \;\; \longrightarrow \;\; -S-S-S_x- \;\; + \cdot S- \tag{1.7}$$

The formation of free biradicals in molten sulfur above 160°C
can also be proven chemically. Below the above temperature, when
sulfur exists predominantly as cyclooctasulfur, addition of halogen
(chlorine, bromine, iodine) and hydrogen sulfide to the melt does
not change its viscosity. However, above 160°C, the addition of
halogens considerably decreases the viscosity of sulfur [271,286,
287]. This result can be explained as being due to the reaction of
halogens with sulfur biradicals leading to the termination of the
growth of polymer chains [286,288]:

$$S_n \cdot + X_2 \longrightarrow X-S_n-X \tag{1.8}$$

The possibility of the existence of branched sulfur chains has
recently been rejected [1] whereas previously such a possibility was
seriously considered [289,290]. Sulfur molecules can exist either
as zig-zag-like chains, or as closed rings [291,292]. Molten sulfur
can be considered as a reversible equilibrium system, with an equilibrium between chains of octasulfur and high-polymer polychain
sulfur [149,274,293,294]:

$$nS_8 \text{(ring)} \rightleftharpoons n-S_8- \text{(chain)} \rightleftharpoons (S_8)_n \text{(chain)} \tag{1.9}$$

It has been noted that, below the critical temperature, biradical
sulfur chains and thermally still uncleaved cyclooctasulfur rings
can form charge-transfer complexes [295]. As mentioned above,
molten sulfur is pale yellow; however, with increasing temperature
the color becomes more intense and the melt finally has a brown-red
color. It is assumed that the changes in color of the melt above
200°C are due to the presence of S_3 and S_4 molecules which even in
a concentration of $\sim$ 2% give intense coloration to polymeric sulfur
[31,296,297]. The high reactivity of hot sulfur melts is also likely

to be due mainly to the presence of small amounts of small sulfur
molecules such as S_3, S_4, and possibly also S_2 [31,296]. Optical
methods have been used to measure the velocity of ultrasound (fre-
quency 4 MHz) in molten sulfur at 120-143°C [298]. Within the above
temperature range, the velocity of ultrasound decreases linearly
with increasing temperature. The absorption of ultrasound with
frequencies 5.8, 12, and 15 MHz increases with increasing temperature,
starting at 160°C, and stops $\sim$ 200°C [299].

The electrical conductivity of liquid sulfur has been measured
in the 113-900°C temperature range. The conductivity is at a minimum
at 170°C which is consistent with the low degree of polymerization
of liquid sulfur at this temperature [300].

The viscosity of liquid sulfur has been studied over a wide
temperature range [271,301-303]. The effect of pressure upon the
viscosity of molten sulfur has also been investigated [304].

The density [305,306], thermal conductivity [307-309], and the
coefficient of expansion of liquid sulfur [305,310,311] have been
measured. Thermomagnetic measurements have been carried out [312,
313]. Molten sulfur shows photoelectric conductivity [314].

X-ray studies of liquid sulfur have been carried out over the
100-340°C temperature range [315]. The x-ray diffraction diagram
of liquid sulfur has maxima corresponding to the following distances
between the planes: 5.50, 3.60, 1.72, and 1.03 Å [280]. In the
130-1100°C range, the absorption spectra (250-200 nm) of liquid and
gaseous sulfur have been measured at different pressures [316-318].

1.3.3. The Gaseous State

At temperatures above 200°C, the gas phase contains considerable
numbers of S_4, S_3, and S_2 molecules. Above 500°C, S_2 molecules are
the prevailing species and their concentration reaches its maximum
at 800-1400°C [185,197,319,320]. Above 1000°C, and especially above
1500°C, the diatomic sulfur molecules dissociate with the formation
of free sulfur atoms [1].

The presence of various molecular forms of sulfur in the vapor
phase determines its color. Sulfur vapor at about the boiling point
of sulfur is orange-red. With increasing temperature, the color
initially becomes redder, but later it becomes less intense again
and at 650°C the color is straw-yellow. The red color of sulfur in
the vapor phase is obviously due to the presence of S_3 and S_4 molecules, similarly as in the liquid state [1,37].

In the 150-400°C range, sulfur vapor consists essentially of
S_6 (60-70%) and S_8, S_7, and S_5 (30-40%) molecules. At higher
temperatures (500-700°C), the concentration of S_n molecules (n = 5,

6,7,8) sharply decreases and there is a strong increase in the concentration of S_4 and S_3 molecules [37].

Mass spectral data indicate that saturated sulfur vapor contains about 20% S_8, 40% S_7, 20% S_6, 6% S_5, and 4% S_4, S_3, and S_2 molecules [32,40,321].

In addition to the "fragments" of cyclooctasulfur, other cyclic molecules were discovered in the vapor phase at low pressure, i.e., S_9, S_{10}, S_{11} [29,70], and S_{12} [40]. However, there are contrasting views concerning the presence of S_n molecules (with n > 8) in sulfur vapor [1,29,37,296,322].

Partial pressures of S_n molecules (n = 1,2...8) in the total pressure of sulfur vapor were calculated at normal pressure. At low temperatures, the highest partial pressure is displayed by S_6, S_7, and S_8 molecules, at the highest temperature by S_2 and S_3 molecules [323].

Equilibrium constants for the equilibria between sulfur molecules with different numbers of atoms were determined at different temperatures and it has been found that saturated sulfur vapor contains all possible types of S_n molecules (n = 1,2,3...8) [40].

There is an indication that, in the vapor phase, S_n molecules with n > 5 are exclusively cyclic whereas S_4 molecules are either cyclic or branched [18,282]. Sulfur vapor is diamagnetic at low temperatures [324]; however, with increasing temperature, sulfur vapor begins to contain some paramagnetic S_2 molecules.

The density of sulfur vapor has been studied at various temperatures [325-331]. The chemical potential of sulfur in the 500-950°C temperature range has been measured [332].

1.4. SOLUBILITY

The solubility of elemental sulfur in various organic solvents is limited (cf. Table 1.2). It is very soluble in carbon disulfide (at 20°C, the solubility is 28.5 wt.%). The solubility of sulfur in carbon disulfide increases with increasing temperature at constant pressure and reaches its maximum at $\sim$ 111°C. On the other hand, the solubility shows a negligible increase at constant temperature with increasing pressure (from 70 to 360 atm) [333]. Further temperature increases from 120 to 220°C lead to a noticeable decrease in the solubility of sulfur because of the formation of an insoluble modification, i.e., S_μ (up to 33%). A sharp decrease in the solubility of sulfur is observed above 180°C [334]. The difference in solubility in carbon disulfide serves as the basis for the extraction of S_π from the plastic modification of sulfur; molecules of S_π in carbon disulfide solution are eight-membered chains [18,149].

Table 1.2.   Solubility of Sulfur in Organic Solvents

Solvent	Formula	Temp. (°C)	Wt.%	Ref.
Acetic acid	CH_3COOH	20	0.032	372
Acetone	CH_3COCH_3	15	0.045	334,373
Aniline	$C_6H_5NH_2$	15	0.700	344,374
		100	18.000	
Benzene	C_6H_6	20	1.768	370,375,376
		100	17.500	372
Benzoyl chloride	C_6H_5COCl	0	1.000	377
		134	55.800	
Benzyl chloride	$C_6H_5CH_2Cl$	35	2.570	378
Bromoform	$CHBr_3$	6	3.640	379
Butyl alcohol	C_4H_9OH	30	0.187	380
Carbon disulfide	CS_2	0	18.780	376,381,382
		0	19.300	
		20	28.500	
		46	56.200	
		81	79.400	
Carbon tetrachloride	CCl_4	25	0.831	383
		25	0.872	376,384
Chlorobenzene	C_6H_5Cl	70	20.000	385
Chloroform	$CHCl_3$	15	0.868	350,376,386
		30	1.281	
1,2-Dibromoethane	$BrCH_2CH_2Br$	22	2.400	387
		50	6.400	
p-Dichlorobenzene	$\underline{p}-C_6H_4Cl_2$	97	40.000	376,385
1,2-Dichloroethane	$ClCH_2CH_2Cl$	25	0.826	383
1,2-Dichloroethylene	$ClCH=CHCl$	25	1.254	384
Diethyl ether	$(C_2H_5)_2O$	23	0.283	388
Ethanol	C_2H_5OH	25	0.031	389
Ethyl bromide	C_2H_5Br	25	1.860	390
Ethyl formate	$HCOOC_2H_5$	20	0.152	350

Table 1.2 (continued)

Solvent	Formula	Temp.(°C)	Wt.%	Ref.
Formic acid	$HCOOH$	100	0.035	391
Glycerol	$HOCH_2CHOHCH_2OH$	15	0.140	392
Heptane	C_7H_{16}	0	0.124	383
		54	0.926	
Hexane	C_6H_{14}	0	0.160	393,394
		20	0.250	
		100	2.800	
Isobutyl alcohol	$(H_3C)_2CHCH_2OH$	20	0.110	395
Methanol	CH_3OH	18.5	0.028	396
Methylene iodide	CH_2I_2	10	9.100	397
2-Naphthol	$C_{10}H_7OH$	118	34.000	
		150	70.000	
Octane	C_8H_{18}	25	0.360	383
Pentanol	$C_5H_{11}OH$	110	2.150	398
Pentachloroethane	Cl_3CCHCl_2	25	1.210	383
Phenol	C_6H_5OH	89.5	9.100	382
Pyridine	C_5H_5N	20	1.480	376,399
Quinoline	C_9H_7N	80	16.000	400
1,1,2,2-Tetrachloro-ethane	$Cl_2CHCHCl_2$	100	18.000	383
Tetrachloroethylene	$Cl_2C=CCl_2$	25	1.150	388
Thionyl chloride	$SOCl_2$	0-60	-	401
Toluene	$C_6H_5CH_3$	20	1.827	337,376
Trichloroethylene	$Cl_2C=CHCl$	25	1.600	386
m-Xylene	$m\text{-}C_6H_4(CH_3)_2$	20	1.140	383
		80	10.290	
p-Xylene	$p\text{-}C_6H_4(CH_3)_2$	100	17.850	383

The solubility of sulfur consisting of S_6 and S_{12} molecules has been determined in carbon disulfide at -31 to 46°C [76]. It has been noted that there are no phase transitions in the case of S_6 and S_{12}. The different solubilities of S_6, S_8, and S_{12} in carbon disulfide, due to differences in crystal lattice energies, make it possible to separate these three molecular forms of sulfur. Dilute solutions of sulfur in carbon disulfide are paramagnetic [335]. This indicates that sulfur is present as S_2 in dilute solutions.

In an effort to find suitable solvents for sulfur for cryoscopic and ebullioscopic measurements, it was found that sulfur is poorly soluble in the following organic compounds: heptane, acenaphthene, biphenyl, diphenylmethane, anthracene, dihydroanthracene, phenanthrene, fluorene, chloroform, 1,2-dibromoethane, bromonaphthalene, diethyl ether, ethanol, ethyl acetate, and resorcinol [336]. Sulfur reacts with many of these compounds at their respective boiling points. At temperatures which are close to the melting points of the respective solvents, sulfur is poorly soluble in benzene, nitrobenzene, phenol, thymol, cetyl alcohol, acetic acid, and anethole. Sulfur reacts chemically with some of the above solvents [337].

The solubility of sulfur at 25-140°C has been measured in dimethyl sulfoxide and its mixtures with acetone, water, and ethanol [338]. The solubility of sulfur in 100 g of dimethyl sulfoxide increases with increasing temperature from 0.097 g at 25°C to 4.815 g at 140°C. The solubility of sulfur is considerably lower in mixed solvents. The solubility of sulfur in hexane, heptane, octane, trichloroethylene, and perchloroethylene in the 25-121°C range has been examined. Equations describing the solvent-temperature dependence have been presented [339].

The solubility of sulfur in compounds of general formula R-X where R = hydrophobic radical, X = hydrophilic radical has been determined [340].

The solubility of sulfur in liquid hydrogen sulfide has been studied in the -80°C to +120°C temperature range. At -80°C, the solubility is $5 \cdot 10^{-6}$ moles of S_8 per 1 g liquid hydrogen sulfide and it is somewhat higher at +80°C under pressure ($5 \cdot 10^5$ moles of S_8) [341]. The solubility of sulfur in liquid ammonia has also been determined [342]. These solutions can be used as fertilizers and in rubber vulcanization.

The solubility of sulfur in paraffin has been determined over a wide temperature range [343-347]. With increasing temperature, the solubility increases from 0.50 wt.% at 90°C to 9.86 wt.% at 136°C. Above 150°C sulfur interacts with paraffin and hydrogen sulfide is liberated (cf. Section 3.1.3). Sulfur is soluble in different types of gasoline. In light gasoline the solubility of sulfur is

0.619 wt.% at 20°C [348-350]. In heavy gasoline (ligroin) the
solubility of sulfur is 2.5 and 23.0 g per 100 g solvent at 15°C and
100°C, respectively [351]. Heavy oil dissolves 6.0 and 52.5 g sulfur
in 100 g solvent at 15°C and 100°C, respectively [348,349]. The
solubility of sulfur in rubber has been studied at 40-95°C [352-360],
and at 65-110°C [361,362]. Vulcanized rubber contains 2.8 and 7.1
wt.% free sulfur at 55°C and 95°C, respectively.

Sulfur forms colored solutions with various solvents. Thus,
solutions of sulfur in oleum are light blue to blue. The blue color
of sulfur solutions in oleum is due to the presence of $S_4^{\oplus}$ [363-365]
or $S_8^{\oplus}$ cations [366]. Sulfur forms bluish-green solutions in hexamethylphosphoric triamide. It is assumed that the color is due to
the presence of S_3^- anions [367]. Solutions of sulfur in dimethylformamide are pale blue. It has been shown by UV spectroscopy that
the solution contains three depolymerization products of S_8, probably
S_4, S_3, and S_2 [368].

Solutions of sulfur in organic solvents (carbon disulfide,
benzene, toluene, carbon tetrachloride, acetone, chloroform) are
quite sensitive toward light [369,370]. Upon irradiation of these
solutions, a precipitate of sulfur is formed.

The solubility of sulfur in water has been studied at different
temperatures. At 25°C, it is $(1.9 \pm 0.6) \cdot 10^{-8}$ mole of S_8/kg [371].

The UV absorption curves of solutions of sulfur in chloroform,
ethanol, n-hexane, and methanol are similar, with a maximum at 265 nm
and a minimum at 250 nm [402].

REFERENCES

1. B. Meyer (ed.), Elemental Sulfur: Chemistry and Physics, Wiley-
    Interscience, New York (1965).
2. O. M. Nefedov and M. N. Manakhov, Angew. Chem., 78, 1039 (1966).
3. L. Pauling, Proc. Natl. Acad. Sci. USA, 35, 495 (1949).
4. M. Schmidt, Angew. Chem., 85, 474 (1973).
5. C. K. Ingold, Structure and Mechanism in Organic Chemistry,
    2nd ed., Cornell Univ. Press, Ithaca—London (1969), p. 22.
6. F. Bernardi and C. Zauli, J. Chem. Soc. (London), Ser. A, 2633
    (1968).
7. S. Oae, Organic Chemistry of Sulfur, Plenum Press, New York
    (1977).
8. D. P. Graig and C. Zauli, J. Chem. Phys., 37, 601 (1962).
9. K. Keeton and D. P. Santry, Chem. Phys. Lett., 7, 105 (1970).
10. D. W. J. Cruickshank and B. C. Webster, in: Inorganic Sulfur
    Chemistry, ed. by G. Nickless, Elsevier, Amsterdam (1968), p. 7.
11. D. W. J. Cruickshank, B. C. Webster, and D. F. Mayers, J. Chem.
    Phys., 40, 3733 (1964).
12. W. W. Fogleman, D. J. Miller, H. B. Johassen, and L. C. Cusachs,
    Inorg. Chem., 8, 1209 (1969).

13. K. A. Levison and P. G. Perkins, Theor. Chim. Acta, 14, 206 (1969).
14. G. L. Carlson and L. G. Pedersen, J. Chem. Phys., 62, 4567 (1975).
15. L. C. Cusachs and D. J. Miller, Adv. Chem. Ser., 1 (1972).
16. D. J. Miller and L. C. Cusachs, Chem. Phys. Lett., 3, 501 (1969)
17. S. D. Thompson, D. G. Carroll, F. Watson, M. O'Donnell, and S. P. McGlynn, J. Chem. Phys., 45, 1367 (1966).
18. B. Meyer and K. Spitzer, J. Phys. Chem., 76, 2274 (1972).
19. W. R. Salaneck, N. O. Lipari, A. Paton, R. Zallen, and K. S. Liang, Phys. Rev. B, Solid State, 12, 1493 (1975).
20. R. Bosch and W. Schmidt, Inorg. Nucl. Chem. Lett., 9, 643 (1973)
21. H. C. Whitehead and G. Anderman, J. Phys. Chem., 77, 721 (1973).
22. B. Meyer, Chem. Rev., 64, 429 (1964).
23. B. Meyer, Elemental Sulfur, Elsevier, Amsterdam (1968), p. 240.
24. A. A. Palma and N. V. Cohan, Rev. Mex. Fis., 19, 15 (1970).
25. H. Garcia-Fernandez, Bull. Soc. Chim. France, 265 (1958).
26. D. J. Gibbons, Mol. Cryst., 10, 137 (1970).
27. B. Meyer, Chem. Rev., 76, 367 (1976).
28. J. E. Aken, J. A. Prins, R. Reijuhart, and F. Tuinstra, Mater. Res. Bull., 3, 219 (1968).
29. M. Schmidt, Angew. Chem., 85, 474 (1953).
30. S. Abrahams, Acta Crystallogr., 18, 566 (1965).
31. B. Meyer, Adv. Chem. Ser., 53 (1972).
32. R. Meyer, Z. Chem., 9, 321 (1973).
33. G. Bergson, Arkiv Kemi, 16, 315 (1961).
34. D. M. Gardner and G. K. Fraenkel, J. Am. Chem. Soc., 76, 5291 (1954).
35. D. M. Gardner, Ph.D. Thesis, Columbia Univ., New York (1955).
36. D. M. Gardner and G. K. Fraenkel, J. Am. Chem. Soc., 78, 3279 (1956).
37. N. W. Luft, Monatsh. Chem., 86, 474 (1955).
38. P. W. Schenk and U. Tuemmler, Z. Anorg. Allg. Chem., 315, 271 (1962).
39. A. G. Pinkus, C. B. Concilio, J. L. McAttee, and J. S. Kim, J. Polym. Sci., 40, 581 (1959).
40. M. Schmidt and H. D. Block, Angew. Chem., 79, 944 (1967).
41. J. A. Semlyen, Trans. Faraday Soc., 63, 2343 (1967).
42. N. V. Richardson and P. Weinberger, J. Electron. Spectrosc. Relat. Phenom., 6, 109 (1975).
43. A. Burrie, H. Garcia-Fernandez, H. G. Heal, and R. J. Romsay, J. Inorg. Nucl. Chem., 37, 311 (1975).
44. D. R. Salahub, A. E. Foti, and V. H. Smith, J. Am. Chem. Soc., 99, 8067 (1977).
45. M. Elbanowski and J. A. Wojczak, Phosphorus Sulfur, 5, 107 (1978).
46. M. Elbanowski, UAM Ser. Chim. (Warshava), 86 (1978).
47. F. D. Wayne, Chem. Phys. Lett., 31, 97 (1975).
48. H. S. Liszt, Astrophys. J., 219(2), 454 (1978).
49. F. D. Wayne and E. A. Colbourn, Mol. Phys., 34, 1141 (1977).
50. N. A. Narasimham, V. Sethuraman, and K. V. Apparao, J. Mol. Spectrosc., 59, 142 (1976).

51. N. A. Narasimham, K. V. Apparao, and T. K. Balasubramanian, J. Mol. Spectrosc., 59, 244 (1976).
52. T. Arai and T. Martin, Phys. Status Solidi (B), 81, 185 (1977).
53. J. M. Dyka, L. Golob, and N. Jonathan, J. Chem. Soc. Faraday Trans., 71, 1026 (1975).
54. S. T. Lee, S. Suzer, and D. A. Shirley, Chem. Phys. Lett., 41, 25 (1976).
55. J. Berkowitz, J. Chem. Phys., 62, 4074 (1975).
56. P. A. Freedman, W. J. Jones, and A. Rogstad, J. Chem. Soc. Faraday Trans., 71, 286 (1975).
57. A. G. Hopkins and C. W. Brown, J. Chem. Phys., 62, 1598 (1975).
58. R. Stendel, Z. Naturforsch., 30b, 281 (1975).
59. R. N. Dixon and J. M. V. Hugo, Mol. Phys., 29, 953 (1975).
60. W. C. Swope, Y. P. Lee, and H. F. Schaefer, J. Chem. Phys., 70, 947 (1979).
61. A. L. Suris, L. S. Aslonyan, and S. N. Shorin, Khim. Vys. Energ., 8, 392 (1974).
62. E. Bechtold, L. Wiesberg, and J. H. Block, Z. Phys. Chem. (Frankfurt am Main), 97, 97 (1975).
63. R. R. Smardzewski, J. Chem. Phys., 68, 2878 (1978).
64. T. H. Gee, J. Weston, and E. Ralph, J. Chem. Phys., 68, 1736 (1978).
65. V. E. Bondybey and J. H. English, J. Chem. Phys., 69, 1865 (1978).
66. S. A. Averkiev, L. S. Agroskin, V. G. Aleksandrov, V. N. Bogo-molov, Y. N. Volgin, A. I. Gutman, T. B. Zhukova, V. P. Petra-kovskii, and D. S. Poloskin, Fiz. Tverd. Tela, 20, 434 (1978).
67. N. R. Carlsen and H. F. Schaefer, Chem. Phys. Lett., 48, 390 (1977).
68. C. E. Moore, Atomic Energy Levels as Derived from Analysis of Optical Spectra, Vol. 1, US Government Printing Office, Washington, D.C. (1949-1958).
69. K. Spitzer, Ph.D. Thesis, University of Washington (1973).
70. M. Schmidt, Chemie Unserer Zeit, 7, 11 (1973).
71. J. Berkowitz, J. Chem. Phys., 62, 4074 (1975).
72. J. Kao, Inorg. Chem., 16, 3347 (1977).
73. R. E. Davis, Mechanism of Sulfur Reactions, Elsevier, Amsterdam (1968), p. 85.
74. M. Schmidt, B. Block, H. D. Block, H. Köpf, and E. Wilhelm, Angew. Chem., 80, 660 (1968).
75. M. Schmidt and E. Wilhelm, Chem. Commun., 1111 (1970).
76. M. Schmidt and H. D. Block, Z. Anorg. Allg. Chem., 385, 119 (1971).
77. R. Stendel and H. J. Mäusle, Z. Anorg. Allg. Chem., 457, 165 (1979).
78. J. Kao and N. L. Allinger, Inorg. Chem., 16, 35 (1977).
79. Z. S. Herman and K. Weiss, Inorg. Chem., 14, 1592 (1975).
80. M. Schmidt and E. Wilhelm, Angew. Chem., 78, 1020 (1966).
81. M. Schmidt, G. Knippschild, and E. Wilhelm, Chem. Ber., 101, 381 (1968).
82. M. Schmidt, A. Kutoglu, and E. Hellner, Angew. Chem., 78, 1021 (1966).
83. J. Kawada and E. Hellner, Angew. Chem., 82, 390 (1970).

84. R. Stendel, R. Reinhard, and F. Schuster, Angew. Chem., 89, 756 (1977).

85. R. Stendel and F. Schuster, J. Mol. Struct., 44, 143 (1978).

86. M. Gardner and A. Rogstad, J. Chem. Soc. Dalton Trans., 599 (1973)

87. C. F. Buchholz, Gelens News J. Chem., 3, 7 (1804).

88. D. Detry, J. Drowart, P. Caldtinger, H. Keller, and H. Rickert, Z. Phys. Chem., 55, 314 (1967).

89. U. I. Zahozszky, Angew. Chem., 80, 661 (1968).

90. R. Reinhardt, R. Stendel, and F. Schuster, Angew. Chem., 90, 55 (1978).

91. R. Stendel, J. Steidel, T. Sandow, and F. Schuster, Z. Naturforsch., 33b, 1198 (1978).

92. M. Schmidt, E. Wilhelm, T. Debaerdenacker, E. Hellner, and A. Kutoglu, Z. Anorg. Allg. Chem., 405, 153 (1974).

93. J. Niimura, Kagaku Chem. Jpn., 26, 308 (1971).

94. R. Stendel and D. F. Eggerz, Spectrochim. Acta, A31, 879 (1975).

95. R. Stendel and M. Rebsch, J. Mol. Spectrosc., 51, 189 (1974).

96. I. Buchler, Angew. Chem., 78, 1026 (1966).

97. W. T. King, Spectrochim. Acta, A31, 1421 (1975).

98. R. E. Barletta and W. T. King, Spectrochim. Acta, A33, 247 (1977)

99. R. Stendel, Spectrochim. Acta, A31, 1065 (1975).

100. T. Debaerdenacker and A. Kutoglu, Naturwissenschaften, 60, 49 (1973).

101. J. V. Sefirov and P. M. Sozki, Zh. Strukt. Khim., 17, 745 (1976)

102. K. Krebs and H. Beine, Z. Anorg. Allg. Chem., 355, 113 (1967).

103. E. D. West, J. Am. Chem. Soc., 81, 29 (1959).

104. T. Debaerdenacker and A. Kutoglu, Cryst. Struct. Commun., 3, 611 (1974).

105. O. Erämetsä and L. Niinistö, Suom. Kemi, 42, 471 (1969).

106. K. Neuman, Z. Phys. Chem., 171, 399 (1934).

107. A. M. Kellas, J. Chem. Soc., 903 (1918).

108. H. F. Schaefer and G. D. Palmer, J. Chem. Educ., 17, 473 (1960).

109. M. D. Gernez, C. R. Acad. Sci. (Paris), 74, 804 (1872).

110. F. Feher, Colloquium Section of Inorganic Chemistry, IUPAC, Muenster (1954), p. 112.

111. N. H. Hartshone and M. H. Roberts, J. Chem. Soc., 1097 (1951).

112. W. Reimschüssel, W. Swiatkowski, and Y. Trybulska, J. Therm. Anal., 14, 99 (1978).

113. M. Thackray, Nature, 201, 674 (1964).

114. J. T. Burwell, Z. Kristallogr., 97, 123 (1937).

115. A. G. Pinkus, C. D. Concilio, J. L. McAttee, and J. S. Kim, J. Am. Chem. Soc., 79, 4566 (1957).

116. M. G. Wolf, J. Chem. Educ., 28, 427 (1951).

117. E. D. Eastman and W. C. McGavock, J. Am. Chem. Soc., 59, 145 (1937).

118. D. E. Sands, J. Am. Chem. Soc., 87, 1395 (1965).

119. M. F. Churbanov, I. V. Skripachev, and G. G. Devyatykh, Zh. Neorg. Khim., 21, 2313 (1976).

120. H. Gerding and E. Westrik, Recl. Trav. Chim. Pays-Bas, 62, 68 (1943).

121. P. D. Bartlett and C. Mequerian, J. Am. Chem. Soc., 78, 3720 (1956).

122. G. M. Barrow, J. Chem. Phys., 21, 219 (1953).
123. P. D. Bartlett, A. R. Colter, E. R. Davis, and W. R. Roderick, J. Am. Chem. Soc., 83, 109 (1961).
124. M. L. Chetrorov, Izv. Fiz. Inst., Bulg. Akad. Nauk, 8, 33 (1960).
125. D. W. Scott and J. P. McCullough, J. Mol. Spectrosc., 6, 372 (1961).
126. F. H. Kruse and D. W. Scott, J. Mol. Spectrosc., 20, 276 (1966).
127. D. W. Scott, J. P. McCullough, and F. H. Kruse, Contribution No. 127, Bartlesville Petroleum Research Center, Bartlesville, Oklahoma (1963).
128. O. Erämetsä, Suom. Kemi, B36, 6, 213 (1963).
129. S. Peter, Z. Elektrochem., 57, 289 (1953).
130. P. W. Schenk, Angew. Chem., 65, 325 (1953).
131. A. Smith, W. B. Holmes, and E. S. Hall, Z. Phys. Chem., 52, 602 (1905).
132. A. Smith and W. B. Holmes, Z. Phys. Chem., 54, 257 (1906).
133. A. Smith and C. M. Carson, Z. Phys. Chem., 57, 698 (1907).
134. A. Smith and C. M. Carson, Z. Phys. Chem., 61, 200 (1908).
135. A. Smith and R. H. Brownlee, Z. Phys. Chem., 61, 209 (1908).
136. H. Specker, Z. Anorg. Allg. Chem., 2, 116 (1950).
137. H. Specker, Kolloid-Z., 125, 106 (1952).
138. H. Specker, Angew. Chem., 65, 299 (1953).
139. H. Mueller and E. Schmid, Monatsh. Chem., 85, 719 (1954).
140. C. W. Thompson and N. S. Gingrich, J. Chem. Phys., 31, 1598 (1959).
141. J. J. Trillat and J. Forestier, C. R. Acad. Sci. (Paris), 192, 559 (1931).
142. A. G. Pinkus, Chem. Eng. News, No. 38, 44 (1960).
143. A Ripomonti and G. Vacca, Ric. Sci., 28, 1880 (1958).
144. J. C. Koh and W. J. Klement, J. Phys. Chem., 74, 4280 (1970).
145. E. Beckmann, R. Paul, and O. Liesche, Z. Anorg. Allg. Chem., 103, 189 (1918).
146. M. D. Gernez, C. R. Acad. Sci. (Paris), 82, 115 (1876).
147. P. W. Schenk and U. Thuemmler, Z. Anorg. Allg. Chem., 315, 271 (1962).
148. A. Smith and W. B. Holmes, Z. Phys. Chem., 42, 469 (1903).
149. P. W. Schenk and Thuemmler, Z. Elektrochem., 63, 1002 (1959).
150. W. J. Macknight, J. A. Poulis, and C. H. Massen, J. Macromol. Sci., A1, 699 (1967).
151. J. A. Semlyen, Trans. Faraday Soc., 64, 1396 (1968).
152. D. M. Gardner and G. K. Fraenkel, J. Am. Chem. Soc., 78, 3279 (1956).
153. M. D. Gernez, C. R. Acad. Sci. (Paris), 98, 144 (1884).
154. M. D. Gernez, C. R. Acad. Sci. (Paris), 100, 1584 (1885).
155. M. D. Gernez, J. Phys. Radium, 4, 349 (1885).
156. P. Spica, Z. Kristallogr., 11, 409 (1884).
157. W. Salomon, Z. Kristallogr., 30, 605 (1891).
158. P. Sabatier, C. R. Acad. Sci. (Paris), 100, 1346 (1885).
159. P. Hatmann, Koninkl. Ned. Acad. Wetenschap., B55, 134 (1952).
160. N. H. Hartshone, R. S. Bradley, and M. Thackray, Nature, 173, 400 (1954).

161. N. H. Hartshone and M. Thackray, J. Chem. Soc., 2122 (1957).
162. R. Laitinen and L. Niinistö, J. Therm. Anal., 13, 99 (1978).
163. O. Erämetsä, Suom. Kemi, B32, 15, 233 (1959).
164. E. Korinth and G. Linck, Z. Anorg. Allg. Chem., 171, 312 (1928).
165. E. Korinth, Z. Anorg. Allg. Chem., 174, 57 (1928).
166. O. Erämetsä, Suom. Kemi, B31, 237, 241, 246 (1958).
167. W. Z. Muthmann, Kristallografiya, 17, 336 (1890).
168. M. Henkel, Dissertation, Univ. Giessen, Giessen (1931).
169. A. Smith and C. M. Cerson, Z. Phys. Chem., 77, 661 (1911).
170. H. Stranz and E. Herka, Naturwissenschaften, 48, 596 (1961).
171. E. Korinth, Dissertation, Univ. Jena (1928).
172. K. H. Meyer and Y. Go, Helv. Chim. Acta, 17, 1081 (1934).
173. J. A. Prins, J. Schenk, and P. A. Hosptel, Physica, 22, 770 (1956).
174. J. A. Prins and F. Tuinstra, Physica, 29, 329, 884 (1963).
175. J. A. Prins and N. J. Poulis, Physica, 15, 696 (1949).
176. J. A. Prins, J. Schenk, and L. H. J. Wachters, Physica, 23, 746 (1957).
177. A. Boulle, Ann. Agron., 17, 575 (1947).
178. R. Faivre and G. Chandrean, Proc. Intern. Congr. Pure Appl. Chem., London, 11, 87 (1947).
179. S. R. Das and M. Gosh, Sci. Cult. (Calcutta), 1, 784 (1936).
180. S. R. Das and K. Kay, Sci. Cult. (Calcutta), 2, 650 (1937).
181. S. R. Das, Indian J. Phys., 12, 163 (1938).
182. E. Poulsen, Dissertation, Univ. Honnover, Honnover (1954).
183. M. R. Engel, C. R. Acad. Sci. (Paris), 112, 866 (1891).
184. R. E. Whitfield, Ph.D. Thesis, Harvard Univ., Cambridge, Mass. (1949).
185. F. O. Rice and J. J. Ditter, J. Am. Chem. Soc., 75, 6066 (1953).
186. A. H. W. Aten, Z. Phys. Chem., 88, 321 (1914).
187. M. R. Engel, Bull. Soc. Chim. France, 6, 3, 12 (1891).
188. J. Donohue, A. Caron, and E. Goldish, J. Am. Chem. Soc., 83, 3748 (1961).
189. A. Caron and J. Donohue, J. Phys. Chem., 64, 1767 (1960).
190. C. Frondel and R. E. Whitfield, Acta Crystallogr., 3, 242 (1950).
191. O. Erämetsä, Suom. Kemi, B35, 154 (1962).
192. B. Meyer and E. Schumacher, Helv. Chim. Acta, 43, 1333 (1960).
193. B. Meyer and E. Schumacher, Nature, 186, 801 (1960).
194. O. Skjerven, Z. Anorg. Allg. Chem., 314, 206 (1962).
195. G. Magnus, Pogg Ann., 92, 312 (1854).
196. J. Donohue, A. Caron, and E. Goldish, Nature, 182, 518 (1958).
197. F. O. Rice and C. J. Sparrow, J. Am. Chem. Soc., 75, 848 (1953).
198. F. O. Rice and R. B. Ingalls, J. Am. Chem. Soc., 81, 1856 (1959).
199. A. F. Kapustinskii, A. K. Mal'tsev, and B. V. Mill, Zh. Neorg. Khim., 5, 506 (1960).
200. R. E. Barletta and C. W. Brown, J. Phys. Chem., 75, 4059 (1971).
201. J. Scandellari, Boll. Chim. Farm., 92, 290 (1953).
202. F. O. Rice, Adv. Chem. Ser., 36, 5 (1962).
203. A. F. Kapustinskii, A. K. Mal'tsev, and B. V. Mill, Tr. Mosk. Khim.-Tekhnol. Inst., 35, 73 (1961).

204. A. F. Kapustinskii, A. K. Mal'tsev, and B. V. Mill, Zh. Neorg. Khim., 8, 1559 (1963).
205. B. M. Vul, Fiz. Tverd. Tela, 3, 2264 (1961).
206. N. S. Kandrick, J. E. Miller, and G. W. Crawford, Phys. Rev., 99, 1631 (1955).
207. S. R. Das and M. Gosh, Indian J. Phys., 13, 91 (1939).
208. B. Meyer, Adv. Inorg. Chem. Radiochem., 18, 287 (1976).
209. G. Aminoff, Arkiv Kemi. Min., 7, 20 (1918).
210. A. Hettich and A. Schleede, Z. Phys., 50, 249 (1928).
211. S. Kreutz, Bull. Acad. Polon. Sci., Ser. Sci. Chim., 60 (1916).
212. P. Groth, Elemente der physikalischen und chemischen Kristallographie, München—Berlin (1921), p. 153.
213. V. Rosický, Z. Kristallogr., 58, 113 (1923).
214. J. Novák, Z. Kristallogr., 76, 169 (1930).
215. G. Friedel and R. Weil, C. R. Acad. Sci. (Paris), 190, 243 (1930).
216. P. Niggli, Z. Kristallogr., 58, 490 (1923).
217. H. Strunz, Mineralogische Tabellen, Geest-Portig, Leipzig (1949), p. 27.
218. B. E. Warren and I. T. Burwell, Phys. Rev., 47, 33 (1935).
219. C. S. Venkateswaren, Proc. Indian Acad. Sci., A4, 345 (1937).
220. C. S. Abrahams, Acta Crystallogr., 8, 661 (1955).
221. C. S. Abrahams, Acta Crystallogr., 14, 311 (1961).
222. U. Ventriglia, Periodico Mineral (Rome), 20, 237 (1951).
223. R. B. Cuddeback and H. G. Drickmer, J. Chem. Phys., 19, 790 (1951).
224. J. J. Trillat and J. Forestier, Bull. Soc. Chim. France, 51, 248 (1932).
225. M. C. Neuburger, Z. Kristallogr., 93, 1 (1936).
226. A. S. Cooper, Acta Crystallogr., 15, 578 (1962).
227. V. M. Goldschmidt, Z. Kristallogr., 51, 10, 15 (1913).
228. F. Fairbrother, G. Gee, and Z. T. Merall, J. Polym. Sci., 16, 459 (1955).
229. B. E. Warren, and J. T. Burwell, J. Chem. Phys., 3, 6 (1935).
230. M. Ataka and C. Tanaka, Jpn. Patent 53-110969 (1978); Ref. Zh. Khim., 17, N199P (1979).
231. L. Pauling, The Nature of the Chemical Bond, 2nd ed., Cornell Univ. Press, Ithaca, New York (1948), p. 53.
232. H. A. Skinner, Trans. Faraday Soc., 41, 645 (1945).
233. A. V. Hippel, J. Chem. Phys., 16, 372 (1946).
234. A. M. Bass, J. Chem. Phys., 21, 80 (1955).
235. R. L. Emerald, R. E. Drews, and R. Zallen, Phys. Rev. B: Solid State, 14, 808 (1976).
236. H. J. Bernstein and J. Pouling, J. Chem. Phys., 18, 1018 (1950).
237. H. J. Bernstein and J. Pouling, J. Chem. Phys., 19, 139 (1951).
238. C. MacNeill, J. Opt. Soc. Am., 53, 398 (1963).
239. H. Gerding and E. Westrik, Recl. Trav. Chim. Pays-Bas, 62, 68 (1943).
240. R. Norris, Proc. Indian Acad. Sci., A13, 291 (1941); A16, 287 (1942).
241. G. Gautier and M. Debeau, Spectrochim. Acta, A30, 1193 (1974).

242. F. P. Daly and C. W. Brown, J. Phys. Chem., 80, 480 (1976).
243. M. Goehring, Chem. Ber., 80, 110 (1947).
244. P. Haglund, Z. Chem., 94, 369 (1935).
245. J. Valasck, Phys. Rev., 43, 612 (1933).
246. M. S. Ioffe and V. I. Nefedov, Zh. Strukt. Khim., 15, 424 (1974).
247. K. Grupe, K. Hellwig, and L. Kolditz, Z. Phys. Chem. (Leipzig), 255, 1015 (1974).
248. P. Nielsen, Int. Conf. Garmisch-Partenkirchen, Vol. 1, London (1974), p. 639.
249. W. Spring, J. Chem. Phys., 5, 410 (1907).
250. W. Spring, Recl. Trav. Chim. Pays-Bas, 26, 357 (1907).
251. W. Spring, Bull. Acad. Belg., 2, 88 (1881).
252. R. F. Marchand and T. Scheerer, J. Prakt. Chem., 24, 129 (1841).
253. W. M. Thornton, Proc. R. Soc., Ser. A, 82, 422 (1909).
254. J. Curie, Ann. Chim. Phys., 17, 385 (1889).
255. W. A. Cunningham, J. Chem. Educ., 12, 120 (1935).
256. S. E. Green, Proc. Phys. Soc., 44, 295 (1932).
257. R. N. Mathur and M. B. Nevgi, Z. Phys., 100, 615 (1936).
258. S. S. Bhatnagar and R. N. Mathur, Philos. Mag., 8, 104 (1929).
259. P. Pascal, Ann. Chim. Phys. (Paris), 19, 5 (1910).
260. G. C. Vezzoli, F. Dachill, and R. Roy, J. Polym. Sci., Part A-1, 7, 1557 (1969).
261. B. C. Deaton and F. A. Blum, Phys. Rev., 137, 1131 (1965).
262. T. Baak, Science, 148, 1220 (1965).
263. P. J. Bounds and R. W. Munn, Chem. Phys., 39, 165 (1979).
264. M. M. Antonova, I. T. Brakhnova, and A. B. Borisova, Properties of the Elements [in Russian], Vol. 2, Moscow (1976), p. 1976.
265. M. P. Dokhov, S. N. Sadumkin, A. A. Karashaev, and B. Kh. Uneshev, Zh. Fiz. Khim., 50, 1801 (1976).
266. G. S. Pawley, The Structure and Dynamics of Orthorhombic Sulfur, Rend. Sen. Int. F., s. Enrico Fermi, New York—London (1975), p. 149.
267. T. Luty and G. S. Pawley, Phys. Status Solidi, 69B, 551 (1975).
268. R. P. Rinald and G. S. Pawley, J. Phys. C: Solid State Phys., 8, 599 (1975).
269. C. S. Lu and J. Donohue, J. Am. Chem. Soc., 66, 818 (1944).
270. P. Mondeau-Manval and P. Schneider, C. R. Acad. Sci. (Paris), 186, 151 (1923).
271. R. F. Bacon and R. Fanell, J. Am. Chem. Soc., 65, 639 (1943).
272. W. N. Tuller (ed.), The Sulphur Data Book, McGraw-Hill, New York (1954).
273. V. M. Lekae and L. N. Elkin, Physicochemical and Thermodynamic Constants of Elementary Sulfur [in Russian], Min. Vysch. Sred. Spets. Obrazovaniya RSFSR, Moscow (1964).
274. A. V. Tobolsky and E. Eisenberg, J. Am. Chem. Soc., 81, 780 (1959).
275. R. E. Powell and H. Eyring, J. Am. Chem. Soc., 65, 648 (1943).
276. R. H. Ewell and H. Eyring, J.Chem. Phys., 5, 726 (1937).
277. G. Gee, Trans. Faraday Soc., 48, 515 (1952).
278. N. S. Gingrich, J. Chem. Phys., 8, 29 (1940).

279. H. Krebs, Angew. Chem., 65, 293 (1953).
280. J. A. Prins, Physica, 20, 124 (1954).
281. P. W. Schenk, Z. Anorg. Allg. Chem., 280, 1 (1955).
282. J. Schenk, Physica, 23, 546 (1957).
283. J. E. Mark, Macromolecules, 11, 627 (1978).
284. J. A. Prins and J. Schenk, Plastica, 216 (1953).
285. H. Jakobson and W. Stockmayer, J. Chem. Phys., 18, 1600 (1950).
286. H. Krebs, Silicium, Schwefel, Phosphate, Colloq. Sek. Anorg.
     Chem. Intern. Union Reine und Angew. Chem., Münster, 107
     (1954).
287. R. Fanelly, Ind. Eng. Chem., 38, 39 (1946).
288. J. Schenk and J. A. Prins, Nature, 172, 957 (1953).
289. L. F. C. Parker, Ind. Rubber J., 108, 387 (1945).
290. J. C. Petrick, Trans. Faraday Soc., 32, 347 (1945).
291. H. Krebs and E. F. Weber, Z. Anorg. Allg. Chem., 272, 288
     (1953).
292. G. C. Vezzoli, P. J. Kisatsky, L. W. Doremus, and P. J. Walsh,
     Appl. Opt., 15, 327 (1976).
293. M. Schmidt, Inorg. Macromol. Rev., 1, 101 (1970).
294. O. Skjerven, Kolloid-Z., 152, 75 (1957).
295. T. K. Wiewiorowski, A. Parthasarathy, and B. Slaten, J. Phys.
     Chem., 72, 1890 (1968).
296. B. Meyer and T. V. Oommen, The Color of Liquid Sulfur,
     Plenum Press, New York (1970/1971), p. 2.
297. W. A. Pryor, Mechanisms of Sulfur Reactions, McGraw-Hill,
     New York (1962).
298. M. Baccaredda and E. Butta, Ann. Chim. (Rome), 45, 50 (1955).
299. A. W. Pryor and E. G. Richardson, J. Phys. Chem., 59, 14
     (1955).
300. E. H. Baker and T. G. Davey, J. Mater. Sci., 13, 1951 (1978).
301. R. F. Bacon and R. Fanelli, Ind. Eng. Chem., 34, 1043 (1942).
302. R. Fanelli, J. Am. Chem. Soc., 72, 4016 (1950).
303. C. C. Farr and D. B. McLeod, Proc. R. Soc., Ser. A, 118, 534
     (1928).
304. T. Doi, Rev. Phys. Chem. Jpn., 33, 41 (1963).
305. A. M. Kellas, J. Chem. Soc., 113, 903 (1918).
306. V. Weimann, Kolloid-Z., 6, 250 (1910).
307. H. Braune and D. Moeller, Z. Naturforsch., A 9, 210 (1954).
308. F. Feher and E. Hellwig, Z. Anorg. Allg. Chem., 294, 63 (1958).
309. G. L. Lewis and M. Randall, J. Am. Chem. Soc., 33, 476 (1911).
310. A. Smith, W. B. Holmes, and E. S. Hall, J. Am. Chem. Soc., 27,
     797 (1905).
311. H. Kopp, Liebigs Ann. Chem., 93, 129 (1855).
312. K. Honda, Ann. Phys. (Leipzig), 32, 1027 (1910).
313. L. Neel, C. R. Acad. Sci. (Paris), 194, 2035 (1932).
314. G. Nadzhtakov and N. T. Kashukeev, Dokl. Akad. Nauk Bolg., 7,
     5 (1954).
315. I. G. Poltavtsev and I. V. Titenko, Zh. Fiz. Khim., 49, 301
     (1975).
316. G. Weser, W. W. Warren, and F. Hensen, Ber. Bunsenges. Phys.
     Chem., 80, 1225 (1976).
317. M. Zanini and J. Tauc, J. Non-Cryst. Solids, 23, 349 (1977).

318. G. Weser, F. Hansel, and W. W. Warren, Ber. Bunsenges. Phys. Chem., $\underline{82}$, 588 (1978).
319. H. Staudinger and W. Kreis, Helv. Chim. Acta, $\underline{8}$, 71 (1925).
320. D. A. Peterson, Avail. NTIS. from Gov. Rep. Announce Index (U.S.), $\underline{79(13)}$, 96 (1979).
321. A. Eisenberg, J. Chem. Phys., $\underline{39}$, 1852 (1963).
322. H. Braune and E. Steinbacher, Z. Naturforsch., $\underline{A7}$, 486 (1952).
323. M. Wakihara, J. Nii, T. Uchida, and M. Tahiguchi, Chem. Lett. Jpn., 621 (1977).
324. E. J. Shaw and T. E. Phipps, Phys. Rev., $\underline{38}$, 174 (1931).
325. G. Preuner and W. Schupp, Z. Phys. Chem., $\underline{68}$, 129 (1910).
326. H. Wartenberg, Z. Anorg. Chem., $\underline{56}$, 320 (1908).
327. H. Biltz, Z. Phys. Chem., $\underline{19}$, 416 (1896).
328. H. Biltz and V. Meyer, Z. Phys. Chem., $\underline{4}$, 249 (1889).
329. W. T. Cooke, Proc. R. Soc., Ser. A, $\underline{77}$, 148 (1906).
330. H. Biltz and G. Preuner, Ber., $\underline{34}$, 2490 (1901).
331. G. Preuner, Z. Phys. Chem., $\underline{44}$, 733 (1903).
332. T. A. Ramaharayanan and W. L. Worrell, J. Electrochem. Soc., $\underline{127}$, 1717 (1980).
333. J. G. Roof, Soc. Petrol. Eng. J., $\underline{11(3)}$, 272 (1971).
334. I. T. Tatarenko, Khim. Ind., 39 (1954).
335. C. Courty, C. R. Acad. Sci. (Paris), Ser. C, $\underline{273}$, 193 (1971).
336. W. R. Orndorff and G. L. Terrasse, Am. Chem. J., $\underline{18}$, 173, 192 (1896).
337. J. Prchlík, L. Hlinák, and J. D. Katáková, Paliva, $\underline{34(11)}$, 298 (1954).
338. T. Kawakami, N. Kubota, and H. Terui, Technol. Repts. Iwate Univ., $\underline{5}$, 77 (1971).
339. G. S. Frolov, V. I. Lazarev, I. S. Antsykovich, and T. V. Epishkina, Zh. Prikl. Khim., $\underline{48}$, 1853 (1975).
340. L. E. Hakka, Brit. Patent 1392813 (1975); Ref. Zh. Khim., $\underline{5}$, L50P (1976).
341. J. J. Smith, D. Jensen, and B. Meyer, J. Chem. Eng. Data, $\underline{15}$, 144 (1970).
342. A. Zipp and S. Zipp, Sulphur Inst. J., $\underline{4(1)}$, 2 (1968).
343. O. A. Abrarov, A. A. Zhuravlev, A. Sultanov, and N. Abdulsalyamov, Dokl. Akad. Nauk Uzb. SSR, 25 (1971).
344. H. R. Krujt, Z. Phys. Chem., $\underline{65}$, 485 (1909).
345. S. Brion, C. R. Acad. Sci. (Paris), $\underline{56}$, 876 (1863).
346. H. Sienebeck, Petroleum, $\underline{18}$, 281 (1922).
347. H. L. Dunlap, Chem. Met. Eng., $\underline{34}$, 298 (1927).
348. E. Pelouze, C. R. Acad. Sci. (Paris), $\underline{68}$, 1179 (1869).
349. E. Pelouze, C. R. Acad. Sci. (Paris), $\underline{69}$, 56 (1869).
350. K. Huerre, Union Pharm., $\underline{62}$, 352 (1921).
351. V. Vesselovsky and V. Kalichevsky, Ind. Eng. Chem., $\underline{23}$, 181 (1931).
352. M. Akamatsu, T. Yokoyama, and G. Hashiz, J. Chem. Soc. Jpn., Ind. Chem. Sect., $\underline{74}$, 1628 (1971).
353. H. Daunenberg, Kautschuk, $\underline{3}$, 104 (1927).
354. A. R. Kemp, F. S. Malm, G. G. Winspear, and B. Stiratelli, Ind. Eng. Chem., $\underline{32}$, 1075 (1940).

355. A. R. Kemp, F. S. Malm, and G. G. Winspear, Kautschuk, 18, 99
      (1942).
356. J. Williams, Rubber Chem. Technol., 12, 191 (1939).
357. T. C. Morris, Ind. Eng. Chem., 24, 584 (1932).
358. G. J. Amerongen and R. Houwinj, Rev. Gen. Caoutch. Plast. Ed.
      Gen., 9, 293 (1942).
359. M. L. Selker and A. R. Kemp, Ind. Eng. Chem., 39, 895 (1947).
360. G. J. Veersen, Chem. Weekblad, 45, 573 (1949).
361. B. S. Grishin, I. A. Tutorskii, I. S. Yuzovskaya, and
      Khun Kuang Khok, Vysokomol. Soedin., Ser. B, 17, 723 (1975).
362. B. S. Grishin, I. A. Tutorskii, I. S. Boikacheva, and
      G. V. Kochetkova, Vysokomol. Soedin., Ser. A, 17, 2481 (1975).
363. H. Lux and E. Böhm, Chem. Ber., 98, 3210 (1965).
364. R. A. Beandet, Chem. Commun., 1083 (1971).
365. W. F. Giggenbach, Chem. Commun., 852 (1970).
366. M. Stillings, M. C. R. Symons, and J. G. Wilkinson, J. Chem.
      Soc., A, 3201 (1971).
367. T. Chivers and J. Drummond, Chem. Commun., 1623 (1971).
368. H. Lux, S. Benninger, and E. Bohm, Chem. Ber., 101, 2485 (1968).
369. G. A. Raukin, J. Phys. Chem., 11, 1 (1907).
370. A. Wigand, Z. Phys. Chem., 77, 423 (1911).
371. J. Boulegue, Phosphorus Sulfur, 5, 127 (1978).
372. O. Anders, Z. Phys. Chem., A 164, 145 (1933).
373. H. Garcia-Fernandez, Bull. Soc. Chim. France, 14, 594 (1947).
374. P. I. Kuznetsov, Tr. Azerbaidzhan. Univ., 52 (1939).
375. J. Meyer, Z. Anorg. Chem., 33, 140 (1903).
376. E. Pincovschi, A. Zaharra, and L. Filipescu, Rev. Roum.-Chim.,
      22, 287 (1971).
377. H. Prinz, Liebigs Ann. Chem., 223, 355 (1884).
378. I. Boguskii and V. Jakubovskii (Yakubovskii), Zh. Russ. Khim.
      Ova., 37, 92 (1905).
379. M. Amadori, Gazz. Chim. Ital., 52, 387 (1922).
380. W. J. Kelly and K. B. Ayers, Ind. Eng. Chem., 16, 148 (1924).
381. W. Jacek, Bull. Acad. Crac., A, 26, 33 (1915).
382. A. Cossa, Ber. 1, 138 (1868).
383. J. H. Hildebrand and C. A. Jenks, J. Am. Chem. Soc., 43, 2172
      (1921).
384. K. A. Hofmann, H. Kirmrenther, and A. Thal, Ber., 43, 183 (1910).
385. G. Bruni and C. Pellizola, Atti Accad. Naz. Lincei, Classe Sci.
      Fis., 30, 158 (1921).
386. R. Delaplage, J. Pharm. Chim., 26, 139 (1922).
387. K. Etard, Ann. Chim. Phys. (Paris), 2, 503 (1894).
388. D. H. Wester and A. Bruins, Pharm. Weekblad, 51, 1443 (1914).
389. T. L. Davis and J. W. Hill, J. Am. Chem. Soc., 49, 3114 (1927).
390. J. N. Brönsted, Z. Phys. Chem., 55, 371 (1906).
391. C. Vulpius, Arch. Pharm. (Weinheim, Ger.), 13, 229 (1878).
392. A. M. Ossendowski, J. Pharm. Chem., 26, 162 (1907).
393. J. S. Lewis, J. Chem. Soc., 761 (1929).
394. A. Grünert and G. Tölg, Talanta, 18, 881 (1971).
395. V. I. Alekseev, Liebigs Ann. Chem., 235, 305 (1886).

396. C. A. Lobry, Z. Phys. Chem., <u>10</u>, 782 (1892).
397. J. W. Retgert, Z. Anorg. Allg. Chem., <u>3</u>, 343 (1893).
398. A. Gerardin, Ann. Chim. Phys. (Paris), <u>5</u>, 129 (1865).
399. W. Langenbeck and H. C. Rhiem, Kautschuk, <u>12</u>, 156 (1936).
400. D. L. Hammick and W. E. Holt, J. Chem. Soc., 1995 (1926).
401. K. A. Kleinedienst and M. L. McLaughlin, J. Chem. Eng. Data, <u>24</u>, 203 (1979).
402. H. L. Friedman and M. Kerker, J. Colloid Interface Sci., <u>8</u>, 80 (1953).

2
Methods of Preparation and Chemical Properties of Different Modifications of Sulfur

A large amount of data concerning the high and diversified reactivity of sulfur is available [1-10].

Because sulfur can undergo various allotropic transformations, depending on the experimental conditions, it can form reactive particles containing from one to several million atoms. The cyclic molecule of S_8 is thermodynamically most stable. In addition to S_8, other cyclic molecules, i.e., S_6, S_7, S_9, S_{10}, S_{11}, S_{12}, S_{18}, and S_{20} can be studied. It seems that isolation and investigation of cyclic molecules of sulfur with a small number of sulfur atoms (S_3, S_4, S_5) and of S_{13}, S_{14}, S_{15}, and S_{17} will become a topic of investigation in the more distant future. The highly reactive monatomic sulfur and the molecular diatomic sulfur have, however, been studied in sufficient detail.

2.1. MONATOMIC SULFUR

Monatomic sulfur is formed by heating sulfur vapor above $1500°C$ at reduced pressure [11], as well as by the action of an electric discharge upon sulfur vapor [8,10-12], and by the decay of S_9 occurring upon moderate heating according to the following scheme [10]:

$$S_9 \longrightarrow S_8 + [S_1] \qquad (2.1)$$

Atomic sulfur cannot be obtained under the conditions used for cleavage of organic disulfides because the process which takes place under these conditions is $8S \rightarrow S_8$ [13,14].

33

A convenient method for the preparation of monatomic sulfur is the photolysis of certain sulfur compounds in the gas phase at low temperatures. The most commonly used method is photochemical decomposition of carbonyl sulfide in an inert gas atmosphere or the photochemical reaction of carbonyl sulfide with mercury vapor [8, 10-12,15-23]:

$$OCS \xrightarrow{h\nu} CO + [S_1^*] \qquad (2.2)$$

In order to generate monatomic sulfur in its triplet or singlet state from carbonyl sulfide, it is necessary to increase the energy by 59.5 kcal/mole and 85.9 kcal/mole, respectively [24]. The wavelength commonly used for the photolysis is 3325 Å or shorter. This leads to the simultaneous formation of sulfur atoms in both triplet and singlet states.

Atomic sulfur may also be obtained by the photolysis of ethylene sulfide [10,12,23], organic isothiocyanates [25,26], and carbon disulfide [23,27].

Monatomic sulfur is highly reactive even at room temperature. In the singlet excited state sulfur atoms can stereospecifically add to C=C bonds and insert themselves into C-H bonds. In the triplet state, sulfur atoms react only with carbon-carbon multiple bonds [10,12,15-29]:

$$-\overset{|}{\underset{|}{C}}-H + S\,(^1D_2) \rightarrow -\overset{|}{\underset{|}{C}}-SH \qquad (2.3)$$

$$\rangle C=\overset{|}{C}H + S\,(^1D_2) \rightarrow \underset{\diagdown S \diagup}{\rangle C-\overset{|}{C}H} \rightarrow \rangle C=\overset{|}{C}-SH \qquad (2.4)$$

$$\rangle C=C\langle + S\,(^3p_2) \rightarrow \underset{\diagdown S \diagup}{\rangle C-C\langle} \qquad (2.5)$$

Reactions 2.3 and 2.5 occur in the gas phase and, as a rule, at elevated pressures.

2.2. DIATOMIC SULFUR

Recently, considerable attention has been given to studies of the simplest form of molecular sulfur, i.e., S_2, an analog of molecular oxygen O_2. Thermal excitation can lead to the formation of S_2 molecules only at temperatures higher than $500°C$ and at reduced pressures [8-10,30,31]. Molecules of S_2 can be isolated from the gas phase as a blue [9,32] or purple modification [33-35] if suitable conditions are used.

At room temperature, S_2 molecules are quite unstable. However, solutions of diatomic sulfur and of other labile oligoatomic molecules of sulfur can be kept for many days at low temperature [36-42]. At room temperature, S_2 molecules can be stabilized using phosphine-iridium or π-cyclopentadienylniobium complexes such as [Ir(S_2)(Ph_2P-CH_2-CH_2-PPh_2)$_2$]Cl [39,43] or (π-C_5H_5)$_2$Nb(S_2)Cl [44].

In preparative chemistry S_2 can be obtained in high yields (90%) by the photolysis of dichloro- or dibromo-disulfane S_2Hal_2 [8,9,45,46], or by gently heating the cyclic allotropic modification S_{10} [10]:

$$S_{10} \longrightarrow S_8 + [S_2]. \tag{2.6}$$

Diatomic and other simple sulfur molecules may be obtained as a result of the complete or partial loss of sulfur on heating unsaturated cyclic oligosulfides:

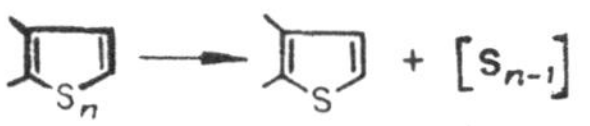

$\longrightarrow$ $+ [s_{n-1}]$ (2.7)

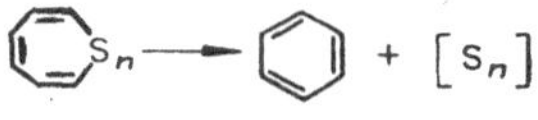

$\longrightarrow$ $+ [s_n]$ (2.8)

The high reactivity of diatomic sulfur is due to the possibility of existence of its molecules in excited triplet and singlet states [8,10,47]. It is assumed [39] that S_2 molecules participate in the interaction of sulfur with carbon-carbon double bonds under certain conditions, e.g.:

$$>C=C< + [S_2] \longrightarrow >\underset{S-S}{C-C}< \longrightarrow 2 >C=S \tag{2.9}$$

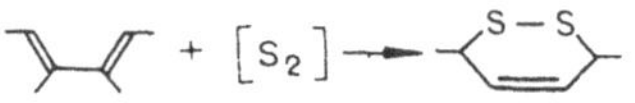

$+ [S_2] \longrightarrow$ (2.10)

2.3. OLIGO- AND POLYATOMIC SULFUR

During the thermal excitation of sulfur it is very difficult, from the practical point of view, to choose conditions under which only one of the possible allotropic modifications can participate. The chemical reactivity of sulfur upon thermal activation depends on the temperature and its molecular structure, and can be approximately expressed in the following way [10,48-54]:

cyclic-S_8 (< 95°C, low activity) $\xrightarrow{\ > 95°C\ }$ linear polymers (low activity) $\xrightarrow{\ > 187°C\ }$ linear oligomers (increased activity) $\xrightarrow{>200°C}$ increased content of molecules with a small number of sulfur atoms (high reactivity).

In order to obtain a certain molecular form of sulfur, one can
use photolysis of its compounds or suitable chemical transformations.
Thus, e.g., the photolysis of S_3Cl_2 and S_4Cl_2 gives the allotropic
modifications S_3 and S_4, respectively [8,9]; both these forms possess
high chemical reactivity:

$$S_3Cl_2 \xrightarrow{h\nu} S_3 + Cl_2 \tag{2.11}$$

$$S_4Cl_2 \xrightarrow{h\nu} S_4 + Cl_2 \tag{2.12}$$

Some allotropic modifications of sulfur which are less stable
than S_8 can be obtained by a redox reaction between oligosulfanes
and dichloro-oligosulfanes in dilute solutions [9,55-57]. This
method was used to obtain and characterize the individual cyclic
molecules S_6, S_{10}, S_{12}, S_{18}, and S_{20}. However, this method has
limited use because of the difficulties connected with the synthesis
and isolation of the individual oligosulfanes:

$$H_2S_n + S_mCl_2 \longrightarrow S_{n+m} + 2HCl \tag{2.13}$$

$$n = 4, \; m = 2; \; n = 6, \; m = 4; \; n + m = 12$$

A convenient method for the preparation of various allotropic
modifications of sulfur is based on the interaction of bis(π-cyclo-
pentadienyl)titanium(IV) pentasulfide with the corresponding dichloro-
sulfanes [58-60]:

$$
\begin{array}{c}
SCl_2 \\
\hline
\end{array} \longrightarrow \tfrac{1}{2}S_6 + \tfrac{1}{4}S_{12} + X \tag{2.14}
$$

$(\pi-C_5H_5)_2TiS_5$

$S_2Cl_2 \longrightarrow S_7 + X \tag{2.15}$

S_4Cl_2 + trace HCl $\longrightarrow S_9 + X \tag{2.16}$

S_6Cl_2 + trace HCl $\longrightarrow S_{11} + X \tag{2.17}$

$SO_2Cl_2 \longrightarrow \tfrac{1}{2}S_{10} + X + SO_2 \tag{2.18}$

$$X = (\pi-C_5H_5)_2TiCl_2$$

Reaction 2.14 gives basically the allotropic form S_6 (87%),
and also a small amount of S_{12} (11%) [58]. The cyclic allotropic
form S_7 has been obtained and characterized for the first time in
an analogous manner [Eq. (2.15)] [58]. The x-ray diffraction
data indicate that S_7 possesses a specific structure in which the
individual sulfur atoms have nonequivalent electronic environments.
Because of this, the -S-S- bonds in S_7 are not energetically equivalent [59]. S_7 polymerizes in the presence of light, and upon
heating, and when kept for some time it slowly changes into a diffi-
cult-to-dissolve crystalline modification whose structure has not
been determined so far.

"Odd" allotropic modifications of sulfur, i.e., S_9 and S_{11}, were obtained using reactions (2.16) and (2.17), respectively [9,60]. These allotropic modifications are stable in the dark at lower temperatures. Treatment of the bis(π-cyclopentadienyl)titanium(IV) pentasulfide with sulfur chloride [Eq. (2.18)] gave the allotropic modification S_{10}.

Attempts to obtain the cyclic modification S_5 using reaction (2.19) have been unsuccessful, obviously because of the high instability of this modification of sulfur [9]:

$$(\pi\text{-}C_5H_5)_2MoS_4 + SCl_2 \longrightarrow (\pi\text{-}C_5H_5)_2MoCl_2 + [S_5] \qquad (2.19)$$

In saturated benzene solutions exposed to light S_6 is partially converted to S_{12}. The latter form was also obtained upon extraction of a cooled melt of ordinary sulfur (at $\sim 120°C$) [61]. This indicates that in the vicinity of the melting point of sulfur, other modifications of sulfur are capable of existence in the liquid sulfur (in addition to S_8).

Thus, the allotropic forms of sulfur known at the present time are S_n with n = 6-12, 18, and 20. Among these allotropic modifications, S_{12} is a relatively stable molecule with the highest melting point (148°C). The stability of S_{12} is confirmed by the corresponding calculations [62]. S_9 is somewhat more stable than S_6; S_6, S_9, S_{10}, and S_{11} undergo ready polymerization above 60°C [10]. There are indications that transformations of the cyclic molecules S_6, S_7, etc. into cyclo-S_8 occurring in solutions of the corresponding allotropic modification of sulfur in an organic solvent in the presence of an initiator (light, atmospheric oxygen, triethylamine, a mixture of sulfur dioxide or hydrogen sulfide with triethylamine) follows the following scheme: ring opening $\rightarrow$ polymerization $\rightarrow$ depolymerization [8,24,63,64]. It is also known that the activity of S_{12} as a Lewis acid in the reaction with diphenyl-o-tolylphosphine is higher than that of S_8, but lower than that of S_6 [65]. However, so far there is insufficient complete data available which would make it possible to systematically compare the stability and reactivity of S_8 and other allotropic modifications of sulfur S_n. Almost nothing is known about the structure of the S_5-S_{12} molecules in the gas phase.

2.4. REACTIONS OF SULFUR IN AQUEOUS MEDIA

The course of the reactions of elemental sulfur with organic compounds depends considerably on the medium in which they take place [10]. A universal solvent like water is rarely used in the reactions of organic compounds with elemental sulfur.

On heating in water the ring of the S_8 molecules undergoes cleavage and sulfide, polysulfide, and oxygen-containing anions are formed [66]. The detailed mechanism of this process is unknown.

It is feasible to carry out reactions of sulfur with certain aromatic amines [67] or aliphatic diamines [68-70] in aqueous or aqueous–ethanolic media in order to decrease the possibility of side reactions occurring at the same time. The interaction of sulfur with aliphatic mono- and diamines in aqueous media, in the absence of hydrogen sulfide acceptors and without their subsequent introduction, is associated with the transformation of sulfur into thiosulfate and sulfide anions, with the participation of hydroxide ions in the reaction [69]:

$$4S + 6OH^{\ominus} \rightarrow [S_2O_3]^{2-} + 2S^{2-} + 3H_2O \qquad (2.20)$$

The reaction of alkylamines with sulfur in an aqueous medium gives a mixture of hydrogen hydropolysulfides and the thiosulfates of the corresponding amines [69]:

$$4RNH_2 + 2(m+1)S + 3H_2O \rightarrow 2[RNH_3]^{\oplus} \cdot [HS_m]^{\ominus} \qquad (2.21)$$
$$+ [RNH_3]_2^{\oplus} \cdot [S_2O_3]^{2-}$$

Diamines give an analogous reaction with the participation of both amino groups [69] and formation of the corresponding hydropolysulfides and thiosulfates. When reaction (2.22) is carried out in the absence of water, e.g., in benzene, no thiosulfates are formed and the principal reaction products are hydropolysulfides [68,69]:

$$3H_2N(CH_2)_n NH_2 + 2(m+1)S + 3H_2O \rightarrow$$
$$[H_3N(CH_2)_n NH_3]^{2+} S_m^{2-} + [H_3N(CH_2)_n NH_3]^{2+} [S_2O_3]^{2-} \qquad (2.22)$$

When considering reactions of sulfur in an aqueous medium in the presence of ammonium hydroxide or hydroxides of alkaline earth metals it is important to remember that the following reactions first take place [66]:

$$\frac{1}{2}S_8 + 5NH_4OH \xrightarrow{250-350^{\circ}C} (NH_4)_2 SO_4 + 3NH_4SH + H_2O \qquad (2.23)$$

$$\frac{1}{2}S_8 + 5NaOH \xrightarrow{250-350^{\circ}C} Na_2SO_4 + 3NaSH + H_2O \qquad (2.24)$$

$$\frac{1}{2}S_8 + 4NaOH \xrightarrow{100^{\circ}C} Na_2S_2O_3 + 2NaSH + H_2O \qquad (2.25)$$

$$\frac{1}{2}S_8 + 4NH_4OH \xrightarrow{100^{\circ}C} (NH_4)_2 S_2O_3 + 2NH_4SH + H_2O \qquad (2.26)$$

$$\frac{1}{2}S_8 + \frac{5}{2}Mg(OH)_2 \xrightarrow{225^{\circ}C} MgSO_4 + \frac{3}{2}Mg(SH)_2 + H_2O \qquad (2.27)$$

$$\tfrac{3}{8}S_8 + 2Ca\,(OH)_2 \xrightarrow{225^\circ C} CaSO_3 + Ca\,(SH)_2 + H_2O \qquad (2.28)$$

$$HS^\ominus + \tfrac{n}{8}S_8 \rightarrow HS_{n+1}^\ominus. \qquad (2.29)$$

2.5. REACTIONS OF ELEMENTAL SULFUR IN LIQUID AMMONIA AND WITH AMINES

Liquid ammonia and amines are widely used in the reactions of elemental sulfur with organic compounds (see Chapter 7), because these bases are excellent activating agents [10]. Anhydrous ammonia reacts with sulfur, and this reaction leads to the final formation of the cyclic compounds HNS_7 and S_4N_4 [71-76].

$$2NH_3 + S_8 \rightarrow NH_4^\oplus + H_2N - S_8^\ominus \qquad (2.30)$$

$$NH_3 + H_2N - S_8^\ominus \rightarrow NH_2SH + H_2N - S_7^\ominus \qquad (2.31)$$

$$H_2N - S_7^\ominus + H_2N - S - S_n^\ominus \rightarrow HNS_7 + H_2NSH + S_n^{2-} \qquad (2.32)$$

$$64NH_3 + 5S_8 \longrightarrow 4N_4S_4 + 24\,(NH_4)_2S \qquad (2.33)$$

The conversion of toluene into benzonitrile by its reaction with sulfur in liquid ammonia is worth mentioning [77]:

$$C_6H_5CH_3 + 3S + NH_3 \rightleftharpoons C_6H_5C\!\!\diagup_{\!NH_2}^{\!S} + 2H_2S \rightleftharpoons C_6H_5CN + 3H_2S \qquad (2.34)$$

The interaction of sulfur with amines may be utilized both for the catalysis of reactions of sulfur with other organic compounds (cf. Section 2.6) and for the synthesis of the products of the reaction of sulfur with amines (cf. Chapter 7).

In most cases, sulfur does not react appreciably with tertiary amines at low temperatures [10,70,78-81]. Yellow solutions of sulfur in pyridine when kept for some time become darker and contain paramagnetic particles, according to ESR spectra [78,79]. It is assumed that pyridine reacts with the cyclic molecules of S_8 by cleaving the ring and by stabilizing the "active" linear S_8 molecules which then readily react with ammonia or amines to give compounds of the ammonium polysulfide type as the final products.

It should be noted that the action of tertiary amines upon sulfur depends on the basicity of the amine (trialkylamines > 4-methylpyridine > pyridine) and also on the steric hindrance at the nitrogen atom in the amine regardless of its basicity (pyridine > 2,6-dimethylpyridine) [70]. One cannot exclude that interaction of a tertiary amine with sulfur in the presence of light may lead to the formation of charge-transfer complexes which can be partially ionic in the excited state [70]:

$$R_3N + S_8 \rightarrow (R_3N^{\oplus}S_8^{\ominus})^* \qquad (2.35)$$

Sufficiently basic primary and secondary amines activate sulfur very easily and thus may undergo various reactions with sulfur. Solutions of sulfur in amines are colored (dark yellow, orange-red, green) and thus one may assume that donor—acceptor-type complexes are formed in the first phase of their interaction (2.36). Later active particles of both ions and ion-radicals are formed [70,78, 82,83]. Both Raman spectra [8] and the electric conductivity of sulfur solutions in amines [70,83] indicate these solutions contain polysulfides and other ions of different structures whose concentration changes slowly with time. The dissociation of sulfur in amines is a reversible process and acidification of these solutions precipitates elemental sulfur. Upon addition of sulfide-ion acceptors (HgO, PbO) to solutions of sulfur in amines, sulfur imides are formed [10,74,75]:

$$
\begin{array}{c}
\cdot S_8^{\ominus} + \cdot \overset{\oplus}{N}HR_2 \\
\Updownarrow \\
S_8 + HNR_2 \rightleftarrows [S_8 \leftarrow NHR_2] \rightleftarrows HS-S_7-NR_2 \\
-HNR_2 \Big\Updownarrow HNR_2 \\
PbS\downarrow + R_2N-S_7-NR_2 \underset{-H_2O}{\overset{PbO}{\leftarrow}} \quad R_2\overset{\oplus}{N}H_2S^{\ominus}-S_7-NR_2
\end{array}
\qquad (2.36)
$$

Other possible directions of the reactions of sulfur with amines under different conditions are discussed in Chapter 7.

2.6. CATALYSIS OF CHEMICAL REACTIONS OF SULFUR

Various catalysts may be used in the reactions of elemental sulfur with organic compounds. The S-S bonds in sulfur molecules are readily cleaved by nucleophiles ($N^{\ominus}$) or by electrophiles ($E^{\oplus}$), as well as by free radicals (R·) [6,7,65,84-91]. The polysulfide chains formed are much more reactive than the relatively stable S_8 molecules. Because of this, the above-mentioned types of reagents may be used as catalysts in the reactions of elemental sulfur:

$$S_8 + N^{\ominus} \rightarrow N-S_7-S^{\ominus} \qquad (2.37)$$

$$S_8 + E^{\oplus} \rightarrow E-S_7-S^{\oplus} \qquad (2.38)$$

$$S_8 + R\cdot \rightarrow R-S_7-S \qquad (2.39)$$

In the presence of nucleophilic agents, many reactions of elemental sulfur occur under mild conditions [4,92-106]. Nucleophilic attack on the sulfur molecules starts by cleavage of the S_8 ring with the formation of dipolar particles. Thus, triarylphosphines react with sulfur according to the following scheme [65,79,89,90].

Elemental sulfur reacts analogously with organomagnesium compounds [7] and with organometallics containing two metals, e.g., Ph$_3$MLi (M = Ge, Sn, Pb) [6]:

$$Ar_3P + S_8 \rightarrow Ar_3P^{\oplus} \rightarrow S - S_6 - S^{\ominus} \qquad (2.40)$$

$$Ar_3P + \overset{\ominus}{S} -\!- S_6 - S \leftarrow \overset{\oplus}{P}Ar_3 \rightarrow Ar_3PS + Ar_3P^{\oplus} \rightarrow S - S_5 - S^{\ominus} \qquad (2.41)$$

$$Ar_3P + \overset{\ominus}{S} - S \leftarrow \overset{\oplus}{P}Ar_3 \rightarrow 2Ar_3PS \qquad (2.42)$$

Important catalysts of the reactions of elemental sulfur are primary, secondary, and often also tertiary amines which facilitate the ionization of carbon or sulfur acids and thus sharply increase their reactivity in their reactions with elemental sulfur. Because of the efficient catalytic effect of amines, many processes which take place at high temperatures in the absence of any amine may be carried out at 40-60°C or even at room temperature in the presence of amines [10,92,107-110]. In this respect, the contributions made by Mayer and Gewald [10,92,107,108] and by Asinger (cf. Chapters 5 and 7) are especially important.

Another method for the introduction of sulfur into organic compounds in the presence of basic catalysts is via the Willgerodt-Kindler reaction, i.e., the reaction of alkyl aryl ketones with elemental sulfur and secondary amines [111]:

$$Ar-\underset{O}{C}-(CH_2)_n\, CH_3 + S + HNR_2 \xrightarrow{130°C} Ar-(CH_2)_{n+1}-C\!\!\begin{array}{c} \nearrow S \\ \searrow NR_2 \end{array} \qquad (2.43)$$

Under suitable conditions [10,92,111] aromatic and heterocyclic methyl and chloromethyl derivatives, alcohols, aldehydes, and other organic compounds also give the Willgerodt-Kindler reaction (cf. Section 3.3). The base-catalyzed introduction of sulfur into 1-alkynes is of preparative importance [93-95]:

$$RC\equiv CH \xrightarrow[-H^+]{B} RC\equiv C^{\ominus} \xrightarrow{S_8} RC\equiv C-S_n-S^{\ominus}$$
$$\qquad\qquad\qquad -s_n \downarrow \uparrow +s_n$$
$$RC=CH \rightleftharpoons RCH=C=S \rightleftharpoons RC\equiv C-SH \underset{\overset{H^{\oplus}}{\rightleftharpoons}}{} RC\equiv C-S^{\ominus} \qquad (2.44)$$
$$\underset{S}{\diagdown\diagup}$$

In the reactions of aromatic nitriles with sulfur, the catalytic activity of trialkylamines increases with increasing length of the alkyl substituents. Reaction (2.45) is five times faster in the presence of tri-n-octylamine than in the presence of triethylamine [96]:

$$2 \text{ ArCN} + S \xrightarrow{R_3N} \underset{N-S}{\overset{Ar}{\underset{\|}{\text{N}}}}\!\!\!\!\!\!\text{Ar} \tag{2.45}$$

$$\text{Ar} = C_6H_5, \; \underline{p}\text{-}CH_3C_6H_4, \; \beta\text{-}C_{10}H_7$$

The catalytic action of gaseous ammonia in the reactions of organic sulfides and disulfides with elemental sulfur in absolute ethanol leading to the formation of the corresponding oligo- and polysulfides is worth mentioning [97]:

$$R_2S_n + S_m \underset{-NH_3}{\overset{+NH_3}{\rightleftarrows}} R_2S_{n+m} \tag{2.46}$$

The reaction of carbonyl compounds with sulfur and ammonia [98] is similar (for details, see Chapter 6, Asinger's reaction). In addition to amines and ammonia, other basic catalysts may be used in various cases; these catalysts can form the following ions upon dissociation: S^{2-}, OH^-, CN^-, I^-, $S_2O_3^{2-}$, ROS_2^-, $R_2NCS_2^-$, etc. [13,37, 64,84,99–102].

Sulfide ions are very widely used as catalysts for sulfur reactions, e.g., in the preparation of alkali polysulfides from elemental sulfur or polysulfanes [84], hydropolysulfides of aliphatic amines, $(R_3N)_2 \cdot H_2S_n$ (n = 6,7,9) [103], and diamines, $(CH_2)_n(NH_2)_2 \cdot H_2S_m$ (n = 2,3,4; m = 3,5,6,7) [104] (cf. Section 7.9), in the reactions of the corresponding amines with sulfur and hydrogen sulfide, in the preparation of dialkyl polysulfides, and in the reaction of sulfur with dialkyl disulfides [84]. Nickel(II) dithiobenzoate reacts with elemental sulfur (2.47) only when the reaction is catalyzed by sulfide anions. This reaction leads to the formation of the corresponding complex [105]:

$$2C_6H_5C\underset{S}{\overset{S}{<}}\!\!\!\!\diagup Ni \diagdown\!\!\!\!\overset{S}{\underset{S}{>}}CC_6H_5 + \;\; 2/8 \; S_8 \;\; \xrightarrow[CC_6H_6]{[S^{2-}]}$$

$$(C_6H_5\overset{S}{\overset{\|}{C}}\!\!-\!\!S)_2Ni\diagdown\!\!\!\!\overset{S}{\underset{S}{}}\!\!\!\!\diagup Ni(S\!\!-\!\!\overset{S}{\overset{\|}{C}}C_6H_5)_2 \tag{2.47}$$

On interaction with sulfur, tetralin may be dehydrogenated to naphthalene under milder conditions in the presence of sulfide anions [106].

Reactive polysulfide ions S_4^{2-}, S_6^{2-}, and S_8^{2-} can be produced in the electrochemical reduction of elemental sulfur in dimethylformamide or dimethyl sulfoxide [112].

Mineral and organic protic acids (HCl, H_2SO_4, H_3PO_4, $ArSO_3H$), Lewis acids ($AlCl_3$, $AlBr_3$, $FeCl_3$, $SbCl_3$, SbF_5), I_2, SI_2, CuI_2, and other compounds can be used as acidic catalysts in the reactions of elemental sulfur with organic compounds.

Electrophilic attack upon S_8 molecules may be illustrated by their reaction with Lewis acids. The adducts thus formed can readily undergo further reactions, e.g., they can react with aromatic hydrocarbons giving diaryl sulfides (cf. Chapter 3). Aluminium chloride or iodine (and other compounds) are used as catalysts in the synthesis of phenothiazine and its N-substituted derivatives (cf. Section 7.2.2), in the Ferrario reaction (cf. Chapter 6) and in some modifications of the Willgerodt-Kindler reaction (cf. Section 7.11):

$$S_8 + AlCl_3 \rightarrow \overset{\oplus}{S} - S_6 - S \rightarrow \overset{\ominus}{AlCl_3} \qquad (2.48)$$

The catalysts aluminium chloride, iron(III) chloride, and iodine are very important in the synthesis of iminothionyl chlorides, e.g., in the reaction of N, N-dichloroalkylamines with elemental sulfur (cf. Section 7.9) which does not take place in the absence of a catalyst [113]:

$$\begin{array}{c} R-NCl_2 \\ \vdots \\ \overset{\oplus}{S}-S_n-S \rightarrow Cat^{\ominus} \end{array} \xrightarrow[-[\oplus S - S_n \rightarrow Cat^{\ominus}]]{} [RN-SCl] \rightarrow \underset{\underset{Cl}{|}}{RN} = SCl_2 \xrightarrow{RNCl_2} etc. \qquad (2.49)$$

Hydrochloric acid, sulfuric acid, and iodine may be used as catalysts in the reaction of aniline with sulfur (cf. Section 7.2.1).

When studying the reactions of elemental sulfur, one has to remember that oxidizing agents oxidize S_8 molecules to sulfur cations [7,10,114,115]. Thus in oleum, for example, sulfur undergoes gradual oxidation with the formation of the cations S_4^{2+}, S_8^{2+}, and S_{16}^{2+}. The S_8^{2+} ion is in equilibrium with the radical cation $S_4^{+\cdot}$, and S_{16}^{2+} is in equilibrium with $\cdot S_8^{+}$ [114,115]. So far cations of sulfur have not been used in organic reactions.

In some special cases, copper, zinc, platinium, activated charcoal, zinc sulfide, and potassium fluoride have been used as catalysts in sulfur reactions (cf. Chapters 4 and 7). Many reactions of sulfur with organic compounds occur under mild conditions when hydrogen sulfide acceptors are used (HgO, MgO, PbO, mercuric amides and imides, amines of Hg salts) (cf. [116] and Chapter 7).

Accelerators of rubber vulcanization which are widely used are the dithiocarbamates and benzothiazolethiolates of zinc, lead, and other heavy metals [84]. (The vulcanization of rubber is discussed in detail in Chapter 3.)

Free radicals cause ready cleavage of the cyclic S_8 molecule with the formation of polysulfides. Radicals formed during the decomposition of azo compounds attack S_8 even at 112°C [85]. Aliphatic radicals formed from triazenes react quantitatively with sulfur [88]:

$$C_tH_5NH-N=NR+HOH \longrightarrow C_6H_5NH_2+ RN=NOH \qquad (2.50)$$

$$RN=NOH \longrightarrow R\cdot+\cdot OH+N_2 \qquad (2.51)$$

$$R\cdot+S_8 \longrightarrow RS\cdot_n \longrightarrow R-S_m-R \qquad (2.52)$$

Elemental sulfur inhibits the polymerization of various vinylic monomers because radicals formed from these monomers react preferentially with the cyclic S_8 molecules [7,117-121].

In reactions occurring in the 160-220°C temperature range and in the pyrolysis of some sulfur-containing compounds, sulfur reacts in the form of diradicals, $\cdot S_n \cdot$, and, because of this, it serves as an initiator in various radical processes [122] such as dehydrogenation by sulfur and the high-temperature uncatalyzed introduction of sulfur into various organic compounds.

REFERENCES

1. A. Senning (ed.), Sulfur in Organic and Inorganic Chemistry, Vols. 1-3, Marcel Dekker, New York (1972).
2. R. F. Gould (ed.), Advances in Chemistry Series 110, Sulfur Research Trends, American Chemical Society, Washington D.C. (1972).
3. G. Nickless (ed.), Inorganic Sulfur Chemistry, Elsevier, Amsterdam—London—New York (1968).
4. M. J. Janssen (ed.), Organosulfur Chemistry, Wiley-Interscience, New York—London—Sidney (1967).
5. N. Kharasch (ed.), Organic Sulfur Compounds, Pergamon Press, Oxford—London—New York—Paris (1961).
6. B. Meyer (ed.), Elemental Sulfur, Chemistry and Physics, Wiley-Interscience, New York (1965).
7. W. A. Pryor, Mechanisms of Sulfur Reactions, McGraw-Hill, New York—San Francisco—Toronto—London (1962).
8. T. Chivers and J. Drummond, Chem. Soc. Rev., 2, No. 2, 233 (1973).
9. M. Schmidt, Angew. Chem., 85, 474 (1973).
10. R. Mayer, Z. Chem., 13, 321 (1973).
11. B. Meyer, Elemental Sulfur, Ref. [3], p. 241.
12. O. P. Strausz, The Chemistry of Atomic Sulfur, Ref. [1], Vol. 2.
13. O. Foss, Acta Chem. Scand., 4, 404 (1950).
14. Ch. C. Price and Sh. Oae, Sulfur Bonding, Ronald Press, New York (1962) p. 44.
15. O. P. Strausz, The Addition of Sulfur Atoms to Olefines, Ref. [2], p. 137.

16. O. P. Strausz, The Reactions of Atomic Sulfur, Ref. [4], p. 11.
17. A. R. Knight, O. P. Strausz, and H. E. Gunning, J. Am. Chem. Soc., 85, 2349 (1963).
18. A. R. Knight, O. P. Strausz, S. M. Malm, and H. E. Gunning, J. Am. Chem. Soc., 86, 4243 (1964).
19. H. E. Gunning, Trans. R. Soc. Can., Sect. 1-3, 2, 293 (1964).
20. H. A. Wieba, A. R. Knight, O. P. Strausz, and H. E. Gunning, J. Am. Chem. Soc., 87, 1443 (1965).
21. K. S. Sidhu and E. M. Lown, J. Am. Chem. Soc., 88, 254 (1966).
22. W. J. MacKnight, Ann. N.Y. Acad. Sci., 159 (Pt. 1), 267 (1969).
23. H. E. Gunning, The Reaction of Atomic Sulfur, Ref. [6], p. 265.
24. R. E. Davis, Mechanisms of Sulfur Reactions, Ref. [3], p. 85.
25. V. Schmidt and Ch. Osterroht, Angew. Chem., 77, 455 (1965).
26. O. M. Nefedov and M. N. Manakov, Angew. Chem., 78, 1039 (1966).
27. W. D. McGrath, T. Morrow, and D. N. Dempster, Chem. Commun., 516 (1967).
28. O. P. Strausz, H. E. Gunning, A. S. Denes, and I. G. Csizmadia, J. Am. Chem. Soc., 94, 8317 (1972).
29. R. B. Klemm and D. D. Davis, Int. J. Chem. Kinet., 5, No. 3, 375 (1973).
30. W. Klemm and H. Kilian, Z. Phys. Chem., B49, 279 (1941).
31. H. Braune, S. Peter, and V. Neveling, Z. Naturforsch., 6A, 32 (1951).
32. B. Mayer, Adv. Chem. Ser., 110, 53 (1972).
33. F. O. Rice and C. J. Sparrow, J. Am. Chem. Soc., 75, 848 (1953).
34. G. Skandellari, Boll. Chim. Farm., 92, 290 (1953).
35. L. Brewer and G. D. Brabson, J. Chem. Phys., 44, 3274 (1966).
36. L. Brewer, C. D. Brabson, and B. Mayer, J. Chem. Phys., 42, 1385 (1965).
37. H. Lux, S. Benninger, and E. Böhm, Chem. Ber., 101, 2485 (1968).
38. T. V. Oommen, Diss. Abstr. Int., 31B, 3904 (1971).
39. A. P. Ginsberg and W. E. Lindsell, J. Chem. Soc., D, 232 (1971).
40. W. D. Bonds and J. A. Ilbers, J. Am. Chem. Soc., 94, 3413 (1972).
41. T. Chivers and J. Drummond, J. Chem. Soc., D, 1623 (1971).
42. H. Lux and H. Anslinger, Chem. Ber., 94, 1161 (1961).
43. A. P. Ginsberg, W. E. Lindsell, and W. E. Silverthorn, Trans. N.Y. Acad. Sci., 303 (1971).
44. P. M. Treichel and G. P. Werber, J. Am. Chem. Soc., 90, 1753 (1968).
45. B. Mayer, T. Stroyer-Hansen, D. Jensen, and T. V. Oommen, J. Am. Chem. Soc., 93, 1034 (1971).
46. B. Mayer, T. V. Oommen, and D. Jensen, J. Phys. Chem., 75, 912 (1971).
47. D. J. Meschi and A. W. Searcy, J. Chem. Phys., 51, 5134 (1969).
48. A. V. Tobolsky and W. J. MacKnight, Polymeric Sulfur and Related Polymers, Wiley-Interscience, New York (1965).
49. T. K. Wiewiorowski and F. J. Touro, J. Phys. Chem., 70, 3528 (1966).

50. A. Eisenberg and L. A. Telter, J. Phys. Chem., 71, 2332 (1967).
51. A. T. Ward, J. Phys. Chem., 72, 4133 (1968).
52. A. Eisenberg, Macromolecules, 2, 44 (1969).
53. R. E. Harris, J. Phys. Chem., 74, 3102 (1970).
54. B. Mayer and K. Spitzer, J. Phys. Chem., 76, 2274 (1972).
55. M. Schmidt, Inorg. Macromol. Rev., 1, 101 (1970).
56. M. Schmidt and E. Wilhelm, Inorg. Nucl. Chem. Lett., 1: 39 (1965).
57. M. Schmidt and E. Wilhelm, Angew. Chem. Int. Ed. Engl., 5, 964 (1966).
58. M. Schmidt, B. Block, H. D. Block, H. Köpf, and E. Wilhelm, Angew. Chem., 80, 660 (1968); Angew. Chem. Int. Ed. Engl., 7, 632 (1968).
59. I. Kawada and E. Hellner, Angew. Chem., 82, 390 (1970).
60. M. Schmidt and E. Wilhelm, J. Chem. Soc., D, 111 (1970).
61. M. Schmidt and H. D. Block, Angew. Chem., 79, 944 (1967).
62. F. Lin, Diss. Abstr. Int., 31B, 7222 (1971).
63. P. D. Bartlett, G. Lohaus, and C. D. Weis, J. Am. Chem. Soc., 80, 5064 (1958).
64. P. D. Bartlett, A. K. Colter, R. E. Davis, and W. R. Roderick, J. Am. Chem. Soc., 83, 109 (1961).
65. F. Feher and D. Kurz, Z. Naturforsch., 24B, 1089 (1969).
66. Ref. [7], p. 11.
67. B. Jaeckel, Chem. Ber., 83, 578 (1950).
68. H. Krebs and K.-H. Müller, Z. Anorg. Allg. Chem., 281, 187 (1955).
69. M. G. Voronkov and A. Ya. Legzdyn', Zh. Org. Khim., 3, 465 (1967).
70. R. E. Davis and H. F. Nakshbendi, J. Am. Chem. Soc., 84, 2085 (1962).
71. O. Ruff and E. Geisel, Ber., 38, 2659 (1905).
72. A. P. Zipp and S. G. Zipp, Sulfur Inst. J., 4(1), 2 (1968).
73. J. W. Bergstrom, J. Am. Chem. Soc., 48, 2319 (1926).
74. M. Becke-Goehring, Angew. Chem., 73, 589 (1961).
75. H. Jenne and M. Becke-Goehring, Chem. Ber., 91, 1950 (1958).
76. Borchers, Gebr., A.-G: Brit. Patent 753,667 (1956); C. A., 51, 8127h (1957).
77. M. G. Toland, J. Org. Chem., 27, 869 (1962).
78. N. W. H. M. Wu, Diss. Abstr. Int., 30B, 4550 (1970).
79. W. G. Hodgson, Sh. A. Buckler, and G. Peters, J. Am. Chem. Soc., 85, 543 (1963).
80. H. Krebs and E. F. Weber, Z. Anorg. Allg. Chem., 272, 288 (1953).
81. C. R. Platzmann, Bull. Chem. Soc. Jpn., 4, 257 (1929).
82. F. P. Daly and C. W. Brown, J. Phys. Chem., 77, 1859 (1973).
83. T. Chivers and I. Drummond, J. Chem. Soc. Dalton Trans., 631 (1974).
84. H. Krebs, Die Katalytische Aktivierung der Schwefels, Westdeutscher Verlag, Köln und Opladen (1958).
85. A. J. Parker and N. Kharash, Chem. Rev., 59, 583 (1959).
86. C. H. Wang and S. G. Cochen, J. Am. Chem. Soc., 81, 3005 (1959).

87. E. S. Tinyakova, B. A. Dolgoplosk, and M. P. Tikhomolova, Zh. Obshch. Khim. 25, 1387 (1955).
88. W. Ya. Andakushkin, B. A. Dolgoplosk, and N. I. Radchenko, Zh. Obshch. Khim., 26, 2972 (1956).
89. P. D. Bartlett and G. Meguerian, J. Am. Chem. Soc., 78, 3710 (1956).
90. P. D. Bartlett, E. Cox, and R. E. Davis, J. Am. Chem. Soc., 83, 103 (1961).
91. M. Schmidt, Angew. Chem., 73, 394 (1961).
92. R. Mayer and K. Gewald, Angew. Chem., 79, 298 (1967).
93. H. Spies, K. Gewald, and R. Mayer, J. Prakt. Chem., 313, 804 (1971).
94. R. Mayer and B. Gebhardt, Chem. Ber., 97, 1298 (1964).
95. M. Schmidt and V. Potschka, Naturwissenschaften, 50, 302 (1963).
96. W. Mack, Angew. Chem., 79, 1106 (1967).
97. B. Holmberg, Ber., 43, 220 (1910).
98. F. Asinger, M. Thiel, and F. Pallas, Liebigs Ann. Chem., 602, 37 (1957).
99. F. Asinger, H. Offermanns, D. Neuray, and P. Mueller, Monatsh. Chem., 101, 1295 (1970).
100. US Patent, 2,496,319 (1950); Chem. Zentralbl., II, 1511 (1950).
101. M. Goehring, W. Helbing, and J. Appel, Z. Anorg. Allg. Chem., 254, 185 (1947).
102. P. D. Bartlett and R. E. Davis, J. Am. Chem. Soc., 80, 2513 (1958).
103. F. Feher and M. Baudler, Z. Anorg. Allg. Chem., 258, 132 (1949).
104. H. Krebs and K. H. Müller, Z. Anorg. Allg. Chem., 281, 187 (1955).
105. W. Hieber and R. Brück, Z. Anorg. Allg. Chem., 269, 13 (1952).
106. A. Jennen, C. R. Acad. Sci. (Paris), 241, 1581 (1955).
107. K. Gewald, J. Prakt. Chem., 32, 26 (1966).
108. K. Gewald and E. Schinke, Chem. Ber., 99, 2712 (1966).
109. M. Chaykovsky and M. Lin, J. Med. Chem., 16, 188 (1973).
110. M. Nakanishi and T. Tahara, J. Med. Chem., 16, 214 (1973).
111. R. Wegler, E. Kühle, and W. Schäfer, Angew. Chem., 70, 351 (1958).
112. R. Bonnatere and G. Cauquis, J. Chem. Soc., Chem. Commun., 293 (1972).
113. L. N. Markovskij, G. S. Fed'yuk, E. S. Levchenko, and A. V. Kirsanov, Zh. Org. Khim., 9, 2502 (1973).
114. R. J. Gillespie and P. K. Ummat, Inorg. Chem., 11, 1674 (1972).
115. W. Giggenbach, Inorg. Chem., 10, 1327, 1333 (1971).
116. P. S. Pishchimuka, Zh. Obshch. Khim., 21, 1689 (1951).
117. K. U. Ingold and B. P. Roberts, Free-Radical Substitution Reactions, Wiley-Interscience, New York—London—Sydney—Toronto (1971).
118. A. Krakay, Ber., 11, 1261 (1878).
119. P. D. Bartlee and G. Kwart, J. Am. Chem. Soc., 74, 3969 (1952).
120. J. L. Kice, J. Am. Chem. Soc., 76, 6274 (1954).

121.   J. L. Kice, J. Polym. Sci., $\underline{19}$, 123 (1956).
122.   A. W. Horton, J. Org. Chem., $\underline{14}$, 761 (1949).

3
The Action of Sulfur upon Hydrocarbons

3.1. ALKANES

3.1.1. Methane and the Lower Alkanes

The following reactions may take place when alkanes are treated with sulfur:

1) Introduction of sulfur with the formation of thiols, sulfides, and polysulfides which can then undergo further decomposition leading to unsaturated compounds and hydrogen sulfide.
2) "S-Dehydrogenation" leading to alkenes, alkadienes, acetylene, or benzene which can then further react incorporating sulfur and undergoing polymerization with the formation of hydrogen sulfide.
3) Processes leading to the introduction of large amounts of sulfur into the molecules and exhaustive S-dehydrogenation. The final products are macromolecular tar-like or carbon-like substances, sulfur-containing dyes, polymeric carbon sulfides, etc.
4) Formation of unsaturated sulfur heterocycles (thiophenes, thiophthenes, tetrahydrothiophenes, 1,2-dithiol-3-thiones).
5) "S-Oxidation" with the formation of carbon disulfide.

All the above-mentioned reactions are mainly important from the industrial point of view and form the basis of industrial processes for the synthesis of carbon disulfide (cf. Section 3.1.4), hydrogen sulfide (cf. Section 3.1.3), 1,3-butadiene, and thiophene and its homologs.

The reactions studied in most detail are those of elemental sulfur with methane [1-9]. The reaction of sulfur with methane has

49

been studied over a wide temperature range (550-1000°C) and at
different pressures (30-400 atm), both with and without various
catalysts [1,2,10-14]. The principal reaction products are carbon
disulfide, hydrogen sulfide, and hydrogen:

$$CH_4 + 4S \longrightarrow CS_2 + 2H_2S \qquad (3.1)$$

$$CH_4 + 2S \longrightarrow CS_2 + 2H_2 \qquad (3.2)$$

With added water vapor, the reaction can be selectively directed
towards the formation of hydrogen sulfide [13]. The kinetics of
the reaction of methane with sulfur in the gaseous phase at 600-700°C
has been studied [15]. The thermodynamic parameters of this reaction
within the 25-1220°C temperature range and at a pressure of 1-5 atm
have been calculated [16]:

$$CH_4 + 4S + 2H_2O \longrightarrow 4H_2S + CO_2 \qquad (3.3)$$

A regular change in the relative reactivity of ethane and
ethylene in the reaction with elemental sulfur in the presence of
metal sulfides as catalysts has been established [17,18]. The
principal reaction products obtained in the reaction of sulfur with
a mixture of lower alkanes (C_1-C_3) at 450-650°C and 6-12 atm are
hydrogen sulfide and carbon disulfide [5,19]. When lower alkanes
are heated with sulfur above 220°C they undergo dehydrogenation with
the formation of hydrogen sulfide. This process can be carried out
under controlled conditions which stop the conversion at the stage
of alkene formation. The reaction of sulfur with alkanes is first
order with respect to the hydrocarbon and its rate does not depend
on the sulfur vapor pressure. However, when going from ethane to
butane, the reaction rate increases 400 times and it shows an even
higher increase for branched hydrocarbons [20]. The reaction
products are alkenes with the same number of carbon atoms. The
dehydrogenation of ethane in the presence of sulfur gives not only
ethylene but also acetylene when an appropriate $S:C_2H_6$ ratio is
chosen [21]. Alkenes are the products of the reaction of sulfur
with alkanes, starting with methane and ending with heptane, carried
out at ∿ 1000°C and in the presence of catalysts such as alumina
(in the case of methane) or vanadium pentoxide [22]. Methane gives
ethylene and the higher alkanes give the corresponding higher
alkenes. All these compounds are then converted into carbon disulfide.

At 650-750°C and 0.04-0.1 s contact time, ethane and propane
are dehydrogenated by sulfur vapor to the corresponding alkenes
(65-95% yield). At higher temperatures (880°C) and with longer
contact time (2-3 s), the yield of ethylene from ethane is only 50%
[21]. A synthetic procedure for propene has been worked out on the
basis of the dehydrogenation of propane with sulfur as the selective
dehydrogenating agent (51% yield) [23,24]. The reaction is carried

out at 700-730°C and 17-18 atm, with a contact time of 120 ms. Atomic sulfur is very reactive toward hydrocarbons. Dehydrogenation of the C_3H_8 petroleum fraction by sulfur at 170°C has been performed [25].

The basic reaction of sulfur atoms in their singlet state, $S(^1D_2)$, with alkanes is a concerted one-step attack on the C-H bond with the formation of a thiol. The insertion occurs in the gas phase and is not selective with respect to the various types of C-H bonds. It takes place at an equal rate for ethane, propane, and cyclopropane. Parallel with this insertion, deactivation of excited sulfur atoms takes place:

$$S(^1D_2) + RH \begin{array}{l} \xrightarrow{K_1} RSH \\ \xrightarrow{K_2} S(^3P) + RH \end{array}$$

$$(3.4)$$
$$(3.5)$$

The probability of this incorporation (3.4) increases with increasing wavelength. Thus, for the reaction of atomic sulfur with ethane the k_1/k_2 ratio is 0.71 when the effective wavelength of the exciting radiation in the photolysis of carbonyl sulfide is 2490 Å, whereas it is only 0.31 when the wavelength is 2288 Å. In the case of propane, this ratio is 0.16 at 2288 Å, and for cyclopropane it is 0.50 at 2490 Å. The absolute value of k_1 is $> 2 \cdot 10^{10}$ liter·mole^{-1}·s^{-1} [26,27] and the activation energy of reactions (3.4) and (3.5) is ~ 1 kcal/mol or somewhat higher. The reaction of methane with $S(^1D_2)$ leads to the formation of excited methanethiol which then decomposes. The first step of the reaction (insertion of sulfur) is accompanied by a large decrease of enthalpy. The final reaction products are CH_3SH, H_2S, CH_3SSCH_3, CH_3SCH_3, CS_2, C_2H_6, H_2, and elemental sulfur [26,28]:

$$\begin{array}{ll} S(^1D_2) + CH_4 \longrightarrow CH_3SH^* & \Delta H \simeq -82.7 \text{ kcal} \\ & \\ CH_3SH^* - \begin{cases} \rightarrow CH_3 + SH & \Delta H \simeq -7 \text{ kcal} \\ \rightarrow CH_3S + H & \Delta H \simeq +8 \text{ kcal} \\ \rightarrow CH_2 + H_2S & \Delta H \simeq +1 \text{ kcal} \\ \rightarrow CS + 2H_2 & \Delta H \simeq -19 \text{ kcal} \end{cases} \end{array}$$

$$(3.6)$$

3.1.2. Higher Alkanes

Low-molecular-weight alkanes react with sulfur considerably slower and at higher temperatures than do high-molecular-weight alkanes. Thus, hexane does not react with sulfur even on heating at 210°C for twenty-four hours in a sealed tube [29,30]. Nevertheless, certain experimental evidence shows that heptane and iso-octane when shaken with sulfur at 60-90°C liberate hydrogen sulfide and form thiols and disulfides [31]. Disulfides can eliminate hydrogen sulfide and may then be transformed into alkenes which undergo polymerization.

Interaction of lower alkanes with sulfur above 220°C leads to their dehydrogenation with the formation of hydrogen sulfide. The dehydrogenation of butane by sulfur is a chain process and involves radicals [20]. The radicals are formed from butane by the rupture of C–H bonds according to the following scheme. Olefins are intermediate products formed in the reaction of sulfur with butane, pentane, and heptane at ∿ 1000°C in the presence of vanadium pentoxide. Later in the process these olefins are converted into carbon disulfide [22]. Isobutane is dehydrogenated by sulfur vapor at 650–760°C and gives isobutene in a high yield [21] which decreases at higher temperatures. A mixture of mono-, di- and polysulfides is obtained from the reaction of sulfur with 2,4-dimethylpentane at 160–180°C in the presence of di-tert-butyl peroxide (total yield 72%) [32]:

$$CH_3CH_2CH_2CH_3 + S \xrightarrow{S_8 \to 8S} CH_2CH_2\dot{C}HCH_3 + HS\cdot$$

$$CH_3CH_2CH_2CH_3 + HS\cdot \longrightarrow CH_3CH_2\dot{C}HCH_3 + H_2S$$

$$CH_3CH_2\dot{C}HCH_3 - \left[\begin{matrix} \to \dot{C}H_3 + CH_3CH=CH_2 \\ \to CH_3\dot{C}H_2 + CH_2=CH_2 \end{matrix} \right. \tag{3.7}$$

$$CH_3CH=CH_2 + S \longrightarrow CH_3CH=\dot{C}H + HS\cdot$$

$$CH_3CH=\dot{C}H + \dot{C}H_3 \longrightarrow CH_3CH=CHCH_3$$

A radical chain mechanism for the high-temperature reaction of higher alkanes with sulfur catalyzed by amines has been proposed [33]. The reaction starts by the cleavage of the S_8 rings to give bi-radicals, $\cdot S_2\cdot$, which then react with a hydrocarbon and yield sulfides according to the following scheme:

$$S_8 \longrightarrow 4\cdot SS\cdot$$

$$R—CH_3 + \cdot SS\cdot \longrightarrow R—CH_2\cdot + HSS\cdot$$

$$HSS\cdot \longrightarrow HS\cdot + \cdot S\cdot$$

$$R—CH_3 + HS\cdot \longrightarrow R—CH_2\cdot + H_2S$$

$$R—CH_2\cdot + \cdot SS\cdot \longrightarrow R—CH_2—SS\cdot$$

$$R—CH_2—SS\cdot + R—CH_2\cdot \longrightarrow R—CH_2—SS—CH_2—R \tag{3.8}$$

$$R—CH_2\cdot + \cdot S\cdot \longrightarrow R—CH_2—S\cdot$$

$$R—CH_2—SS\cdot + R—CH_2—SS\cdot \longrightarrow R—CH_2—S_4—CH_2—R$$

$$R—CH_2—SS\cdot + R—CH_2—S\cdot \longrightarrow R—CH_2—S_3—CH_2—R$$

$$R—CH_2\cdot + R—CH_2\cdot \longrightarrow R—CH_2—CH_2—R$$

Friedmann was the first to observe the formation of thiophenes in the reaction of sulfur with alkanes [34,35]. He was able to show that heating octane with sulfur in a sealed tube at 270–280°C produces a small amount of substituted thiophenes, $C_8H_{12}S$ (according to the author, these compounds are 2,5-diethylthiophene, 4-isopropyl-3-methylthiophene, and 3,4-dimethylthieno[2,3-b]thiophene,

$C_8H_2S_2$). The reaction is accompanied by the isomerization of octane to 2,3,4-trimethylpentane which further reacts with sulfur according to the scheme:

$$CH_3CH_2CH_2CH_2CH_2CH_2CH_2CH_3 \longrightarrow CH_3-CH-CH-CH-CH_3 \longrightarrow$$
$$\underset{CH_3\ \ CH_3\ \ CH_3}{\qquad\qquad}$$

$$\xrightarrow[-3H_2S]{+4S} \ \ \underset{S\ \ CH_3}{CH_3}\boxed{}\overset{CH-CH_3}{} \ \xrightarrow{+3S} \ \ \underset{S\ \ S}{CH_3}\boxed{}CH_3 +2H_2S$$

$$(3.9)$$

Butane or heptane react very slowly with sulfur in a steel bomb at 300-350°C to give small amounts of sulfur-containing compounds, probably thiophenes [36]. The interaction of sulfur with pentane and isooctane in an autoclave at 275-285°C yields a complex mixture of products consisting of sulfides, thiophenes, tetrahydrothiophenes, thieno[2,3-b]thiophenes, and bithiophenes [37,38]. The reaction of butane with sulfur serves as the basis for an industrial preparation of thiophene in $\sim$ 70% yield. The process is carried out at 570°C, the contact time is two seconds and the ratio of the starting materials is (wt.) 1:1 [39-43]. In addition to thiophene the reaction products include 1,3-butadiene, butenes, hydrogen sulfide, carbon disulfide, and lower hydrocarbons. Another industrial synthesis of thiophene is based on the interaction of butane and its dehydrogenation products with sulfur vapor at 700°C and atmospheric pressure [44-48]. The principal product obtained in the high-temperature reaction of butane with sulfur on irradiation with a ruby laser (100 MW) is carbon disulfide [49]:

$$\underset{CH_3\ \ CH_3}{\overset{CH_2-CH_2}{|\ \ \ \ |}} \ \xrightarrow{+4S} \ \ \boxed{}_S + 3H_2S$$

$$(3.10)$$

The interaction of pentane with sulfur in the vapor phase at 400-600°C gives a mixture containing methylthiophene, thieno[2,3-d]-1,2-dithiol-3-thione, and 2H-thiapyran-2-thione [50].

Heating alkanes with sulfur in the presence of aluminium chloride, gallium chloride, or iron(III) chloride under pressure leads to the formation of thiols and sulfides [51]. The halides accelerate this reaction. The reaction of butane or pentane with sulfur in the presence of aluminium chloride under pressure and at 125-135°C is accompanied by isomerization and in the case of butane gives dibutyl sulfide, 2-methylpropanethiol and diisobutyl sulfide. In the case of pentane, the products are 3-methylbutanethiol and the corresponding sulfides and disulfides. Together with these compounds a small amount of 4,5-dimethyl-1,2-dithiol-3-thione is formed [52].

Sulfuration products of C_4-C_6 alkanes obtained at 400-850°C including thiophene, methylthiophenes, thiophenol, etc., have been suggested as possible corrosion inhibitors [53]. Sulfuration of hexadecane (cetane) at 200-240°C gives sulfides, disulfides, and hydrogen sulfide [54,55]. The amount of sulfides increases with increasing temperature. In the presence of aluminium chloride, the reaction gives hexadecanethiol, octanethiol, and a mixture of gaseous products containing methane, propane, isobutane, isopentane, and hydrogen sulfide [56].

Two competing reactions take place when sulfur reacts with higher alkanes. At 240°C, the principal reaction is dehydrogenation leading to alkenes, cycloalkanes, arenes, and asphaltenes. The process occurring at lower temperatures (140°C) is the insertion of sulfur with the formation of polysulfides [57,58]. The reaction of sulfur with paraffin, vaseline, mineral oils, etc. is used for the preparation of hydrogen sulfide (cf. Section 3.1.3). The sulfuration of petroleum gases, paraffinic hydrocarbons, kerosene fractions, heavy oils and other heavy fractions of petroleum at 700-1300°C represents an industrial method for the preparation of carbon disulfide (cf. Section 3.1.4).

The interaction of polyethylene with sulfur at 200-250°C is accompanied by the incorporation of sulfur into C-H bonds with subsequent cross-linking between the polyethylene chains. The maximum amount of sulfur incorporated does not depend on temperature and is 31-37% [59-61]. The modification of the polyethylene surface by atomic sulfur in the singlet state is irreversible and involves insertion of sulfur atoms into the C-H bonds on the polyethylene surface leading to the formation of mercapto groups [62].

3.1.3. Preparation of Hydrogen Sulfide from Hydrocarbons

The interaction of elemental sulfur with paraffins (and some other high-boiling organic substances) represents a convenient method for the preparation of hydrogen sulfide. This method has been successfully used both in the laboratory and in pilot-scale experiments [29,63-91]. The aim of the first investigations of the reactions of sulfur with paraffinic hydrocarbons was the preparation of hydrogen sulfide whereas no attention was paid to other reaction products. In 1870 Champion and Pallot [64] and one year later Galletly [65-69] heated paraffin with sulfur and obtained a continuous stream of hydrogen sulfide which stopped after the removal of the heat source. Hydrogen sulfide formed in this manner is rather pure and contains only a small amount of carbon disulfide [74,78]. A disadvantage of this reaction is the foaming and spattering of the reaction mixture [70,71]. To ensure a quiet and uniform reaction it is advisable to add pumice [70], asbestos [78, 83], kaolin [91], kieselguhr [89], or other fillers. A temperature increase in the reaction of paraffin with sulfur above 200°C increases the yield of hydrogen sulfide and is close to the theoretical yield at 300°C [41,81]. However, considerable carbonization is observed. The final reaction product of the sulfuration of paraffin is carbon sulfide and not carbon, as assumed earlier [74, 78,83]. Carbon sulfide is a substance with a composition of $(C_5S)_n$ [76] or $(C_2S)_n$ [92]. As a rule the carbon-like end products of the sulfuration of paraffin contain more than 40% of nonextractable sulfur, i.e., the atomic ratio $C:S \geq 4$.

The interaction of higher paraffinic hydrocarbons with sulfur at 750-1300°C gives hydrogen sulfide and asphalt-like products [31, 93-95]. When choosing the temperature conditions for the reaction of paraffin with sulfur one has to keep in mind that, whereas the formation of hydrogen sulfide starts at 150°C, at 230°C a noticeable amount of carbon disulfide is also formed [76]. The catalytic effect of metals (Fe, Al, Zn, Cu, Co, Ni, Pb, Ag), carbon black, charcoal, cadmium sulfide, anhydrous aluminium chlorides, and other substances upon the temperature at which the reaction of sulfur with paraffin begins has been investigated [84]. Aluminium chloride and carbon black have been found to be the best catalysts.

The reaction of paraffin with sulfur at 500-800°C in the presence of kaolin gives a 57% yield of hydrogen sulfide [96]. In addition to paraffin, vaseline [74], mazut [96], and mineral oils of various origins [77,86,97] may be used as starting materials for the preparation of hydrogen sulfide. A very convenient method of obtaining hydrogen sulfide consists of the dropwise addition of mineral oil to molten sulfur preheated to 350-400°C [72].

The reaction of sulfur with methane has been suggested as an industrial method for the preparation of hydrogen sulfide [12,28]. At 600-650°C and in the presence of various catalysts, sulfur quantitatively reacts with methane and water vapor giving hydrogen sulfide and carbon dioxide according to the general scheme [13]

$$CH_4 + 4S + 2H_2O \longrightarrow 4H_2S + CO_2 \tag{3.11}$$

To prevent the formation of carbon disulfide, the reaction is carried out in two steps. The first step involves the interaction of sulfur with methane at 600-650°C to give hydrogen sulfide and carbon disulfide. In the second step, carbon disulfide is converted to hydrogen sulfide and carbon dioxide by treatment with water vapor at 300-350°C. When heated with sulfur to 450-500°C at 200-400 atm, methane is completely converted to hydrogen sulfide and carbon disulfide [7]. Hydrogen sulfide is formed during the preparation of carbon disulfide from methane in the presence of alumina [1,2, 98]. At temperatures up to 700°C, the predominant reaction is

$$CH_4 + 2S_2 \longrightarrow CS_2 + 2H_2S \tag{3.12}$$

At higher temperatures (850-1200°C), no formation of hydrogen sulfide is observed and the only product is carbon disulfide. In

order to increase the yield of hydrogen sulfide from methane, it is
advisable that the latter be preheated to about ∿ 250°C. Molten
sulfur is then added under 4 atm pressure and the mixture heated to
500-700°C [99]. The reaction products obtained from the interaction
of sulfur with hydrocarbons (consisting basically of the lower
alkanes, C_1-C_3) at 450-650°C and 6-12 atm include carbon disulfide,
hydrogen sulfide, and sulfur [5,19].

3.1.4. Preparation of Carbon Disulfide

Lampadius in 1796 was the first person to obtain carbon disulfide by treating carbon with sulfur vapor [106]. Carbon disulfide
was manufactured using this method until the middle of this century
when a new process based on natural gas was introduced [100,101].
However, the classical method for the preparation of carbon disulfide
is still important today [102-105]. The best temperature for the
preparation of carbon disulfide from its elements is 830-900°C. The
addition of sodium, copper, magnesium, or lead salts (1.6-2.9% of
the carbon weight) [107-109], alkali metal carbonates, or silver
nitrate [110-112] considerably increases the reaction rate and the
yield of carbon disulfide. Lower alkanes (C_1-C_5), and natural gas
form carbon disulfide by heating with sulfur above 450°C.

The industrial method for the preparation of carbon disulfide
by treating sulfur vapor with hydrocarbon gases at 1000°C was
initially suggested in 1930 [128-130]. Starting from 1940, special
attention has been given to the development of an industrial process
whereby carbon disulfide is produced from methane or natural gas and
sulfur at 800-1000°C [3,11,63,131-144].

The thermodynamic data for the synthesis of carbon disulfide
from methane and sulfur at atmospheric and higher pressures (7 atm)
have been determined [136]. A thermodynamic study of the formation
of carbon disulfide from its elements has also been carried out
[121,127]. For the reaction C(graphite) + $S_2 \rightleftharpoons CS_2$ the heat of
formation of carbon disulfide at ∿ 1000°C has been found to be
-3.44 kcal/mole [121,127], -3.25 kcal/mole [123], -5.57 kcal/mole [125],
and -6.10 kcal/mole [126]. The above differences in the heat of
formation are probably due to the various methods used for its determination.

Continuous methods of preparation of carbon disulfide from
charcoal and sulfur vapor at 800-1000°C have been patented [113].
Carbon disulfide can be obtained from coke and sulfur at 600°C.
The desorption of sulfur from the surface of carbon has been studied
at 1200°C [114].

The basic physical chemistry of the formation of carbon disulfide from its elements has been studied in detail [4,115-126].

Berthelot has shown [115,116] that the reaction leading to the formation of carbon disulfide from its elements is a reversible process:

$$C + S_2 \rightleftharpoons CS_2 \qquad (3.13)$$

The conversion of methane into carbon disulfide has been studied most thoroughly [4,8,144]. The kinetics, thermodynamics, and mechanism of the reaction have been investigated [3,9,14,22], and the following equations represent the course of the reaction:

$$CH_4 + 2S_2 \rightleftharpoons CS_2 + 2H_2S \qquad (3.14)$$

$$CH_4 + {}^2/_3 S_6 \rightleftharpoons CS_2 + 2H_2S \qquad (3.15)$$

$$CH_4 + {}^1/_2 S_8 \rightleftharpoons CS_2 + 2H_2S \qquad (3.16)$$

$$CH_4 + S_2 \rightleftharpoons CS_2 + 2H_2 \qquad (3.17)$$

$$CH_4 + {}^1/_3 S_6 \rightleftharpoons CS_2 + 2H_2 \qquad (3.18)$$

$$CH_4 + {}^1/_4 S_8 \rightleftharpoons CS_2 + 2H_2 \qquad (3.19)$$

It is assumed that at least three reactions (3.14, 3.15, and 3.16) occur simultaneously because all the three types of sulfur molecules are present in the vapor. At 500-700°C, these reactions lead to an almost complete conversion of methane to carbon disulfide.

In order to obtain carbon disulfide, it is usually recommended to use hydrocarbon mixtures containing methane, ethane, and propane [137-140]. Methods based on the use of hydrocarbon mixtures containing hydrocarbons with five or more carbon atoms have also been described [131,141]. There are patents for the preparation of carbon disulfide from sulfur and paraffinic hydrocarbons containing five or more carbon atoms at 700-1300°C [145,146], petroleum gases [147], the kerosene fraction of petroleum [148], and heavy oil and other heavy petroleum distillates [149]. The reaction of gaseous or liquid hydrocarbons with liquid or gaseous sulfur has been carried out at 450-700°C and 2-24 atm [5,150]. The reaction products consist of carbon disulfide, hydrogen sulfide, and unreacted sulfur.

In order to decrease the amount of by-products, especially hydrogen sulfide, carbon disulfide may be obtained by the interaction of C_1-C_6 alkanes with sulfur vapor and sulfur dioxide at 700-850°C under pressure (2-20 atm) or at atmospheric pressure [151].

When a gaseous mixture of hydrocarbons (C_2-C_4) and sulfur is irradiated with a ruby laser (100 MW), carbon disulfide is the principal reaction product [49]. The reaction leading to the formation of carbon disulfide in a laser plasma with the participation of atomic carbon and sulfur (S or S_2) can be represented as follows:

$$C + S \rightarrow CS \xrightarrow{+S} CS_2 \tag{3.20}$$

$$C + S_2 \longrightarrow CS_2 \tag{3.21}$$

A method for the preparation of carbon disulfide by introducing sulfur and benzene vapors at 600°C and 400°C respectively, into the oxygen plasma stream of a high-frequency plasmatron has been developed. The yield of carbon disulfide is 86% [152].

In order to make the process of carbon disulfide preparation more efficient, a mixture of sulfur vapor and hydrocarbons is brought into contact with fluidized graphite through which an electric current is being passed, thus heating the mixture to 800-1000°C [153-155]. Catalytic processes of carbon disulfide preparation patented during the past 25 years make it possible to obtain a 90-98% yield of carbon disulfide [7,11,134,136,142]. The catalytic gas-phase reactions of sulfur with methane have been investigated at 550-700°C (with aluminium acetate or chromic acid as the catalyst) [11,14]. Silica gel and chromium-activated alumina are highly active in the reaction of a sulfur-methane mixture (molar ratio 0.5 or 0.6) at 600°C [148]. The following catalysts for the temperature range of 450-700°C have been suggested: silica gel and copper, zirconium, and thorium phosphates [107,108,111,112,157]. Tungsten, molybdenum, chromium [135,158], zirconium dioxide on silica gel [157,159,160], and aluminium acetate and chromic acid [11] have been used as catalysts in the preparation of carbon disulfide from methane and sulfur. Catalytic reactions leading to the formation of carbon disulfide from sulfur and a hydrocarbon mixture containing more than 90% methane at 450-700°C and 3-30 atm have been described [150,161, 162].

A method of preparation of carbon disulfide from sulfur vapor and hydrocarbons in the presence of liquid sulfur has been proposed [1,99,132,133]; the reaction temperature and pressure can vary within a wide range. The following substances have been recommended as catalysts for this process: Cu, Fe, Zn, Mo, Ni, Co, Th, their oxides and sulfides, and activated charcoal. There is a patent for a catalytic process leading to the preparation of carbon disulfide by treating a mixture of methane, ethane, propane, and butane with sulfur vapor at 800-1000°C in the presence of copper, silver, or gold sulfides [131]. The following substances have also been recommended as possible catalysts in the sulfuration of paraffinic hydrocarbon mixtures giving carbon disulfide at 550-1000°C: W, Cr, Mo [135], Al_2O_3 [19,107], $MgSiO_3$ or MgO sensitized by 10% of Ni, Mo and Co oxides [138], and silica gel [163]. The catalytic activity of sulfidized chromium, molybdenum, nickel, and vanadium oxides deposited on alumina or silica has been investigated in a flow system at 450-700°C in the reaction of sulfur with pentane or octane [164]. Carbon disulfide is the principal reaction product.

3.2. ALKENES

3.2.1. Ethylene

The course of the reaction of ethylene with sulfur depends chiefly on the reaction conditions. When ethylene is passed through boiling sulfur considerable amounts of hydrogen sulfide and carbon disulfide are obtained together with traces of thiophene [165]. The reaction is accompanied by considerable carbonization. The formation of thiophene can be represented by the following equation:

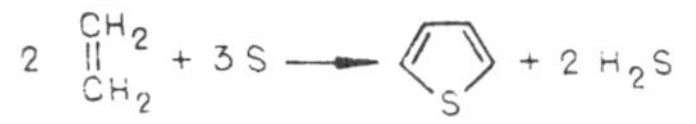

$$2 \, \overset{CH_2}{\underset{CH_2}{\|}} + 3\,S \longrightarrow \langle\!\!\langle \underset{S}{} \rangle\!\!\rangle + 2\,H_2S \qquad\qquad (3.22)$$

It has been shown [166] that thiophene is also formed from ethylene when it is passed over hot pyrites. In this case nascent sulfur participates in the reaction.

When ethylene is passed through molten sulfur at 325°C, ethanethiol, hydrogen sulfide, and small amounts of carbon disulfide and diethyl sulfide are obtained [167,168]. When using pyrites at 350°C instead of sulfur, about 1% of thiophene is obtained together with hydrogen sulfide and ethanethiol. Ethylene passed through molten sulfur heated to 350-400°C gives a bluish-black dye containing 61% sulfur [169]. Ethylene passing through a contact zone with sulfur vapor at high temperature and flow rate is converted into carbon disulfide [42,130].

The reaction of ethylene with sulfur at 175°C under pressure leads to the formation of sulfuration products whose hydrogenation yields alkanethiols [170,171]. Macromolecular sulfuration products may be prepared by treating ethylene with molten sulfur [172]. It has been stated that the prolonged slow passage of ethylene through molten sulfur at 150°C did not result in a reaction [173]. This result may be explained as being due to the use of an incorrect experimental technique which did not allow the isolation of non-volatile reaction products.

In spite of these data it has been reported [174] that ethylene undergoes a vigorous reaction with sulfur at temperatures above its melting point. The primary reaction products of this reaction, carried out under pressure and with xylene as the solvent, are an insoluble elastomer and a partially volatile oil soluble in xylene. The following structures have been suggested for the volatile fraction of the oil:

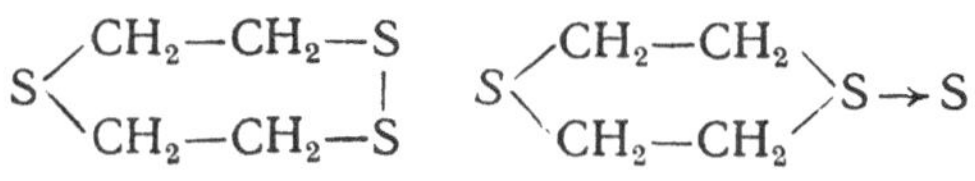

$$S\!\!\begin{array}{c} \diagup CH_2{-}CH_2{-}S \\ \diagdown CH_2{-}CH_2{-}S \end{array} \qquad S\!\!\begin{array}{c} \diagup CH_2{-}CH_2 \\ \diagdown CH_2{-}CH_2 \end{array}\!\!\!S \to S$$

The structure proposed for the nonvolatile component is

$$
\begin{array}{c}
\text{S} \quad \text{S} \\
\text{S}_{\searrow} \quad {}_{\nearrow}\text{C}_2\text{H}_4\text{--S--C}_2\text{H}_4{}_{\searrow} \quad {}_{\nearrow}\text{S} \\
\quad\quad \text{S} \quad\quad\quad\quad\quad\quad\quad \text{S} \\
\text{S}^{\nearrow} \quad {}^{\searrow}\text{C}_2\text{H}_4\text{--S--C}_2\text{H}_4{}^{\nearrow} \quad {}^{\searrow}\text{S} \\
\text{S} \quad \text{S}
\end{array}
$$

However, the first of these compounds is not the 1,2,5-trithiepane described in the literature [175] but probably its dimer.

The insoluble polymer seems to consist of ethylene groups linked by sulfide or polysulfide bridges. This reaction is catalyzed efficiently by the sulfuration products formed in the reaction [176-178]. The reaction of ethylene with sulfur at 300°C has been studied using the EPR method. The reaction proceeds by a radical mechanism and is initiated by sulfur biradicals [179]. Kinetics of both the homogeneous reaction of ethylene with sulfur within the 360-480°C range and the heterogeneous process in the presence of cobalt, tungsten, molybdenum, and cadmium sulfides have been investigated [180]. In both cases, the main reaction products are carbon disulfide, ethanethiol, and thiophene. The reactions of ethylene with atomic sulfur in its singlet or triplet state proceed in different directions [181-184]:

$$
CH_2{=}CH_2 + S\,({}^3P) \rightarrow H_2C\underset{\diagdown S \diagup}{\text{---}}CH_2 \tag{3.23}
$$

$$
CH_2{=}CH_2 + S\,({}^1D_2) \rightarrow \left[H_2C\underset{\diagdown S \diagup}{\text{---}}CH_2 \right]^* \tag{3.24}
$$

$$
\downarrow
$$

$$
CH_2{=}CH_2 + S\,({}^1D_2) \rightarrow CH_2{=}CH\text{---}SH \tag{3.25}
$$

$$
CH_2{=}CH_2 + S\,({}^1D_2) \rightarrow [CH_2{=}CH_2]^* + S\,({}^3P) \tag{3.26}
$$

The molar ratio of ethylenethiol and thiirane formed in the reaction of ethylene with sulfur $S({}^1D_2)$ is 49 to 51. Ethylenethiol can be formed not only by the insertion of sulfur into the C-H bonds but also by the isomerization of thiirane [185]. The formation of episulfide was also observed when the sulfur surface was irradiated with a laser (K_2F, λ 249 nm, energy 30 MJ) in an atmosphere of ethylene. The yield of episulfide was proportional to the irradiation energy to the power of 3.5. Hydrogen, acetylene, ethylenethiol, etc. are by-products of the reaction [186]. A quantum-chemical calculation of the reaction of ethylene with sulfur leading to thiirane has been carried out [187].

3.2.2. Higher Alkenes

The reaction of sulfur with higher alkenes has attracted the interest of many researchers. This is due to the fact that this reaction is involved in the mechanism of vulcanization of rubber macromolecules and to its relevance to the structure of vulcanized rubber. In 1886 it was shown [188] that heating mineral oils with sulfur resulted in reaction of the latter with only the unsaturated hydrocarbons and led to the formation of sulfuration products (thiols and hydrogen sulfide).

Boiling hexene for a long time with sulfur did not give any reaction [189]. However, if the reaction is carried out in a sealed tube at 210°C for twenty-four hours, almost all the olefin reacts with the formation of a large amount of hydrogen sulfide. It has been assumed [189] that addition of one or several sulfur atoms to the double bond takes place. Further heating the cyclic sulfides or polysulfides so formed leads to the liberation of hydrogen sulfide. The higher the number of sulfur atoms bonded to the olefin molecule, the more pronounced is the formation of hydrogen sulfide:

$$C_6H_{12} + 2S \rightarrow C_6H_{12} \Big\langle \begin{smallmatrix} S \\ | \\ S \end{smallmatrix} \qquad\qquad (3.27)$$

$$C_6H_{12} + 6S \rightarrow C_6H_{12} \Big\langle \begin{smallmatrix} S_5 \\ | \\ S \end{smallmatrix} \rightarrow 6C + 6H_2S \qquad (3.28)$$

The reaction of 2-hexene with sulfur under pressure at 270–280°C gives sulfuration products which have the following composition: $C_6H_{12}S$, $C_{12}H_{24}S$, $C_{18}H_{34}S_2$, and $C_{24}H_{40}S_3$, as well as an asphalt-like residue, $C_{24}H_{20}S_2$ [190]. It is assumed that a dithiol is the intermediate reaction product which then loses a molecule of hydrogen sulfide to give an unsaturated thiol:

$$\begin{array}{ccc} | & | & | \\ CH & CH\!-\!SH & CH\!-\!SH \\ \| \;{\scriptstyle +H_2S+S} & | & | \\ CH \xrightarrow{} & CH\!-\!SH \rightarrow & CH \qquad +H_2S \\ | & | & \| \\ CH_2 & CH_2 & CH \\ | & | & | \end{array} \qquad (3.29)$$

This thiol can be further converted into an unsaturated sulfide by the removal of a molecule of hydrogen sulfide from two thiol molecules with the formation of a sulfide $(C_nH_{2n-1})_2S$, or by addition of the thiol to the double bond of the starting olefin with the formation of a sulfide $(C_nH_{2n})_2S$. Analogous reaction products were isolated from the reaction of sulfur with 1-octene [190].

2-Octene reacts with sulfur, on heating in a sealed tube, in an entirely different manner and gives 3,4-dimethyl-2,5-dipentyl-thiophene:

$$H_3C-CH \atop H_{11}C_5-CH \| \quad +3S+ \quad {CH-CH_3 \atop CH-C_5H_{11}} \| \quad \longrightarrow \quad {H_3C \atop H_{11}C_5}\diagdown\diagup{CH_3 \atop \diagdown S \diagup C_5H_{11}} \quad +2H_2S \qquad (3.30)$$

Among other reaction products, a thiophene homolog $C_8H_{12}S$, a thieno-[2,3-b]thiophene homolog $C_8H_8S_2$, and a compound $C_{24}H_{38}S_2$ have been found.

In a study of the interaction of sulfur with alkenes under conditions which are intermediate between vulcanization and dehydrogenation (180°C) it has been found that the active agent is hydrogen sulfide formed during the reaction [167,168]. Hydrogen sulfide adds to the starting olefin and gives a saturated thiol which further reacts with the olefin to give a saturated sulfide:

$$RCH=CHR' \xrightarrow[S]{+H_2S} \underset{\underset{SH}{|}}{RCH_2CHR'} \xrightarrow[S]{RCH=CHR'} {RCH_2CH{\diagup R' \atop \diagdown} \atop RCH_2CH{\diagdown \atop \diagup R'}}S \qquad (3.31)$$

Sulfur plays a dual role, being both a reagent and a catalyst of the thiylation.

Many authors have pointed out the important role of hydrogen sulfide in the sulfuration of unsaturated compounds [191-194]. When alkenes are heated with sulfur to $\sim$ 190°C, a vigorous evolution of hydrogen sulfide is observed. Sometimes this is erroneously taken as the initial stage of the reaction. Actually, alkenes begin to interact with sulfur at lower temperatures (120-160°C) and, at a very low rate, even at 40°C [195-197].

The reaction of sulfur with alkenes under vulcanization conditions (120-140°C) and in the presence of accelerators (2-mercapto-benzo-1,3-thiazole, zinc oxide, etc.) [198-200] gives volatile and non-volatile sulfuration products in various ratios. The volatile reaction products possess a structure in which two alkene molecules are joined in the α-position by one or several sulfur atoms. Thus, the sulfuration of 2-butene gives bis-2-buten-1-yl sulfide and a small amount of the analogous disulfide:

$$CH_3CH=CHCH_3 \xrightarrow{+2S} 2CH_3CH=CHCH_2SH \xrightarrow[-H_2S]{} (CH_3CH=CHCH_2)_2S \qquad (3.32)$$

The sulfuration of 2-methylpropene under the conditions used in the vulcanization of rubber and in the absence of catalysts leads to the formation of cross-linkages between the alkene molecules and thus to polysulfides [201-204]. The latter are thermally quite unstable and above 140°C they quickly decompose with the formation of hydrogen sulfide, thiols, and sulfides.

2-Methyl-2-butene and 2-methyl-1-butene interact with sulfur
to yield a mixture of the isomeric symmetric and unsymmetric sulfides and disulfides containing allyl-type moieties:

$$-CH_2-\underset{\underset{CH_3}{|}}{C}=CH-CH_3 \quad -\underset{\underset{CH_3}{|}}{CH}-\underset{\underset{CH_3}{|}}{C}=CH_2 \quad -CH_2-CH=\underset{\underset{CH_3}{|}}{C}-CH_3$$

$$-CH_2-\underset{\underset{C_2H_5}{|}}{C}=CH_2$$

It has been assumed that the reaction of sulfur with alkenes is a
free-radical chain process [199,200]. Under the reaction conditions,
an intermediate allyl-type radical is formed:

$$CH_2=\underset{\underset{CH_3}{|}}{C}-\overset{\cdot}{CH}-CH_3$$

This radical further reacts with sulfur to give the above-mentioned
free radicals which then react with sulfur to give sulfides or polysulfides. The reaction of 2-methyl-2-butene with sulfur at 142°C
carried out under pressure [205] yields a mixture of liquid polysulfides of the type R-S$_n$-R' where R and R' are alkenyl groups of
the allylic type and n changes from 2 to 6 depending on the duration
of heating. A compound $C_5H_6S_3$ has also been isolated; this compound
was thought to be 3-methylthiophene-2,5-dithiol (1) which is tauto-
meric with 3-methyltetrahydrothiophene-2,5-dithione (2) [205]:

$$\underset{(\underline{1})}{HS-\overset{\overset{\displaystyle CH-\underset{\underset{S}{\diagdown}}{C}-CH_3}{\|}}{\underset{\diagup}{C}}\quad\overset{\|}{C}-SH}\rightleftarrows\underset{(\underline{2})}{S=\overset{\overset{\displaystyle H_2C-CH-CH_3}{|\quad|}}{\underset{\diagup}{C}}\quad C=S}$$

(1)   (2)

Later it was established that this compound is actually 4,5-dimethyl-
1.2-dithiol-3-thione [205-207].

$$\underset{S}{\overset{\displaystyle H_3C-\underset{\underset{\diagdown}{\|}}{C}-C=S}{\underset{\diagup}{\underset{\displaystyle H_3C-C}{}}\quad S}}$$

   The direction of the reactions of 2-methyl-1-butene, 2-methyl-
2-butene, 2-pentene, and 2,3-dimethyl-2-butene with excess sulfur
at 170°C under pressure is greatly influenced by the structure of
the starting alkene [190,206,208,209]. In all cases, the liquid
volatile sulfuration products are unsaturated sulfides, R-S$_n$-R',
where R is an alkyl group, R' an alkenyl group, and n = 3, 2, or 1.
Thus, in the case of 2-methyl-2-butene, R' can be:

$$-CH_2-CH=C-CH_3 \qquad -CH_2-C=CH-CH_3$$
$$\qquad\qquad | \qquad\qquad or \qquad\qquad |$$
$$\qquad\qquad CH_3 \qquad\qquad\qquad\qquad CH_3$$

It seems that the sulfides are secondary reaction products as disulfides are the first sulfuration products obtained under milder conditions.

The following reaction mechanism has been suggested for the formation of alkyl alkenyl polysulfides during the interaction of sulfur with alkenes [206]. The first stage of sulfuration is the free-radical formation of β,γ-unsaturated alkylpolysulfanes:

$$R^1-C=CH-CH_2-R \xrightarrow{S_n} R^1-C=CH-CH-S_nH \xrightarrow[S]{R^1R^2C=CH-CH_2R}$$
$$\quad | \qquad\qquad\qquad\qquad | \qquad\quad |$$
$$\quad R^2 \qquad\qquad\qquad\qquad R^2 \qquad R$$

$$\qquad\qquad R^1R^2C=CH-CH-R$$
$$\qquad\qquad\qquad\qquad\qquad |$$
$$\longrightarrow \qquad\qquad\qquad S_n \qquad\qquad\qquad\qquad\qquad (3.33)$$
$$\qquad\qquad\qquad\qquad\qquad |$$
$$\qquad\qquad R-CH_2CH_2-CR^1R^2$$

$$R,R^1,R^2 = H, alkyl; n \leqslant 8.$$

Addition of the latter compound to the starting alkene leads to the formation of alkyl alkenyl polysulfides where the number of sulfur atoms depends on the reaction conditions.

The sulfuration of alkenes containing the groupings $R'-CH=C(R)-CH_3$ or $R'-CH_2-C(R)=CH_2$ gives 1,2-dithiol-3-thiones [205,210-255]. The reaction can be represented as follows:

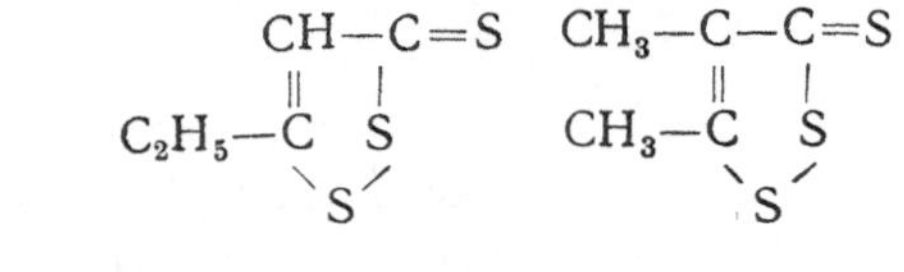

$$
\begin{array}{c}
R-C-CH_3 \\
R'-C-H \\
\\
R-C=CH_2 \\
R'-CH_2
\end{array}
\xrightarrow{+5\,S}
\begin{array}{c}
R \diagdown \qquad \diagup S \\
\qquad \diagup \diagdown S \\
R' \diagdown S \diagup
\end{array}
+ 2\ H_2S
\qquad (3.34)
$$

Thus, the reaction of sulfur with 2-pentene and 2-methyl-2-butene produces the corresponding crystalline derivatives of 1,2-dithiol-3-thione:

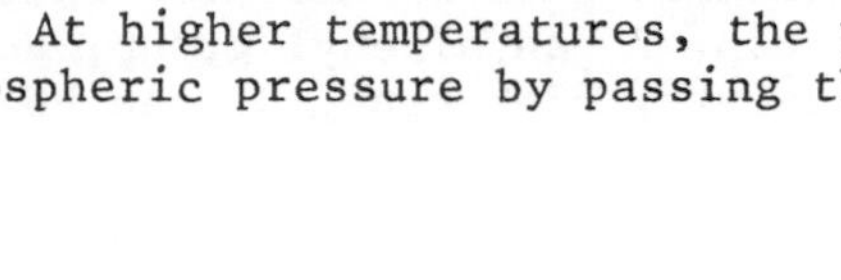

$$
\begin{array}{cc}
CH-C=S & CH_3-C-C=S \\
\|\ | & \qquad \|\ | \\
C_2H_5-C\ \ S & CH_3-C\ \ S \\
\diagdown S \diagup & \diagdown S \diagup
\end{array}
$$

The formation of 1,2-dithiol-3-thiones occurs at $160-220°C$ and consequently, it has to be carried out under pressure in the case of the lower alkenes. At higher temperatures, the process can be performed even at atmospheric pressure by passing the gaseous alkene through molten sulfur. Because sulfur reacts with unsaturated compounds by numerous pathways at the same time, the 1,2-dithiol-3-thiones formed in 20-40% yield are not the only reaction products. It has been assumed [218,219,257,258] that thiols are the intermediate products in the formation of the 1,2-dithiols. The thiols are then converted to trisulfides. The latter give the 1,2-dithiol-3-thiones upon thermal decomposition in the presence of sulfur:

$$2 \begin{array}{c} CH-CH_3 \\ \| \\ R-CH \end{array} \xrightarrow{+S} 2 \begin{array}{c} CH-CH_2SH \\ \| \\ R-CH \end{array} \xrightarrow[-H_2S]{+2S} \begin{array}{c} CH-CH_2S_3\ CH_2-CH \\ \| \qquad\qquad \| \\ R-CH \qquad\quad CH-R \end{array}$$

$$\xrightarrow[>170^\circ C]{+6\,S} 2\ \ R-\!\!\boxed{}^{S}_{S-S} + 3\,H_2S \tag{3.35}$$

The reaction of alkenes with sulfur in the gaseous phase at 500°C affords 1,2-dithiol-3-thiones in 50% yield [259].

A patent exists describing the preparation of copolymers, used in the manufacture of lacquers, based on the reaction of sulfur with propene or butene at 150-200°C [260].

The reaction of sulfur with alkenes at 140°C gives, together with alkenyl polysulfides, cyclic polysulfides containing two or more sulfur atoms per double bond [261,262]:

$$\begin{array}{c} C_6H_{13}-CH-CH_2 \\ S\big<\qquad\quad\big>S \\ H_2C\!-\!\!-\!\!-CH-C_6H_{13} \end{array} \qquad \begin{array}{c} S-C_8H_{16} \\ \diagup \qquad \diagdown \\ H_{16}C_8 \qquad S \\ \diagdown \qquad \diagup \\ S-C_8H_{16} \end{array} \qquad \begin{array}{c} CRR'\!-\!\!-\!CR''R''' \\ \diagup \qquad\qquad \diagdown \\ S \qquad\qquad\quad S \\ \diagdown \qquad\qquad \diagup \\ CR'R'''\!-\!\!CR'R' \end{array}$$

The principal products of this reaction, obtained in the presence of diethylamine, are tert-alkanethiols, dialkyl disulfides, and a small amount of the corresponding polysulfides [263]. The interaction of sulfur with 2-alkylpropenes, $H_3C-C(R)=CH_2$, above 100°C in tetrahydrothiophene-S,S-dioxide and in the presence of dimethyl-formamide gives compounds of the general formula [264]

$$(\overline{S-S-C-C(R)}=C-S-)_n\ X \\ \qquad\quad \| \\ \qquad\quad S$$

where R is an alkyl group, X a salt-forming group, and n the valence of the group X.

A continuous process for the preparation of 1,2-dithiol-3-thiones involving the reaction of alkenes (isobutene and diisobutylene) with excess sulfur at 450-550°C and 0.5-5.0 atm has been described [265]. The reaction of alkenes (isobutene, its dimer, and its trimer) with sulfur at 180-200°C in the presence of sulfur dioxide which decreases the yield of hydrogen sulfide and accelerates the formation of the corresponding 1,2-dithiol-3-thione derivatives ($C_4H_4S_3$, $C_8H_{12}S_3$, $C_{12}H_{20}S_3$) is the subject of a patent [266].

The reaction of sulfur in the vapor phase, diluted with hydrogen sulfide or carbon disulfide, with alkenes at 600-750°C and at 1-10 bar leads to the formation of high-purity carbon disulfide (99.9%) [267]. Under these conditions, propene is converted into carbon disulfide in quantitative yield.

Sulfuration of 2-methyl-2-butene under vulcanization conditions in the presence of zinc soap as a catalyst [268] gives disulfides according to the scheme:

$$R-H \xrightarrow{S} R-S- \xrightarrow{Zn} (RS)_2 Zn \xrightarrow{S} R-S-S-R + ZnS \qquad (3.36)$$

where R is an allylic group. In this case, the formation of disulfides occurs with the participation of the catalyst. The presence of a methyl group bonded to an sp^2 carbon atom increases the reactivity of the alkene toward sulfur. Polysulfides are the principal reaction products obtained by the interaction of alkenes with sulfur at 20°C in the presence of zinc oxide or zinc propionate [269].

Reactions of alkenes with atomic sulfur in its singlet or triplet state proceed in different directions [182-184]. Sulfur atoms in the triplet state S(^{3}P) obtained by the photolysis of carbonyl sulfide in the presence of excess carbon dioxide react with alkenes to give only the corresponding thiirane [270-272]:

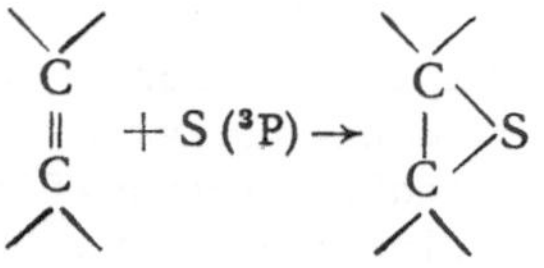

$$(3.37)$$

An increase of the alkyl group length (for alkyl groups bonded to an sp^2 carbon atom) inhibits the above reaction. At the same time, electron-donating groups decrease the energy of activation of reactions of unsaturated compounds with S(^{3}P) atoms (Table 3.1) [270-274]. Atomic sulfur in the singlet state, S($^1D_2$), can add to the double bond, but it can also insert itself into the C-H bonds:

1)

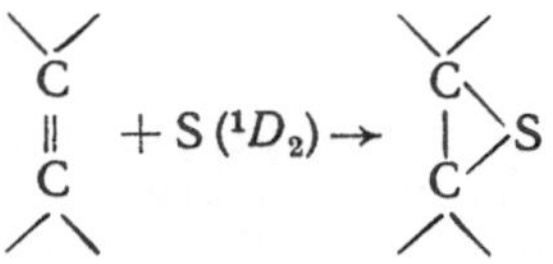

$$(3.38)$$

Table 3.1. Relative Activation Energies of Addition of $S(^3P)$ to Alkenes

Compound	E_a (ethylene) $-$ E_a, kcal/mole	Compound	E_a (ethylene) $-$ E_a, kcal/mole
$H_2C=CH_2$	0.0	$(H_3C)_2C=CHCH_3$	3.0
$H_2C=CHCH_3$	1.1	$(H_3C)_2C=C(CH_3)_2$	3.4
$H_3CCH\overset{c}{=}CHCH_3$	2.1	$H_2C=CHF$	−0.7
$H_3CCH\overset{t}{=}CHCH_3$	2.0	$H_2C=CHCl$	−0.5
$H_3CC=CH_2$ $\quad |$ $\quad CH_3$	2.4	$H_2C=CF_2$	−1.7
		$F_2C=CF_2$	−1.4

$$2) \qquad \underset{/\,\backslash}{\overset{\backslash/}{\underset{C}{\overset{C}{\underset{\|}{\,}}}}} + S(^1D_2) \rightarrow \overset{\backslash/}{\underset{C-SH}{\overset{C}{\underset{\|}{\,}}}} \qquad\qquad (3.39)$$

The reaction products obtained from $S(^1D_2)$ and alkenes are shown in Table 3.2. Vinylic-type thiols can be formed not only by the insertion of sulfur into C-H bonds but also by the isomerization of the originally formed thiirane [185]:

Table 3.2. Products Formed in the Reaction of $S(^1D_2)$ with Alkenes and Their Derivatives

Unsaturated compound	Molar ratio (%)		
	$>$C=CH-SH	$>$C=CHCH$_2$-SH	$\overset{\backslash}{C}-\overset{/}{C}$ with S
$H_2C=CH_2$	49	–	51
$H_2C=CHCH_3$	19	19	62
$H_2C=C(CH_3)_2$	12	32	56
$H_3CCH=CHCH_3$	0	32	68
$H_2C=CHCH_2CH_3$	12	29	59
$(H_3C)_2C=CHCH_3$	0	42	58
$(H_3C)_2C=C(CH_3)_2$	–	50	50
$H_2C=CHF$	32	–	68
$HFC=CHF$	0	–	100

$$CH_2-CH_2 \rightarrow CH_2=CH-SH \qquad (3.40)$$
$$\underset{S}{\diagdown \diagup}$$

3.3. THE WILLGERODT REACTION

The interaction of alkenes, alkadienes, and cycloalkenes with sulfur and aqueous ammonia or amines (a modification of the Willgerodt reaction) has not been studied in detail [174,275-287]. When propene or 2-methylpropene is heated with sulfur and a concentrated aqueous solution of ammonia below 200°C, the product is an oil consisting mainly of the corresponding di- and trisulfides. At higher temperatures (210-220°C), the amide of the corresponding acid is formed in good yield, according to the scheme:

$$\underset{R'}{\overset{R}{\diagdown}}C=CH_2 + 2S + NH_3 + H_2O \rightarrow \underset{R'}{\overset{R}{\diagdown}}CHCONH_2 + 2H_2S \qquad (3.41)$$
$$(70-75\%)$$

where R = H, R' = CH_3; or R = R' = CH_3. Under analogous conditions, 1,3-butadiene gives a 10-20% yield of butanoamide, whereas cyclohexene and vinylcyclohexene give tar-like products. Refluxing diisobutylene (2,4-dimethylpent-1-ene) with sulfur and morpholine gives a 14% yield of dithio-oxalo dimorpholide, $(C_5H_8NOS)_2$ [285]. A patent describes the preparation of alkanoic acids by the reaction of sulfur with C_4-C_7 alkenes in the presence of ammonia and water at 300-360°C under pressure [286].

Under the conditions of the Willgerodt reaction, arylalkenes and arylalkynes are converted to amides. Phenylacetylene and sulfur react in the presence of ammonium polysulfide at 200-220°C to give phenylacetamide [288,289]:

$$C_6H_5C\equiv CH \xrightarrow[S]{(NH_4)_2SS_n} C_6H_5CH_2CONH_2 \qquad (3.42)$$
$$(\sim 50\%)$$

Heating phenylacetylene with sulfur and morpholine gives phenylthioacetomorpholide [289-292]:

$$C_6H_5C\equiv CH \xrightarrow[S]{HN(CH_2CH_2)_2O} C_6H_5CH_2\overset{\overset{\displaystyle S}{\|}}{C}N(CH_2CH_2)_2O \qquad (3.43)$$
$$(60\%)$$

The latter is also formed in an analogous reaction with styrene [288,289]:

$$C_6H_5CH=CH_2 \xrightarrow[S]{HN(CH_2CH_2)_2O} C_6H_5CH_2\underset{\underset{\displaystyle S}{\|}}{C}N(CH_2CH)_2O \qquad (3.44)$$

Kinetic data have been obtained for the reaction of styrene with sulfur in the presence of morpholine, hexene-1,6-diamine, poly-ethyleneamide and diphenylguanidine in p-xylene at 100-130°C [291].

The reaction of 3-phenylpropene or 1-phenylpropene with ammonium polysulfide at 205°C gives 3-phenylpropanoamide [292]:

$$C_6H_5CH_2CH=CH_2- \atop C_6H_5CH=CHCH_3-} \xrightarrow{(NH_4)_2SS_n} C_6H_5CH_2CH_2CONH_2 \qquad (3.45)$$

Under analogous conditions, 2-phenylpropene forms 3-phenylpropano-amide in low yield ($\sim$ 3%) [21]. When 1-phenylpropene or 1-(4-methyl-phenyl)-1-propene are refluxed with sulfur and morpholine, dithio-oxalo dimorpholide $(C_5H_8NOS)_2$ is obtained (5% yield) [285]. Heating 1-phenylpropyne with sulfur, aqueous ammonia, and pyridine at 150°C gives 3-phenylpropanoamide [290,291]:

$$C_6H_5C \equiv CCH_3 + S + NH_3 + H_2O \rightarrow C_6H_5CH_2CH_2CONH_2 + H_2S \qquad (3.46)$$

The principal reaction products obtained from the interaction of sulfur and concentrated aqueous ammonia with alkylbenzenes (toluene, ethylbenzene, cumene, p-xylene) at 235-260°C are amides. including some with a lower number of carbon atoms (cf. Table 3.3). Benzamide is formed in all cases. The reaction can be depicted as follows:

$$C_6H_5CH_3 + 3S + NH_3 + H_2O \rightarrow C_6H_5CONH_2 + 3H_2S \qquad (3.47)$$
$$C_6H_5CH(CH_3)_2 + 3S + NH_3 + H_2O \rightarrow C_6H_5\underset{\underset{CH_3}{|}}{C}H CONH_2 + 3H_2S \qquad (3.48)$$

It is known that the classical Willgerodt reaction occurs without any change in the carbon skeleton. However, ethylbenzene, cumene, and p-xylene give benzamide as one of the products under the con-ditions of the Willgerodt reaction. The mechanism of the shortening of the side chain has not yet been elucidated.

Table 3.3. Products Obtained in the Reaction of Alkyl-
           benzenes with Sulfur and Aqueous Ammonia at
           235-260°C

Alkylbenzene	Reaction products	Overall yield (%)
Toluene	Benzamide	15-20
Ethylbenzene	Benzamide + phenylacetamide	20
Cumene	Benzamide + phenylacetamide + 2-phenylpropanoamide	20
p-Xylene	Benzamide + terephthalamide	60

3.4. ALKADIENES AND ALKAPOLYENES

Alkadienes with conjugated double bonds react with sulfur above 350°C via an intermolecular thiophene ring closure according to the scheme:

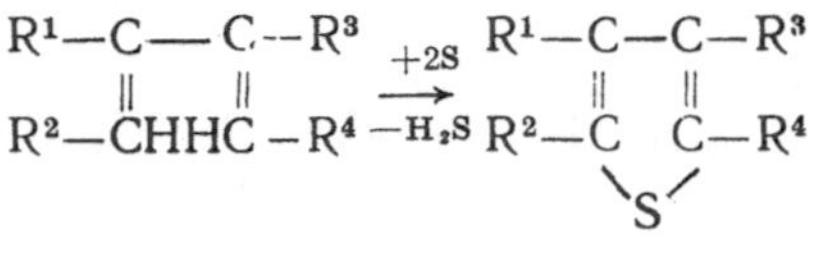

$$R^1-C-C-R^3 \quad \underset{-H_2S}{\overset{+2S}{\longrightarrow}} \quad R^1-C-C-R^3$$

(3.49)

The formation of thiophenes in the sulfuration of 1,3-butadiene was first observed by Steinkopf [293]. He passed isoprene vapor over red-hot pyrites. The yield of 3-methylthiophene was so low that this reaction remains interesting only from a theoretical point of view.

The possibility of obtaining thiophenes by the reaction of 1,-3-alkadienes with sulfur has been investigated [44,45]. When passed through molten sulfur at 350°C, 1,3-alkadienes are converted to thiophene and its homologs which do not contain any isomers or dihydrothiophenes. The side products in this reaction are carbon disulfide and nonvolatile substances. In the case of 1,3-butadiene, the yield of thiophene is only 6%. However, isoprene can be converted into 3-methylthiophene in 40% yield. An industrial procedure for the synthesis of thiophene from sulfur and 1,3-butadiene has been published [39-43]. The originality of this method consists of the fact that the starting 1,3-butadiene is obtained by dehydrogenation of butane or butene with sulfur. The reaction is a stepwise process:

$$\text{Butane} \overset{S}{\rightarrow} \text{butene} \overset{S}{\rightarrow} \text{1,3-butadiene} \overset{S}{\rightarrow} \text{thiophene} \qquad (3.50)$$

The by-products are thiophenol, homologs of thiophene, and macromolecular sulfur-containing compounds.

A patent describes the preparation of thiophene and its homologs by treating 1,3-alkadienes with sulfur at its boiling point (445°C) [46]. The yield of thiophene from 1,3-butadiene is 41%, and the yield of 3-methylthiophene from isoprene is 47%. The reaction between 1,3-butadiene and sulfur proceeds smoothly at 160°C in xylene under pressure and gives a red oil which resembles the sulfuration products of alkenes [294]. The reaction of isoprene with sulfur gives a low yield of a compound whose structure has been assigned as shown below [218,221]:

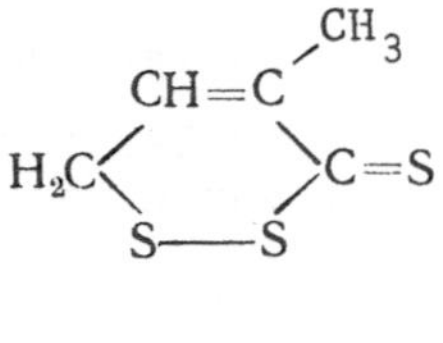

Presumably, this compound should be 4,5-dimethyl-1,2-dithiol-3-thione. A patent is available describing the preparation of solid thermoplastic copolymers of 1,3-butadiene and isoprene with sulfur, whose content in the copolymer can vary from 20 to 80% [295].

The interaction of 1,3-pentadiene with sulfur at 400-600°C gives a mixture containing methylthiophene, thieno[2,3-d]-1,2-dithiol-3-thione, and 2H-thiapyran-2-thione [50]. When 2,4-hexadiene vapor is passed through molten sulfur at 420°C, 2,5-dimethylthiophene is formed [296]:

$$\begin{array}{ccc} \text{CH—CH} & & \text{CH—CH} \\ \| \quad \| & \xrightarrow[-\text{H}_2\text{S}]{+2\text{S}} & \| \quad \| \\ \text{H}_3\text{C—CH HC—CH}_3 & & \text{H}_3\text{C—C} \quad \text{C—CH}_3 \\ & & \diagdown\text{S}\diagup \end{array} \qquad (3.51)$$

The reaction of 1,3-butadiene with atomic sulfur in its triplet state gives vinylthiirane, thiophene, and hydrogen sulfide [271]:

$$2 \; \begin{array}{c} \text{H}_2\text{C=CH} \\ | \\ \text{H}_2\text{C=CH} \end{array} + 3\text{S}(^3\text{P}) \longrightarrow \text{H}_2\text{C=CH—CH} - \text{CH}_2 + \bigcirc_\text{S} + \text{H}_2\text{S} \qquad (3.52)$$

Sulfur can add to 1,3-alkadienes either in the 1,4- or in the 1,2-positions [297,298]:

$$\qquad (3.53)$$

$$\longrightarrow 2 \; \diagup\text{C=S} \qquad (3.54)$$

The reaction of sulfur with alkadienes containing isolated double bonds has been studied under the conditions of the vulcanization of rubber [193,194,201,203,267,279,299-302]. The reaction starts above 100°C and occurs with a moderate rate at 140°C (the usual vulcanization temperature for rubber). By-products are formed above 140°C [201,203,282,299].

The interaction of sulfur with diisoprenes (dihydromyrcene, geraniolene) leads to cross-linking of these molecules via polysulfide bridges, similarly to alkenes. However, because the molecules of these hydrocarbons contain two double bonds, intramolecular cyclization is also possible leading to the formation of the sulfur-containing rings. The reaction of sulfur with 2,6-dimethyl-2,6-octadiene (dihydromyrcene, 3) and 2,6-dimethyl-1,5-octadiene (geraniolene, 4) gives two principal reaction products: a low-boiling unsaturated monosulfide, $C_{10}H_{18}S$, and a mixture of high-boiling, thermally unstable polysulfides of the R-S_n-R' type. The reaction is a free-radical process and can occur either with participation

of the α-methylene groups or with the participation of the double
bonds [201,203,282,299]:

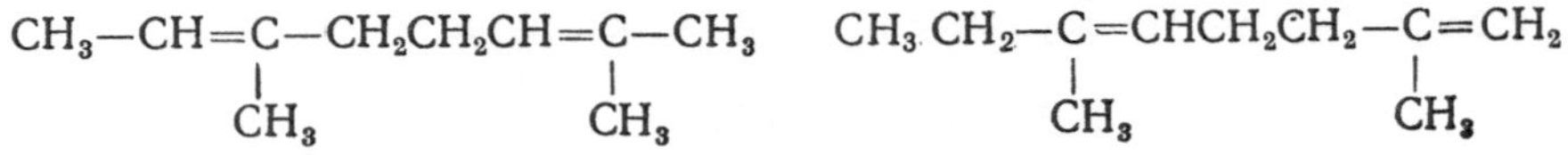

$$CH_3-CH=C-CH_2CH_2CH=C-CH_3 \qquad CH_3\ CH_2-C=CHCH_2CH_2-C=CH_2$$
$$\qquad\quad | \qquad\qquad\quad | \qquad\qquad\qquad\qquad | \qquad\qquad\qquad |$$
$$\qquad\ CH_3 \qquad\qquad\ CH_3 \qquad\qquad\qquad CH_3 \qquad\qquad\ CH_3$$

$$\underline{3} \qquad\qquad\qquad\qquad\qquad\qquad \underline{4}$$

Unstable polysulfides formed in the reaction of diisoprenes with
sulfur decompose quickly above 140°C to give hydrogen sulfide,
thiols, and sulfides [193,194]. The S-S bonds undergo cleavage with
the formation of thiyl radicals which subsequently give thiols and
cyclic sulfides.

Similarly to diisoprenes, the products obtained in the reaction
of sulfur with polyisoprenic hydrocarbons (squalene [300,301] and
natural rubber [279,300,302]) are thermally unstable saturated polysulfides and bis-tertiary cyclic sulfides. The latter are mainly
secondary reaction products.

The principal products obtained in the reaction of sulfur with
2,6-dimethyl-2,6-octadiene at 110-140°C in the presence of diethylamine are cyclic sulfides of the type shown below [303-307]:

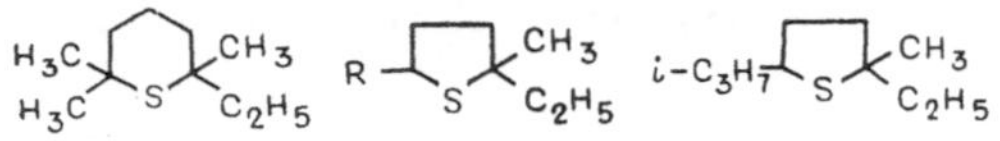

$$\underset{(3\%)}{\overset{H_3C}{\underset{H_3C}{\rangle}}\!\!\underset{S}{\bigcirc}\!\!\underset{C_2H_5}{\overset{CH_3}{\langle}}} \qquad \underset{(25-50\%)}{R\!-\!\underset{S}{\bigsqcup}\!\!\underset{C_2H_5}{\overset{CH_3}{\langle}}} \qquad \underset{(\sim 1.5\%)}{i\text{-}C_3H_7\!-\!\underset{S}{\bigsqcup}\!\!\underset{C_2H_5}{\overset{CH_3}{\langle}}}$$

where R = -CH(CH₃)₂, -C=CH₂, -C-SH   and cross-linked sulfides
$$\qquad\qquad\qquad\qquad\qquad | \qquad\quad |$$
$$\qquad\qquad\qquad\qquad\ CH_3 \qquad CH_3$$

R—Sₙ-R' where R = R' = $\underset{H_3C}{\overset{H_3C}{\rangle}}\!\!\underset{S}{\bigcirc}\!\!\underset{C_2H_5}{\overset{CH_3}{\langle}}$.

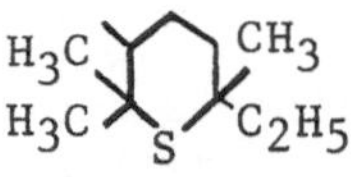

2,6-Dimethyl-2,4,6-octatriene reacts with sulfur at 210°C with the
formation of a substituted 5-(2-thienyl)-1,2-dithiol-3-thione :
[247,308]:

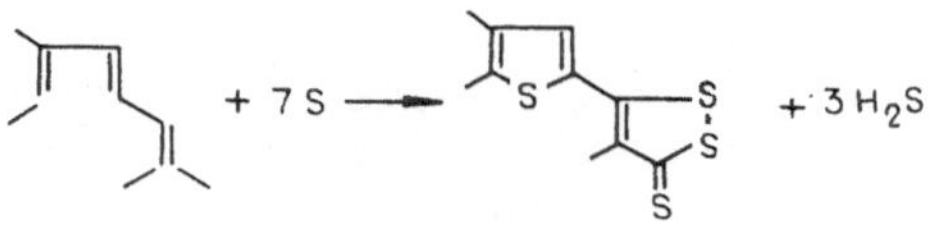

$$\qquad\qquad + 7\,S \longrightarrow \qquad\qquad + 3\,H_2S \qquad\qquad\qquad (3.55)$$

The reaction of excess S₈ with oligomeric styryllithium and polyiso-
prenyllithium in benzene at 25°C leads to the formation of symmetrical dialkyl polysulfides of the type [309] R—Sₙ—R where n = 3, 4.
Heating crystalline poly-4-methylpent-1-ene with sulfur at 250-320°C
markedly increases the induction period of polymer oxidation [310].

High-quality varnishes for surface coatings are obtained by heating polyalkadienes (by-products in the manufacture of butadiene-based rubber) with sulfur at 140°C and subsequent addition of a benzo[b]-furan resin and heating to 280-300°C [311].

3.5. VULCANIZATION

The hot vulcanization of rubber is the most important practical application of the sulfuration of unsaturated compounds. A considerable number of investigations of the interaction of sulfur with alkenes were carried out as studies of model systems whose goal was to elucidate the mechanism of vulcanization and the structure of the vulcanized rubber (see 3.2.2). The elucidation of the direction and of the mechanism of the reaction of sulfur with such a complex unsaturated hydrocarbon as rubber is a very difficult task. One of the earliest questions was whether the changes of the physical properties of raw rubber occurring during its vulcanization were due to a chemical reaction, or whether they were caused by a physicochemical process. The first period of investigation of vulcanization was dominated by physicochemical ideas [312]. In the middle of the nineteenth century, a number of studies demonstrated the possibility of the chemical interaction of sulfur with rubber [313-316].

The first chemical theory of vulcanization was proposed by Weber [317,318]. He stated that vulcanization consisted of a stepwise formation of addition products between sulfur and rubber, starting from $(C_{10}H_{16})_{10}S$ (the lower limit) and ending with polyisoprene sulfide, $(C_{10}H_{16})_{10}S_{20}$ (the upper limit).

Later, the gradual formation of a series of sulfides was rejected [319-322]. However, it was still assumed that vulcanization of rubber gave the individual sulfides, $(C_5H_8)_nS$ where n = 1 to 10 [323-328]. At the beginning of this century, it was still thought that rubber was a polymer of dimethylcyclooctadiene and that vulcanization with sulfur led to the cyclic sulfides (5 and 6):

$$
\begin{array}{cc}
\begin{array}{c}
- C(CH_3)\,CH_2CH_2CH- \\
\quad | \qquad\qquad\quad | \\
\quad S \qquad\qquad\quad S \\
\quad | \qquad\qquad\quad | \\
-CHCH_2CH_2(CH_3)C- \\
\\
5
\end{array}
&
\begin{array}{c}
\overset{\displaystyle CH_3}{\underset{|}{}} \qquad\qquad \overset{\displaystyle CH_3}{\underset{|}{}} \\
-C-CH_2 - CH_2- CH-C-CH_2-CH_2 -CH- \\
\quad| \qquad\qquad\qquad | \quad | \qquad\qquad\qquad | \\
\qquad\qquad\qquad\qquad S \quad S \\
\quad| \qquad\qquad\qquad | \quad | \qquad\qquad\qquad | \\
-CH-CH_2-CH_2-C-\ CH-CH_2-CH_2-C- \\
\qquad\qquad\qquad\quad | \qquad\qquad\qquad\qquad | \\
\qquad\qquad\qquad\ CH_3 \qquad\qquad\qquad\ CH_3 \\
\\
6
\end{array}
\end{array}
$$

It was also assumed that, during the vulcanization process, sulfur adds to the double bonds and that vulcanized rubber is a mixture of sulfide (7) and a disulfide (8) [329]:

$$\left[\begin{matrix} CH_3-C-CH_2-CH_2-\overset{|}{CH} \\ \parallel \qquad\qquad\qquad >S \\ CH-CH_2-CH_2-\underset{|}{C}-CH_3 \end{matrix}\right]_n \qquad \left[\begin{matrix} CH_3-C-CH_2-CH_2-\overset{|}{CH} \\ S\triangleleft| \qquad\qquad\qquad >S \\ CH-CH_2-CH_2-\underset{|}{C}-CH_3 \end{matrix}\right]_n$$

$$\underset{7}{} \qquad\qquad\qquad\qquad \underset{8}{}$$

It was also assumed [330,331] that vulcanization is a chemical reaction involving the addition of sulfur to a few or all of the double bonds of rubber and that it leads to two stable products, soft rubber and ebonite (hard rubber). The assumption was that soft rubber is obtained when sulfur only adds to the terminal double bonds of the polyisoprene molecule, whereas in the case of ebonite addition occurs to all double bonds:

$$\begin{matrix} S \qquad\qquad\qquad\qquad\qquad S \\ \diagup\diagdown \qquad\qquad\qquad\qquad \diagup\diagdown \\ C-C-C-C-C-C-C=C\ldots C=C-C-C-C-C-C \end{matrix} \quad \text{soft rubber}$$

$$\begin{matrix} S \qquad\quad S\quad S \qquad\qquad\quad S \\ \diagup\diagdown \quad \diagup\diagdown \diagup\diagdown \qquad\qquad \diagup\diagdown \\ C-C-C-C-C-C\ldots C-C-C-C-C-C \end{matrix} \quad \text{ebonite}$$

In agreement with this, it was found that the content of sulfur in soft rubber corresponds to the composition $(C_5H_8)_{200}S_2$. The later steps of the vulcanization process were represented as the addition of one, two, or three sulfur atoms to the polyisoprene double bonds with the formation of cyclic structures [317,332-336]:

$$\left[\begin{matrix} \qquad\quad CH_3 \\ -CH-CH_2-C- \\ \underset{|}{S}\text{———}\underset{|}{S} \end{matrix}\right] \quad \left[\begin{matrix} \qquad\qquad CH_3 \\ -CH-CH_2-CH_2-C- \\ |\text{————}S\text{————}| \end{matrix}\right]$$

$$\left[\begin{matrix} CH_3 \\ -CH-C- \\ \diagdown\diagup \\ S \end{matrix}\right] \quad \left[\begin{matrix} CH_3 \\ -CH-C- \\ \diagdown\diagup \\ S \\ \parallel \\ S \end{matrix}\right] \quad \left[\begin{matrix} \quad CH_3 \\ -CH-C- \\ \underset{|}{S}\quad\underset{|}{S} \\ \diagdown\diagup \\ S \end{matrix}\right]$$

$$\left[\begin{matrix} CH_3 \\ -CH-C- \\ \underset{|}{S}\text{—}\underset{|}{S} \end{matrix}\right] \quad \left[\begin{matrix} \qquad CH_3 \\ -CH-C- \\ \diagdown\diagup \\ S \\ \diagup\diagdown \\ -CH-C- \\ \underset{|}{CH_3} \end{matrix}\right] \quad \text{etc.}$$

These structures were later rejected. Nevertheless, one cannot exclude the possibility of the formation of tetrahydrothiophene (9) [334,337-339] or 1,4-dithiane structures (10) [334,340] during vulcanization:

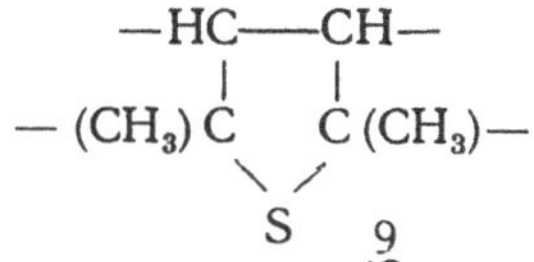

$$
\begin{array}{cc}
\begin{array}{c}
-\mathrm{HC}\!-\!\mathrm{CH}- \\
\quad | \qquad | \\
-(\mathrm{CH_3})\,\mathrm{C} \quad \mathrm{C}\,(\mathrm{CH_3})- \\
\qquad \backslash \;/ \\
\mathrm{S} \qquad \underset{\sim}{9}
\end{array}
&
\begin{array}{c}
\mathrm{S} \\
/\;\backslash \\
-\mathrm{HC} \quad \mathrm{CH}- \\
\quad | \qquad | \\
-(\mathrm{CH_3})\,\mathrm{C} \quad \mathrm{C}\,(\mathrm{CH_3})- \\
\qquad \backslash \;/ \\
\mathrm{S} \qquad \underset{\sim}{10}
\end{array}
\end{array}
$$

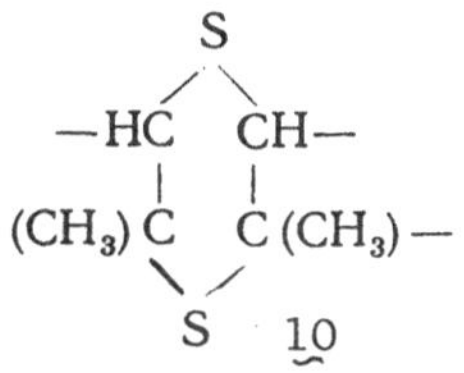

All these data are interesting from the historical point of view and illustrate the evolution of concepts concerning the mechanism of rubber vulcanization. It has been shown that several consecutive or parallel reactions take place during vulcanization and that the mechanism of these reactions is different (radical or ionic) [341-343]. Primary reactions occurring with the participation of sulfur, hydrocarbon rubber and a vulcanization accelerator give labile intermediates with no cross-linkages between the polyisoprene molecules. The vulcanizations proper represent the next series of reactions. The primary reactions of the polymeric chains of rubber during vulcanization with sulfur occur with the participation of double bonds or adjacent methylene groups. Activated sulfur can be present either in ionic or in radical form and, consequently, it can react with polyisoprene via an ionic or a radical mechanism:

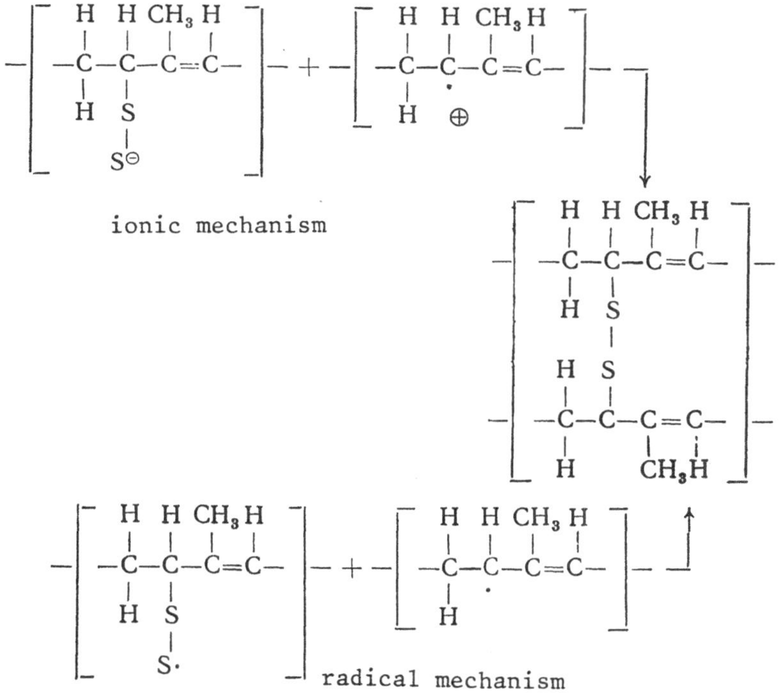

ionic mechanism

(3.56)

radical mechanism

Sulfide and polysulfide bridges formed predominantly at α-methylene carbon atoms [202,208,344-348] lead to linear and cyclic structures [282,349,350] as with model compounds (cf. Section 3.2.2).

In addition to polysulfide bridges of the type $C-S_n-C$ formed during the interaction of rubber with sulfur biradicals, $\cdot S_n \cdot$, mercapto radicals, $HS \cdot$, and polymercapto radicals $HS_n \cdot$, polyprene macromolecules can be cross-linked even via C-C bonds [342,351-353]. These bonds can be formed by the interaction of polymer radicals with one another or with the starting rubber molecules:

$$\text{a)} \left[-CH_2-CH=\overset{\overset{\displaystyle CH_3}{|}}{C}-CH_2-\right] + \cdot S_n \cdot \rightarrow \left[-CH_2-CH=\overset{\overset{\displaystyle CH_3}{|}}{C}-\overset{\displaystyle CH}{\underset{\displaystyle \cdot}{}}-\right] + \cdot S_n H \tag{3.57}$$

$$\text{b)} \left[-CH_2-CH=\overset{\overset{\displaystyle CH_3}{|}}{C}-\overset{\displaystyle CH}{\underset{\displaystyle \cdot}{}}-\right] + S_8 \rightarrow \left[-CH_2-CH=\overset{\overset{\displaystyle CH_3}{|}}{C}-\overset{\displaystyle CH}{\underset{\overset{\displaystyle |}{\cdot S_n}}{}}-\right] + S_{8-n} \tag{3.58}$$

$$\text{c)} \left[-CH_2-CH=\overset{\overset{\displaystyle CH_3}{|}}{C}-\overset{\displaystyle CH}{\underset{\overset{\displaystyle |}{\cdot S_n}}{}}-\right] + \left[-CH_2-CH=\overset{\overset{\displaystyle CH_3}{|}}{C}-CH_2-\right] \rightarrow$$

$$\rightarrow \begin{array}{c} \left[-CH_2-CH=\overset{\overset{\displaystyle CH_3}{|}}{C}-\overset{\displaystyle CH}{|}-\right] \\ \underset{\overset{\displaystyle |}{S_n}}{} \\ \left[-CH_2-\overset{\displaystyle CH}{|}-\overset{\overset{\displaystyle CH_3}{|}}{C}-CH_2-\right] \end{array} \quad \xrightarrow{\;+\left[-CH_2-CH=\overset{\overset{\scriptstyle CH_3}{|}}{C}-CH_2-\right]\;}$$

$$\rightarrow \begin{array}{c} \left[-CH_2-CH=\overset{\overset{\displaystyle CH_3}{|}}{C}-CH-\right] \\ \underset{\overset{\displaystyle |}{S_n}}{} \\ \left[-CH_2-\overset{\displaystyle CH}{|}-\overset{\overset{\displaystyle CH_3}{|}}{CH}-CH_2-\right] \end{array} + \left[-CH_2-CH=\overset{\overset{\displaystyle CH_3}{|}}{C}-\overset{\displaystyle CH}{\underset{\displaystyle \cdot}{}}-\right] \tag{3.59}$$

The vulcanization of rubber often leads to the initial formation of thiols which, by interaction with sulfur or an accelerator, yield di- and polysulfides [355]:

$$2R-S_n H \xrightarrow[-H_2 S]{+S} R-S_n-S_n-R \tag{3.60}$$

Cross-linking can also occur without the formation of rubber thiols [356-362]. When cross-linking occurs at the double bonds, the sulfur can add either in the form of mercapto or polymercapto groups [363, 364]. This can take place as an intermolecular process with the formation of S-H bridges or intramolecularly with simultaneous cyclization [342,365]. The subsequent reactions which can lead to a loss of sulfur from polysulfide bonds and the formation of new ring systems have not yet been elucidated. It is likely that vulcanized rubber can contain so-called vicinal bridges [366]. In this case, each pair of such bridges behaves as a single cross-linkage. Such cross-linkages are formed according to the scheme:

$$2-\left[-CH_2-CH=\overset{\overset{\displaystyle CH_3}{|}}{C}-CH_2-\right]-+2\cdot S_n\cdot \rightarrow \begin{array}{l} -\left[-CH_2-CH-\overset{\overset{\displaystyle CH_3}{|}}{\underset{\underset{\displaystyle S_n}{|}}{C}}-CH_2-\right]- \\[4pt] -\left[-CH_2-\underset{}{CH}-\underset{\underset{\displaystyle CH_3}{|}}{\overset{\overset{\displaystyle |}{S_n}}{C}}-CH_2-\right]- \end{array} \qquad (3.61)$$

The process of natural rubber vulcanization with sulfur at 100-180°C in the presence of amines in a ratio from 1:1 to 10:1 has been studied [367,368]. The formation of cross-linkages in the course of natural rubber vulcanization in the presence of accelerators has been examined. The number of cross-linkages may be greatly changed by varying the amount of sulfur [369,370]. Organic accelerators are widely used in rubber vulcanization. These accelerators considerably decrease the time needed for vulcanization and substantially lower the temperature of the reaction. Also, the use of accelerators makes it possible to decrease the amount of sulfur necessary for achieving the optimal properties of vulcanized rubber [371,372]. The mechanism of the action of vulcanization accelerators has been studied in detail [341,354,365,373-379]. Organic accelerators of rubber vulcanization can be divided into the following basic groups: dithiocarbamates, thiuram disulfides, thiazoles, sulfenamides, guanidines, aldehyde-amines, and xanthates. The literature dealing with organic accelerators of rubber vulcanization is extensive and it is not possible to give a complete list of references. The subject is well covered by reviews and monographs [342,373,380-393].

3.6. ALKYNES

The reaction of acetylene with sulfur at 300°C was initially studied by Meyer and Sandmeyer [165,166]. The principal reaction products were hydrogen sulfide, carbon disulfide, traces of thiophene, and carbon. The formation of thiophene may be represented as follows: ·

$$\begin{array}{c} CH \\ ||| \\ CH \end{array} + S + \begin{array}{c} CH \\ ||| \\ CH \end{array} \longrightarrow \left\langle\!\!\!\bigcirc\!\!\!\right\rangle_S \qquad (3.62)$$

The yield of thiophene from acetylene has been considerably improved
by replacing elemental sulfur with pyrites [293,394-397]. More
recently, the products of this reaction have been found to be
hydrogen sulfide, butane, 1,3-butadiene, acetaldehyde, acetone, 2-
and 3-methylthiophene, 2,3-dimethylthiophene, 3-ethylthiophene, a
considerable amount of thiophene, traces of benzene, and acetylene
and other acetylenic hydrocarbons. The mechanism of thiophene
formation may be represented by the scheme:

$$\begin{array}{c} \text{HC} \quad \text{CH} \\ \text{III} + \text{III} \\ \text{HC} \quad \text{CH} \end{array} \longrightarrow \begin{array}{c} \text{HC} - \text{CH} \\ \text{II} \quad \text{II} \\ \text{HC} \quad \text{CH} \end{array} \xrightarrow{+\text{S}} \begin{array}{c} \text{HC} \quad \text{CH} \\ \text{HC} \diagdown_{\text{S}} \diagup \text{CH} \end{array} \qquad (3.63)$$

Condensation of the above biradical with another acetylene molecule
and sulfur (or hydrogen sulfide) gives "α,β-thioxene" and 3-ethyl-
thiophene. The formation of the methylthiophenes may be due to the
presence of methane which is obviously liberated during the reaction
[398,399]. The appearance of oxygen-containing compounds among the
reaction products must be ascribed to the presence of moisture in
the acetylene used. According to other data, no traces of thiophene
are formed when acetylene is passed over strongly heated sulfur
[400,401]. In this case the reaction products were carbon disulfide,
hydrogen sulfide, thieno[2,3-b]thiophene (11) and probably also
benzo[b]thiophene (12). At 275°C, acetylene ignites in the presence
of sulfur vapor and burns with a yellow smoky flame. The reaction
of acetylene with sulfur at 290-390°C gives carbon disulfide, hydrogen
sulfide, carbon, thieno[2,3-b]thiophene (11), traces of thiophene, and
traces of benzenethiol [402]:

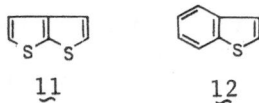

11   12

    According to other data [403], at temperatures below the
boiling point of sulfur the latter reacts with acetylene to give
about 5% of thiophene and large amounts of hydrogen sulfide and
carbon disulfide. At the optimum reaction temperature, i.e., 500°C,
75% of the acetylene is converted to liquid sulfuration products
consisting of 77% carbon disulfide, 12% thiophene, and 6% thieno-
[2,3-b]thiophene [403]. Thiophene is formed by the direct interaction of acetylene with sulfur and not by a secondary reaction with
carbon disulfide [404]. The reaction of acetylene with sulfur vapor
gives liquid thieno[2,3-b]thiophene (11) or thieno[3,4-b]thiophene
(13) and solid thieno[3,2-b]thiophene (14) [405]. The isomeric bithiophenes are formed as well as thieno[2,3-b]thiophene [406]. A
patent describes the preparation of carbon disulfide by the action
of sulfur on alkynes, especially in the presence of catalysts [407].
It has been observed [408] that, upon contact with aluminium oxide,
acetylene reacts with sulfur vapor with the formation of thiophene
and its homologs:

13   14

Tetrathiooxalic acid is formed [409] on condensation of acetylene with sulfur dissolved in aromatic amines:

$$HC\equiv CH \xrightarrow{+4S} HS-\underset{\underset{S}{\|}}{C}-\underset{\underset{S}{\|}}{C}-SH \qquad (3.64)$$

This acid then forms dithiooxanilide (15) with aniline and reacts with N,N-dimethylaniline to give a compound whose structure has been claimed to be (16):

$$C_6H_5NHC-C-NHC_6H_5$$
$$\underset{S}{\overset{\|}{}}\;\underset{S}{\overset{\|}{}}$$

15

$$\begin{array}{c} S \\ \diagup\diagdown \\ C=C-C_6H_4-N(CH_3)_2 \\ | \\ S \\ | \\ C=C-C_6H_4-N(CH_3)_2 \\ \diagdown\diagup \\ S \end{array}$$

16

First attempts to observe a reaction between sulfur and higher alkynes were unsuccessful [42]. However, conditions were later found under which 2-alkynes would react with sulfur to form 1,2-dithiol-3-thiones in low yield, together with other sulfuration products [410]. Acetylene, when heated with sulfur to 450°C, also forms 1,2-dithiol-3-thione in low yield [411]:

$$R-\underset{\underset{}{C}}{\overset{C-CH_3}{\underset{\|}{\|}}} \xrightarrow{+4S} R-\underset{S\diagdown S}{\boxed{}}S + H_2S \qquad (3.65)$$

Phenylacetylene reacts with sulfur and carbon disulfide at room temperature with the formation of 4-phenyl-1,3-dithiol-2-thione [412-414]:

$$C_6H_5C\equiv CH \xrightarrow{S_8, CS_2} C_6H_5C\equiv C-SH \xrightarrow{CS_2} C_6H_5\underset{S}{\overset{S}{\boxed{}}}=S \qquad (3.66)$$

The reaction passes through an intermediate step involving the formation of an unsaturated thiol [415] which then undergoes cyclization in the presence of carbon disulfide to give 4-phenyl-1,3-dithiol-2-thione. Because of the prototropic equilibrium between the 1-alkyne-1-thiol and the corresponding thioketene [297, 417-422], the former can rearrange to phenylthiirene:

$$C_6H_5C\equiv C-SH \rightleftharpoons C_6H_5CH=C=S \rightleftharpoons C_6H_5-\underset{S}{\overset{|}{C}}=CH \qquad (3.67)$$

A method for the preparation of divinyl sulfide from the reaction of acetylene with elemental sulfur at 80-120°C in dimethyl sulfoxide or hexamethylphosphoric triamide in the presence of alkali and water has been developed [423,424]:

$$CH\equiv CH + S \xrightarrow[KOH]{H_2O} CH_2=CH-SH + CH_2=CH-SOH$$
$$\downarrow CH\equiv CH$$
$$CH_2=CH-S-CH=CH_2$$
$$(80\%) \qquad\qquad\qquad (3.68)$$

The absolute rate constants and the Arrhenius parameters for the reaction of $S(^3P_{2,1,0})$ with alkynes have been measured using photolysis and vacuum UV spectroscopy (Table 3.4) [425].

Six possible resonance structures have been proposed for the product of the addition of sulfur to acetylene, of which (18) and (19) were claimed to be the main structures. Analogously, resonance structures (23) to (33) were suggested for the product from the addition of sulfur to propene [425]:

17   18   19   20   21   22

23   24   25   26

Table 3.4. Rate Constants for the Reactions
            of $S(^3P)$ Atoms with Alkynes (25°C)

Substrate	K, 10^8 liter·mole^{-1}·s^{-1}
Acetylene	2.3 ± 0.4
Propyne	48 ± 2
1-Butyne	33 ± 2
2-Butyne	160 ± 20
2-Pentyne	180 ± 20

$$CH_3C \equiv C-SH$$

$$\underset{\underset{\sim}{27}}{} \qquad \underset{\underset{\sim}{28}}{} \qquad \underset{\underset{\sim}{29}}{} \qquad \underset{\underset{\sim}{30}}{}$$

$$\underset{\underset{\sim}{31}}{} \qquad \underset{\underset{\sim}{32}}{} \qquad \underset{\underset{\sim}{33}}{}$$

Addition of atomic sulfur in its triplet or singlet state to alkynes gives very labile thiirenes [426]:

$$R-C \equiv C-R' \xrightarrow{S} \underset{\diagdown S \diagup}{R-C=C-R'} \tag{3.69}$$

Thiirenes easily react with alkynes to give thiophenes and solid polymeric materials. The addition of atomic sulfur to alkynes is an electrophilic process:

$$\underset{\diagdown S \diagup}{-C=C-} + RC \equiv CR' \longrightarrow \underset{S}{\boxed{}}\underset{R'}{\overset{R}{}} \tag{3.70}$$

3.7. CYCLOALKANES AND THEIR AROMATIC DERIVATIVES

Interaction of elemental sulfur in its singlet state with cyclopropane, cyclobutane, and cyclopentane takes place at room temperature with the formation of the corresponding cycloalkanethiols as the principal reaction products [297,427,428]:

$$\underset{H_2C-CH_2}{\overset{(CH_2)_n}{\diagup \diagdown}} + S\,(^1D_2) \rightarrow \underset{H_2C-CH-SH}{\overset{(CH_2)_n}{\diagup \diagdown}} \tag{3.71}$$

where n = 1,2,3.

The dehydrogenation of cyclohexane and its homologs to the corresponding aromatic hydrocarbons has not been observed in their reactions with sulfur. However, the validity of these observations is doubtful.

Thus, heating cyclohexane with sulfur in a sealed tube at 240-280°C gives benzenethiol, dibenzothiophene and probably also benzene, but not cyclohexene and cyclohexadiene as erroneously stated by the authors [429,430]:

$$\text{(cyclohexane)} \xrightarrow{+4S} \text{(benzenethiol)}\!-\!SH + 3\,H_2S \qquad (3.72)$$

$$2\,\text{(benzenethiol)}\!-\!SH \xrightarrow{S} \text{(dibenzothiophene)} + 2\,H_2S \qquad (3.73)$$

Under analogous conditions, methylcyclohexane reacts with sulfur to give 2-methylbenzenethiol, and 1,3-dimethylcyclohexane gives 2,6-dimethylbenzenethiol and 1,2-bis(3-methylphenyl)ethane (34) [429]:

$$2\,\underset{\substack{\text{1,3-dimethyl-}\\\text{cyclohexane}}}{\text{(structure)}} \xrightarrow[-7\,H_2S]{+7S} \underset{34}{\text{(structure)}\!-\!CH_2\!-\!CH_2\!-\!\text{(structure)}} \qquad (3.74)$$

The interaction of sulfur with 2-cyclohexyl-2-butene at 195–205°C yields 2-phenylbutane [431]. It is assumed that 2-cyclohexyl-2-butene isomerizes during the reaction to 2-cyclohexenylbutane [29–31] which is then dehydrogenated by sulfur to the corresponding alkylbenzene:

$$\underset{\overset{|}{CH_3}}{\text{(cyclohexyl)}}\!-\!C\!=\!CH\!-\!CH_3 \longrightarrow \underset{\overset{|}{CH_3}}{\text{(cyclohexenyl)}}\!-\!CH\!-\!CH_2\!-\!CH_3 \xrightarrow[-2\,H_2S]{+2S} \underset{\overset{|}{CH_3}}{\text{(phenyl)}}\!-\!CH\!-\!CH_2\!-\!CH_3 \qquad (3.75)$$

1,2-Dimethylenecyclohexane (35) when treated with sulfur forms 3,4,5,6,7,8-hexahydrobenzo[d]-(1,2)dithiine (36), 2,3,4,5,6,7-hexahydrobenzo[c]thiophene (37) and 3,4,5,6-tetrahydrobenzo[c]-thiophene (38) [432]:

$$\underset{35}{\text{(structure)}} \xrightarrow{S} \begin{cases} \underset{38}{\text{(structure)}} \\ \underset{37}{\text{(structure)}} \longrightarrow \underset{36}{\text{(structure)}} \end{cases} \qquad (3.76)$$

Hydrindene undergoes ready dehydrogenation with sulfur to form indene [34]:

$$\text{(hydrindene)} + S \longrightarrow \text{(indene)} + H_2S \qquad (3.77)$$

Tetralin [433–436] and its derivatives [437–441] undergo ready dehydrogenation with sulfur at 200–320°C to give the corresponding naphthalenes:

$$\text{(tetralin)} + 2\,S \longrightarrow \text{(naphthalene)} + 2\,H_2S \qquad (3.78)$$

The reaction is considerably accelerated and the yield of naphthalenes increased by the presence of organic sulfides [433,435] or
amines [433,434]. The catalytic activity of amines increases in the
order 4-methylamine < diphenylamine < triethanolamine < N-ethyl-N-
benzylaniline.

The dehydrogenation of tetralin by sulfur in the presence of
$Na_2S \cdot 9H_2O$ takes place at 140°C and not at 180°C as is the case in
the absence of this catalyst [434]. The catalytic activity may be
explained as being due to the heterolytic cleavage of the covalent
S-S bond in the cyclic S_8 molecule with the formation of polysulfide
ions. These ions undergo homolytic cleavage with the formation of
free radicals which then serve as initiators of dehydrogenation.

Heating tetralin with sulfur gives transparent, low-viscosity
copolymers which find use in the manufacture of varnishes [260].

Decalin reacts with sulfur at 170-290°C to give a high yield
of naphthalene. The reaction rate considerably increases in the
presence of diphenyl sulfide, dinaphthyl sulfide, or di(3-methyl-
butyl) sulfide [435]. 1-Phenyldecalin undergoes ready sulfur
dehydrogenation with the formation of 1-phenylnaphthalene [57]. The
latter was also obtained by dehydrogenating 1-phenyltetralin using
sulfur [444]:

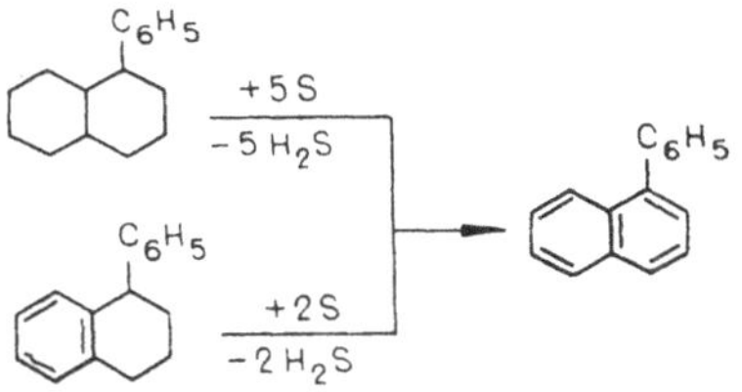

(3.79)

Heating cyclooctane with sulfur to 210-220°C gives xylenes
[445,446]. It seems that the reaction is preceded by the isomerization of cyclooctane to dimethylcyclohexane which is then
dehydrogenated by sulfur to xylene:

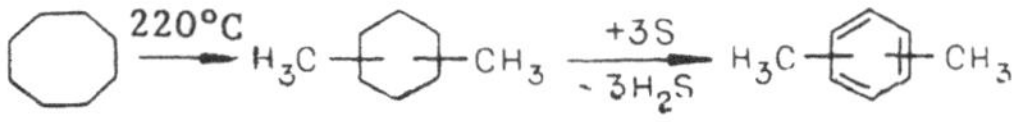

(3.80)

Tetrahydroacenaphthene, when heated with sulfur at 180-200°C, undergoes an almost quantitative dehydrogenation to acenaphthene [447].
Acenaphthene can undergo further dehydrogenation to acenaphthylene
which can then interact with excess sulfur to give diacenaphthyleno-
thiophene [448,449]:

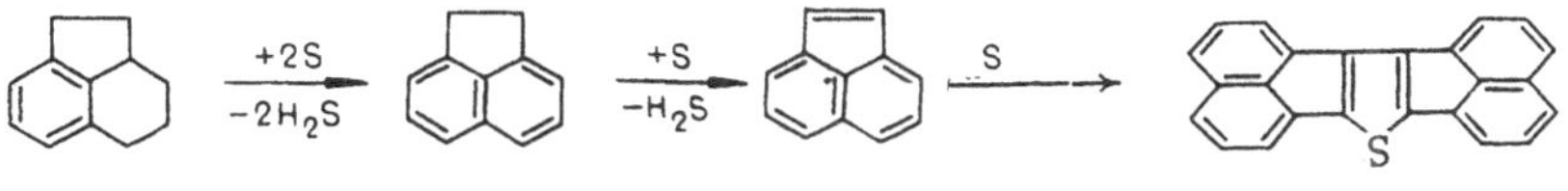

(3.81)

The reaction of sulfur with 1,1-pentamethylenebicyclo[4.1.0]-heptane at 190–210°C gives biphenyl [450]:

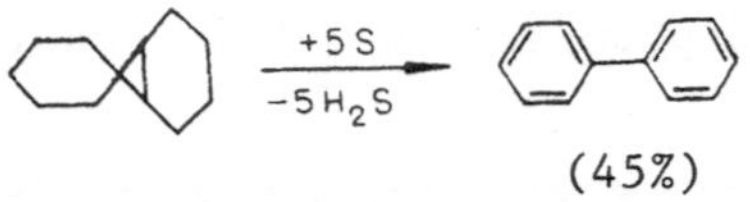

$$\text{(structure)} \quad \xrightarrow[-5\,H_2S]{+5\,S} \quad \text{(biphenyl)} \qquad (3.82)$$

(45%)

Dodecahydrochrysene is dehydrogenated by sulfur at 200°C to give octahydrochrysene [451]:

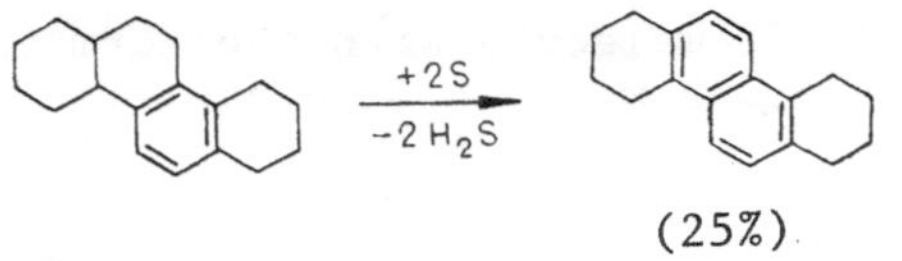

$$\text{(structure)} \quad \xrightarrow[-2\,H_2S]{+2\,S} \quad \text{(structure)} \qquad (3.83)$$

(25%)

Dehydrogenation of 4a,11-dimethyloctahydrochrysene with sulfur at 250°C is accompanied by demethylation and yields 5-methylchrysene [452–454], and only a low yield of 4,5-dimethylchrysene is obtained [455]. Perhydroazulene is easily dehydrogenated by sulfur to azulene [456]:

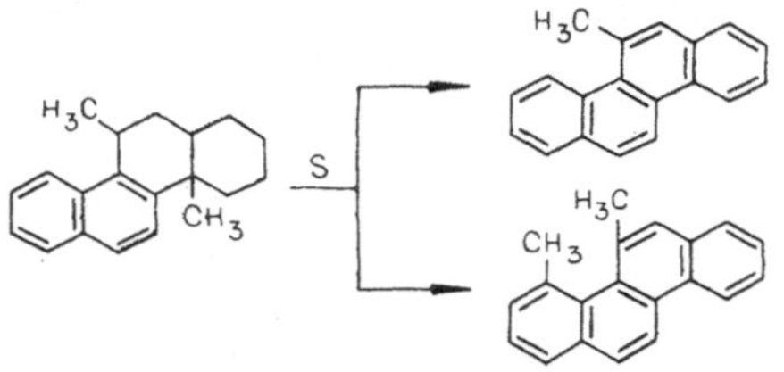

$$\text{(structure)} \xrightarrow{S} \text{(products)} \qquad (3.84)$$

$$\text{(structure)} \quad \xrightarrow[-5\,H_2S]{+5\,S} \quad \text{(azulene)} \qquad (3.85)$$

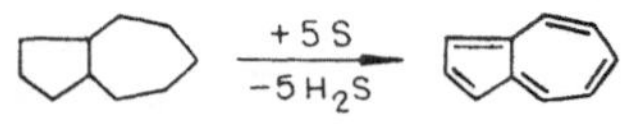

3.8.   CYCLOALKENES

The dehydrogenation of various unsaturated hydroaromatic compounds by sulfur is widely used and represents a convenient method for the determination of their structure [430,457,458,460,461].

Cyclohexene reacts with sulfur at 150°C to give cyclohexanethiol, dicyclohexyl sulfide, dicyclohexyl disulfude and saturated polysulfides of polymeric character [167,168,173,268,462]:

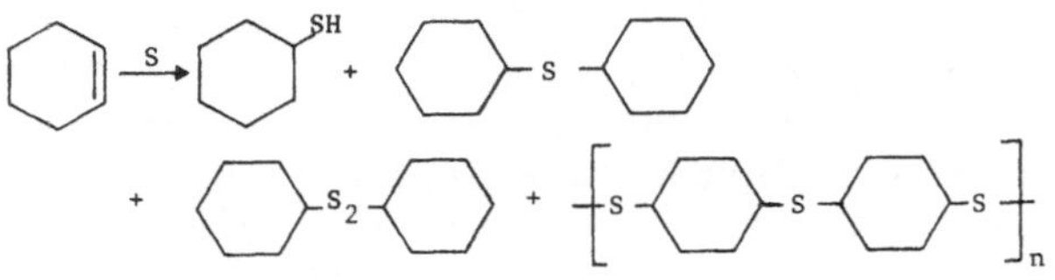

$$\text{(structures)} \qquad (3.86)$$

The reaction of S_8 with excess cyclohexene in the presence of $Fe_3(CO)_{12}$ (58°C, 16 h) in an inert gas atmosphere affords 1,2,3,4-tetrathiadecalin (39), 2,3,4,5,6-pentathiabicyclo[5.4.0]undecane (40) (total yield 5%) and complexes (41,42,43) [463]:

$$(3.87)$$

$$\underset{41}{S_2Fe_2(CO)_6} \quad + \quad \underset{42}{S_2Fe_3(CO)_9} \quad +$$

1-Methylcyclohexene behaves similarly. In the presence of sulfur, it adds hydrogen sulfide and forms the corresponding thiol and sulfide [193]. Thus, in contrast to alkenes, cyclohexene and its analogs have a tendency to form saturated thiols upon treatment with sulfur.

Sulfuration of cyclohexene under the conditions of rubber vulcanization also leads to nonvolatile polysulfides having the composition $(C_6H_{10}S_4)_n$ where n $\sim$ 3.5 [201,203,282,299,464]. The polysulfides obtained are thermally quite unstable and above $140°C$ undergo rapid decomposition with the formation of hydrogen sulfide, thiols, and sulfides. The reaction of sulfur with cycloalkenes is a free-radical process [201,203,299]:

$$(3.88)$$

A polar chain mechanism has also been suggested for the reaction of sulfur with cyclohexene and methylcyclohexene. In this mechanism, intermediate formation of a polythiosulfenyl cation $RS_n^{\oplus}$ and subsequent addition to the double bond are assumed [304].

The interaction of sulfur with cyclohexene in the presence of diethylamine at $140°C$ gives dicyclohexyl sulfide, dicyclohexyl disulfide, and saturated polymeric products of the type:

$$C_6H_{11}-S_n-C_6H_{10}-S_n-C_6H_{11}$$

It is assumed that the sulfur-introducing agent in this case is diethylammonium hydrosulfide which is formed in the reaction of sulfur with diethylamine [263].

The reaction of cyclohexene with sulfur has been investigated in
the presence of various catalysts, such as zinc dithiocarbamate,
tetramethylthiuram disulfide, tetramethylthiuram sulfide and zinc
oxide [465]. On irradiation of a solution of cyclohexene and sulfur
in carbon disulfide, compounds (44) with n = 1-3 are formed [466].
Irradiation of S_8 is believed to afford molecules S_3 and S_4 which
attack the alkene C=C bond:

$$\text{(3.89)}$$

1-Methylcyclohexene heated with sulfur at 190-205°C in xylene
gives benzo[c]-1,2-dithiol-3-thione [235,282,467,468]. Under these
conditions, 1-methylcyclohexene derivatives react with sulfur in an
analogous manner [467]:

$$\text{(3.90)}$$

The reaction between sulfur and 1-phenylcyclohexene or its derivatives
gives biphenyls [469-471]:

$$\text{(3.91)}$$

(70%)

The reaction of sulfur with cycloheptatriene (45) in sulfolane
at 70°C in the presence of pyridine as a catalyst has been investigated and found to afford the trithiabicyclodecadiene (46) in 21%
yield. The latter was also obtained from (47) and cyclohexasulfur
in the absence of a catalyst in 11% yield [472]:

$$+ 3 \text{ S} \longrightarrow \qquad \text{(3.92)}$$

The reaction of dicyclopentadiene with sulfur in the presence of
amines at 115-125°C gives the corresponding trithiolanes (47) [473]:

$$+ 3 \text{ S} \longrightarrow \qquad \text{(3.93)}$$

At 140°C dicyclopentadiene dissolves liquid sulfur to form sulfur-
dicyclopentadiene polymers [474].

Dihydronaphthalenes 1-phenyl-, 1-methyl-, and 1-ethyl-3,4-dihydronaphthalene are readily dehydrogenated with sulfur at 250-280°C to yield naphthalene and its derivatives [475,476]:

$$\underset{(91\%)}{\text{(structure)} \xrightarrow[-H_2S]{+S} \text{(structure)}} \tag{3.94}$$

On heating with sulfur to 200°C in xylene, 1-methyl-3,4-dihydronaphthalene forms 1H-naphtho[2,1-c]-1,2-dithiole-1-thione [467]:

$$\text{(structure)} \xrightarrow[-3H_2S]{+6S} \text{(structure)} \tag{3.95}$$

Under analogous conditions, 2-methyl-3,4-dihydronaphthalene is converted to 3H-naphtho[1,2-c]-1,2-dithiol-3-thione [467]. Sulfur, pretreated with ammonia in dimethylformamide, adds stereospecifically to bicyclo[2.2.1]heptene (norbornene) at 110°C to yield exo-3,4,5-trithiatricyclo[5.2.1.0^{2,6}]decane (48) [477]:

$$\text{(structure)} + 3S \xrightarrow[110°C]{NH_3, DMF} \text{(structure)} \tag{3.96}$$

$$\underset{48 \quad (86\%)}{}$$

Heating sulfur with norbornene (49) at 100°C for six hours in the presence of sodium 2,6-dimethylphenoxide leads to the trithiatricyclodecane (48) in 74% yield. Simultaneously the trithiatetracyclotridecene (50) was isolated in low yield [478]:

$$\text{(structure)} \xrightarrow{3\ S} \begin{cases} \text{(structure)} & 48 \\ \text{(structure)} & 50 \end{cases} \tag{3.97}$$

Irradiation (λ 350 nm) of a mixture of norbornene (49) and sulfur in carbon disulfide leads to the formation of 3,4,5-trithiatricyclo-[5.2.1.0]decane (48) and 2,3-epithianorbornene (51) in 77 and 8% yields, respectively (3.98) [466]. The reaction of norbornadiene (52) with sulfur at 100°C in the presence of 2,5-bis-octyldithio-1,3,4-thiadiazole (3.99) leads to the episulfide (53) [479]:

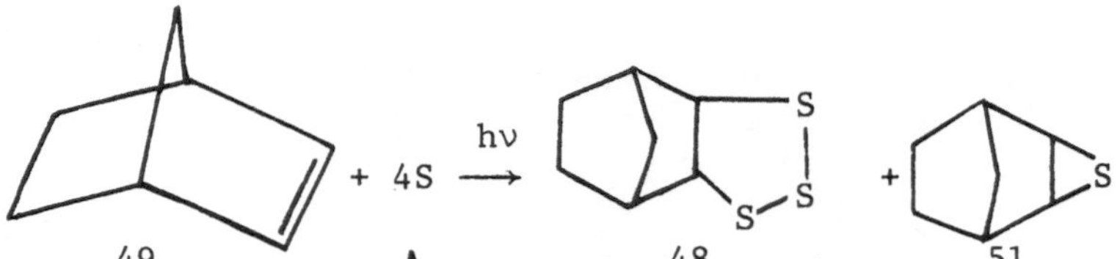

$$+ 4S \xrightarrow{h\nu} \qquad + \qquad \qquad (3.98)$$

49   48   51

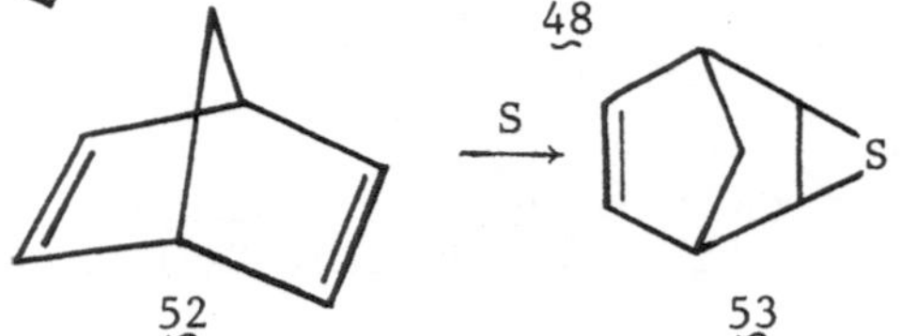

$$\xrightarrow{S} \qquad\qquad (3.99)$$

52   53

Similarly, the reaction of sulfur with 5-ethylidenebicyclo[2,2,1]-heptene gives 8-ethylidene-exo-3,4,5-trithiatricyclo[5.2.1.0^{2,6}]-decane. Only the endocyclic double bond participates in the reaction with sulfur [477]:

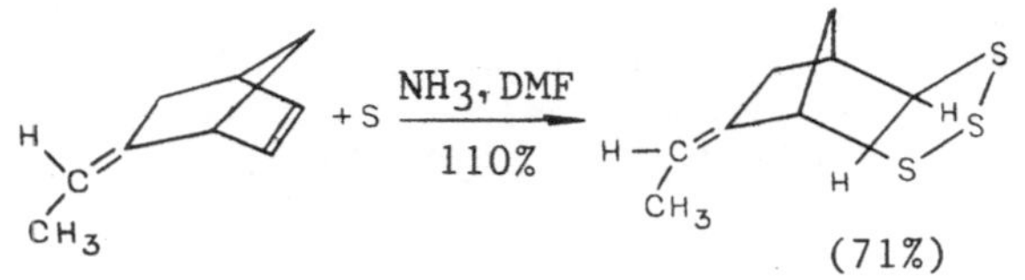

$$+ S \xrightarrow[110\%]{NH_3, DMF} \qquad\qquad\qquad (3.100)$$

(71%)

The reaction of sulfur with the cyclopentadiene trimer (54) at 100°C in the presence of 2,5-bis(octyldithio)-1,3,4-thiadiazole affords the trithiolane (55) in 85% yield [479]:

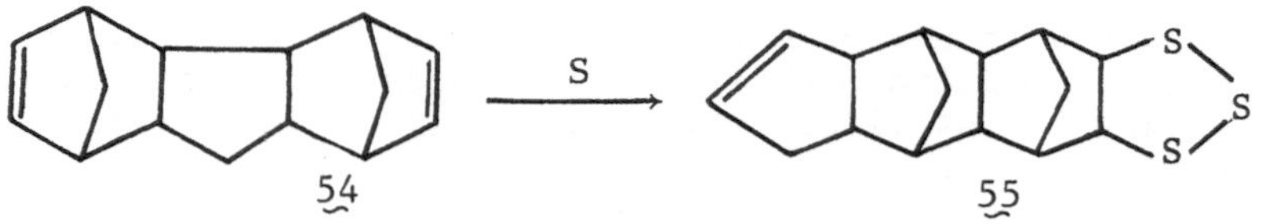

$$\xrightarrow{S} \qquad\qquad (3.101)$$

54   55

When sulfur reacts with the dihydro-derivatives of pentacene, hexacene, and heptacene in boiling nitrobenzene or trichlorobenzene, these compounds are dehydrogenated to the corresponding aromatic hydrocarbons. Further introduction of sulfur leads to the formation of high-melting sulfur-containing compounds whose structures have not been established [480].

3.9. TERPENES AND TERPENOIDS

    The introduction of sulfur into terpenes has been of interest to researchers for a long time, especially from the point of view of obtaining technically useful products, such as synthetic resins [459,481-483], gold-plating mixtures [484-497], accelerators for rubber vulcanization [498], oils for metal machining [499], varnishes [481], etc. [500,501].

Sulfurated turpentine oils have been used as medicines for a very long time. In the nineteenth century, they were used in folk medicine for the treatment of animals [496]. Later on, the sulfur-containing terpene derivatives have been used as their complexes with gold salts as gold-plating agents for china and ceramics [484-497]. In 1858, turpentine was distilled with sulfur [461] to give hydrogen sulfide and asphalt. It was established in 1870 that the reaction of turpentine with sulfur leads to the direct introduction of one, two, or more sulfur atoms into the terpene molecule [497]:

$$C_{10}H_{16}+2S \rightarrow C_{10}H_{14}S+H_2S \tag{3.102}$$

$$C_{10}H_{16}+4S \rightarrow C_{10}H_{12}S_2+2H_2S \tag{3.103}$$

When rosin (colophony) is heated with sulfur to 400°C colophthaline, $C_{11}H_{10}$, is obtained [503]. Heating the resin oil with sulfur affords retene [504]. The sulfuration of rosin gives a hydrocarbon which, according to its molecular composition, corresponds to pimanthrene, $C_{16}H_{14}$ [505].

Vesterberg [506] was the first to use sulfur for the dehydrogenation of terpenoids to aromatic compounds. He studied the dehydrogenation of abietic acid to retene. Dehydrogenation with sulfur has been used for the preparation of retene and pimanthrene from unsaturated resin oils [504,505,507-510], rosin [503,511], copal [512,513], and abietic [514-518] and pimaric acids [511,519]:

$$\tag{3.104}$$

The dehydrogenation of sesqui- and polyterpenes with sulfur, as well as the dehydrogenation of steroids, has been systematically used and improved [461]. It has been shown that hydrocarbons belonging to the naphthalene series, e.g., cadalene [461,517,519] and eudalene [461,517] can be obtained by the dehydrogenation with sulfur of certain sesquiterpenes which therefore must be hydrogenated naphthalenes.

The sulfur dehydrogenation of abietic, (+)-pimaric, and agathenedicarboxylic acids to phenanthrene derivatives has been used as proof of the structure of diterpenes. This reaction established that triterpenes are hydrogenated derivatives of picene [461,514, 517,520].

The reaction of sulfur with linalool and its acetate at 160°C gives $C_{10}H_{18}OS_6$ and $C_{12}H_{20}O_2S_3$ respectively [332]. These compounds were assigned the structure of "thioozonides" containing the grouping:

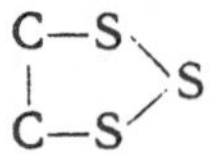

$$
\begin{array}{c}
C-S \\
| \quad\;\; \diagdown S \\
C-S \diagup
\end{array}
$$

However, these compounds are probably derivatives of 1,2-dithiol-3-thione because compounds containing the three-carbon grouping $-CH=\underset{|}{C}-CH_3$ or $-CH_2-\underset{|}{C}=CH_2$ react with sulfur to give the corresponding derivatives of the 1,2-dithiol [218,228,235,247,442,443,521-526]:

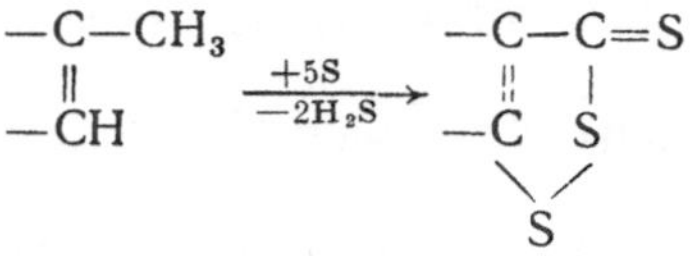

$$
\begin{array}{c}
-C-CH_3 \\
\parallel \\
-CH
\end{array}
\xrightarrow[-2H_2S]{+5S}
\begin{array}{c}
-C-C=S \\
\parallel \quad\;\; | \\
-C \quad\; S \\
\diagdown \diagup \\
S
\end{array}
\tag{3.105}
$$

Prolonged heating of pinene and Russian turpentine with sulfur leads to strong resinification and the formation of a red-brown oil, together with a very small amount of hydrogen sulfide [489,491,492]. This oil yields a cyclic monosulfide (56) and a cyclic disulfide (57), to which the following structures were assigned:

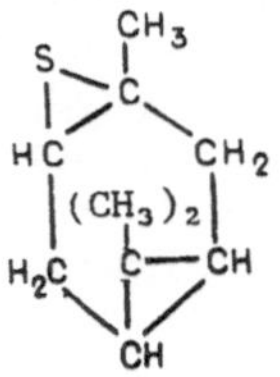

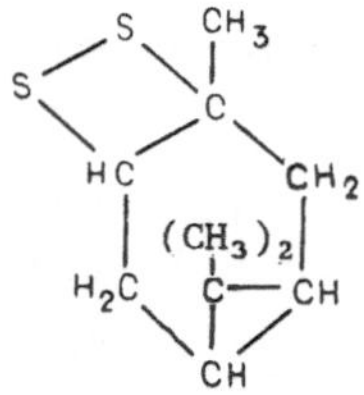

When α-pinene is refluxed with sulfur [527] polysulfides of the following type are formed:

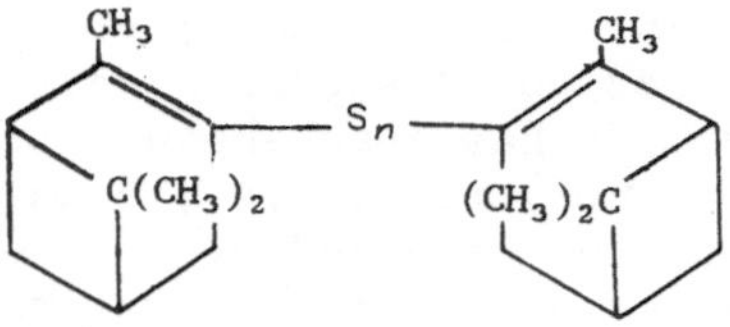

where n = 1-4.

The sulfuration of American turpentine, containing mainly α-pinene, leads to the formation of a liquid with the formula $C_{20}H_{24}S_5$ and of a solid with a composition corresponding to $C_{10}H_{12}S_4$ [266,528]. The formation of the first compound can be explained as follows:

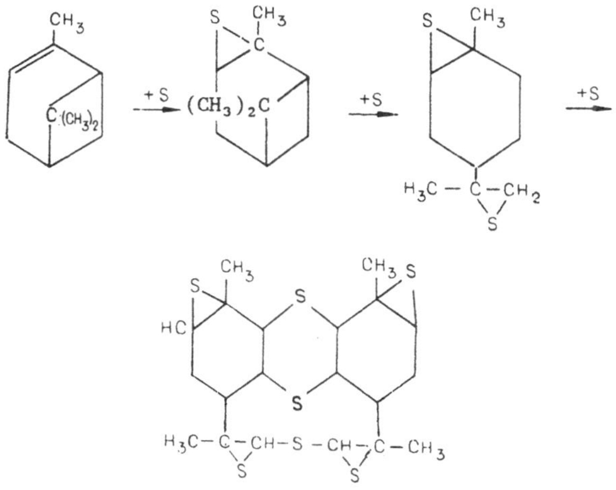

(3.106)

According to other data [529], the reaction of α-pinene with sulfur gives mainly polymerization products and only traces of a sulfide, $C_{10}H_{16}S$. A method for the preparation of α-pinenemercaptan by heating α-pinene with sulfur in the presence of hydrogen and 3-7% of a catalyst (WS_2 or NiS) at 20-100 atm and 160-210°C has been developed [530].

Heating α-limonene with sulfur at 160°C leads to the formation of a small amount of p-cymene and of a compound to which was erroneously assigned the structure of a thiirane:

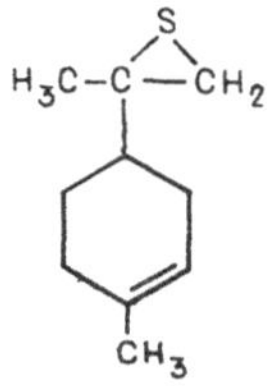

Similar sulfuration products containing only one sulfur atom were obtained from the reactions of many other terpenes with sulfur [529-532]. Later it was established that the reaction of sulfur with limonene [533] gives 4-isopropyltoluene, p-cymene, the optically active sulfides (58) and (59), the optically inactive cyclic sulfide (60), and the polysulfides (61) to (63). It is likely that the latter compounds are the primary reaction products:

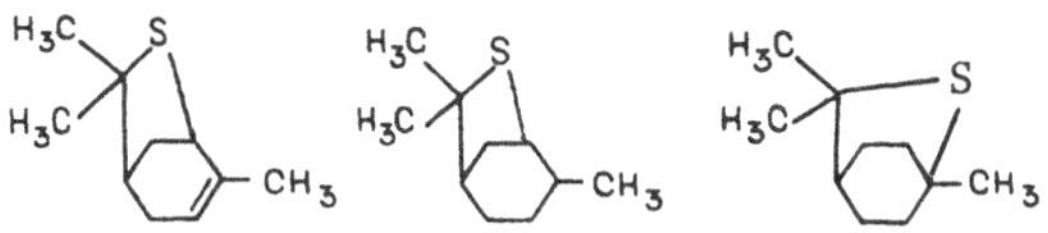

58   59   60

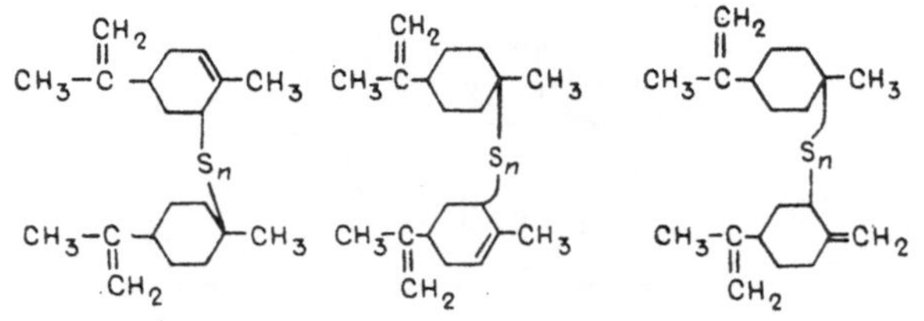

61   62   63

In their reaction with sulfur, dipentene, α-terpineol and its acetate give the same sulfide, $C_{10}H_{18}S$, which has been assigned the following structure [531,532]:

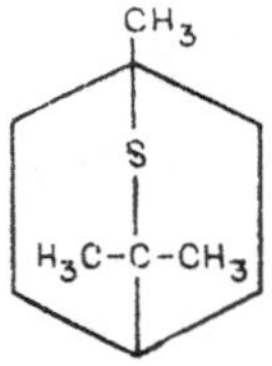

The reaction of sulfur with elemol [534,535] leads to a compound $C_{14}H_{18}S$ which has been assigned a thianaphthene structure:

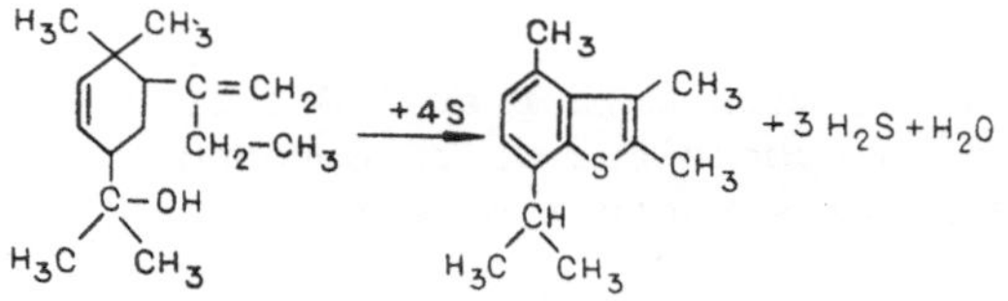

(3.107)

It has been noted in the dehydrogenation of the isomeric sesqui-terpenes bisabolene (64) and zingiberene (66) that the number and position of double bonds in the starting compounds have a consid-erable effect upon the structure of the reaction products. Bisabo-lene forms 2-methyl-6-(4-methylphenyl)heptane (65) on treatment with sulfur, whereas zingiberene gives the naphthalene derivative cadalene (67) [520]:

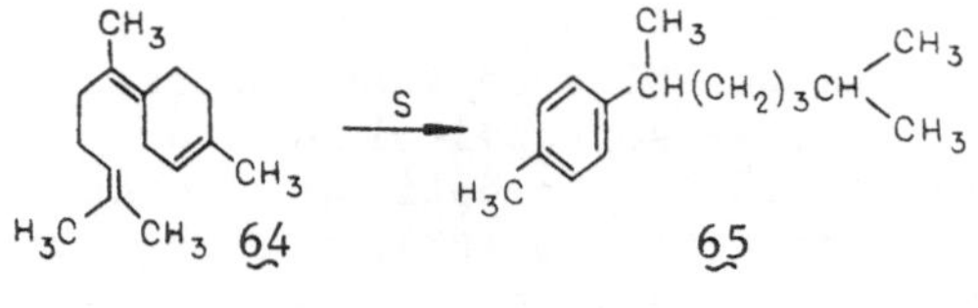

(3.108)

64   65

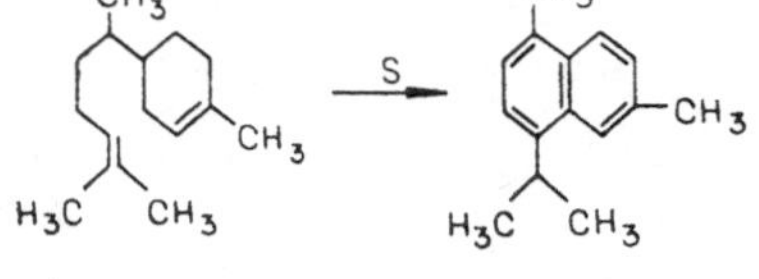

(3.109)

66   67

α-Terpinene undergoes dehydrogenation on heating with sulfur, and gives p-cymene (50% yield) [461]. Dehydrogenation of ionene (68) with sulfur gives a 10% yield of 1,6-dimethylnaphthalene (69) [536]:

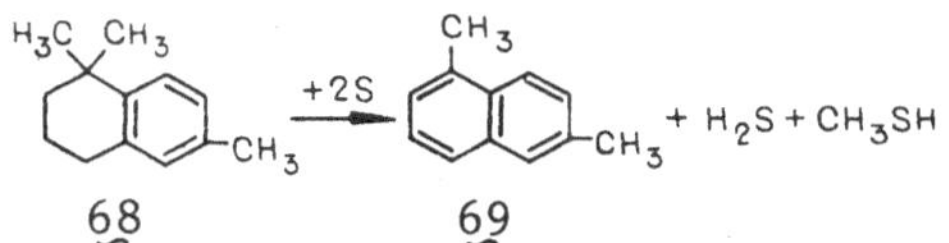

$$+2S \longrightarrow + H_2S + CH_3SH \qquad (3.110)$$

68 69

Heating β-cadinene (70) with sulfur gives the corresponding naphthalene derivative, cadalene (67) [461]:

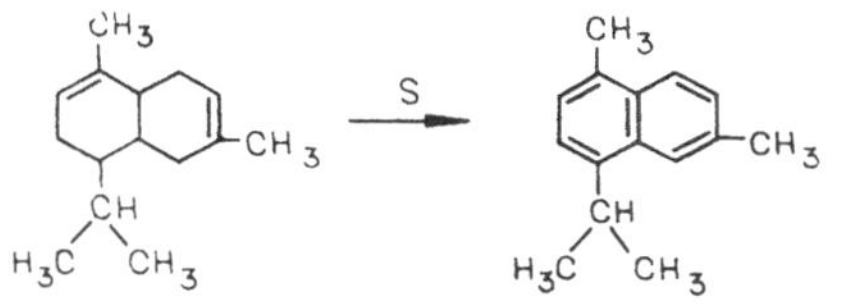

$$\xrightarrow{\text{S}} \qquad (3.111)$$

70 67

α-Eudesmol (71) is dehydrogenated by sulfur to give eudalene (7-isopropyl-1-methylnaphthalene) (73) via the intermediate formation of dihydroeudesmene (72) [537]:

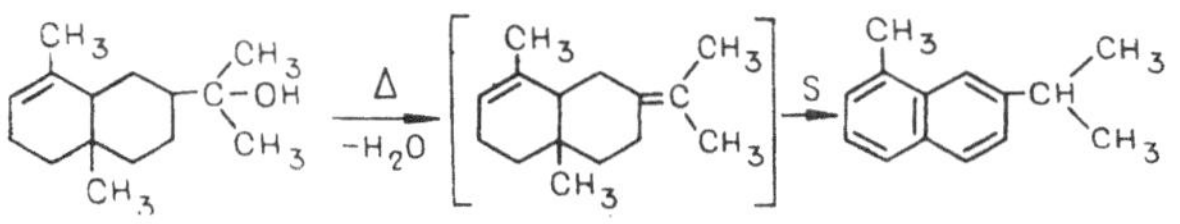

$$\xrightarrow[-H_2O]{\Delta} \left[\quad \right] \xrightarrow{\text{S}} \qquad (3.112)$$

71 72 73

The bicyclic sesquiterpene machilol forms 1-ethyl-7-isopropyl-naphthalene upon dehydrogenation with sulfur [538].

α- and β-selinene, (74) and (75) respectively, are dehydrogenated to eudalene (73) when treated with sulfur. During this reaction elimination of the angular methyl group takes place [537]:

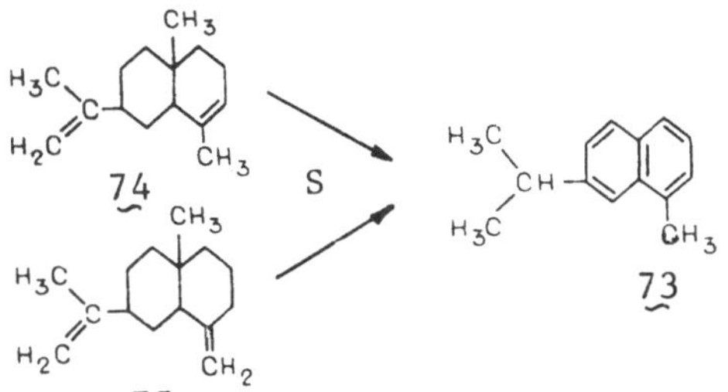

$$\xrightarrow{\text{S}} \qquad (3.113)$$

74 75 73

The dehydrogenation of the tricyclic sesquiterpene copaene (76) with sulfur gives exclusively cadalene (67) [539]:

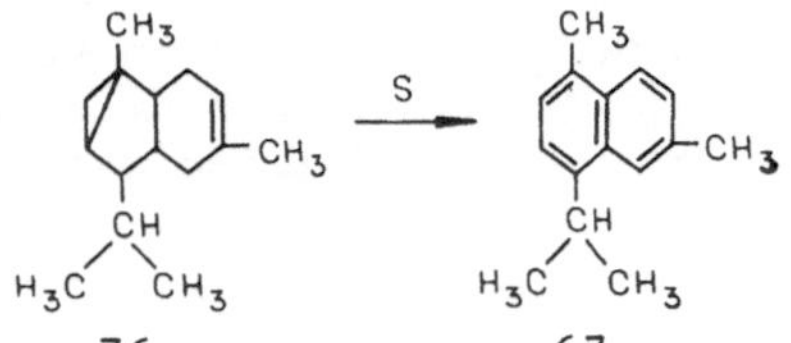

76 67 (3.114)

The dehydrogenation of guaiene (77) with sulfur at 200-260°C for two to three hours leads to the isolation of guaiazulene (78) as the main reaction product together with small amounts of 3,5,8-trimethylazuleno[6,5-b]thiophene (79), 3,6,9-trimethylazuleno[4,5-b]-thiophene (80), 2-methylthioguaiazulene (81), and 2,2'-biguaiazulene (82) [540]:

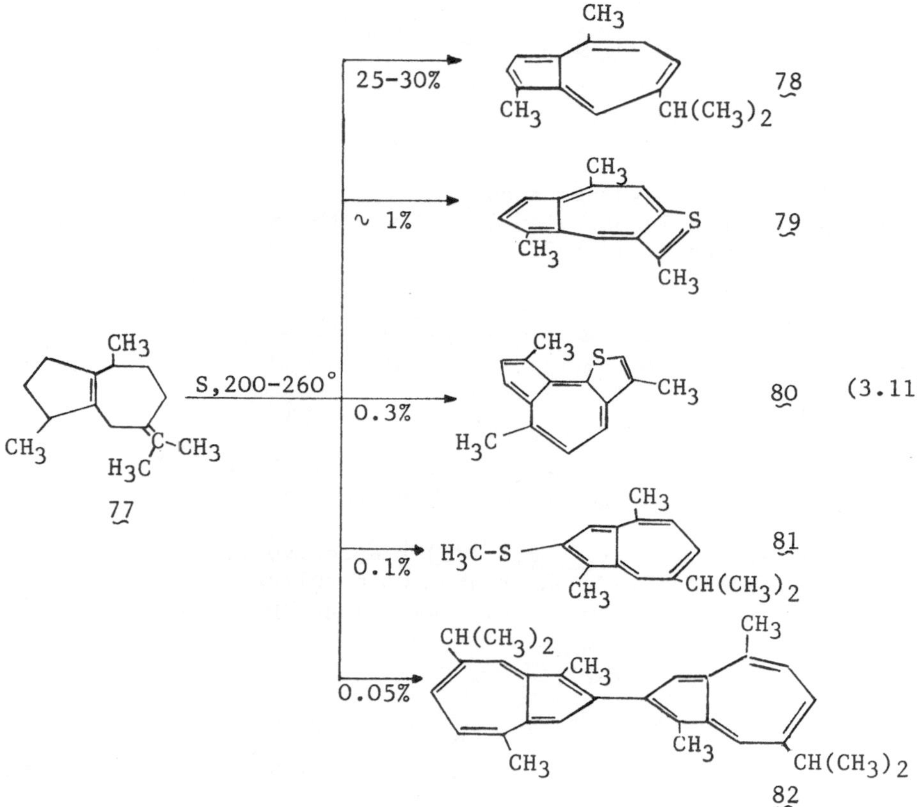

77 78 79 80 81 82 (3.115)

The sulfuration of dipentene and pulegone leads to the initial formation of a viscous oil from which a compound corresponding to $C_{10}H_6S_4$ can be isolated. This compound contains condensed thiophene and 1,2-dithiol-3-thione rings [218,219,230,304]:

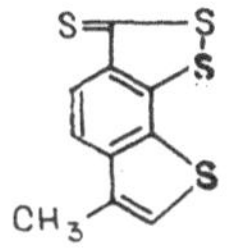

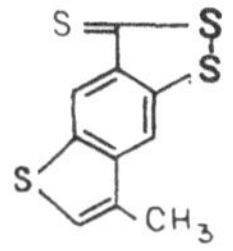

The reaction of carvone with sulfur gives a phenolic compound, $C_{10}H_6OS_4$ [230], with the following possible structures:

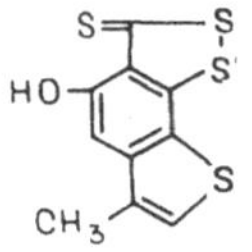

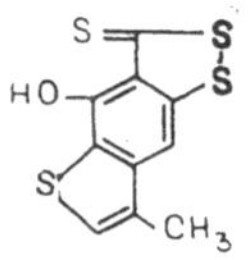

The reaction of 3-carene with sulfur above 100°C gives a sulfuration product containing 1 to 40% sulfur depending on the duration of heating [541].

3.10. AROMATIC HYDROCARBONS

Aromatic hydrocarbons are very stable to sulfur. They begin to react with sulfur only at 220-250°C. The inertness of these hydrocarbons with respect to sulfur can be shown by their method of purification. The hydrocarbon in question is heated with 1-3% sulfur at temperatures up to 300°C under pressure [457,542,543].

When a mixture of benzene and sulfur vapor is passed through a red-hot tube, a white crystalline substance, $C_{12}H_8S_3$, is formed, together with hydrogen sulfide. This substance is probably thianthrene, contaminated with sulfur [544]. When benzene is heated with sulfur in a sealed tube at 400-500°C, hydrogen sulfide is formed together with a small amount of thianthrene [545]. Thianthrene is also obtained by the reaction of benzene with sulfur in the presence of lead(II) oxide [546]. Benzene reacts with sulfur at 350°C to give benzenethiol, diphenyl sulfide, thianthrene, and hydrogen sulfide [547]. Benzenethiol is the primary reaction product, and the other products are formed by its subsequent interaction with sulfur and benzene. The results of the reaction of sulfur with benzene to give thiophenol, diphenyl sulfide and polysulfide as the main reaction products, have been summarized [548].

When aromatic hydrocarbons are heated with sulfur in the presence of Friedel-Crafts catalysts, a vigorous reaction occurs at 80-140°C [549]. Thus, in the presence of aluminium chloride, benzene reacts with sulfur at 80°C with the formation of hydrogen sulfide, thianthrene, diphenyl sulfide, and benzenethiol [549-555]:

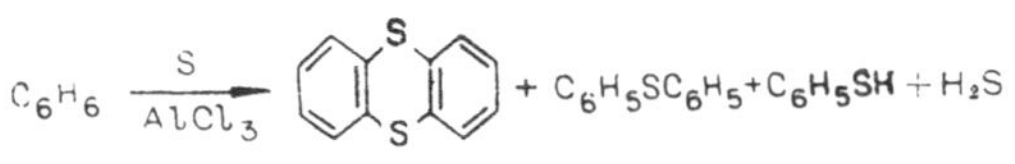

$$C_6H_6 \xrightarrow[AlCl_3]{S} \text{[thianthrene]} + C_6H_5SC_6H_5 + C_6H_5SH + H_2S \qquad (3.116)$$

The relative yields of diphenyl sulfide and thianthrene at a constant C_6H_6:S ratio depend on the amount of the catalyst. An increase in the benzene content in the reaction mixture leads to a decrease in the yield of thianthrene and to an increase in the yield of diphenyl sulfide [555]. The best yield of thianthrene can be obtained using the molar ratio $AlCl_3$:C_6H_6:S = 1:2:4 [552]. At the molar ratio $AlCl_3$:C_6H_6:S = 0.5:1:1, the principal reaction product is diphenyl sulfide [552] formed according to the scheme:

$$4S + 2AlCl_3 + 4C_6H_6 \rightarrow 2(C_6H_5)_2S \cdot AlCl_3 + 2H_2S \qquad (3.117)$$

The following scheme has been suggested for the reaction of sulfur with benzene using aluminium chloride as a catalyst:

$$S_n \cdot AlCl_3 + C_6H_6 \cdot AlCl_3 \rightarrow C_6H_5S_nH \cdot AlCl_3 + AlCl_3 \qquad (3.118)$$

$$C_6H_5S_nH \cdot AlCl_3 + C_6H_6 \cdot AlCl_3 \rightarrow (C_6H_5)_2S \cdot AlCl_3 + H_2S_{n-1} \qquad (3.119)$$

$$(C_6H_5)_2S \cdot AlCl_3 + S_n \rightarrow C_6H_4 {\Large \langle} \begin{smallmatrix} S \\[4pt] S \end{smallmatrix} {\Large \rangle} C_6H_4 \cdot AlCl_3 + H_2S_{n-1} \qquad (3.120)$$

The reaction of sulfur with benzene and its homologs in the presence of aluminium chloride has been patented as a method for the preparation of intermediates for polyvinyl chloride plasticizers [556].

The interaction of sulfur with biphenyl in the presence of anhydrous aluminium chloride leads to dibenzothiophene [553,557-561]. At 230-240°C, this reaction is complete within eight to ten hours and gave a 60-80% yield of dibenzothiophene [557-561]:

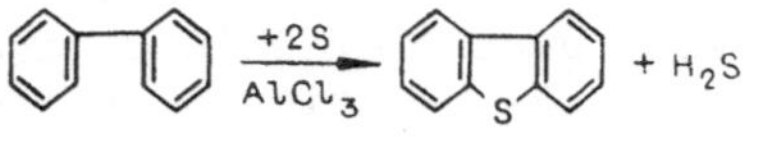

$$(3.121)$$

Sulfur and naphthalene vapors react in a red-hot iron tube to give [2,3-b;2',3'-d]dinaphthothiophene, 1,8-epidithionaphthalene, and hydrogen sulfide [562,563]. The reaction of sulfur with naphthalene catalyzed by inorganic halides and the subsequent reduction of the sulfuration product gave naphthalenethiol in good yield [564]. A reaction of sulfur with naphthalene and other aromatic hydrocarbons in the presence of sulfuric acid, aromatic sulfonic acids, or phosphoric acid has been patented [565-568]:

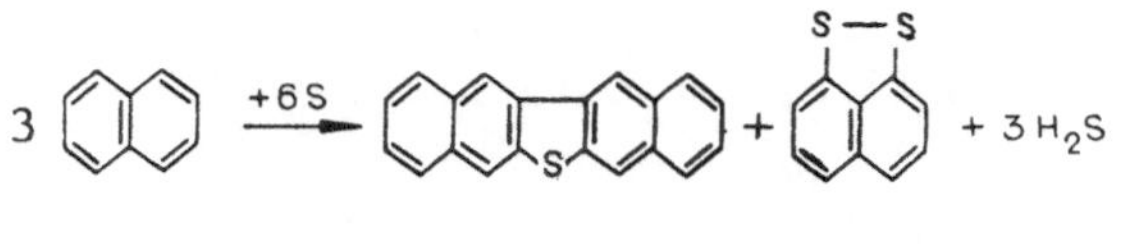

$$(3.122)$$

The sulfuration products so obtained have been recommended for the impregnation of fabrics, for the preparation of varnishes, and for the tanning of animal pelts. There is also a patent describing the preparation of an ebonite substitute by heating naphthalene with sulfur in the presence of sulfuric acid followed by the subsequent addition of formalin [569].

Acenaphthene heated with sulfur at 290°C undergoes dehydrogenation with subsequent trimerization leading to the formation of decacyclene (tris-perinaphthylenebenzene). Bis(1,8-naphthylene)-thiophene is a secondary reaction product [570-573]:

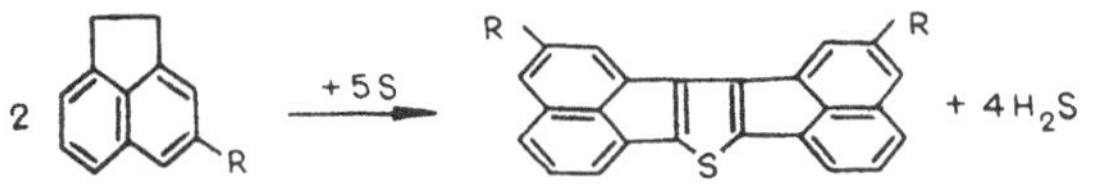

$$2 \quad \text{[structure]} \quad \xrightarrow{+5S} \quad \text{[structure]} \quad + 4H_2S \qquad (3.123)$$

In its reaction with sulfur, 4-benzylacenaphthene reacts according to the above scheme ($R = CH_2-C_6H_5$) giving tribenzyltrinaphthylenebenzene (14% yield) and also some dibenzyldinaphthylenethiophene [574]. A sulfur-containing dye may also be obtained by heating acenaphthene with sulfur [575].

Indene reacts with sulfur at 180-250°C to give a number of sulfur-containing compounds: $C_{18}H_{12}S$, $C_{36}H_{22}S$, $C_{36}H_{24}S$, $C_{27}H_{20}S$, $C_{36}H_{24}S_3$, $C_{45}H_{22}S$, $C_{45}H_{24}S$, and $C_{45}H_{32}S$ [575,577]. Among these compounds, only a thiophene derivative $C_{18}H_{12}S$ has been identified so far:

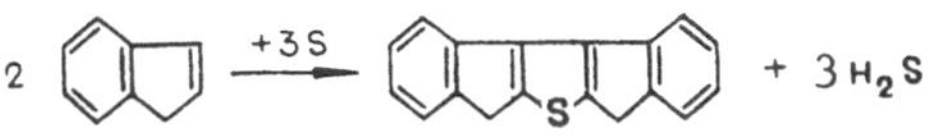

$$2 \quad \text{[structure]} \quad \xrightarrow{+3S} \quad \text{[structure]} \quad + 3H_2S \qquad (3.124)$$

Fluorene when heated with sulfur at 300°C gives 9,9'-bifluorene (83) [578], and in the presence of excess sulfur the product is bifluorenylidene (84) [579-582]:

$$2 \begin{matrix} C_6H_4 \\ | \\ C_6H_4 \end{matrix}\!\!>\!\!CH_2 \xrightarrow[-H_2S]{+S} \begin{matrix} C_6H_4 \\ | \\ C_6H_4 \end{matrix}\!\!>\!\!CH-CH\!\!<\!\!\begin{matrix} C_6H_4 \\ | \\ C_6H_4 \end{matrix} \xrightarrow[-H_2S]{+S} \begin{matrix} C_6H_4 \\ | \\ C_6H_4 \end{matrix}\!\!>\!\!C=C\!\!<\!\!\begin{matrix} C_6H_4 \\ | \\ C_6H_4 \end{matrix} \qquad (3.125)$$

83   84

The reaction of phenanthrene with sulfur at 340-360°C in a stream of carbon dioxide gives a 15% yield of tetrabenzo[a,c,h,i] thianthrene [583]. Anthracene heated with sulfur at 350-360°C gives various dyes [584,585]. The reaction of sulfur with technical anthracene has been patented as a method for the preparation of an asphalt substitute [587]. The addition of 25% of anthracene

and 2.5% of sulfur to pitch and subsequent heating gives a product
with properties which are identical with those of the starting
pitch. Thus, this technique can be used to increase the yield of
the pitch [588]. The action of sulfur upon the distillates obtained
from coal tars containing anthracene, phenanthrene, and other aromatic hydrocarbons transforms these hydrocarbons into pitch-like
substances which have been patented as additives improving the
lubricating characteristics of oils [589]. The reaction of sulfur
with anthracene at 400°C gives sulfur-containing polymers [590].

On passing the vapor of various aromatic hydrocarbons (benzene,
naphthalene, anthracene, biphenyl) over molten sulfur at 240-260°C,
reddish-brown sulfur-containing dyes are obtained. The sulfur
content in these dyes is almost constant and each of the aromatic
rings is bonded to four or five sulfur atoms [591,592].

Sulfur dissolved in sulfuryl chloride undergoes condensation
with aromatic hydrocarbons (naphthalene, anthracene, or their mixtures) to give plastics [593]. The interaction of anthracene with
sulfur at 300°C in argon leads to the formation of paramagnetic
polymers which exhibit semiconducting properties [594,595]. Pentacene readily reacts with excess sulfur in boiling 1,2,4-trichloro-
benzene and gives pentacene hexasulfide (85) [596]. When hexacene
reacts with sulfur the hexasulfide (86) or (87) is obtained. However, no octasulfide is formed. Because of the low solubility of
the hexasulfides (86) and (87) it is not possible to differentiate
between these two structures using NMR spectra [596].

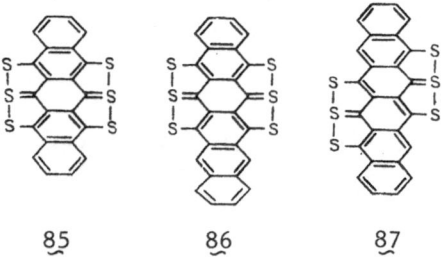

85   86   87

3.11. ARYLALIPHATIC HYDROCARBONS

3.11.1. Arylalkanes

The presence of an aromatic substituent on an aliphatic hydrocarbon chain considerably increases its reactivity toward sulfur.
Arylalkanes react with sulfur much more easily than the corresponding alkanes.

When sulfur and toluene vapors are passed through a tube heated
to 800°C, a complex mixture of condensation and sulfuration products
is formed. This mixture contains bibenzyl, stilbene, tetraphenyl-

thiophene, 1,2,3,4-tetraphenylbutane, 2,2'-dimethylbiphenyl, 2-phenylbenzo[b]thiophene and [1]benzothieno[3,2-b]-[1]benzo-thiophene [586,597-599]. Toluene heated with sulfur in a sealed tube at 250-300°C gives stilbene and tetraphenylthiophene [599]. A free-radical mechanism for the reaction of toluene with sulfur has been proposed [600]:

$$C_6H_5CH_3 + \cdot S \cdot \rightarrow HS \cdot + C_6H_5CH_2 \cdot$$

$$2C_6H_5CH_2 \cdot \rightarrow C_6H_5CH_2CH_2C_6H_5 \xrightarrow{\cdot S \cdot}$$

$$\rightarrow C_6H_5CH_2\dot{C}HC_6H_5 \rightarrow C_6H_5CH_2CH-CHCH_2C_6H_5$$
$$\downarrow \cdot S \cdot \qquad\qquad\qquad | \qquad | $$
$$\qquad\qquad\qquad\qquad C_6H_5 \quad C_6H_5$$

$$C_6H_5CH=CHC_6H_5 \xrightarrow{\cdot S \cdot} C_6H_5CH-CHC_6H_5$$
$$\qquad\qquad\qquad\qquad\qquad | $$
$$\qquad\qquad\qquad\qquad\qquad S \cdot$$

(3.126)

As seen from Table 3.5 below, substituted thiophenes, 1,2-di-thiol-3-thiones and sulfides are the main products of the reaction of alkylbenzenes with sulfur.

Table 3.5. The Reaction of Alkylbenzenes with Sulfur

Alkylbenzenes	Reactions and reaction conditions	Reaction products	Yield, %	References
	S, 400°C		12	601
	S, 115-130°C, AlCl3			547, 553
			27	602

Table 3.5.(continued)

Alkylbenzenes	Reactions and reaction conditions	Reaction products	Yield, %	References
H₃C—⟨benzene⟩—CH₃	S, 300-350°C 20-30 atm	[⟨benzene⟩—CH=CH]ₙ		604, 605
⟨benzene⟩ with CH₃, CH₃ (o-xylene)	S, NH₃, 450-600°C Metal oxides	C_6H_6 $C_6H_5CH_3$ $C_6H_5C{\equiv}N$ $\underline{m}-H_3CC_6H_4C{\equiv}N$ $\underline{m}-C_6H_4(CN)_2$	12 3 3 8 7	605, 606
H₃C—⟨benzene⟩—CH₃ with CH₃	S, AlCl₃	H₃C, CH₃ thianthrene, H₃C, CH₃	15-30	553
⟨benzene⟩—CH₂CH₃	S, 240°C	C_6H_5 thiophene C_6H_5 C_6H_5 thiophene C_6H_5	35 12	600, 603
	S, 600-700°C	$C_6H_5CH=CH_2$	85-95	607- 613
⟨benzene⟩ with CH₃, H₃C, CH₃ (mesitylene)	S, AlCl₃	CH₃ CH₃ H₃C—S—CH₃ CH₃ CH₃; CH₃ CH₃ H₃C—S,S—CH₃ CH₃ CH₃		603
RO—⟨benzene⟩—CH(CH₃)CH₃	S, 190-200°C, mercuri-acet-amide	RO—⟨benzene⟩=S, S, S ring R = H, Alk, Ph, PhCO	11-88	618
R^2 R^1 R^3—⟨benzene⟩—CH(CH₃)CH₃ R^4	S, 190-200°C, mercuri-acet-amide	R^2 R^1 R^3—⟨benzene⟩=S, S, S R^4 ring R^1,R^2,R^3,R^4 = H, Me, OMe, OEt, i-Pr	15-60	617

Table 3.5.(continued)

Alkylbenzenes	Reactions and reaction conditions	Reaction products	Yield, %	References
R^3, R^2, R^1, $CH_2CH_2CH_3$ (trialkyl benzene)	S, 186–189°C mercapto-benzothiazole	(thiophene structure) $R^1,R^2,R^3 = Alk$	11–37	618
C_6H_5–$(CH_2)_3CH_3$	S, 195–200°C	(2-phenylthiophene)	5	627
C_6H_5–$CHCH_2CH_3$ | CH_3	S, 190–240°C	(2-phenylthiophene)	4	626
	S, 190–200°C Fe, activated charcoal	(2-phenylthiophene)	4	626
	S, 180–200°C mercuri-acet-amide	(thiophene-thione structure)	1	628
		(2-phenylthiophene)	6	
	S, 200–210°C morpholine	(thiophene-thione structure)	2	628
		(2-phenylthiophene)	15	
C_6H_5–$C(CH_3)_3$	S, 180–200°C	A product has not been isolated	–	621
C_6H_5–$CH_2C(CH_3)_3$	S, 200°C	A product has not been isolated	–	621
C_6H_5–$(CH_2)_2CH(CH_3)_2$	S, 200–245°C	C_6H_5–(thiophene)–CH_3	27	621

2-(2-Naphthyl)propane reacts with sulfur to give a 57% yield of 4-(2-naphthyl)-1,2-dithiol-3-thione [620]. 1,2-Diphenylpropane when heated with sulfur gives 4,5-diphenyl-1,2-dithiol-3-thione [620]:

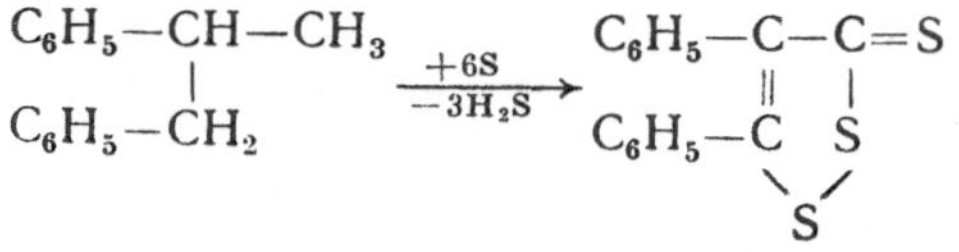

$$
\begin{array}{c}
C_6H_5-CH-CH_3 \\
| \\
C_6H_5-CH_2
\end{array}
\xrightarrow[-3H_2S]{+6S}
\begin{array}{c}
C_6H_5-C-C=S \\
\parallel\quad | \\
C_6H_5-C\quad S \\
\diagdown\diagup \\
S
\end{array}
\qquad (3.127)
$$

(27%)

In the absence of catalysts, isopropyl, tert-butyl, and tert-pentyl groups bonded to the aromatic ring are stable towards sulfur [621]. This makes it possible to obtain retene (7-isopropyl-1-methylphenanthrene) [622] and 7-tert-butylnaphthalene [623] by dehydrogenating the respective tetrahydro-derivatives with sulfur.

Tetraphenylethylene is the principal product from the reaction of diphenylmethane with sulfur at 250-290°C [629-640]:

$$(C_6H_5)_2CH_2 + 2S + H_2C(C_6H_5)_2 \rightarrow (C_6H_5)_2C=C(C_6H_5)_2 + 2H_2S \qquad (3.128)$$

(64%)

Together with tetraphenylethylene, 1,1,2,2-tetraphenylethane and thiobenzophenone are also formed. The proposed mechanism of the reaction [630,631] involves the initial formation of diphenylmethane-thiol:

$$2\,(C_6H_5)_2\,CH_2 \xrightarrow{+2S} 2\,(C_6H_5)_2\,CHSH \xrightarrow[-H_2S]{+S}$$

$$\rightarrow (C_6H_5)_2\,CHSSCH\,(C_6H_5)_2 \xrightarrow{t°} (C_6H_5)_2\,C=S + (C_6H_5)_2\,CH_2 + S$$

$$(C_6H_5)_2\,CS + (C_6H_5)_2\,CH_2 \xrightarrow{t} (C_6H_5)_2\,CH\cdot + (C_6H_5)_2\,CHS\cdot \qquad (3.129)$$

$$2\,(C_6H_5)_2\,CH\cdot \rightarrow (C_6H_5)_2\,CH-CH\,(C_6H_5)_2$$

$$2\,(C_6H_5)_2\,CHS\cdot \rightarrow (C_6H_5)_2\,CHSSCH\,(C_6H_5)_2$$

$$(C_5H_5)_2\,CHCH\,(C_6H_5)_2 \xrightarrow[-H_2S]{S} (C_6H_5)_2\,C=C\,(C_6H_5)_2$$

The dehydrogenation of diphenylmethanethiol with sulfur has been carried out experimentally (this reaction is accelerated by the presence of secondary amines) [630]. Most of the thiobenzophenone formed according to the above scheme reacts further with diphenyl-methane (which serves as an electron acceptor) to give rise to the formation of two different types of radicals which subsequently give the dimers 1,1,2,2-tetraphenylethane and dibenzhydryl disulfide.

Depending on their basicity, amines catalyze the interaction of sulfur with diphenylmethane [634,638]. The reaction is also accelerated in the presence of 1,3-benzothiazole-2-thiol [634,639] or bis-1,3-benzothiazol-2-yl disulfide [634,636,637,639]. When a

mixture of diphenylmethane, fluorene, and sulfur is heated to 240-
300°C diphenylmethylenefluorene (88) is obtained [641]. Tetraphenylethylene, 1,1,2,2-tetraphenylethane, bifluorenylidene, and 9,9'-
bifluorene are also formed in this reaction:

$$(C_6H_5)_2CH_2 + H_2C{<}^{C_6H_4}_{C_6H_4} \xrightarrow{2S} {}^{C_6H_5}_{C_6H_5}{>}C{=}C{<}^{C_6H_4}_{C_6H_4} + 2H_2S \qquad (3.130)$$

$$88$$

Even prolonged heating of triphenylmethane with sulfur at 300°C
does not lead to any noticeable interaction [642]. However, triphenylmethyl radicals easily react with sulfur giving bis-triphenyl-
methyl polysulfides [643,644]:

$$2(C_6H_5)_3C\cdot + S_n \rightarrow (C_6H_5)_3C-S_n-C(C_6H_5)_3 \qquad (3.131)$$

1,2-Diphenylethane (bibenzyl) is dehydrogenated with sulfur to
give stilbene [645] which further reacts with sulfur to give tetraphenylthiophene [641,646-648]:

$$C_6H_5CH_2CH_2C_6H_5 \xrightarrow[-H_2S]{+S} C_6H_5CH{=}CHC_6H_5 \qquad (3.132)$$

$$2\,C_6H_5CH{=}CHC_6H_5 \xrightarrow[-2H_2S]{+3S} \underset{C_6H_5 \quad S \quad C_6H_5}{\overset{C_6H_5 \qquad C_6H_5}{\text{thiophene}}} \qquad (3.133)$$

1,2,3,4-Tetraphenylbutane is also formed [647]. A free-radical
mechanism for the reaction of bibenzyl with sulfur [600], analogous
to that described above for the reaction of toluene with sulfur,
has been suggested.

2,2'-Dimethylbiphenyl reacts with sulfur at 250°C to give
phenanthrene in good yield [622,649]:

$$\underset{H_3C\ CH_3}{\text{biphenyl}} + 2S \longrightarrow \text{phenanthrene} + 2H_2S \qquad (3.134)$$

1-Methylnaphthalene heated with sulfur in a sealed tube at
300-310°C gives a mixture of 1,2-bis-1'-naphthylethane (89), 1,2-
bis-1'-naphthylethylene (90), and picene (91) [624]:

$$(3.135)$$

$$89 \qquad 90 \qquad 91$$

Under analogous conditions, 2-methylnaphthalene gives 1,2-bis-2'-naphthylethane, 1,2-bis-2'-naphthylethylene, and two compounds with the formulas $C_{22}H_{14}S$ (92) and $C_{22}H_{14}S_2$ (93) respectively [624]. The following structures were later formulated for the above two compounds [600]:

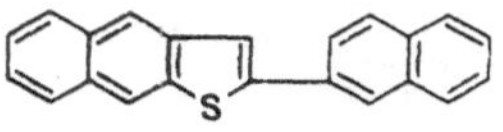

92 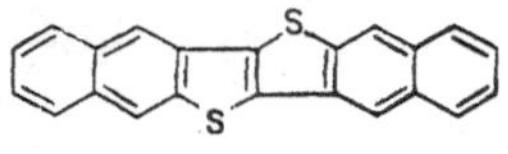   93

The sulfuration of 1-methylacenaphthene at 200°C gives acenaphtheno[2,1-c]-[1,2]-dithiol-3-thione [650]:

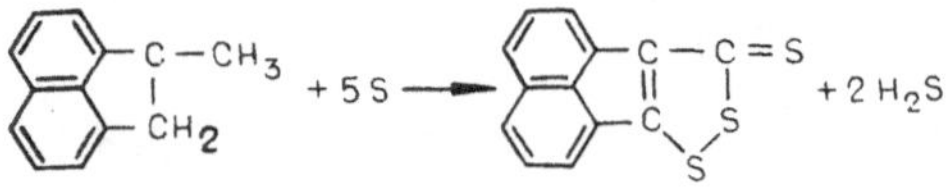

(3.136)

When polyarylethanes with the general formula $(CH_2CH_2C_6H_3(R))_n$ (where R = H or alkyl) are heated with sulfur, formation of polymers takes place involving the removal of hydrogen atoms from the $-CH_2-CH_2-$ groups and formation of new chemical bonds [651].

3.11.2. Arylalkenes and Arylalkadienes

The presence of a phenyl group on an olefinic chain substantially increases its reactivity toward sulfur, especially in those cases in which the phenyl group is bound to sp^2-carbon. In 1891 it was shown that heating stilbene with sulfur is useful for the preparation of tetraphenylthiophene [652,653]:

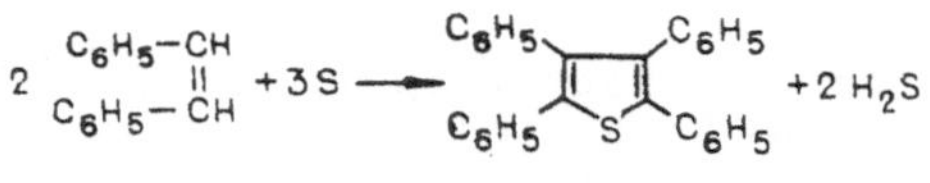

(3.137)

The reaction of styrene with sulfur occurs at 190-230°C in an analogous manner and 2,4- and 2,5-diphenylthiophenes are obtained as the reaction products [173,654]:

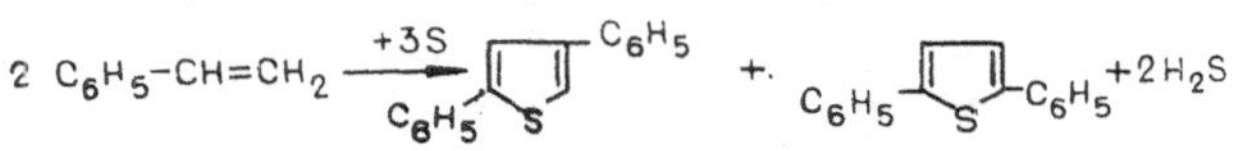

(3.138)

2,5-Diphenylthiophene is formed to a minor extent. The hydrogen sulfide formed in this reaction reduces some of the styrene to ethylbenzene [654,655]:

$$C_6H_5\,CH{=}CH_2 + H_2S \longrightarrow C_6H_5{-}C_2H_5 + S \qquad (3.139)$$

Styrene reacts more readily with sulfur than stilbene [655]. Stilbene may also be reduced by hydrogen sulfide during the sulfuration process. However, the product is toluene and not bibenzyl as one might expect:

$$C_6H_5CH\!=\!CHC_6H_5+2H_2S \longrightarrow 2C_6H_5CH_3+2S \qquad (3.140)$$

Under milder conditions, styrene as well as some other unsaturated aromatic compounds can react with sulfur without the formation of hydrogen sulfide. It has been stated [294,654,657,658] that the reaction starts by the formation of an unstable adduct with sulfur which, upon further heating, decomposes to give hydrogen sulfide and diphenylthiophene. A systematic investigation of the sulfuration reaction of phenylalkenes and their derivatives has shown [621,626, 627,660-664] that the reaction of sulfur with α,β-dialkylstyrenes gives 3-phenylthiophene or its homologs:

$$
\begin{array}{c}
C_6H_5-C=C-R \\[2pt]
\quad | \qquad | \\
R''-CH_2 \; CH_2-R'
\end{array}
+ 3S \rightarrow
\begin{array}{c}
C_6H_5-C-C-R \\[2pt]
\quad \| \quad \| \\
R''-C \quad C-R' \\
\quad \diagdown_{S}\diagup
\end{array}
+ 2H_2S \qquad (3.141)
$$

where R, R', R'' = alkyl.

All possible methyl substituted 3-phenylthiophenes as well as some more complex 3-phenylthiophene homologs have been obtained by this method [626,660-662,664]. The reaction of sulfur with β-ethyl-α-propylstyrene and β-isobutyl-α-methylstyrene gives 2-ethyl-5-methyl-3-phenylthiophene and 2-isopropyl-4-phenylthiophene respectively:

$$
\begin{array}{c}
C_6H_5-C=CH \\[2pt]
\quad | \qquad \diagdown \\
C_2H_5CH_2 \quad CH_2CH_3
\end{array}
\xrightarrow[-2H_2S]{+3S}
\begin{array}{c}
C_6H_5-C-CH \\[2pt]
\quad \| \quad \| \\
C_2H_5-C \quad C-CH_3 \\
\quad \diagdown_{S}\diagup
\end{array}
\qquad (3.142)
$$

$$
\begin{array}{c}
C_6H_5-C=CH \\[2pt]
\quad | \qquad | \\
CH_3 \; CH_2CH(CH_3)_2
\end{array}
\xrightarrow[-2H_2S]{+3S}
\begin{array}{c}
C_6H_5-C-CH \\[2pt]
\quad \| \quad \| \\
HC \quad C-CH(CH_3)_2 \\
\quad \diagdown_{S}\diagup
\end{array}
\qquad (3.143)
$$

However, the yields of the thiophene derivatives decrease with increasing length of the side chain in the starting styrenes.

It has been suggested [659,660,663,664] that the formation of 3-phenylthiophenes occurs via dehydrogenation of the starting α,β-dialkylstyrene with sulfur to a 2-phenyl-1,3-butadiene homolog. The reaction scheme is as follows:

$$
\begin{array}{c}
C_6H_5-C=C-R \\[2pt]
\quad | \qquad | \\
R''-CH_2 \; CH_2-R'
\end{array}
\xrightarrow[-H_2S]{+S}
\begin{array}{c}
C_6H_5-C-C-R \\[2pt]
\quad \| \quad \| \\
R''-CH \; CH-R'
\end{array}
\xrightarrow[-H_2S]{+2S}
\begin{array}{c}
C_6H_5 \quad\quad R \\
\quad \boxed{} \\
R'' \quad S \quad R'
\end{array}
\qquad (3.144)
$$

To confirm this scheme, the reaction of 2-methyl-3-phenyl-1,3-butadiene with sulfur at 180-230°C has been studied. This reaction gives 3-methyl-4-phenylthiophene [44,46,243,626,662,664]:

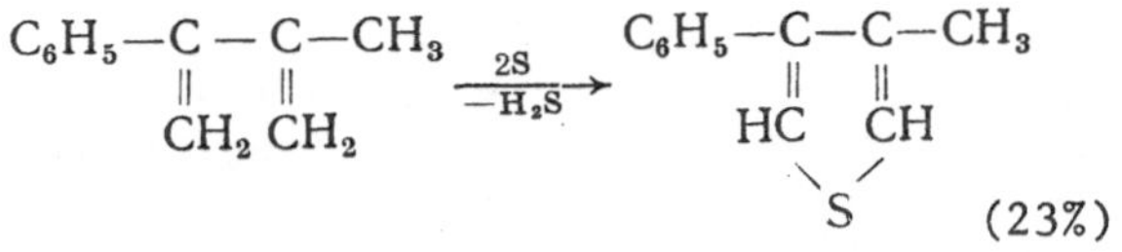

$$
\underset{\substack{| \\ CH_2}}{C_6H_5-C} \underset{\substack{| \\ CH_2}}{- C} -CH_3 \xrightarrow[-H_2S]{2S} \underset{\substack{\diagdown \quad \diagup \\ S}}{\underset{HC \quad CH}{C_6H_5-C-C-CH_3}} \qquad (3.145)
$$
$$(23\%)$$

A method for the synthesis of 2-phenylthiophene and its homologs has also been formulated. The method is based on the reaction of sulfur with the isomeric phenylbutenes at 190-200°C [627,664,665]:

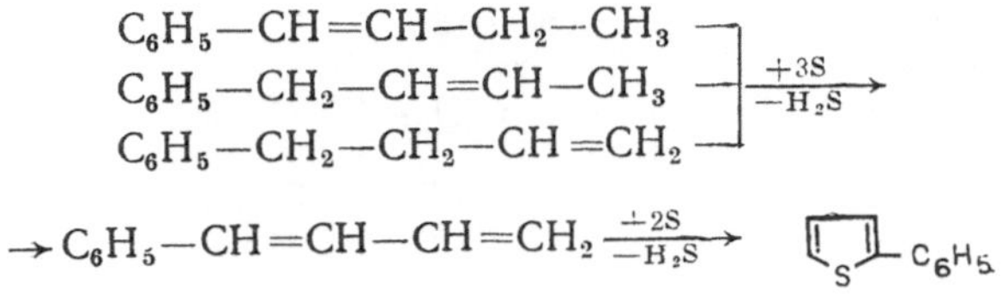

$$
\left.\begin{array}{l} C_6H_5-CH=CH-CH_2-CH_3 \\ C_6H_5-CH_2-CH=CH-CH_3 \\ C_6H_5-CH_2-CH_2-CH=CH_2 \end{array}\right\} \xrightarrow[-H_2S]{+3S} \qquad (3.146)
$$

$$
\rightarrow C_6H_5-CH=CH-CH=CH_2 \xrightarrow[-H_2S]{+2S} \quad \underset{S}{\boxed{}}-C_6H_5
$$

In this case it is also assumed that 1-phenyl-1,3-butadiene is formed as an intermediate. The sulfuration of the latter then gives an 8.5% yield of 2-phenylthiophene.

All the isomeric methyl-2-phenylthiophenes have been synthesized by the reaction of sulfur with 1-phenyl-1-pentene, 3-methyl-1-phenyl-1-butene, and 2-methyl-1-phenyl-1-butene [621]:

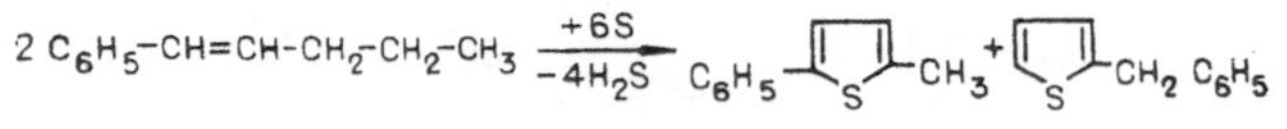

$$
2\,C_6H_5-CH=CH-CH_2-CH_2-CH_3 \xrightarrow[-4H_2S]{+6S} C_6H_5-\underset{S}{\boxed{}}-CH_3 + \underset{S}{\boxed{}}-CH_2\,C_6H_5 \qquad (3.147)
$$

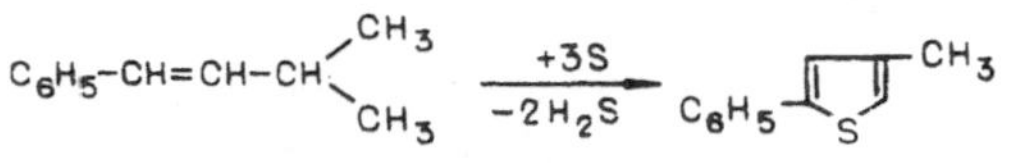

$$
C_6H_5-CH=CH-CH\underset{\diagdown CH_3}{\overset{\diagup CH_3}{}} \xrightarrow[-2H_2S]{+3S} C_6H_5-\underset{S}{\boxed{}}-CH_3 \qquad (3.148)
$$

$$
C_6H_5-CH=\underset{\substack{| \\ CH_3}}{C}-CH_2-CH_3 \xrightarrow[-2H_2S]{+3S} \underset{S}{\boxed{}}\overset{CH_3}{\underset{C_6H_5}{}} \qquad (3.149)
$$

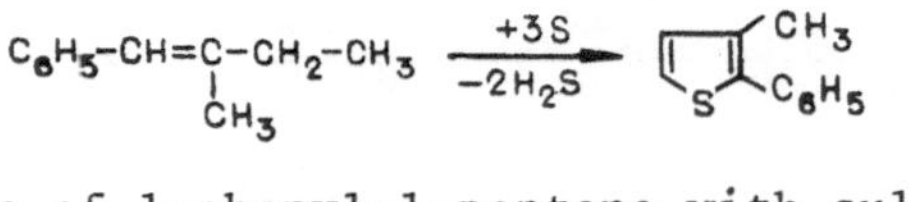

The reaction of 1-phenyl-1-pentene with sulfur produces a small amount of 2-benzylthiophene, together with 2-methyl-5-phenylthiophene as the principal reaction product. Thus, an intramolecular ring closure to the thiophene system is observed in the reaction of sulfur with phenylalkenes containing at least four carbon atoms in the straight chain, regardless of the position of the double bond.

The following reaction mechanism has been proposed (shown with 1-phenyl-2-butene):

$$
C_6H_5-CH_2-CH=CH-CH_3 \xrightarrow{+S} C_6H_5-\underset{\substack{| \\ SH}}{CH}-CH=CH-CH_3 \longrightarrow
$$

$$\longrightarrow SH^{\ominus} + \left[C_6H_5-CH \overset{\frown}{=} CH=CH-CH_3 \right]^{\oplus} \longrightarrow H_2S$$

$$+ \quad C_6H_5-CH=CH-CH=CH_2 \xrightarrow{+S} \underset{C_6H_5}{\underset{S}{\bigcirc}} \xrightarrow[-H_2S]{+2S} \underset{C_6H_5}{\underset{S}{\bigcirc}} \qquad (3.150)$$

When the reaction between sulfur and phenylalkenes was studied with compounds containing more than two, but less than four, chain carbon atoms, phenylthiophenes were not obtained. However, a new reaction course leading to 4-(or 5-)phenyl-1,2-dithiol-3-thiones was discovered [211,213,666]:

$$R-CH=C-CH_3 \xrightarrow[-2H_2S]{+5S} \underset{R}{\overset{R'}{\bigcirc}}\overset{S}{\underset{S}{}} \qquad (3.151)$$
$$\qquad\qquad \overset{|}{R'}$$

$$R=H, \; C_6H_5; \quad R'=H, \; CH_3, \; C_6H_5$$

The following mechanism for the reaction of phenylpropenes with sulfur has been suggested (shown in the case of 1-phenylpropene):

$$C_6H_5-CH=CH-CH_3 \xrightarrow{+S} C_6H_5-CH=CH-CH_2SH \xrightarrow[-H_2S]{+2S}$$

$$\rightarrow C_6H_5-CH=CH-C\underset{SH}{\overset{\diagup S}{\diagdown}} \xrightarrow[-H_2S]{+2S} \underset{S\diagdown S}{\overset{H_5C_6}{\bigcirc}}=S \qquad (3.152)$$

Studies of the reaction of sulfur with unsaturated aromatic hydrocarbons (allylbenzene, propenylbenzene, α-methylstilbene) resulted in an alternative mechanism for the formation of 1,2-dithiol-3-thiones [216,219,521,526]:

$$\text{Olefin} \xrightarrow[<180°C, \text{ fast}]{+ S} \text{thiol} \xrightarrow[\text{very fast}]{+ S} \text{disulfide} \longrightarrow$$

$$\xrightarrow[\text{very fast}]{+ S} \text{trisulfide} \xrightarrow[\text{slow}]{>160°C} \text{1,2-dithiol-3-thione}$$
$$\qquad\qquad\qquad\qquad\qquad\qquad\qquad\qquad (3.153)$$

It has been shown that there are two basic types of compounds which can react with sulfur to give 1,2-dithiol-3-thiones [228-230]:

$$C_6H_5-CH=CH-CH_3 \quad \text{and} \quad C_6H_5-C=CH_2$$
$$\qquad\qquad\qquad\qquad\qquad\qquad\qquad\qquad \overset{|}{CH_3}$$

The effect of substituents in the benzene ring on the formation of 1,2-dithiol-3-thiones has been studied [228-230].

On heating 1-(α-methylvinyl)naphthalene with sulfur (190°C, 3 h), 1-methylnaphtho[2,1-b]thiophene (94) and 4-(1-naphthyl)-1,2-dithiol-3-thione (95) are formed.

$$(3.154)$$

94 (29%)   95 (11%)

Under the same conditions, 2-(α-methylvinyl)naphthalene gives 3-methylnaphtho[1,2-b]thiophene and 4-(2-naphthyl)-1,2-dithiol-3-thione in 30% and 2.3% yield, respectively [667]. 1-Methylace-naphthylene reacts similarly with sulfur [235,668]:

$$(3.155)$$

In a study of the reaction of anethole (1-p-methoxyphenyl-propene) with sulfur it was assumed that sulfur reacts in the form of cyclooctasulfur when forming the corresponding 1,2-dithiol-3-thione [669]:

$$CH_3O\ C_6H_4CH{=}CHCH_3 \xrightarrow[-2H_2S]{+5/8S_4}
\begin{array}{c} CH_3O\ C_6H_4-C{=}CH \\ |\qquad | \\ S\qquad C{=}S \\ \diagdown S \diagup \end{array}$$

$$(3.156)$$

There is a patent describing the synthesis of α-methylstyrene copolymers by the use of sulfur at 150-250°C. These copolymers are light in color, possess a low viscosity [260], and find use in the manufacture of varnishes. The reaction of styrene with sulfur leads to the formation of hard, thermoplastic copolymers [295] with a sulfur content of 20 or 80%.

The double bond in thermochromic ethylenes undergoes cleavage in the presence of sulfur at 280°C to give the corresponding thio-ketones [670-677]:

$$\diagup \overset{\diagup}{C}{=}C\overset{\diagdown}{\diagdown} \underset{-2S}{\overset{+2S}{\rightleftarrows}} 2 \overset{\diagdown}{\diagup}C{=}S$$

$$(3.157)$$

The mechanism of this reaction has been suggested to involve the addition of two sulfur atoms to the free valences of the bi-radicaloid form of the starting unsaturated compound. The cyclic disulfide thus formed further decomposes with the formation of two thioketone molecules [675,676,678]:

$$\begin{array}{c} R\diagdown \qquad \diagup R'' \\ C{=}C \\ R'\diagup \qquad \diagdown R''' \end{array} \rightarrow
\begin{array}{c} R\diagdown \qquad \diagup R'' \\ C{-}C \\ R'\diagup | \quad | \diagdown R''' \end{array} \xrightarrow{+2S}$$

$$\rightarrow \begin{array}{c} R \\ \diagdown \\ R' \diagup \end{array} \!\!\! C\!\!-\!\!C \!\!\! \begin{array}{c} \diagup R'' \\ \diagdown \\ \diagdown R''' \end{array} \underset{S-S}{\overset{\mid\ \ \mid}{}} \rightarrow \begin{array}{c} R \\ \diagdown \\ R' \diagup \end{array} \!\!\! C\!\!=\!\!S + \begin{array}{c} R'' \\ \diagdown \\ R''' \diagup \end{array} \!\!\! C\!\!=\!\!S \qquad (3.158)$$

In support of this scheme, the capability of triphenylmethyl radicals to react with sulfur with the formation of bis-triphenyl-methyl polysulfides has been cited [679]:

$$2\,(C_6H_5)_3\,C\cdot \xrightarrow{+nS} (C_6H_5)_3\,C\!-\!S_n\!-\!C\,(C_6H_5)_3\,. \qquad (3.159)$$

However, no such reaction takes place with diphenylmethylenefluorene (96), benzylidenefluorene (97), and tetraarylethylenes (98):

$$(C_6H_5)_2\,C\!=\!C \!\!\! \begin{array}{c} \diagup C_6H_4 \\ \mid \\ \diagdown C_6H_4 \end{array} ; \quad C_6H_5CH\!=\!C \!\!\! \begin{array}{c} \diagup C_6H_4 \\ \mid \\ \diagdown C_6H_4 \end{array}$$

$$\underset{96}{} \qquad\qquad \underset{97}{}$$

$$(C_6H_4R)_2 C\!=\!C\,(C_6H_4R)_2$$

$$\underset{98}{}$$

$$R = H,\ Ph,\ OMe,\ NMe_2\,.$$

This seems to indicate that the principal factor responsible for the cleavage of the double bond according to scheme (3.157) is the stability of the thiocarbonyl compounds formed.

REFERENCES

1. H. O. Folkins and E. Miller, Food Machinery and Chem. Corp. U.S. Patent 2,661,267 (1953); C. A., 48, 7859 (1954); 2,666,690 (1954); C. A., 48, 5452 (1954).
2. G. Gevidalli, Combust. Flame, 7, 5 (1953).
3. R. A. Fischer and J. M. Smith, Ind. Eng. Chem., 42, 704 (1950).
4. D. Stull, Ind. Eng. Chem., 41, 1968 (1949).
5. D. R. Olsen, U.S. Patent 3,250,595 (1966); C. A., 65, 1381 (1966).
6. A. R. Clark, Chem. Process. (London), 48, 442 (1967).
7. I. Varga and P. Benedek, Magy. Kem. Folyoirut, 56, 36 (1950)..
8. I. P. Kuznetsov, V. M. Lekaj, N. G. Vilesov, and A. A. Ivlev, Chem. Prom. (Moscow), 41, 823 (1965).
9. I. P. Kuznetsov, A. G. Kasatkin, V. M. Lekaj, L. N. Elkin, and N. G. Vilesov, Tr. Mosk. Khim.-Tekhnol. Inst., 47, 80 (1964).
10. M. de Simo, U.S. Patent 2,187,393 (1940); C. A., 34, 3453 (1940).
11. R. E. Stantou, U.S. Patent 2,788,261 (1957); Chem. Zentralbl., 12244 (1958).
12. B. T. Brooks and J. Humphrey, Ind. Eng. Chem., 9, 740 (1917).
13. R. F. Bacon and E. S. Boe, Ind. Eng. Chem., 37, 469 (1945).
14. G. Nabor and J. M. Smith, Ind. Eng. Chem., 45, 1272 (1953).
15. A. Draganescu, I. Bica, C. Petrescu, M. Pavlovschi, S. Serbau, and N. Marjanov, Rev. Roum. Chim. (Bukharest), 18, 1859 (1973).

16. Z. Leszyuski and S. Kubica, Rev. Roum. Chim. (Bukharest),
     18, 107 (1973).
17. V. A. Zazhygalov, Ukr. Khim. Zh., 44, 1050 (1978).
18. G. A. Komashko, M. Y. Rubanik, and V. A. Zazhygalov, Catalysts
     for the Preparation and Conversion of Sulfur [in Russian],
     Nauka, Novosibirsk (1979), p. 100.
19. H. O. Folkins, E. Miller, and H. Henning, U.S. Patent 2,709,639
     (1955); Chem. Zentralbl., 10849 (1957).
20. W. M. Bryce and C. Hinshelwood, J. Chem. Soc., 3379 (1949).
21. R. E. Morningstar, Diss. Abstr., 18, 1368 (1958).
22. W. J. Thomas and R. F. Strickland, Trans. Faraday Soc., 53, 972
     (1957).
23. D. H. Cortez, J. S. Land, and L. J. Van Nice, AIChE Symp. Ser.,
     69(127), 134 (1973).
24. D. H. Cortez and J. S. Land, Oil Gas J., 70, 62 (1972).
25. C. Giavarini, Rass. Chim., 27, 169 (1975).
26. A. R. Knight, O. P. Strausz, S. M. Malm, and H. E. Gunning,
     J. Am. Chem. Soc., 86, 4243 (1964).
27. R. J. Donovan, L. J. Kirsch, and D. Husaiu, Nature, 222, 1164
     (1969).
28. P. Fowles, M. de Sorgo, A. J. Yarwood, O. P. Strausz, and H. E.
     Gunning, J. Am. Chem. Soc., 89, 1352 (1967).
29. C. Engler and H. Hofer, Das Erdöl, Vol. 1, Leipzig (1912), p. 530.
30. J. Spanier, Dissertation, Karlsruhe (1910).
31. A. S. Velikovskii and S. M. Pavlova, Neftyanoe Khozyaistvo, 23,
     33 (1945).
32. G. G. Wanless, P. E. Wei, and J. J. Rehner, Rubber Chem.
     Technol., 35, 118 (1962).
33. J. Koru, H. W. Priuzler, and D. Pape, Erdöl, Kohle, Erdgas,
     Erdgas Petrochem., 19, 651 (1966).
34. W. Friedman, Ber., 49, 1344 (1916).
35. W. Friedman, Petroleum, 16, 1299 (1916).
36. K. Baker and E. Reid, J. Am. Chem. Soc., 51, 1566 (1929).
37. W. Friedman, Pet. Refinery, 20(10), 55 (1941).
38. W. Friedman, J. Inst. Petrol. (London), 37, 40 (1951).
39. R. C. Hansford, H. E. Rasmussen, and A. N. Sachanen, U.S. Patent
     2,450,659 (1948); C. A., 43, 1066 (1949).
40. H. E. Rasmussen and F. E. Ray, Chem. Ind. (London), 60, 593,
     620 (1947).
41. R. C. Hansford, H. E. Rasmussen, C. Meyers, and A. N. Sachanen,
     U.S. Patent 2,450,658 (1948); C. A., 43, 1066 (1949).
42. H. E. Rasmussen, R. C. Hansford, and A. N. Sachanen, Ind. Eng.
     Chem., 38, 376 (1946).
43. H. E. Rasmussen and R. C. Hansford, U.S. Patent 2,450,685-6-7
     (1948); C. A., 43, 1067 (1949).
44. A. F. Shepard, A. L. Henne, and T. J. Modley, J. Am. Chem. Soc.,
     56, 1355 (1934).
45. L. Lepage and Y. Lepage, J. Heterocycl. Chem., 15, 1185 (1978).
46. D. D. Coffman, U.S. Patent, 2,410,401 (1946); C. A., 41, 2086
     (1947).
47. R. C. Hansford, H. E. Rasmussen, and C. G. Meyers, Swed. Patent
     124,817 (1945); C. A., 44, 2414 (1950).

48. R. D. Caeser and P. D. Brauton, Ind. Eng. Chem., 44, 122 (1952).
49. F. P. Miknis and J. P. Biscar, High Temp. Sci., 4, 48 (1972).
50. R. L. Hodgson, U.S. Patent 3,350,408 (1965); C. A., 68, 114429 (1968).
51. J. R. Geigi, Brit. Patent 783,037 (1957); C. A., 52, 3860 (1958).
52. A. Hopff, R. Roggero, and G. Valkanas, Rev. Chim. (Bucarest), 7, 921 (1962); C. A., 61, 4201 (1964).
53. W. Bilenberg, Braunkohlenarhiv, 4, 40 (1923).
54. J. Korn. H. W. Prinzler, and D. Pape, Erdöl Kohl, 19(9), 651 (1966).
55. P. Nikolinski and V. Mircheva, Tr. Khim. Tekhnol. Inst., Sofia, 13, 185 (1966).
56. P. Nikolinski and V. Mircheva, Tr. Khim. Tekhnol. Inst., Sofia, 15, 21 (1968).
57. P. L. Bocca, U. Petrossi, and V. Piconi, Chim. Ind. (Milan), 55(5), 425 (1973).
58. U. Petrossi, P. L. Bocca, and F. Pacor, Ind. Eng. Chem. Prod. Res. Develop., 11, 214 (1972).
59. B. A. Dogadkin and A. A. Dontsov, Vysokomol. Soedin., 3(11), 1746 (1961).
60. B. A. Dogadkin and A. A. Dontsov, Vysokomol. Soedin., 7(11), 1841 (1965).
61. B. A. Dogadkin and A. A. Dontsov, Dokl. Akad. Nauk SSSR, 138(6), 1349 (1961).
62. D. A. Olsen and A. J. Osteraas, J. Polym. Sci., 1913 (1969).
63. C. M. Thacker and E. Miller, Ind. Eng. Chem., 36, 182 (1944).
64. P. Champion and H. Pallot, C. R. Acad. Sci. (Paris), 70, 620 (1870).
65. J. Galletly, Chem. News, 24, 62 (1871).
66. J. Galletly, Z. Chem., 471 (1871).
67. J. Galletly, Pharm. J. Transl., 2, 506 (1871).
68. J. Galletly, Bull. Soc. Chim. France, 16, 234 (1871).
69. J. Galletly, Am. J. Pharm., 1, 513 (1871).
70. J. Galletly, Chem. News, 40, 154 (1879); Bull. Soc. Chim. France, 35, 435 (1879).
71. W. Johnston, Chem. News, 40, 167 (1879).
72. A. Lidov, Zh. Russ. Fiz.-Khim. Ova., 13, 514 (1881).
73. S. J. Taylor, Chem. News, 47, 145 (1883); Ber., 16, 1094 (1883).
74. E. Prothiere, Pharm. Ztg., 48, 78 (1903).
75. M. A. Rakusin, Pet. Z., 18, 581 (1922).
76. H. Siebeneck, Petroleum, 18, 281 (1922).
77. E. Bindschedler and E. W. Rugeley, U.S. Patent 1,565,894 (1925); C. A., 20, 483 (1926).
78. A. Henwood, R. M. Garey, W. Goldberg, and E. Field, J. Franklin Inst., 199, 685 (1925).
79. W. E. Ebaugh, Denison Univ., 21, 403 (1926).
80. A. Henwood, U.S. Patent 1,623,942 (1927); C. A., 21, 1692 (1927).
81. H. L. Dunlap, Chem. Met. Eng., 34, 298 (1927).
82. R. Brieger, Pharm. Ztg., 74, 1339 (1929).
83. H. Gfeller and K. Schaeffer, Schweiz. Apoth. Ztg., 67, 109 (1929).
84. E. D. Scudder and R. E. Lyons, Proc. Indian Acad. Sci., 40,

185 (1931).

85. M. Raffo and G. Rossi, Gazz. Chim. Ital., $\underline{44}$, 104 (1914).
86. F. J. Nellensteyn and D. Thoeues, Chem. Weekbl., $\underline{29}$, 582 (1932).
87. C. Lefevre and C. Desgres, C. R. Acad. Sci. (Paris), $\underline{198}$, 1432 (1934).
88. E. E. Aynsley and P. L. Robinson, Gazz. Chim. Ital. $\underline{65}$, 14 (1935).
89. A. Romwalter, Math., Naturw. anz. Ungar. Acad. Wiss., $\underline{61}$, 122 (1942).
90. F. B. Oberhauser, F. A. Herrera, M. S. Minoz, H. F. Torres, and G. Wiehr, Rev. Quim. Farm. (Rio de Janeiro), $\underline{8}$, 12 (1951).
91. F. Vigide and A. Hermida, Inform. Quim. Anal., $\underline{13}$, 61 (1959).
92. H. F. Faylor, Mem. Proc. Manchester Lit. Phil. Soc., $\underline{79}$, 99 (1935).
93. O. Lange, Die Schwefelfarbstoffe, 2. Aufl., Leipzig (1925).
94. L. G. Gurvich, Scientific Fundamentals of Petroleum Processing, Moscow—Leningrad (1940), p. 71.
95. E. Graefe, Angew. Chem., $\underline{34}$, 509 (1921).
96. F. Gomez and A. Lago-Hermida, Inform. Quim. Anal., $\underline{13}$, 61 (1959)
97. H. C. H. Jensen, Fr. Patent 1,080,492 (1954).
98. N. G. Vilesov, V. I. Kilko, V. Y. Skripto, A. D. Biba, V. B. Tkachenko, and V. G. Bolshunov, USSR Patent 649,648 (1979); C. A., $\underline{92}$, P96608c (1980).
99. R. W. Timmerman, A. G. Draeger, and J. W. Getz, U.S. Patent 2,857,250 (1958); C. A., $\underline{53}$, 4675 (1959).
100. O. Kausch, Der Swefelkohlenstoff, Berlin (1929).
101. V. A. Kozlov, Der Swefelkohlenstoff, Moscow (1933).
102. V. S. Smurov and B. S. Aronovich, Erhalten Schwefelkohlenstoff (1966), p. 36.
103. I. L. Iselev, G. A. Ivanov, and D. P. Alferova, Chem. Technol. Schwefelkohlenstoff, 9 (1970).
104. M. Iwanai and Y. Kagaku, Plant. Process, $\underline{14(1)}$, 106 (1972).
105. S. R. Reddy, A. K. Gayen, and S. C. Naik, Chem. Age India, $\underline{20}$, 497 (1969).
106. W. A. Lampadius, Philos. Mag., $\underline{20(1)}$, 131 (1805); Ann. Chim. Phys., $\underline{49(1)}$, 243 (1804).
107. H. O. Folkins, C. A. Porter, E. Miller, and H. Henning, U.S. Patent 2,568,121 (1951); C. A., $\underline{46}$, 2249 (1952).
108. H. O. Folkins, U.S. Patent 2,616,793 (1952); C. A., $\underline{47}$, 1906 (1953).
109. F. Koref, Z. Anorg. Allg. Chem., $\underline{66}$, 73 (1910).
110. C. M. Thacker, U.S. Patent 2,428,727 (1947); C. A., $\underline{42}$, 1397 (1948).
111. H. O. Folkins and E. Miller, U.S. Patent 2,536,680 (1951); C. A., $\underline{45}$, 3134 (1951).
112. H. O. Folkins and E. Miller, U.S. Patent 2,565,215 (1951); C. A., $\underline{45}$, 10521 (1951).
113. Y. Moriwake, U.S. Patent 3,402,021 (1964); Australian Patent 266,166 (1964); Finland Patent 41,384 (1964).
114. B. R. Puri, A. K. Balwar, and R. S. Hazra, J. Indian Chem. Soc., $\underline{44}$, 975 (1967).
115. M. Berthelot, Ann. Chim. Phys., $\underline{23}$, 209 (1881).
116. M. Berthelot, Ann. Chim. Phys., $\underline{28}$, 126 (1893).
117. J. Thomsen, Ber., $\underline{16}$, 2616 (1883).

118. G. N. Lewis and W. N. Lacey, J. Am. Chem. Soc., 57B, 719 (1924).
119. W. J. Huff and J. C. Holtz, Ind. Eng. Chem., 19, 1268 (1927);
      20, 226 (1928).
120. E. Terres and H. Wasermann, Z. Angew. Chem., 45, 795 (1932).
121. P. C. Cross, J. Chem. Phys., 3, 168, 825 (1935).
122. H. Guerin, M. Bastick, J. Bastick, and J. Adam-Gironne,
      C. R. Acad. Sci. (Paris), 228, 87 (1949).
123. H. Guerin and J. Adam-Gironne, Bull. Soc. Chim. France, 78,
      607 (1949).
124. R. Lepsoe, Ind. Eng. Chem., 30, 96 (1938).
125. S. A. Amironova and I. E. Vilijnski, Tr. Ural. Inst. Khim.,
      111 (1955).
126. K. Wieland, Rev. Inst. France Petrole, 13, 635 (1958).
127. K. K. Kelley, U.S. Bur. Mines Bull., 406 (1937).
128. L. G. Farbenindustri, Ger. Patent 476,598 (1929); Chem.
      Zentralbl., II, 487 (1929).
129. W. Fletscher, T. Wheeler, and J. McAnley, Brit. Patent 331,734
      (1930); Chem. Zentralbl., II, 2442 (1930).
130. T. Wheeler and W. Francis, Brit. Patent 391,994 (1930); C. A.,
      26, 3342 (1932).
131. M. De Simo, U.S. Patent 2,187,393 (1940); C. A., 34, 3454 (1940).
132. E. Dittrich and J. Varga, Ger. Patent 697,186 (1940); C. A., 35,
      6400 (1941).
133. E. Dittrich and J. Varga, Ger. Patent 723,396 (1942); C. A., 37,
      5204 (1943).
134. L. Preismann, U.S. Patent 2,474,067 (1949); C. A., 43, 7201
      (1949).
135. V. Sborgi and E. Giovannini, Chim. Ind., 31, 391 (1949).
136. G. Cavidalli, Rev. Inst. Franc. Petrol. Anu. Combust. Lignides,
      7, 5 (1953).
137. M. M. Marisic, U.S. Patent 2,636,810 (1953); C. A., 47, 10186
      (1953).
138. K. W. Guebert, U.S. Patent 2,712,982-3-4-5 (1955); Chem.
      Zentralbl., 9482 (1957).
139. Fr. Patent 1,123,679 (1956); Chem. Zentralbl., 9346 (1958).
140. Indian Patent 53,968 (1957); Chem. Zentralbl., 9595 (1961).
141. A. A. Banks, Brit. Patent 943,737 (1963); C. A., 60, 5262
      (1964).
142. R. W. Timmermann and H. C. Kutz, U.S. Patent 2,888,329 (1959);
      Chem. Zentralbl., 7030 (1961).
143. R. C. Forney and J. M. Smith, Ind. Eng. Chem., 43, 1841 (1951).
144. A. R. Clark, Chem. Proc. Eng. (London), 13, (1967).
145. A. A. Banks, Brit. Patent 939,209 (1958).
146. A. A. Banks, Ger. Patent 1,122,046 (1962); C. A., 57, 4315
      (1962).
147. A. A. Banks, Ger. Patent 1,122,047 (1962); C. A., 57, 4316
      (1962).
148. S. Robinson, U.S. Patent 2,593,345 (1952); C. A., 46, 7320
      (1952).
149. E. Herault, Rev. Prod. Chim., 64, 217 (1961).
150. Z. Leszczyuski, I. Kubica, and I. Strzelecki, Pol. Patent
      52,227 (1965); C. A., 68, 59106 (1968).

151. Fr. Patent 2,053,879 (1969); C. A., 76, 74421 (1972).
152. S. N. Shorin, A. L. Suris, and D. Hebecker, Ger. Patent 2,501,237 (1976); C. A., 85, 162795v (1976).
153. B. S. Aranovich, B. A. Borodulya, and B. L. Gansha, USSR Patent 321,467 (1969); C. A., 76, 74411 (1972).
154. A. A. Peliks and E. S. Shapiro, Chemiefares, 30 (1973).
155. B. A. Borodulya, L. M. Vinogradov, B. L. Gansha, and M. V. Graben'ka, Izv. Akad. Nauk Belor. SSR, Ser. Khim., 81, 137 (1973).
156. C. M. Thacker, U.S. Patent 2,330,934 (1943); C. A., 38, 1245 (1944); U.S. Patent 2,411,236 (1946); C. A., 41, 982 (1947).
157. H. O. Folkins and E. Miller, U.S. Patent 2,709,154 (1955); C. A., 49, 11970 (1955).
158. V. Sborgi and E. Giovannini, Ital. Patent 457,263 (1950); C. A., 46, 5278 (1952).
159. H. O. Folkins, U.S. Patent 2,668,752 (1954); C. A., 48, 7859 (1954).
160. H. O. Folkins, U.S. Patent 2,712,985 (1955); C. A., 49, 16374 (1955).
161. C. J. Wenzke, U.S. Patent 3,079,233 (1960); C. A., 58, 13668 (1963).
162. D. Porter, U.S. Patent 2,882,130 (1959); C. A., 53, 13526 (1959)
163. Neth. Patent 111,340 (1954).
164. W. J. Thomas and S. C. Naik, Trans. Inst. Chem. Eng., 48, 129 (1970).
165. V. Meyer and T. Sandmeyer, Ber., 16, 2176 (1883).
166. V. Meyer, Ber., 18, 217 (1885).
167. S. O. Jones, Thesis, J. Hopkins Univ. (1936).
168. S. O. Jones and E. E. Reid, J. Am. Chem. Soc., 60, 2452 (1938).
169. J. D. Palmer and S. G. Lloyd, J. Am. Chem. Soc., 52, 3388 (1930).
170. W. A. Lazier, F. K. Signaigo, and J. H. Werntz, U.S. Patent 2,402,643 (1946); C. A., 40, 5764 (1946).
171. F. K. Signaigo, U.S. Patent 2,402,456 (1946); C. A., 40, 5767 (1946).
172. E. Tengler, Ger. Patent 728,642 (1942); C. A., 38, 431 (1944).
173. K. H. Meyer and W. Hohenemser, Helv. Chim. Acta, 18, 1061 (1935).
174. H. E. Westlake, M. G. Mayberry, M. H. Whitlock, J. R. West, and G. J. Haddan, J. Am. Chem. Soc., 68, 748 (1946).
175. E. Fromm and H. Jörg, Ber., 58, 304 (1925).
176. W. Brown, Bull. Univ. Pittsburgh, 39, 73 (1943).
177. G. A. Komashko, V. A. Zazhigalov, V. A. Kuznetsov, and S. V. Gerei, All-Union Conference of the Mechanisms of Catalytic Reactions [in Russian], Moscow (1974).
178. V. A. Zazhigalov, G. A. Komashko, S. V. Gerei, and M. Y. Rubanik, Neftekhimiya, 17, 84 (1977).
179. V. A. Zazhigalov, I. V. Bacherikova, and G. A. Komashko, Ukr. Khim. Zh., 44, 886 (1978).
180. G. A. Komashko, V. A. Zazhigalov, S. V. Gerei, and M. Y. Rubanik, Catalysis and Catalysts [in Russian], Kiev (1975), p. 22.
181. O. P. Strausz and H. E. Gunning, J. Am. Chem. Soc., 84, 4080

(1962).

182. H. E. Gunning and O. P. Strausz, Adv. Photochem., No. 4, 143, (1966).

183. O. P. Strausz, Adv. Chem. Ser., 110 (1972).

184. B. Meyer (ed.), Elemental Sulfur: Chemistry and Physics, Wiley-Interscience, New York (1965), p. 265.

185. E. M. Lown, H. S. Sandhu, H. E. Gunning, and O. P. Strausz, J. Am. Chem. Soc., 90, 7164 (1968).

186. D. R. Betteridge and J. T. Yardley, Chem. Phys. Lett., 62, 570 (1979).

187. I. G. Czizmadia, Chem. Biochim. Reactiv. Proc. Int. Symp., Jerusalem, 303 (1974).

188. E. Jacobson, Ger. Patent 38,416 (1886); Ber., 20, 184 (1887).

189. R. Spamier, Dissertation, Karlsruhe (1910).

190. W. Friedmann, Petroleum, 11, 693 (1916).

191. Denmark Patent 51,564 (1935); Brit. Patent 453,921 (1935); Chem. Zentralbl., I, 2278 (1937).

192. Fr. Patent 787,810 (1935); C. A., 30, 1068 (1936).

193. R. F. Naylor, J. Polym. Sci., 1, 305 (1946).

194. R. F. Naylor, Rubber Chem. Technol., 20, 353 (1947).

195. W. H. Hoffert and K. Wendtner, J. Inst. Petrol., 35, 171 (1949).

196. D. Spence, Kolloid.-Z., 10, 299 (1912).

197. D. Spence and J. Goung, Kolloid.-Z., 13, 265 (1913).

198. T. K. Hansen and L. M. Kinnard, Brit. Patent 696,439 (1953); C. A., 51, 15587 (1957).

199. R. T. Armstrong, U.S. Patent 2,446,072 (1947); C. A., 42, 7786 (1948).

200. R. T. Armstrong, J. R. Little, and K. D. Doak, Ind. Eng. Chem., 36, 628 (1944).

201. E. H. Farmer, Trans. Faraday Soc., 38, 356 (1942).

202. E. H. Farmer, Rubber Chem. Technol., 15, 765 (1942).

203. E. H. Farmer and F. W. Shipley, J. Polym. Sci., 1, 293 (1946).

204. E. H. Farmer, Rubber Chem. Technol., 20, 341 (1947).

205. M. L. Selker and A. R. Kemp, Ind. Eng. Chem., 39, 895 (1947).

206. A. S. Broun, M. G. Voronkov, and K. P. Katkova, Zh. Obshch. Khim., 20, 726 (1950).

207. W. Cleve, Fortsch. Forsch., 24, 7 (1948).

208. M. L. Selker and A. R. Kemp, Ind. Eng. Chem., 36, 22 (1944).

209. W. Friedmann, Ber., 49, 50 (1916).

210. M. G. Voronkov, Dissertation, Leningrad (1947).

211. M. G. Voronkov and A. S. Broun, Dokl. Akad. Nauk SSSR, 59, 1437 (1948).

212. M. G. Voronkov, Vestn. Leningr. Univ., 2, 146 (1948).

213. A. S. Broun and M. G. Voronkov, Nauchn. Byull. Leningr. Univ., 21, 8 (1948).

214. M. G. Voronkov, A. S. Broun, and G. B. Karpenko, Zh. Obshch. Khim., 19, 395 (1949).

215. A. S. Broun, M. G. Voronkov, and K. P. Katkova, Zh. Obshch. Khim., 20, 726 (1950).

216. B. Böttcher and A. Lüttringhaus, Office Pub. Board, Rept., 1707, (1944); Ind. Eng. Chem., 39, 895 (1947).

217. A. Lüttringhaus, Angew. Chem., 59, 244 (1947).

218. B. Böttcher and A. Lüttringhaus, Liebigs Ann. Chem., <u>557</u>, 89 (1947).
219. A. Lüttringhaus, H. B. König, and B. Böttcher, Liebigs Ann. Chem., <u>560</u>, 201 (1947).
220. A. Lüttringhaus and W. Cleve, Liebigs Ann. Chem., <u>575</u>, 112 (1951).
221. B. Böttcher, Ger. Patent 869,799 (1953); Chem. Zentralbl., 4413 (1953).
222. B. Böttcher, Ger. Patent 855,865 (1952); Chem. Zentralbl., 1594 (1954).
223. O. Gandin and R. Pottier, C. R. Acad. Sci. (Paris), <u>224</u>, 479 (1947).
224. O. Gandin and N. Lozac'h, C. R. Acad. Sci. (Paris), <u>224</u>, 577 (1947).
225. N. Lozac'h, C. R. Acad. Sci. (Paris), <u>225</u>, 686 (1947).
226. O. Gandin, Fr. Patent 941,543 (1949); C. A., <u>44</u>, 9749 (1950).
227. N. Lozac'h, Bull. Soc. Chim. France, 850 (1949).
228. J. Schmitt and A. Lespagnol, C. R. Acad. Sci. (Paris), <u>230</u>, 551 (1950).
229. J. Schmitt and A. Lespagnol, C. R. Acad. Sci. (Paris), <u>230</u>, 1774 (1950).
230. J. Schmitt and A. Lespagnol, Bull. Soc. Chim. France, 459 (1950).
231. D. R. Stevens and W. C. Sternes, U.S. Patent 2,535,705 (1950); C. A., <u>45</u>, 2193 (1951).
232. D. R. Stevens and A. C. Whitaker, U.S. Patent 2,535,706 (1950); C. A., <u>45</u>, 3424 (1951).
233. C. Djerassi and A. Lüttringhaus, Chem. Ber., <u>94</u>, 2305 (1961).
234. B. Böttcher and F. Bauer, Chem. Ber., <u>84</u>, 458 (1951).
235. N. Lozac'h and L. Legrand, C. R. Acad. Sci. (Paris), <u>232</u>, 2330 (1951).
236. O. Gandin, U.S. Patent 2,556,963 (1951); C. A., <u>46</u>, 3079 (1951).
237. N. Lozac'h and J. Teste, C. R. Acad. Sci. (Paris), <u>234</u>, 1891 (1952).
238. R. S. Airs and V. W. David, U.S. Patent 2,653,910 (1953); C. A., <u>48</u>, 10333 (1954).
239. D. R. Stevens and S. C. Camp, U.S. Patent 2,658,900 (1953); C. A., <u>48</u>, 3024 (1954).
240. H. Horold and R. C. Kaufhold, Ger. Patent 11,644 (1956); C. A., <u>53</u>, 5664 (1959).
241. J. Teste and N. Lozac'h, Bull. Soc. Chim. France, 1492 (1954).
242. O. Gandin, U.S. Patent 2,688,620 (1954); C. A., <u>49</u>, 13296 (1955)
243. J. Schmitt and M. Suguet, Bull. Soc. Chim. France, 84 (1955).
244. J. Schmitt, R. Fallard, and M. Suguet, Bull. Soc. Chim. France, 1147 (1956).
245. L. A. Hamilton and P. S. Landis, U.S. Patent 2,995,569 (1957); C. A., <u>52</u>, 2971 (1958).
246. E. K. Field, U.S. Patent 2,857,399 (1958); C. A., <u>53</u>, 8164 (1959).
247. N. Lozac'h and G. Mollier, Bull. Soc. Chim. France, 1389 (1959).
248. P. S. Landis and L. A. Hamilton, J. Org. Chem., <u>25</u>, 1742 (1960).
249. J. Raoul and J. Vialle, Bull. Soc. Chim. France, 1033 (1960).

250. M. Francois, Neth. Patent 6,400,267 (1964); C. A., 62, 564 (1965).

251. T. A. Danilova, Khim. Tekhnol. Topl. Masel, 10, 34 (1965).

252. A. Grandin, C. Boulton, and J. Vialle, Bull. Soc. Chim. France, 4555 (1968).

253. R. L. Hodgson and E. J. Smutny, U.S. Patent 3,394,146 (1968); C. A., 69, 67362 (1968).

254. J. P. Brown, J. Chem. Soc. C., 1077 (1968).

255. R. S. Spindt, D. R. Stevens, and W. E. Baldwin, J. Am. Chem. Soc., 73, 3693 (1951).

256. B. A. Alink and S. Onde, U.S. Patent 3,935,215 (1976); Ref. Zh. Khim., 19, N175P (1976).

257. R. Wegler, E. Kühle, and W. Schäfer, Angew. Chem., 70, 351 (1958).

258. F. Wessely and A. Siegel, Monatsh. Chem., 82, 607 (1951).

259. B. Buatier and J. Moyne, Fr. Patent 2,236,416 (1975); Ref. Zh. Khim., 24, 0296P (1976).

260. R. Weithöner and K. Brockhausen, Ger. Patent 929,448 (1955).

261. L. Bateman, R. W. Glazebrook, C. G. Moore, M. Porter, G. W. Ross, and R. W. Saville, J. Chem. Soc., 2838 (1958).

262. S. Kambara and N. Yamazaki, J. Soc. Rubber Ind. Jpn., 30, 856 (1957).

263. C. G. Moore and R. W. Saville, J. Chem. Soc., 2089 (1954).

264. J. P. Brown, Brit. Patent 1,010,637 (1965); C. A., 64, 4858 (1966).

265. Ger. Patent 1,275,068 (1968); C. A., 69, 86999 (1968).

266. D. R. Stevens and S. C. Camp, U.S. Patent 2,786,829 (1957); C. A., 51, 12148 (1957).

267. Fr. Patent 1,573,969 (1968); C. A., 72, 45624 (1970).

268. C. M. Hull, S. R. Olsen, and W. G. France, Ind. Eng. Chem., 38, 1282 (1946).

269. E. H. Farmer, J. F. Ford, and J. A. Lyons, J. Appl. Chem., 4, 554 (1954).

270. H. A. Wiebe, A. K. Knight, O. P. Strausz, and H. E. Gunning, J. Am. Chem. Soc., 87, 1443 (1965).

271. K. S. Sidhu, E. M. Lown, O. P. Strausz, and H. E. Gunning, J. Am. Chem. Soc., 88, 254 (1966).

272. U.S. Patent 3,334,036 (1967); C. A., 68, 39455 (1968).

273. E. M. Lown, E. L. Dedio, O. P. Strausz, and H. E. Gunning, J. Am. Chem. Soc., 89, 1056 (1967).

274. O. P. Strausz, W. B. O'Callaghau, E. M. Lown, and H. E. Gunning, J. Am. Chem. Soc., 93, 559 (1971).

275. M. A. Naylor and A. W. Anderson, J. Am. Chem. Soc., 75, 5395 (1953).

276. M. Carmack and L. F. De Tar, U.S. Patent 2,495,567 (1949); C. A., 44, 7868 (1950).

277. W. Friedman, Ber., 49, 1551 (1916).

278. L. E. Rindish and B. Eastman, Fr. Patent 936,208 (1948).

279. G. F. Bloomfield, J. Soc. Ind. Chem., 68, 66 (1949).

280. R. F. Naylor, J. Chem. Soc., 1532 (1947).

281. W. H. Hoffert and K. Wendtner, J. Inst. Petrol. (London), 35, 171 (1949).

282. E. H. Farmer and F. W. Shipley, J. Chem. Soc., 1519 (1947).
283. J. A. King and F. H. McMillan, J. Am. Chem. Soc., 68, 525, 632, 2335 (1946).
284. R. Wegler, E. Kühle, and W. Schäfer, Angew. Chem., 70, 351 (1958).
285. F. H. McMillan and J. A. King, J. Am. Chem. Soc., 69, 1207 (1947).
286. M. A. Naylor, U.S. Patent 2,744,134 (1956); C. A., 51, 4618 (1957).
287. M. Carmack and M. A. Schpilmann, Organic Reactions, Vol. 3, New York (1946), p. 83.
288. L. F. De Tar and M. Carmack, J. Am. Chem. Soc., 68, 2025 (1946).
289. M. Carmack and L. F. De Tar, J. Am. Chem. Soc., 68, 2029 (1946).
290. C. Willgerodt and B. Albert, J. Prakt. Chem., 84, 387 (1911).
291. I. N. Zafranskii, A. N. Semenova, and K. E. Zhukova, in: The Chemistry and Technology of Polymers, Krasnoyarsk (1972).
292. J. A. King and F. H. McMillan, J. Am. Chem. Soc., 70, 4143 (1948).
293. W. Steinkopf, Liebigs Ann. Chem., 403, 11 (1914).
294. H. E. Westlake, Chem. Rev., 39, 219 (1946).
295. W. P. Baker, Fr. Patent 1,357,934 (1962); C. A., 62, 18218 (1965).
296. B. A. Arbuzov and E. G. Kataev, Dokl. Akad. Nauk SSSR, 96, 983 (1954).
297. R. Mayer, Z. Chem., 13, 321, (1973).
298. A. J. Ginsberg and W. E. Lindsell, Chem. Commun., 232 (1971).
299. E. H. Farmer, J. Soc. Chem. Ind., 66, 86 (1947).
300. G. F. Bloomfield, J. Polym. Sci., 1, 312 (1946).
301. G. F. Bloomfield, J. Chem. Soc., 1546, 1548 (1947).
302. G. F. Bloomfield, J. Soc. Chem. Ind., 67, 14 (1948).
303. R. W. Glazebrook and R. W. Saville, J. Chem. Soc., 2094 (1954).
304. L. Bateman, C. G. Moore, and M. Porter, J. Chem. Soc., 2856 (1958).
305. L. Bateman, R. W. Glazebrook, C. G. Moore, and R. W. Saville, Rubber Chem. Technol., 30, 397 (1957).
306. L. Bateman, R. W. Glazebrook, and C. G. Moore, J. Chem. Soc., 2846 (1958).
307. L. Bateman, R. W. Glazebrook, and C. G. Moore, J. Appl. Polym. Sci., 1, 257 (1959).
308. J. Mollier, Bull. Soc. Chim. France, 561 (1953).
309. J. F. Bascato, J. M. Catala, E. Franta, and J. Brossas, Makromol. Chem., 180, 1571 (1979).
310. T. G. Khanlarov, S. A. Sadykov, and I. I. Rasulov, Uch. Zap. Azerb. Univ., Ser. Khim. Nauk, 54 (1978).
311. M. Jentus, L. Nowak, and H. Krawczyk, Pol. Patent 45,941 (1962).
312. T. Hancock, Personal Narrative of the Origin and Progress of Caoutchuc India Rubber Manufacture in England, London (1920).
313. B. V. Byzov, Zh. Russ. Fiz. Khim. Ova., 51, 1 (1921).
314. M. Payer, C. R. Acad. Sci. (Paris), 34, 2, 453 (1852).
315. H. L. Terry, J. Soc. Chem. Ind., 11, 970 (1892).
316. B. Ungar, Z. Anal. Chem., 24, 167 (1885).

317. C. O. Weber, Kolloid-Z., 1, 33, 65 (1906).
318. C. O. Weber, Chem. India Rubber, London (1912).
319. G. Hübner, Gummi Ztg., 24, 213 (1909).
320. F. W. Hinrichsen and E. Stern, Chem. Ztg., 33, 756 (1909).
321. D. F. Twiss, Chem. Ind. (London), 36, 782 (1917).
322. D. Spence and J. Goung, Kolloid-Z., 11, 28 (1912).
323. V. B. Margaritov, Usp. Khim., 10, 224 (1941).
324. N. Bacon, J. Phys. Chem., 32, 801 (1928).
325. J. Williams, Proc. Rubb. Technol. Conf. London, 24, 304 (1938).
326. T. Midgley, A. L. Heune, A. F. Shepard, and M. W. Renoll, J. Am. Chem. Soc., 56, 1325 (1934).
327. F. Horie and K. Morikana, Chem. Ind. Jpn., 37, 5 (1934).
328. A. H. Scott, H. L. Cartiss, and A. T. Pherson, J. Res. Bur. Stand., 11, 173 (1933).
329. G. Hübner, Chem. Ztg., 33, 648 (1909).
330. J. T. Blake, Ind. Eng. Chem., 22, 737, 748 (1930).
331. C. G. Boygs and J. T. Blake, Ind. Eng. Chem., 22, 748 (1930).
332. H. Erdmann, Liebigs Ann. Chem., 362, 133 (1908).
333. J. R. Brown and E. A. Hauser, Ind. Eng. Chem., 30, 1291 (1938).
334. G. J. Amerongen and R. Houwink, J. Prakt. Chem., 161, 261 (1943).
335. E. A. Hauser and J. R. Brown, Ind. Eng. Chem., 31, 1225 (1939).
336. A. Rossem, India Rubber J., 92, 845 (1936).
337. H. J. Prins, Chem. Weekbl., 66, 64 (1918).
338. T. Midgley, A. L. Heune, and A. F. Shepard, J. Am. Chem. Soc., 56, 1326 (1934).
339. T. Midgley, A. L. Heune, and A. F. Shepard, J. Am. Chem. Soc., 54, 2953 (1932).
340. H. E. Westlake, C. R. Acad. Sci. (Paris), 39, 219 (1946).
341. B. A. Dogadkin and V. A. Shershnev, Usp. Khim., 30, 1013 (1961).
342. W. Hofmann, Vulkanisation und Vulkanisations Hilfsmittel (1968).
343. J. R. Shelton and E. T. McDonel, Int. Congress of Caoutchouc and Rubber, Washington (1959).
344. M. L. Selker and A. R. Kemp, Ind. Eng. Chem., 40, 1470 (1948).
345. J. W. Baker, W. S. Nathau, and C. W. Shoppee, J. Chem. Soc., 1847 (1935).
346. F. Henrich, Ber., 32, 668 (1899).
347. E. Farmer and S. Michael, Rubber Chem. Technol., 16, 465 (1943).
348. J. W. Baker and M. L. Hemming, J. Chem. Soc., 191 (1942).
349. G. J. Veersen, Riv. Gin. Caoutch., 28, 411 (1951).
350. G. F. Bloomfield, Proc. Sec. Rubber Technol. Conf. London, 79 (1948).
351. B. A. Dogadkin and S. N. Tarasova, Kolloid-Z., 15, 347 (1953).
352. B. A. Dogadkin and S. N. Tarasova, Rubber Chem. Technol., 24, 883 (1954).
353. C. W. Hull, L. A. Wienlands, J. R. Olsen, and H. C. France, Rubber Chem. Technol., 21, 553 (1948).
354. B. A. Dogadkin, Usp. Khim., 24(7), 801 (1955).
355. B. C. Barton, Ind. Eng. Chem., 42, 671 (1950).
356. Y. J. Cunneeu, J. Chem. Soc., 134 (1947); Appl. Chem., 2, 353

(1952).

357. C. W. Hull, Ind. Eng. Chem., 40, 513 (1948).
358. R. J. Frory, J. Polym. Sci., 4, 22 (1949).
359. K. Alder, Ber., 76, 27, 40 (1943).
360. N. Rabjohn, J. Am. Chem. Soc., 70, 1181 (1948).
361. Fr. Patent 976,073 (1952); C. A., 47, 2531 (1953).
362. J. Alphen, Chem. Weekbl., 47, 733 (1951).
363. L. Bateman, Proc. 3d Rubber Technol. Conf. Cambridge (England), 289 (1954).
364. C. Harris, Ber., 38, 1195 (1905).
365. J. Alphen, Angew. Chem., 66(7), 193 (1954).
366. L. Bateman, C. G. Moore, M. Porter, and B. Saville, "Chemistry of Vulcanisation," in: The Chemistry and Physics of Rubber-like Substances, ed. by L. Bateman, London—New York (1963), p. 487.
367. K. Khasimoto and Y. Sumijsi, Jpn. Patent 47-5102 (1977); Ref. Zh. Khim., 8, t566P (1978).
368. P. K. Bandyopadhyay and S. Banerjee, Angew. Makromol. Chem., 64, 59 (1977).
369. R. Bakule and A. Havranek, Rubber Chem. Technol., 51, 72 (1978).
370. D. A. Chapman, J. Elastom. Plast., 10, 129 (1978).
371. K. L. Miller, U.S. Patent 4,124,754 (1978); Ref. Zh. Khim., 12, S265P (1979).
372. H. J. Petterkau and W. Göbel, Ger. Patent 2,645,920 (1978); Ref. Zh. Khim., 4, S309P (1979).
373. G. A. Bloch, Org. Vulcanization Accelerator Caoutchouc, Moscow—Lenigrad (1972).
374. B. Dolgoplosk and E. Tinyakova, in: Vulcanisation Rubber (1954).
375. G. A. Bloch, Gummi Asbest, 11, 670 (1957); 12, 672 (1957).
376. G. A. Bloch, Rubber Chem. Technol., 31(5), 1035 (1958).
377. D. Shelton and E. MacDonel', Caoutchouc Rubber, 52 (1960).
378. E. Tinyakova, E. Khrennikova, B. Dolgoplosk, V. Reikh, and T. Zhuravleva, Zh. Obshch. Khim., 26, 2476 (1956).
379. B. Dogadkin, M. Fildshtein, and D. Pevzner, in: Vulcanisation Rubber (1954).
380. D. Graig, Rubber Chem. Technol., 30, 1291 (1957).
381. G. S. Whitby, Synthetic Rubber (1954).
382. A. Broud, Kunststoffe, 43(1), 8 (1953).
383. C. S. Mills, Rubber Plastics Age (London), 33(3), 116 (1952).
384. R. W. Culloucm, Rubber World, 135, 5723 (1957).
385. W. F. Fischer, R. F. Neu, and R. L. Zapp, Rubber Age (New York), 81(4), 633 (1957).
386. P. K. Bandyopadhyay and S. Banerjee, J. Appl. Polym. Sci., 23, 185 (1979).
387. G. A. Bloch, Dissertation, Kiev (1960).
388. P. Thiollet, Rev. Gen. Cautchouc, 5, 22 (1945).
389. W. Hofmann, Gummi-Asbest-Kunstst., 32, 158 (1979).
390. A. K. Bhowmick, R. Mukhopadhyay, and S. K. De, Rubber Chem. Technol., 52, 725 (1979).
391. C. D. Trivette, E. Morita, and E. J. Joung, Rubber Chem. Technol., 35(5), 1361 (1962).
392. N. Kharasch, S. Y. Potempa, and H. Wehrmeister, Chem. Rev., 39, 269 (1946).

393. P. K. Bandyopadhyay and S. Banerjee, Kautsch. Gummi Kunstst., 32, 961 (1979).

394. W. Steinkopf and G. Kirchhoff, Austria Patent 72,291 (1916); Chem. Zentralbl., 869 (1917).

395. W. Steinkopf, Die Chemie des Thiophens, Leipzig (1941).

396. W. Steinkopf and G. Kirchhoff, Ger. Patent 252,375 (1911); Chem. Zentralbl., 1707 (1912).

397. W. Steinkopf, Liebigs Ann. Chem., 428, 123 (1922).

398. R. Meyer, Ber., 51, 1571 (1918).

399. R. Meyer and H. Wesche, Ber., 50, 422 (1917).

400. J. Capelle, Bull. Soc. Chim. France, 4(3), 150 (1908).

401. O. W. de Coninck, Bull. Soc. Chim. Belg., 305 (1908).

402. C. T. Bhatt, K. S. Nargund, D. D. Kanga, and M. S. Shak, J. Univ. Bombay, 3, 159 (1934).

403. J. B. Peel and P. L. Robinson, J. Chem. Soc., 2068 (1928).

404. V. A. Briscoe, J. B. Peel, and P. L. Robinson, J. Chem. Soc., 2857 (1928).

405. F. Challenger and J. B. Harrison, J. Inst. Petrol. Technol., 21, 135 (1935).

406. J. Bruce, F. Challenger, H. B. Gibson, and W. E. Allenby, J. Inst. Petrol. Technol., 34, 226 (1948).

407. J. Komlos, A. Komlos, and F. F. Engelke, Brit. Patent 265,994 (1927); Chem. Zentralbl., I, 2949 (1927).

408. A. E. Chichibabin, Zh. Russ. Fiz. Khim. Ova., 47, 703 (1915).

409. F. Cousanino and A. Cruto, Gazz. Chim. Ital., 51(1), 177 (1921).

410. F. Challenger, E. A. Mason, E. C. Holdsworth, and R. Emmott, J. Soc. Chem. Ind. (London), 714 (1952).

411. N. Lozac'h, M. Denis, G. Mollier, and J. Teste, Bull. Soc. Chim. France, 1016 (1953).

412. R. Mayer, B. Hunger, K. Pronsa, and A. K. Müller, J. Prakt. Chem., 35, 294 (1967).

413. R. Mayer, B. Gebhardt, J. Fabiau, and A. K. Müller, Angew. Chem., 76, 143 (1964).

414. R. Mayer and B. Gebhardt, Ber., 97, 1298 (1964).

415. M. Schmidt and V. Potschka, Naturwissenschaften, 50, 302 (1963).

416. H. E. Wijers, P. P. Montiju, L. Bransdma, and J. E. Arens, Recl. Trav. Chim. Pays-Bas, 84, 1284 (1965).

417. P. J. W. Schnijl, L. Brandsma, and J. E. Arens, Recl. Trav. Chim. Pays-Bas, 85, 889 (1966).

418. H. E. Wijers, L. Brandsma, and J. E. Arens, Recl. Trav. Chim. Pays-Bas, 86, 670 (1967).

419. H. E. Wijers, C. H. D. Ginkel, L. Brandsma, and J. E. Arens, Recl. Trav. Chim. Pays-Bas, 87, 38 (1968).

420. L. Brandsma and P. J. W. Schnijl, Recl. Trav. Chim. Pays-Bas, 88, 513 (1969).

421. J. Meijer and L. Brandsma, Recl. Trav. Chim. Pays-Bas, 91, 578 (1972).

422. R. Raap and R. G. Micetisch, Can. J. Chem., 46, 1057, 2251 (1968).

423. B. A. Trofimov, N. K. Gusarova, S. V. Amosova, and M. G. Voronkov, Proceedings of the XIIIth Scientific Conference on the Chemistry and Technology of Organosulfur Compounds, Riga (1974).

424. B. A. Trofimov, N. K. Gusarova, S. V. Amosova, and M. G. Voronkov, USSR Patent 568,638 (1977); C. A., 87, 167538q P (1977).

425. A. Roodselaar, I. Safarik, O. P. Strausz, and H. E. Gunning, J. Am. Chem. Soc., 100, 4068 (1978).

426. O. P. Strausz, J. Font, E. L. Dedio, P. Kebarle, and H. E. Gunning, J. Am. Chem. Soc., 89, 4805 (1967).

427. H. E. Gunning, U.S. Patent 3,230,16; (1966); C. A., 64, 11086 (1966).

428. A. R. Knight, O. P. Strausz, and H. E. Gunning, J. Am. Chem. Soc., 85, 1207 (1963).

429. W. Friedmann, Petrol. Z., 11(10), 978 (1916).

430. W. Friedmann, Brennst. Chem., 8, 257 (1927).

431. A. S. Broun and B. V. Ioffe, Nauchn. Byull. Leningr. Univ., 20, 11 (1947).

432. R. Jahn and U. Schmidt, Chem. Ber., 108(2), 630 (1975).

433. A. Jenner and M. Hens, C. R. Acad. Sci. (Paris), 242, 786 (1956).

434. A. Jenner, C. R. Acad. Sci. (Paris), 241, 1581 (1955).

435. F. Wessely and F. Grill, Monatsh. Chem., 77, 282 (1947).

436. M. T. Bogert and D. Davidson, J. Am. Chem. Soc., 56, 185 (1934).

437. N. G. Bromby, A. T. Peters, and F. M. Rowe, J. Chem. Soc., 144 (1943).

438. S. G. Sengupta, Science Culture, 2, 589 (1937).

439. S. G. Clemo and H. G. Dickenson, J. Chem. Soc., 255 (1937).

440. M. T. Bogert, D. Davidson, and A. M. Apfelbaum, J. Am. Chem. Soc., 56, 959 (1934).

441. S. G. Sengupta, J. Prakt. Chem., 151, 82 (1938).

442. B. Böttcher, Fr. Patent 871,802 (1942).

443. B. Böttcher, Swiss Patent 232,596 (1944).

444. R. Weiss and K. Woidich, Monatsh. Chem., 46, 453 (1925).

445. V. Markovnikov, Ber., 20, 1850 (1887).

446. E. Kleiber, Fr. Patent 651,284 (1928); C. A., 23, 3600 (1929).

447. J. Braun, E. Hahn, and J. Seemann, Ber., 55, 1687 (1922).

448. K. Oziewouski, Ber., 36, 962 (1903).

449. P. Rehlander, Ber., 36, 1583 (1903).

450. A. N. Kost and J. J. Grandberg, Zh. Obshch. Khim., 25, 2064 (1955).

451. J. Braun and G. Irmisch, Ber., 85, 883 (1932).

452. M. S. Newman, J. Am. Chem. Soc., 62, 870 (1940).

453. L. F. Fieser and L. M. Joshel, J. Am. Chem. Soc., 62, 1211 (1940).

454. W. E. Bachmann and R. O. Edgerton, J. Am. Chem. Soc., 62, 2550 (1940).

455. M. S. Newmann, J. Am. Chem. Soc., 62, 2295 (1940).

456. L. Ruzicka, Helv. Chim. Acta, 14, 1104 (1931).

457. Fr. Patent 711,723 (1930); C. A., 26, 1615 (1932).

458. I. P. Labunskii, Dissertation (1951).

459. W. B. Preff, U.S. Patent 1,349,914 (1920); C. A., 14, 3339 (1920).

460. P. A. Plattner, Chimia, 55, 131 (1942).

461. L. Ruzicka, Forschr. Chem. Phys. Phys. Chem., 19A, 57, 345 (1928).

462. G. W. Ross, J. Chem. Soc., 2856 (1958).

463. N. S. Nametkin, V. D. Tyurin, I. V. Petrosyan, A. V. Popov, B. I. Kolobkov, and A. M. Krapivin, Izv. Akad. Nauk SSSR, Ser. Khim., 2841 (1980).

464. R. T. Armstrong, U.S. Patent 1,082,106 (1914); C. A., 8, 830 (1914).

465. J. R. Wolfe, Rubber Chem. Technol., 41, 1339 (1968).

466. S. Inone, T. Tezuka, and S. Oae, Phosphorus Sulfur, 4, 219 (1978).

467. L. Legrand, G. Mollier, and N. Lozac'h, Bull. Soc. Chim. France, 327 (1953).

468. M. Ebel, L. Legrand, and N. Lozac'h, Bull. Soc. Chim. France, 161 (1963).

469. J. R. Sherwood and W. F. Short, J. Chem. Soc., 1641 (1932).

470. J. R. Sherwood, W. F. Short, and R. Stansfield, J. Chem. Soc., 1832 (1932).

471. J. R. Sherwood, W. F. Short, and J. Woodcock, J. Chem. Soc., 322 (1936).

472. H. Fritz and C. D. Weis, Tetrahedron Lett., 1659 (1974).

473. G. J. Jayne, H. F. Askew, and D. R. Woods, Ger. Patent 2,533,327 (1974); C. A., 85, 110706x (1976).

474. B. K. Berdoloi, K. Binoy, and E. M. Pearce, Adv. Chem. Ser., 31 (1978).

475. O. Brunner and F. Grof, Monatsh. Chem., 64, 28, 76 (1934).

476. R. Weiss and K. Woidich, Monatsh. Chem., 46, 453 (1925).

477. T. C. Shields and A. N. Kurtz, J. Am. Chem. Soc., 91, 5415 (1969).

478. A. S. Hay, U.S. Patent 4,033,982 (1977); C. A., 87, 102342b (1977).

479. J. Emsley, D. W. Griffiths, and G. J. Jayne, J. Chem. Soc., Perkin, 1, 228 (1979).

480. C. Marschalk, Bull. Soc. Chim. France, 1122 (1939).

481. K. Jellis, Khim. Sint. Smol, 2, 1678 (1942).

482. W. B. Preff, U.S. Patent 1,349,909 (1920); C. A., 14, 3338 (1920).

483. H. A. Winkelman, U.S. Patent 1,841,235 (1932); C. A., 26, 1832 (1932).

484. P. P. Budnikov, Izv. Ivano-Voznesensk. Politekh. Inst., 6, 211 (1922).

485. P. P. Budnikov, Tr. Gos. Inst. Silikatov, No. 2 (1923).

486. P. P. Budnikov and E. A. Shilov, Izv. Ivano-Voznesensk. Politekh. Inst., 4, 110 (1921).

487. P. P. Budnikov and E. A. Shilov, Zh. Russ. Fiz. Khim. Ova., 59, 685 (1924).

488. P. P. Budnikov, Z. Angew. Chem., 35, 653 (1922).

489. P. P. Budnikov, C. R. Acad. Sci. (Paris), 196, 1898 (1933).

490. P. P. Budnikov and E. A. Shilov, Ber., 55, 3848 (1922).

491. P. P. Budnikov and E. A. Shilov, J. Am. Chem. Soc., 6, 1000 (1923).

492. P. P. Budnikov and E. A. Shilov, Ber. Ceram. Ges., 12, 298 (1931).

493. V. E. Tishchenko and S. A. Smirnov, Zh. Prikl. Khim., 111, 7

(1930).

494. J. G. Gentele, Dingl. Pol. J., 137, 273 (1855).
495. J. G. Gentele, Dingl. Pol. J., 139, 436 (1856).
496. H. Reimsch, J. Prakt. Chem., 13, 136 (1838).
497. H. Schwarz, Dingl. Pol. J., 197, 243 (1870).
498. H. A. Gerdner and L. P. Hart, U.S. Patent 1,963,084 (1934);
      C. A., 28, 4939 (1934).
499. W. H. Kobbe, U.S. Patent 1,844,400 (1932); C. A., 26, 1892
      (1932).
500. H. A. Gardner, U.S. Patent 1,963,846 (1934); C. A., 28, 5261
      (1934).
501. W. B. Pratt, U.S. Patent 1,451,711 (1923); C. A., 17, 2064
      (1923).
502. A. Winkler, Chem. Zentralbl., 337 (1858).
503. P. Curie, Chem. News, 30, 189 (1874).
504. W. Kelbe, Ber., 11, 2174 (1878).
505. H. Morris, Proc. Chem. Soc., 70, 102 (1889).
506. A. Vesterberg, Ber., 36, 4200 (1903).
507. Ger. Patent 43,802 (1887); Ber., 21, 553 (1888).
508. C. O. Henke and G. Etzel, U.S. Patent 1,881,565 (1932); Chem.
      Zentralbl., 2464 (1933).
509. W. Nagel and M. Köruchen, Chem. Umschau, 39, 1 (1932).
510. W. Schulze, Liebigs Ann. Chem., 359, 129 (1908).
511. J. R. Hoskins and W. T. Fadyen, Chem. Eng., 53, 195 (1934).
512. L. Ruzicka, Helv. Chim. Acta, 9, 962 (1926).
513. Scheiber and Wedel, Farbe Lack, 50, 64 (1925).
514. E. Bergmann, C. R. Acad. Sci. (Paris), 212, 529 (1941).
515. T. H. Esterfield and G. Bagley, J. Chem. Soc., 1238 (1904).
516. P. Levy, Z. Angew. Allg. Chem., 81, 145 (1913).
517. L. Ruzicka and J. Meyer, Helv. Chim. Acta, 5, 521 (1922).
518. A. J. Virtanen, Liebigs Ann. Chem., 424, 150 (1921).
519. L. Ruzicka and F. Balas, Helv. Chim. Acta, 6, 677 (1923);
      7, 875 (1924).
520. L. Ruzicka, Vecn. Ann., 468, 133, 143 (1929).
521. B. Böttcher, Belg. Patent 441,366 (1941); Ital. Patent 350,755
      (1941), 408,497 (1945).
522. B. Böttcher and A. Müller, Ger. Patent 700,143 (1940); C. A.,
      35, 7139 (1941).
523. W. Weiss, Ger. Patent 693,207 (1940); Chem. Zentralbl., II,
      2534 (1940).
524. F. Biedebach, Arch. Pharm. (Weinheim, Ger.), 280, 304 (1942).
525. L. C. Holt, U.S. Patent 2,443,823 (1948); C. A., 42, 6108 (1948).
526. B. Böttcher and F. Bauer, Liebigs Ann. Chem., 568, 227 (1950).
527. P. A. Levin, Zh. Prikl. Khim., 39, 677 (1966).
528. W. B. Prett, Ind. Eng. Chem., 15, 178 (1923).
529. A. Nakatasuchi, Chem. Ind. Jpn., 33, 408 (1930).
530. F. Krsnak, J. Koranis, and M. Kalina, Czech. Patent 143,950
      (1973); C. A., 78, 3696c (1973).
531. A. Nakatasuchi, Chem. Ind. Jpn., 35, 376 (1932).
532. A. Kakatasuchi, Chem. Ind. Jpn., 36, 254 (1933).
533. C. G. Moore and M. Porter, Tetrahedron, 6, 10 (1959).
534. L. Ruzicka and M. Pfeiffer, Helv. Chim. Acta, 9, 841 (1926).

535. L. Ruzicka and A. G. Veen, Liebigs Ann. Chem., <u>476</u>, 70 (1929).
536. E. A. Rudolph, Dissertation, Zurich (1925).
537. L. Ruzicka, J. Meyer, and M. Mingazzini, Helv. Chim. Acta,
     <u>5</u>, 361, 363 (1922).
538. S. Takagi, J. Pharm. Soc. Jpn., 514, 1001 (1924); C. A., <u>19</u>,
     1704 (1925).
539. J. Henderson and M. Nab, J. Chem. Soc., 3077 (1926).
540. K. Kohara, H. Tachibana, Y. Masujama, and Y. Otani, Bull. Chem.
     Soc. Jpn., <u>52</u>, 1549 (1979).
541. R. W. Iwett, U.S. Patent 2,691,648 (1954); C. A., <u>49</u>, 2723
     (1955).
542. R. Uloth, Brennst.-Chem., <u>10</u>, 297 (1929).
543. F. Hoffmann and R. Uloth, Brit. Patent 306,421 (1929); Chem.
     Zentralbl., II, 963 (1929).
544. A. Renard, C. R. Acad. Sci. (Paris), <u>112</u>, 49 (1891).
545. F. Schultze, Ber., <u>4</u>, 33 (1871).
546. V. Merz and W. Weith, Ber., <u>4</u>, 384 (1871).
547. H. B. Glass and E. E. Reid, J. Am. Chem. Soc., <u>51</u>, 3428 (1929).
548. S. Oae, J. Jpn. Petrol. Inst., <u>17</u>, 752 (1974).
549. C. Friedel and J. Crafts, C. R. Acad. Sci. (Paris), <u>86</u>, 884
     (1878).
550. J. Krafft and R. E. Lyons, Ber., <u>29</u>, 435 (1896).
551. P. Genwresse, Bull. Soc. Chim. France, 15, 1038 (1898).
552. G. Dougherty and P. D. Hammond, J. Am. Chem. Soc., <u>57</u>, 117 (1935).
553. M. G. Voronkov and F. D. Faitel'son, Khim. Geterotsikl. Soedin.,
     245 (1967).
554. J. Boesehen and D. A. W. Koning, Recl. Trav. Chim. Pays-Bas,
     <u>30</u>, 116 (1941).
555. J. Boesehen, Recl. Trav. Chim. Pays-Bas, <u>24</u>, 6, 209, 221 (1905).
556. Brit. Patent 695,399 (1953); C. A., <u>48</u>, 1062 (1954).
557. E. Tchunkur and E. Himmer, Ger. Patent 579,917 (1933); Chem.
     Zentralbl., II, 2457 (1933).
558. E. Chapiro and P. Gach, Belg. Patent 390,439 (1932); C. A., <u>27</u>,
     2696 (1933).
559. H. Gilman and A. G. Jacoby, J. Org. Chem., <u>3</u>, 108 (1938).
560. A. Bürger, W. B. Wartmann, and R. E. Lutz, J. Am. Chem. Soc.,
     <u>60</u>, 2628 (1938).
561. F. Mutz and H. Pufzer, Pbl. 63936, Fiat Microfilm Reel, C 60,
     PB 17, 657.
562. M. Lanfry, C. R. Acad. Sci. (Paris), <u>152</u>, 1254 (1911).
563. N. N. Vorozhtsov and V. I. Rodionov, Dokl. Akad. Nauk SSSR,
     <u>134</u>, 1085 (1960).
564. W. Lazier, F. Signaigo, and L. Wise, U.S. Patent 2,402,645
     (1946); C. A., <u>40</u>, 5761 (1946).
565. F. Hassler, Ger. Patent 399,063 (1924); Chem. Zentralbl., II,
     1546 (1924).
566. F. Hassler, Ger. Patent 407,994 (1925); Chem. Zentralbl., I,
     2138 (1925).
567. F. Hassler, Ger. Patent 409,713 (1925); Chem. Zentralbl., I,
     2730 (1925).
568. Ger. Patent 242,215 (1910); Chem. Zentralbl., I, 297 (1912).
569. R. C. M. Bayard de la Vingtrie, Fr. Patent 493,569 (1920).

570. W. Fridemann, Brennst.-Chem., 8, 257 (1927).
571. C. Dziewonski, Ber., 36, 962 (1903).
572. D. B. Clapp, J. Am. Chem. Soc., 61, 2733 (1939).
573. P. Rehländer, Ber., 36, 1583 (1903).
574. C. Dziewonski, Bull. Soc. Chim. France, 31, 925 (1904).
575. M. Wyler, U.S. Patent 1,354,890 (1920); Chem. Zentralbl., II, 33 (1921).
576. W. Friedmann, Ber., 49, 683 (1916).
577. R. Kattwinkel, Brennst.-Chem., 15, 141 (1934).
578. C. Grabe and B. Mantz, Liebigs Ann. Chem., 290, 238 (1896).
579. B. Mantz, Dissertation, Geneva (1892).
580. C. Grabe, Ber., 25, 3146 (1892).
581. H. F. Schaeffer, Proc. Indian Acad. Sci., 59, 153 (1950).
582. R. Wessgerber, Ber., 34, 1659 (1901).
583. M. Zander and W. H. Franke, Chem. Ber., 106(8), 2752 (1973).
584. Ger. Patent 480,377 (1929); Chem. Zentralbl., II, 2382 (1929).
585. Ger. Patent 186,990 (1906).
586. A. Renard, Bull. Soc. Chim. France, 5, 278 (1891).
587. Ger. Patent 332,888 (1921); Chem. Zentralbl., II, 828 (1921).
588. K. Frey, Asphalt und Teer, Strassenbautechnik, 33, 26 (1933); Chem. Zentralbl., I, 2026 (1933).
589. F. Schreiber, Ger. Patent 330,970 (1920); Chem. Zentralbl., II, 529 (1921).
590. R. McNeill, D. Weiss, and D. Willis, Austr. J. Chem., 18, 447 (1965).
591. G. D. Palmer and S. J. Lloyd, J. Am. Chem. Soc., 52, 3388 (1930).
592. G. D. Palmer, S. J. Lloyd, W. M. Lure, and N. LeMeistre, J. Am. Chem. Soc., 62, 1005 (1940).
593. O. A. Cherry, U.S. Patent 1,616,741 (1927); Chem. Zentralbl., II, 497 (1928).
594. M. Bornehgo, Mater. Plast. Elastomeri, 29, 34 (1963).
595. S. I. Bass, I. I. Levantovskaya, G. V. Dralyuk, B. M. Kovarskaya and A. A. Berlin, Vysokomol. Soedin., Ser. A, 9, 556 (1967).
596. E. P. Goodings, D. A. Mitchard, and G. Owen, J. Chem. Soc., Perkin Trans., 1310 (1972).
597. A. Renard, C. R. Acad. Sci. (Paris), 109, 699 (1889).
598. A. Renard, Bull. Soc. Chim. France, 3, 958 (1889).
599. L. Aronstein and A. S. Nierop, Recl. Trav. Chim. Pays-Bas, 21, 448 (1903).
600. A. H. Horton, J. Org. Chem., 14, 761 (1949).
601. R. Mayer, E. Hoffmann, and J. Faust, J. Prakt. Chem., 23, 77 (1964).
602. H. Weyland, H. Hahl, and R. Berenda, Ger. Patent 365,169 (1922); Chem. Zentralbl., II, 600 (1933).
603. H. Weyland, H. Hahl, and R. Berenda, U.S. Patent 1,427,122 (1922); U.S. Patent 1,426,340 (1922).
604. K. S. Revell, L. Seed, and A. J. Shipman, Brit. Patent 971,774 (1962); C. A., 61, 16185 (1964).
605. W. H. Lind and J. B. Wilkes, U.S. Patent 3,041,368 (1962); C. A., 57, 13695 (1962).
606. W. G. Tolans, U.S. Patent 2,695,312 (1954); C. A., 50, 5752 (1956).

607. O. H. Smith, U.S. Patent 1,870,876 (1932); C. A., 26, 5580 (1932).
608. Fr. Patent 693,876 (1930); C. A., 25, 1538 (1931).
609. O. H. Smith, Can. Patent 308,122 (1931); C. A., 25, 1266 (1931).
610. O. H. Smith, Brit. Patent 354,308 (1929).
611. W. A. Gibbson and O. H. Smith, U.S. Patent 1,997,967 (1935); Chem. Zentralbl., I, 1962 (1936).
612. C. R. M. Cullough and W. H. Gehrke, U.S. Patent 2,392,289 (1946); C. A., 40, 1877 (1946).
613. Ger. Patent 600,268 (1934); C. A., 28, 6448 (1934).
614. E. K. Fields, J. Am. Chem. Soc., 77, 4255 (1966).
615. E. K. Fields, U.S. Patent 2,816,075 (1957); C. A., 52, 7681 (1958).
616. Brit. Patent 808,064 (1959); C. A., 54, 595 (1960).
617. M. G. Voronkov and T. V. Lapina, Khim. Geterotsikl. Soedin., 342 (1965).
618. M. G. Voronkov and T. V. Lapina, Khim. Geterotsikl. Soedin., 522 (1966).
619. M. G. Voronkov and T. V. Lapina, Khim. Geterotsikl. Soedin., 592 (1970).
620. M. G. Voronkov, T. V. Lapina, and I. S. Minkina, Khim. Geterotsikl. Soedin., 999 (1971).
621. M. G. Voronkov and B. L. Gol'stein, Zh. Obshch. Khim., 20, 1218 (1949).
622. H. Berger, J. Prakt. Chem., 133, 331 (1932).
623. O. Brünner and F. Grof, Monatsh. Chem., 64, 28, 76 (1934).
624. W. Friedmann, Ber., 49, 1354 (1916).
625. A. S. Broun and M. G. Voronkov, Dokl. Akad. Nauk SSSR, 59, 1293 (1948).
626. M. G. Voronkov and A. S. Broun, Zh. Obshch. Khim., 18, 70 (1948).
627. M. G. Voronkov, A. S. Broun, G. B. Karpenko, and B. L. Gol'stein Zh. Obshch. Khim., 19, 1356 (1949).
628. M. G. Voronkov and A. N. Pereferkovich, Khim. Geterotsikl. Soedin., 1133 (1967).
629. J. H. Ziegler, Ber., 21, 779 (1883).
630. R. Moreau, Bull. Soc. Chim. France, 628, 918, 922, 1049, 1164 (1955).
631. R. Moreau, Action du soufre sur le diphenylmethane, Thesis, Paris (1956).
632. I. Tsurugi, H. Fukuda, and T. Nakabayashi, Nippon Kagaku Zasshi, 76, 111, 190 (1955).
633. I. Tsurugi, T. Nakabayashi, and T. Yamanaka, Nippon Kagaku Zasshi, 77, 578 (1956).
634. I. Tsurugi, Rubber Chem. Technol., 31, 762, 769, 773, 779, 788, 800 (1958).
635. I. Tsurugi, Bull. Univ. Osaka Univ., A5, 161, 169, 173 (1957).
636. I. Tsurugi and H. Fukuda, Bull. Univ. Osaka Pref., A6, 145 (1958).
637. I. Tsurugi and H. Fukuda, Nippon Kagaku Zasshi, 78, 362 (1957).
638. I. Tsurugi and T. Nakabayashi, Bull. Univ. Osaka Pref., A6, 135 (1958).
639. I. Tsurugi and T. Nakabayashi, Kogyo Kagaku Zasshi, 60, 365 (1957).
640. I. Tsurugi, Nippon Kagaku Zasshi, 77, 1716 (1956).

641. L. Szperl and T. Wierusz, Chem. Polski, <u>15</u>, 19 (1918).
642. J. Ziegler, Ber., <u>23</u>, 2472 (1890).
643. W. Schlenk, Liebigs Ann. Chem., <u>394</u>, 178 (1912).
644. I. Tsurugi and T. Nakabayashi, J. Org. Chem., <u>25</u>, 1744 (1960).
645. B. Radziezewski, Ber., <u>8</u>, 758 (1875).
646. L. Szperl, Roczki Chem., <u>6</u>, 728 (1926).
647. L. Szperl and M. Gradzateim, Rocz. Chem., <u>12</u>, 478 (1932).
648. L. Szperl and L. Kowalsky, Chem. Polski, <u>16</u>, 112 (1919).
649. W. F. Short, H. Stromberg, and A. E. Wiles, J. Chem. Soc., 319 (1936).
650. A. Smith, W. B. Holmes, and E. S. Hall, Z. Phys. Chem. (Leipzig) <u>52</u>, 602 (1905).
651. I. I. Yukel'son and V. S. Glukhovskii, Vysokomol. Soedin., Ser. B, <u>10</u>, 19 (1968).
652. E. Baumann and E. Fromm, Ber., <u>24</u>, 1441 (1891).
653. E. Baumann and M. Klett, Ber., <u>24</u>, 3307 (1891).
654. E. Baumann and E. Fromm, Ber., <u>28</u>, 890 (1895).
655. E. Baumann and E. Fromm, Ber., <u>28</u>, 895 (1895).
656. E. Fromm and O. Achert, Ber., <u>36</u>, 534 (1903).
657. E. Baumann and E. Fromm, Ber., <u>30</u>, 110 (1897).
658. A. Michael, Ber., <u>28</u>, 1633 (1895).
659. A. S. Broun and M. G. Voronkov, Dokl. Nauchn. Sessii Leningr. Univ. (1946).
660. A. S. Broun and M. G. Voronkov, Zh. Obshch. Khim., <u>17</u>, 1162 (1947).
661. A. S. Broun and M. G. Voronkov, Nauchn. Byull. Leningr. Univ., <u>20</u>, 6 (1948).
662. A. S. Broun, M. G. Voronkov, and F. N. Gol'dburt, Nauchn. Byull. Leningr. Univ., <u>18</u>, 14 (1947).
663. A. S. Broun, M. G. Voronkov, and R. A. Shlyakhter, Nauchn. Byull. Leningr. Univ., <u>18</u>, 11 (1947).
664. A. S. Broun and M. G. Voronkov, Dokl. Akad. Nauk SSSR, <u>50</u>, 1293 (1948).
665. A. J. Kosak, J. F. Polchak, N. A. Steele, and C. M. Selwitz, J. Am. Chem. Soc., <u>76</u>, 4450 (1954).
666. M. G. Voronkov, A. S. Broun, and G. B. Karpenko, Zh. Obshch. Khim., <u>19</u>, 1927 (1949).
667. K. Adachi and J. Tanaka, J. Chem. Soc. Jpn. Chem. Ind. Chem., 1666 (1978).
668. N. Lozac'h and G. Mollier, Bull. Soc. Chim. France, 1243 (1950).
669. H. Krebs and E. F. Weber, Z. Anorg. Allgem. Chem., <u>272</u>, 288 (1963).
670. H. Staudinger and J. Meyer, Helv. Chim. Acta, <u>2</u>, 635 (1919).
671. A. Schönberg, Ber., <u>58</u>, 1793 (1925).
672. A. Schönberg and W. Asker, J. Chem. Soc., 272 (1942).
673. A. Schönberg and W. Asker, Chem. Rev., <u>37</u>, 1 (1945).
674. A. Schönberg, A. F. Ismail, and W. Asker, J. Chem. Soc., 442 (1946).
675. A. Schönberg and S. Nickel, Ber., <u>64</u>, 2323 (1931).
676. H. Cortw, Dissertation, Berlin (1934).
677. K. Glew and R. Schaarschnidt, Ber., <u>72</u>, 1246 (1939).
678. E. Bergmann and L. Engel, Z. Phys. Chem., <u>138</u>, 111 (1930).
679. W. Schlenk, Liebigs Ann. Chem., <u>394</u>, 178 (1912).

4
Organic Halides

Very few studies were made on the interaction of sulfur with organic halogen-containing compounds until the second half of the twentieth century. A few isolated papers were devoted to the reaction of sulfur with carbon tetrachloride and tetrabromide [2,3, 5]. There were also several patents concerning the preparation of synthetic resins by the sulfuration of halogenated aromatic hydrocarbons [8,9,91,151-153]. Furthermore, a number of papers were devoted to studies of the solubility of sulfur and sulfur solutions in various organic halides [92,154-164].

However, quite a few investigations have been made in recent years. These studies have demonstrated numerous possible applications of this reaction for the synthesis of new sulfur-containing compounds, some of which are important from a practical point of view, e.g., fungicides, insecticides [163], corrosion inhibitors [33,151], monomers for the synthesis of new types of polymers [46], and sulfur-containing dyes [92]. The wide-ranging applicability of the reaction of sulfur with organic halides was shown for the first time by Voronkov and co-workers in a series of papers published from 1964 onward [110,111,117,120,125,129,132,133,135,137-139].

The reaction of sulfur with halogenated hydrocarbons can proceed in five principal directions:

1. Catalytic dehydrohalogenation with the formation of the corresponding hydrogen halide and an unsaturated compound.
2. Removal of two geminal or vicinal halogen atoms (X) (with the formation of a carbon—carbon or carbon—sulfur double bond, respectively); the sulfur is converted to S_2X_2 or SX_2.

3. Insertion of one or several sulfur atoms into the C—X bond
    with formation of sulfenyl or polythiosulfenyl halides.
4. Formation of linear sulfides and polysulfides.
5. Cyclization leading to sulfur-containing heterocycles.

The reactions mentioned in directions 4 and 5 are accompanied by the
evolution of HX, H_2S, S_2X_2, or even of free halogen (when X = I).
These reactions can occur as consecutive or parallel processes. In
the above-mentioned cases the sulfur has to be thermally, photo-
chemically, or catalytically activated.

4.1. ALKYL HALIDES

The principal products formed in the reaction between sulfur
and carbon tetrachloride at 220°C are carbon disulfide and disulfur
dichloride [1]:

$$6S + CCl_4 \rightleftharpoons CS_2 + 2S_2Cl_2 \tag{4.1}$$

Traces of thiophosgene, trichloromethanesulfenyl chloride, and
bis-trichloromethyl polysulfides, which are obviously intermediates
in reaction (4.1) are also formed. Reactions (4.2)-(4.7) have been
carried out experimentally [1]:

$$CCl_4 + S \longrightarrow CCl_3SCl \tag{4.2}$$

$$2CCl_3SCl + 2S \longrightarrow CCl_3SSCCl_3 + S_2Cl_2 \tag{4.3}$$

$$CCl_3SSCCl_3 + S \longrightarrow CCl_3SSSCCl_3 \tag{4.4}$$

$$CCl_3SSSCCl_3 \longrightarrow CCl_3SCl + Cl_2C{=}S + S \tag{4.5}$$

$$Cl_2C{=}S + S \longrightarrow Cl(ClS)C{=}S \tag{4.6}$$

$$Cl(ClS)C{=}S + 2S \longrightarrow CS_2 + S_2Cl_2 \tag{4.7}$$

In the presence of catalysts (Al, Fe, Cu and their chlorides),
reaction (4.1) takes place at a lower temperature, i.e., 120°C [2].
Powdered reduced iron seems to be the best catalyst. Introduction
of iodine into the reaction mixture facilitates the formation of
thiophosgene [3], whose yield in reaction (4.1) is negligible both
in the presence and in the absence of iron.

The reaction of chloroform with sulfur and alkali fluoride
(KF) in a polar aprotic solvent (sulfolane) at 120–180°C affords
bis-trifluoromethyl disulfide [4]:

$$2Cl_3CH + 3S + 6KF \rightarrow F_3C{-}S{-}S{-}CF_3 + H_2S + 6KCl \tag{4.8}$$

The interaction of sulfur with carbon tetrabromide leads to a complex mixture of sulfur-containing products. The compounds $C_9Br_4S_4$

and $C_2S_3Br_6$ have been isolated from this mixture, as well as carbon disulfide [5].

Halofluoromethanes of the type F_2CHX and F_2CX_2 (X = Cl, Br) react with sulfur vapor at 500-900°C to give a 90% yield of thiocarbonyl difluoride [6]:

$$2F_2CHX+S \longrightarrow F_2C{=}S+HX \qquad (4.9)$$

$$F_2CX_2+S \longrightarrow F_2C{=}S+SX_2 \qquad (4.10)$$

Trifluoroiodomethane and sulfur interact at 260-270°C under pressure to give a 30% yield of bis-trifluoromethyl disulfide [7]:

$$CF_3I+2S \longrightarrow CF_3SSCF_3+I_2 \qquad (4.11)$$

The polysulfides $(F_3C)_2S_3$ and $(F_3C)_2S_4$ and thiocarbonyl difluoride are also formed in small amounts.

The action of sulfur upon the dihalomethanes and dihaloethanes gives rubber-like polymers [8,9].

1,2-Dichloro-1,1,2-trifluoro-2-iodoethane and pentafluoroiodoethane react with sulfur under pressure to yield the corresponding di- or polysulfides [10,11]:

$$CF_2ClCFCl{-}I \xrightarrow{+nS} (CF_2ClCFCl)_2\,S_n+I_2$$

$$2CF_3CF_2I \xrightarrow{+2S} CF_3CF_2{-}S{-}S{-}CF_2CF_3 + I_2 \qquad (4.12)$$

Upon photochemical activation (λ = 253.7 nm), 1-bromopropane reacts with sulfur at 20°C to give Br_2, HBr, C_3H_6, C_3H_7, S_2, and S_6. Disulfur dibromide is also formed in secondary reactions, with a quantum yield of 0.16 [12].

1,1,1,2,3,3,3-Heptachloropropane reacts with sulfur in the presence of aluminium chloride at 160°C to give a 3,4,5-trichloro-1,2-dithiolium salt [13]:

$$CCl_3CHClCCl_3 \xrightarrow[-HCl]{+S} CCl_3CCl{=}CCl_2 \xrightarrow{+4S}_{AlCl_3}$$

The principal products of the reaction between heptafluoro-1-iodopropane and sulfur in a sealed tube at 250°C are the corresponding di- and trisulfides together with a small amount of the monosulfide and iodine [14]:

$$C_3F_7I \xrightarrow{+6S} C_3F_7SC_3F_7 + C_3F_7S_2C_3F_7 + C_3F_7S_3C_3F_7 + I_2 \qquad (4.14)$$

Results analogous to the above reaction have been obtained from the interaction of sulfur with fluorinated iodobutanes, iodopentanes, and dodecafluoro-1-iodoheptane at 200-270°C under pressure [15-17]:

$$2RI \xrightarrow{+S} R-(S)_n-R+I_2 \qquad (4.15)$$

$R=C_3H_7CH_2CF_2$, $(CF_3)_2CFCH_2CF_2$, $CF_3CFClCH_2CF_2$, $H(CF_2)_6CH_2$

$n = 2, 3\text{-}5$

The reaction of sulfur with polychlorobutanes at 210-220°C gives a 70-72% yield of tetrachlorothiophene [18]:

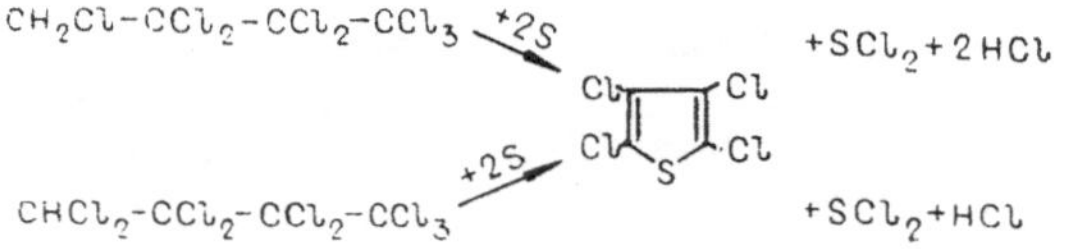

(4.16)

The reaction of sulfur with octafluoro-1,4,-diiodobutane at 250°C leads to the formation of octafluorothiolane in high yield [19]:

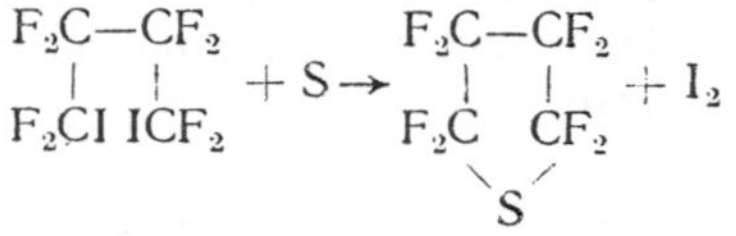

(4.17)

Alkylated phenylacetonitriles (1) react to some extent with elemental sulfur and alkyl halides in the presence of 50% aqueous sodium hydroxide solution and benzyl(triethyl)ammonium chloride (TEBA) [20]:

$$\underset{\mathbf{1}}{PhC(R')HCN} + S + R^2X \xrightarrow[\text{TEBA, DMSO}]{\text{50\% NaOH}} \underset{\mathbf{2}}{PhC(R')(SR'')CN} + \ldots \qquad (4.18)$$

$R' = CH_3$, CH_2CHCH_2, $i\text{-}C_3H_7$, C_6H_5;
$R'' = CH_3$, $n\text{-}C_4H_9$, $X = Cl, Br$

Peculiar transformations take place under these conditions when phenylacetonitrile (3) reacts with sulfur and ethylene dibromide: the desired 2-phenyl-1,3-dithiolane-2-carbonitrile (4) is accompanied *inter alia*, by *trans*-dinitrile (5):

$$PhCH_2CN + S + BrCH_2CH_2Br \xrightarrow[\text{TEBA, DMSO}]{\text{50\% NaOH}}$$

$$\longrightarrow \quad \underset{\underset{\text{4 (8\%)}}{}}{\left[\begin{array}{c}\text{Ph} \quad\quad \text{CN} \\ \diagdown\diagup \\ \text{S} \quad\quad\quad \text{S}\end{array}\right]} \quad + \quad \underset{\underset{\text{5 (14\%)}}{}}{\left[\begin{array}{c}\text{Ph} \quad\quad \text{CN} \\ \diagup\diagdown \\ \text{CN} \quad\quad \text{Ph}\end{array}\right]} \quad + \ldots \qquad (4.19)$$

In the reaction between nitriles (1), sulfur, and alkyl ha-
polysulfides $R''S_nR''$ (mainly n = 2) were formed. Polysulfides (6)
were obtained directly in the reactions of alkyl halides with the
sulfur—sodium hydroxide—dimethyl sulfoxide reagent. The yields of
lides, polysulfides $R''S_nR''$ (mainly n = 2) were formed. Polysulfides
(6) were obtained directly in the reactions of alkyl halides with the

$$RX + S \xrightarrow[\text{DMSO}]{\text{Base}} \underset{\underset{6}{}}{R-S_n-R} \qquad (4.20)$$

Base-activated elemental sulfur in a dipolar aprotic solvent
(DMSO, TEBA, DMF) under mild conditions reacts with haloalkanes,
having electron-withdrawing substituents in the λ-position. Chloro-
acetonitriles react with elemental sulfur in the presence of tertiary amines at room temperature to form cyanodithioformic ammonium
salts. The alkylation of the latter directly in the reaction mixture
leads to esters of cyanodithioformic acid [21,22]:

$$NC-CR_2Cl \xrightarrow{S/R_3N} \underset{\underset{S}{\overset{\|}{}}}{NC-C-S^-} \xrightarrow{+ RX} \underset{\underset{S}{\overset{\|}{}}}{NC-C-SR} \xrightarrow{+ NS-\overset{\overset{\text{S}}{\|}}{C}-SR}$$

$$\longrightarrow NC(RS)C=C(SR)CN \qquad (4.21)$$

$$60\% \ (Z + E)$$

Base-catalyzed sulfuration of various haloacetic acid derivatives by elemental sulfur followed by alkylation leads to 1,1-dithio-
oxalates [23-25]. The reaction proceeds at room temperature without
pressure in dipolar aprotic solvents:

$$\underset{\underset{O}{\overset{\|}{}}}{R-C-CH_2X} \xrightarrow{\text{1) } S_8/\text{Base; 2) R'Y}} \underset{\underset{O}{\overset{\|}{}}}{R-C-C}\diagup^{S}_{\diagdown SR'} \qquad (4.22)$$

$$X = Cl, Br; \ Y = Br, I$$

The reaction of 2-chlorooctane with sulfur in an aqueous
medium at $300°C$ for fifteen minutes gives 2-octanone. This reaction
is promoted by ammonia and hydrogen sulfides [26].

The interaction of polyvinyl chloride with sulfur and benzyl-
mercaptan in a dimethylformamide solution yields a polymer cross-
linked by polysulfide bridges S_n (n = 2-4). The sulfur content of
the polymer increases with temperature and reaction time [27].

Heptafluoroisopropoxypolyfluoroiodoalkanes of the type shown below react with sulfur to form the corresponding disulfides [28]:

$$2(CF_3)_2CFO(CF_2)_m(CH_2)_nI + S \rightarrow$$

$$\rightarrow [(CF_3)_2CFO(CF_2)_m(CH_2)_n]_2S_2 + I_2 \qquad (4.23)$$

4.2. HALOSUBSTITUTED ALKENES AND ALKADIENES

The photochemical reaction of atomic sulfur with vinyl chloride gives the corresponding thiirane and a small amount of 2-chloro-ethylenethiol [29]:

$$CH_2{=}CHCl + S\,(^1D) \rightarrow \underset{\underset{S}{\diagdown\diagup}}{CH{-}CHCl} + CHCl{=}CHSH \qquad (4.24)$$

(47%)

The reaction of singlet-state atomic sulfur $S(^1D)$ with trans-1,2-dichloroethylene gives polymeric products. The photolysis of carbonyl sulfide in the presence of carbon dioxide generates triplet-state atomic sulfur $S(^3P)$, and interaction with cis- or trans-1,2-dichloroethylene gives the corresponding cis- or trans-disubstituted thiirane [29]:

$$HCCl{=}CHCl + S\,(^3P) \rightarrow \underset{\underset{S}{\diagdown\diagup}}{HCCl{-}CHCl} \qquad (4.25)$$

Trichloro- and tetrachloroethylene on heating with sulfur (250°C) under 35-52 atmospheres pressure give a 15-20% yield of tetrachlorothiophene [30]:

$$2Cl_2C{=}CCl_2 + 5S \rightarrow \underset{Cl \quad S \quad Cl}{\overset{Cl \qquad\quad Cl}{\boxed{}}} + 2S_2Cl_2$$

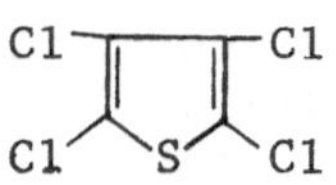

The interaction of 2-chloropropene with sulfur vapor at 550°C gives 4-chloro-1,2-dithiol-3-thione [31]:

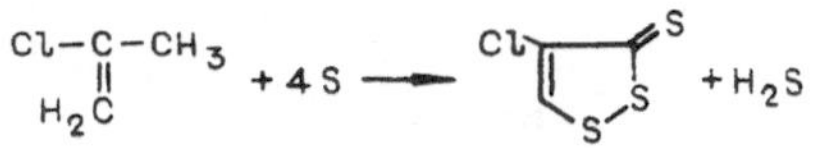

$$\underset{H_2C}{\overset{Cl-C-CH_3}{\|}} + 4S \longrightarrow \underset{S\diagdown S}{\overset{Cl \quad\quad S}{\boxed{}}} + H_2S \qquad (4.26)$$

Polymerization of chloroprene in the presence of sulfur and tetraethylthiuram disulfide leads to the formation of sulfur-modified polychloroprene rubber with good physicomechanical properties [32]. Hexachloropropene interacts with sulfur at 160°C and forms 3,4,5-trichloro-1,2-dithiolium chloride by loss of chlorine atoms from positions 1 and 3 [13,33,34]. The temperature can be lowered in the presence of anhydrous aluminium chloride. Small amounts of hexa-cloroethane and carbon tetrachloride are obtained as by-products:

$$Cl-C-CCl_3 \atop \underset{Cl_2C}{\|} + 3S \rightarrow \underset{Cl}{\overset{Cl}{\bigcirc}} \overset{Cl}{\underset{S}{}}_{S} + S_2Cl_2 \qquad (4.27)$$

$$Cl^-$$

(65%)

1,1,2-Trichloro-3,3,3-trifluoropropene and 1,1,2,3-tetrachloro-3,3-difluoropropene when heated with sulfur in the presence of aluminium chloride react in a similar manner to 1,1,1,2,3,3,3-heptachloropropene and give 3,4,5-trichloro-1,2-dithiolium tetrachloroaluminate:

$$\left.\begin{array}{l} CF_3CCl=CCl_2 \\ CF_2ClCCl=CCl_2 \end{array}\right] \xrightarrow{AlCl_3} Cl-C-CCl_3 \xrightarrow[AlCl_3]{+4S}$$

$$\underset{Cl\cdot C-Cl}{\overset{\|}{}}$$

$$\rightarrow \underset{Cl}{\overset{Cl}{\bigcirc}}\overset{Cl}{\underset{S}{}}_S + S_2Cl_2 \qquad (4.28)$$

$$AlCl_4^-$$

In both cases, the fluorine atoms are replaced by chlorine atoms so that hexachloropropene is the common intermediate [35].

The reaction of sulfur with chloroprene in the presence of a catalyst gives a polymer containing 0.25-2.00% sulfur [36].

The reaction of sulfur with 1,1,2,3,4-pentachloro-1,3-butadiene and with hexa- and heptachlorobutenes occurs at 210-240°C and gives tetrachlorothiophene [18]:

$$\left.\begin{array}{l} HCCl=CCl-CCl=CCl_2 \\ ClH_2ClCCl=CClCCl_3 \\ ClHC=CClCCl_2CCl_3 \\ Cl_2HCCCl=CClCCl_3 \end{array}\right] \xrightarrow{S} \underset{S}{\overset{Cl}{\bigcirc}}\overset{Cl}{} Cl + S_2Cl_2 + HCl \qquad (4.29)$$

Octachloro-1-butene and hexachloro-1,3-butadiene also form tetrachlorothiophene in high yield upon heating with sulfur [37]:

$$\left.\begin{array}{l} Cl_2C=CClCCl=CCl_2 \\ Cl_2C=CClCCl_2CCl_3 \end{array}\right] \xrightarrow{S} \underset{S}{\overset{Cl}{\bigcirc}}\overset{Cl}{} Cl + S_2Cl_2 \qquad (4.30)$$

2,2-Dichloro-1,1-difluoroethylene interacts with sulfur vapor at 450°C to form the corresponding thiirane and thioacyl chloride [38]:

$$CF_2=CCl_2 \xrightarrow{S} F_2\underset{\underset{S}{\diagdown\diagup}}{C-CCl_2} \rightarrow ClF_2C-C\underset{S}{\overset{Cl}{\diagup}} \qquad (4.31)$$

The photochemical reaction of 1,1-difluoroethylene with atomic sulfur gives analogous compounds [39]:

$$F_2C{=}CH_2 + S\,(^1D) \rightarrow F_2C\underset{\diagdown\diagup}{\overset{\hspace{1.5em}}{-}}CH_2 + F_2C{=}CHSH \qquad (4.32)$$
$$S$$

In acetone, the photochemical reaction of 2,2-dichloro-1,1-difluoro-ethylene and of chlorotrifluoroethylene with atomic sulfur at 72°C gives a mixture of β-thiolactones [39,40]:

$$CF_2{=}CClX \xrightarrow{\text{COS}+(CH_3)_2CO} \underset{\substack{|\qquad| \\ S{-}\!\!-\!\!C{=}O \\ (60\%)}}{CF_2{-}C{-}ClX} + \underset{\substack{| \qquad | \\ S{-}\!\!-\!\!-\!\!C{=}O \\ (2{-}5\%)}}{CClX{-}CF_2} \qquad (4.33)$$
$$X{=}Cl,F$$

The reaction of sulfur with chloro- and bromotrifluoroethylene at 450°C in the presence of activated charcoal leads to the corresponding halodifluorothioacetyl fluoride [41,42]:

$$CF_2{=}CFX \xrightarrow{S} XCF_2 C{\overset{\textstyle F}{\underset{\textstyle S}{\big\langle}}} \qquad (4.34)$$
$$X{=}Cl,\ Br$$

Under analogous conditions, sulfur interacts with trifluoro-ethylene with the formation of difluorothioacetyl fluoride [41].

Tetrafluoroethylene reacts with sulfur at 445°C to give tetra-fluorothiirane and higher cyclic polysulfides according to the reaction [43–45]:

$$CF_2{=}CF_2 + \cdot S\cdot_n \rightarrow \cdot S_nCF_2CF_2\cdot \rightarrow CF_2\underset{\diagdown\diagup}{\overset{\hspace{1.5em}}{-}}CF_2 + S_{n-m} \qquad (4.35)$$
$$S_m$$
$$n{=}8,\ m{=}1{-}4$$

The reaction of tetrafluoroethylene with sulfur vapor at 500–600°C in the absence of a catalyst leads to a mixture containing thiocarbonyl difluoride ($\underset{\sim}{7}$), trifluorothioacetyl fluoride ($\underset{\sim}{8}$), and bis-trifluoromethyl disulfide ($\underset{\sim}{9}$) [38]. In the presence of platinum, the reaction of tetrafluoroethylene with sulfur only gives ($\underset{\sim}{7}$) and ($\underset{\sim}{8}$) [46]. In the presence of activated charcoal at 450-500°C, the main product of the reaction of tetrafluoroethylene with sulfur is trifluorothioacetyl fluoride ($\underset{\sim}{8}$) [41,42]:

$$CF_2{=}CF_2 \xrightarrow{S} F_2C{=}S + CF_3 C{\overset{\textstyle F}{\underset{\textstyle S}{\big\langle}}} + CF_3S{-}SCF_3 \qquad (4.36)$$
$$\underset{\sim}{7} \qquad\qquad \underset{\sim}{8} \qquad\qquad \underset{\sim}{9}$$

The interaction of sulfur with tetrafluoroethylene at 250-300°C in the presence of iodine leads to the formation of octafluorothiolane and octafluoro-1,4-dithiane (10) [47,48]. At 300°C, fluorine migrates and (10) isomerizes to pentafluoro-2-trifluoromethyl-1,3-dithiolane (11):

$$\cdot S \cdot _n + CF_2 = CF_2 \rightarrow \cdot S_n CF_2 CF_2 \cdot \xrightarrow{CF_2=CF_2}$$

$$\rightarrow \cdot S_n CF_2 CF_2 CF_2 CF_2 \cdot \rightarrow \begin{array}{c} F_2C - CF_2 \\ | \quad | \\ F_2C \quad CF_2 \\ \diagdown \diagup \\ S \end{array} + S_{n+1}$$

(15%)

(4.37)

$$2 \cdot S_n^{\cdot} + 2 F_2 C = CF_2 \longrightarrow 2 \cdot S_n CF_2 CF_2^{\cdot} \longrightarrow \begin{array}{c} F_2C \diagup S \diagdown CF_2 \\ | \qquad | \\ F_2C \diagdown S \diagup CF_2 \end{array} + \cdot S_{n+2}^{\cdot} + \begin{array}{c} F_2 \diagup S \diagdown F \\ \diagdown \diagdown \diagup \diagdown \\ F_2 \diagdown S \diagup CF_3 \end{array}$$

10 (60%)   11

In the presence of antimony pentafluoride at 20°C, the interaction of tetrafluoroethylene with sulfur leads to trifluoromethyl pentafluoroethyl disulfide and bis-pentafluoroethyl disulfide [49]. Fluorothiocarbonyl is prepared by treating tetrafluoroethylene or chlorotrifluoroethylene with sulfur in carbon disulfide at 400°C [50]. The reaction of tetrafluoroethylene with sulfur in the presence of carbon disulfide or iodine carried out under pressure proceeds as follows [51]:

$$CF_2 = CF_2 + 3S \xrightarrow[250-350°]{CS_2} S \begin{array}{c} \diagup CF_2 - CF_2 \diagdown \\ \diagdown \diagdown S \diagup \end{array} S$$

$$nCF_2 = CF_2 + S \xrightarrow[125-200°]{I_2} -[-(CF_2-CF_2)_n - S_m-]- \qquad (4.38)$$

Tetrafluoroethylene and sulfur react in carbon disulfide under pressure with various unsaturated compounds [48,52]:

$$CF_2 = CF_2 + RCH = CH_2 + S \rightarrow \begin{array}{c} F_2C - CH_2 \\ | \qquad | \\ F_2C \quad CH-R \\ \diagdown \diagup \\ S \end{array} \qquad (4.39)$$

$$R = C_6H_5, CF_3, CH_2OH, OC_2H_5, OCOCH_3$$
$$COOCH_3, CN, Br, Cl$$

$$CF_2=CF_2 + RCH=CR'R'' + S \rightarrow \begin{array}{c} F_2C-CHR \\ | \quad | \\ F_2C \quad CR'R'' \\ \diagdown S \diagup \end{array} \qquad (4.40)$$

R, R', R″ = H, CH₃ etc.

$$CF_2=CF_2 + \begin{array}{c} CH_2 \\ \diagup \diagdown \\ HC \quad CH_2 \\ \| \quad | \\ HC \quad CH_2 \\ \diagdown \diagup \\ CH_2 \end{array} + S \rightarrow \begin{array}{c} CH_2 \\ \diagup \diagdown \\ CF_2-CH \quad CH_2 \\ | \quad | \quad | \\ CF_2 \quad CH \quad CH_2 \\ \diagdown S \diagup \diagdown CH_2 \end{array} \qquad (4.41)$$

$$CF_2=CF_2 + \begin{array}{c} CH=CH \\ | \quad | \\ O=C \quad C=O \\ \diagdown O \diagup \end{array} + S \rightarrow \begin{array}{c} CF_2-CH-C{\diagup\!\!\!\!=O}{\diagdown O} \\ | \quad | \quad \\ CF_2-CH-C{\diagup\!\!\!\!=O}{\diagdown O} \end{array} \qquad (4.42)$$

Under analogous conditions, tetrafluoroethylene and sulfur react with aromatic compounds. Thus, the reaction with benzene gives two isomeric tricyclic compounds [48,52]:

$$2\,CF_2=CF_2 + \bigcirc + S \longrightarrow \qquad \underset{12}{} + \underset{13}{} \qquad (4.43)$$

Thiophene reacts similarly:

$$2\,CF_2=CF_2 + \bigcirc + S \longrightarrow \qquad (4.44)$$

Under the same conditions, naphthalene adds only one molecule of tetrafluoroethylene and one sulfur atom:

$$CF_2=CF_2 + \bigcirc\!\!\bigcirc + S \longrightarrow \qquad (4.45)$$

2-Fluoropropene reacts with sulfur at 550°C to give 4-fluoro-1,2-dithiol-3-thione [31]:

$$\begin{array}{c} F-C-CH_3 \\ \| \\ H_2C \end{array} + 5S \rightarrow \begin{array}{c} FC-C=S \\ \| \quad | \\ HC \quad S \\ \diagdown S \diagup \end{array} + 2H_2S \qquad (4.46)$$

The reaction of sulfur with hexafluoropropene at 300°C leads to the formation of 1,4-dithiane derivatives in high yields [53]:

$$CF_3CF{=}CF_2 + 4S \rightarrow \begin{array}{c} S \\ F_2C \quad CFCF_3 \\ | \quad | \\ CF_3FC \quad CF_2 \\ S \end{array} + \begin{array}{c} S \\ CF_3FC \quad CF_2 \\ | \quad | \\ CF_3FC \quad CF_2 \\ S \end{array} \qquad (4.47)$$

In the gas phase at 425°C, hexafluoropropene reacts with sulfur in the presence of activated charcoal with the formation of hexafluoro-thioacetone and its dimer [41,42,54]. Hexafluorothioacetone is the principal reaction product when the reaction is carried out at 130°C under pressure in nitrobenzene [55-57]:

$$3CF_3CF{=}CF_2 + S \rightarrow \left[\begin{array}{c} CF_3CF{-}CF_2 \\ S \end{array} \right] \rightarrow \begin{array}{c} CF_3 \\ C{=}S \\ CF_3 \end{array} + \begin{array}{c} CF_3 \quad S \quad CF_3 \\ C \quad C \\ CF_3 \quad S \quad CF_3 \end{array} \qquad (4.48)$$

(76%)

2-Trifluoromethylpropene reacts with sulfur at 550°C to afford 4-trifluoromethyl-1,2-dithiol-3-thione [31]:

$$\begin{array}{c} F_3C{-}C{-}CH_3 \\ \| \\ H_2C \end{array} + 5S \rightarrow \begin{array}{c} CF_3{-}C{-}C{=}S \\ \| \quad | \\ HC \quad S \\ S \end{array} + 2H_2S \qquad (4.49)$$

Octafluoroisobutylene reacts with sulfur in tetrahydrothiophene S,S-dioxide in the presence of potassium fluoride at 80°C and under pressure to give a mixture of two sulfuration products [56]:

$$(CF_3)_2C{=}CF_2 + S \rightarrow (CF_3)_3\,CSSSC\,(CF_3)_3$$

$$+ (CF_3)_2\,C{=}C \begin{array}{c} S \\ C{=}C\,(CF_3)_2 \\ S \end{array} \qquad (4.50)$$

(15%)

The corresponding 2,2,5,5-tetrafluoro-3-thiolenes are formed by the reaction of sulfur with hexafluorocyclobutene and 1,2-dichlorotetra-fluorocyclobutene [58]:

$$\begin{array}{c} FC{=}CF \\ \| \quad | \\ F_2C{-}CF_2 \end{array} + S \rightarrow \begin{array}{c} FC{=}CF \\ | \quad | \\ F_2C \quad CF_2 \\ S \end{array} \qquad \begin{array}{c} ClC{=}CCl \\ | \quad | \\ F_2C{-}CF_2 \end{array} + S \rightarrow \begin{array}{c} ClC{=}CCl \\ | \quad | \\ F_2C \quad CF_2 \\ S \end{array} \qquad (4.51)$$

The sulfuration of hexafluorocyclobutene at 350-450°C in the presence of activated charcoal gives hexafluorothiolane-2-thione in very low yield (~ 5%) [41]:

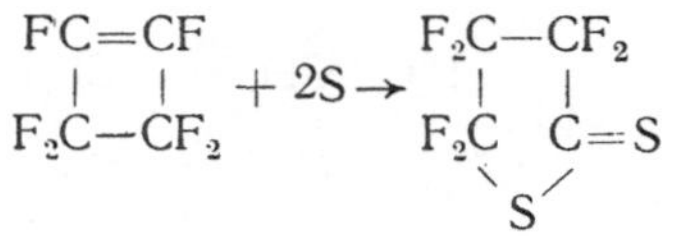

$$
\begin{array}{c}
\text{FC=CF} \\
| \quad | \\
\text{F}_2\text{C—CF}_2
\end{array}
+ 2S \rightarrow
\begin{array}{c}
\text{F}_2\text{C—CF}_2 \\
| \quad \quad | \\
\text{F}_2\text{C} \quad \text{C=S} \\
\diagdown \, \diagup \\
\text{S}
\end{array}
\tag{4.52}
$$

Under pressure and in the presence of potassium fluoride, hexafluorocyclobutene reacts with sulfur at 120°C in a different manner, giving mainly the corresponding 1,2,4,5-tetrathiane, as well as bis-heptafluorocyclobutyl disulfide and bis-pentafluorocyclobuten-1-yl sulfide [56,57]:

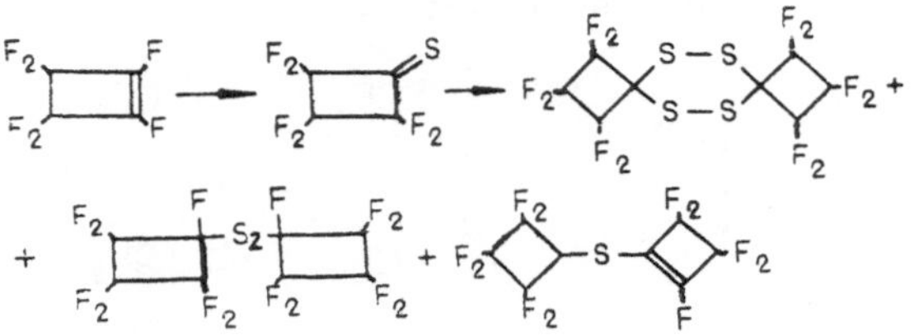

$$\tag{4.53}$$

Perfluoro-4,4-diiodo-1-butene reacts with sulfur at 500°C to give perfluoro-3-butenthioyl fluoride [38]:

$$
\text{CF}_2\text{=CFCF}_2\text{CFI}_2 \xrightarrow{\text{S}} \text{CF}_2\text{=CFCF}_2\text{C}\begin{array}{c}\diagup\text{F}\\\diagdown\!\!\diagdown\text{S}\end{array} + \text{I}_2
\tag{4.54}
$$

Heating octafluoro-1,4-pentadiene with sulfur at 300°C yields octafluoro-3,4-dithiabicyclo[4.1.0]heptane [59]:

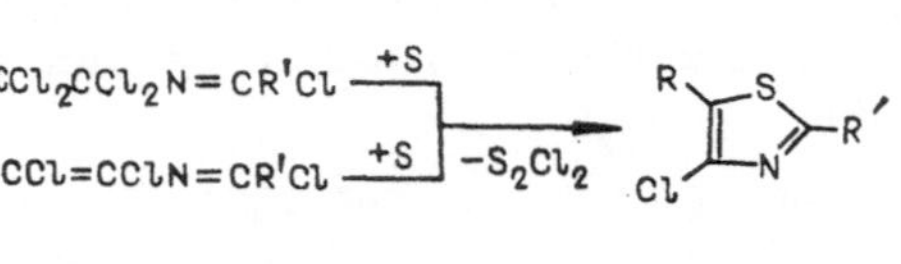

$$
\begin{array}{c}
\text{CF=CF}_2 \\
\diagup \\
\text{F}_2\text{C} \\
\diagdown \\
\text{CF=CF}_2
\end{array}
+ 2S \rightarrow
\begin{array}{c}
\text{CF—CF}_2\text{—S} \\
\diagup \quad \quad \quad | \\
\text{F}_2\text{C} \quad \quad \quad | \\
\diagdown \quad \quad \quad | \\
\text{CF—CF}_2\text{—S}
\end{array}
\tag{4.55}
$$

The reaction of aldimine polychloro-derivatives with sulfur at 200-260°C gives the corresponding thiazole derivatives in 24-91% yields [60].

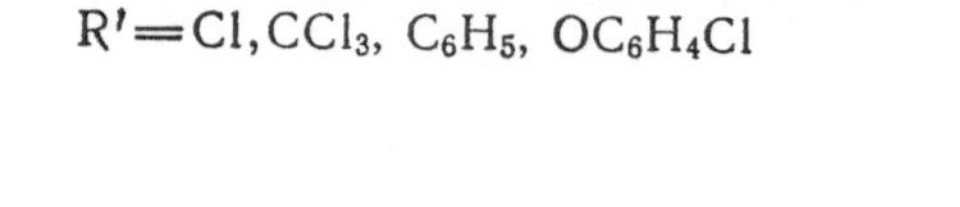

$$\tag{4.56}$$

R = Cl, CN, CCl$_3$, N=CCl$_2$

R' = Cl, CCl$_3$, C$_6$H$_5$, OC$_6$H$_4$Cl

On heating sulfur (220-240°C) with polychloroaldimines of the
structure below, thiazolines are formed [61]. The corresponding
thiazoles are likely to be the intermediates:

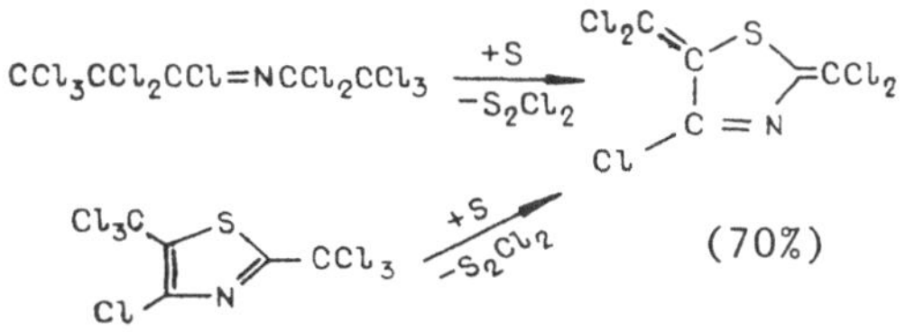

$$CCl_3CCl_2CCl=NCCl_2CCl_3 \xrightarrow[-S_2Cl_2]{+S}$$

(4.57)

(70%)

The reaction of nitrogen-containing octachloro-derivatives
with sulfur at 245-250°C affords chlorothiazolethiazoles [62]:

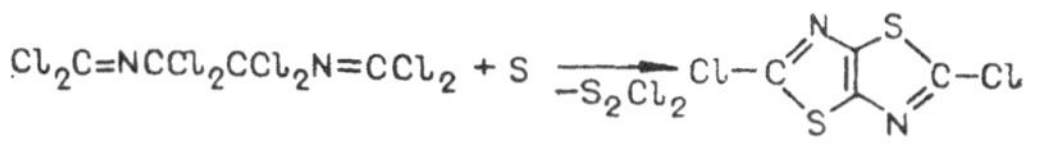

$$Cl_2C=NCCl_2CCl_2N=CCl_2 + S \xrightarrow[-S_2Cl_2]{}$$

(4.58)

Isothiazoles are formed on heating di- or trichloroalkenyl-
nitriles with sulfur [62a]:

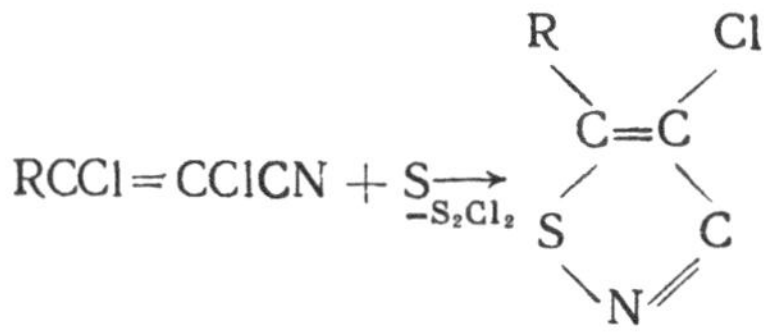

$$RCCl=CClCN + S \xrightarrow[-S_2Cl_2]{}$$

(4.59)

R=Cl, CN

The compounds formed according to schemes (4.56)-(4.59) have
been patented as insecticides and herbicides.

4.3. HALOSUBSTITUTED ALKYNES

Polyfluoro-derivatives of nonterminal acetylenes react with
sulfur at 150-475°C in the presence of iodine at atmospheric or
increased pressure (up to 5000 atm) to give derivatives of 1,2-
dithiet, thiophene, and condensed cyclic compounds [63-65]:

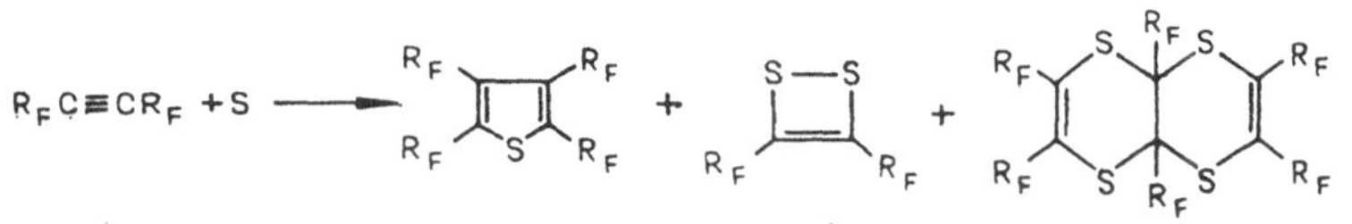

$$R_FC \equiv CR_F + S \longrightarrow$$

(4.60)

14

$$R_F = C_nHF_{2n}, \quad C_nF_{2n}Cl, \quad C_nF_{2n+1}; \quad n=1-12$$

At room temperature, compound (14) dimerizes to yield (15).

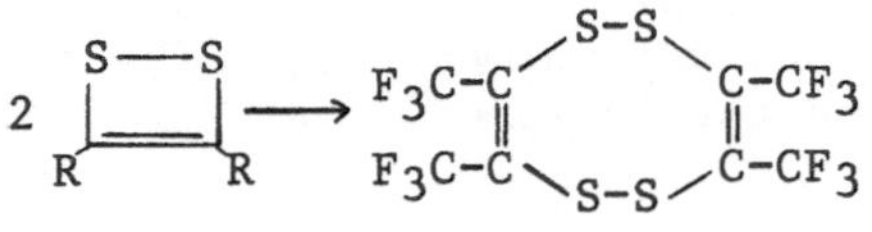

$$2 \quad \underset{R \quad R}{\overset{S—S}{\left|\underset{\|}{\quad}\right|}} \longrightarrow \underset{F_3C-C \underset{S-S}{\diagdown} C-CF_3}{\overset{S-S}{F_3C-C \diagup \overset{\|}{\diagdown} C-CF_3}} \tag{4.61}$$

15

The interaction of hexafluoro-2-butyne with sulfur and carbon disulfide at 200°C gives the cyclic trithiocarbonate (16) and the related compound (17) [66]:

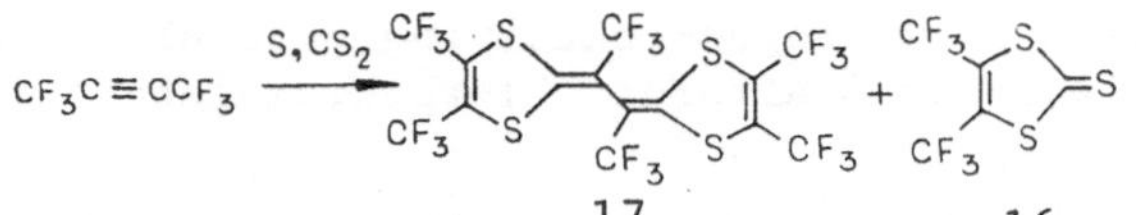

$$CF_3C \equiv CCF_3 \xrightarrow{S,CS_2} \underset{CF_3}{\overset{CF_3}{\diagup}} S \diagdown C = C \diagdown \underset{CF_3}{\overset{CF_3}{S}} + \underset{CF_3}{\overset{CF_3}{\diagup}} S \diagdown S = S \tag{4.62}$$

17   16

In the presence of catalysts capable of generating fluoride ions (CsF), sulfur reacts with perfluoroalkynes via an intermediate enethiolate anion with the formation of tetrasubstituted thiophenes similar to that shown below [67]:

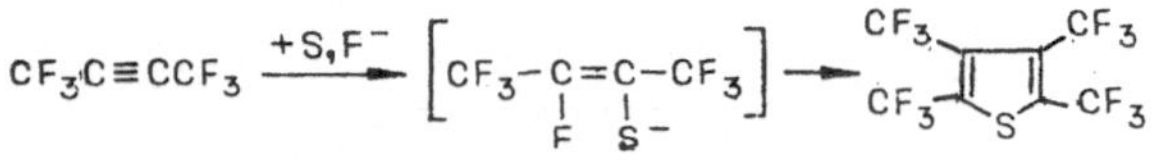

$$CF_3C \equiv CCF_3 \xrightarrow{+S,F^-} \left[\underset{F}{\overset{}{CF_3-C}} = \underset{S^-}{\overset{}{C-CF_3}} \right] \longrightarrow \underset{CF_3 \overset{}{\diagdown}_S \diagup CF_3}{\overset{CF_3 \quad CF_3}{\boxed{}}} \tag{4.63}$$

Atomic sulfur $S(^1D)$ generated by photolysis of carbonyl sulfide reacts with hexafluoro-2-butyne to give tetrafluorothiophene (yields up to 50%) and the corresponding thiirene [68].

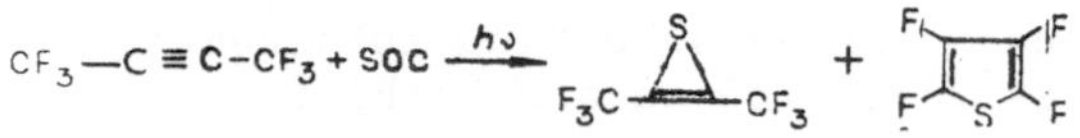

$$CF_3-C \equiv C-CF_3 + SOC \xrightarrow{h\nu} \underset{F_3C \diagdown CF_3}{\overset{S}{\triangle}} + \underset{F \overset{}{\diagdown}_S \diagup F}{\overset{F \quad F}{\boxed{}}} \tag{4.64}$$

The reaction of atomic sulfur with a mixture of acetylene and hexafluoro-2-butyne gives only 2,3-bis-trifluoromethylthiophene; thioketene seems to be an intermediate in this reaction [68]:

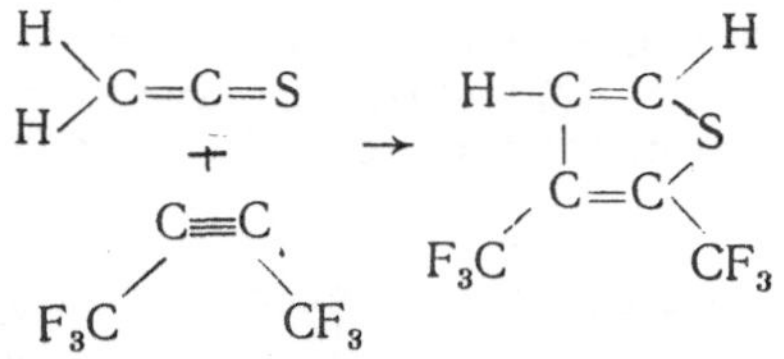

$$\underset{H}{\overset{H}{\diagdown}}C=C=S \quad \longrightarrow \quad \underset{F_3C \quad CF_3}{\overset{H-C=C \diagdown}{\underset{C=C \diagdown}{\overset{H}{\diagup}}S}} \tag{4.65}$$

$$\underset{F_3C \quad CF_3}{\overset{+}{C \equiv C}}$$

4.4. HALOSUBSTITUTED CYCLOALKANES

Hexachlorocyclohexane reacts with sulfur at 270-290°C to give a mixture of p-dichlorobenzene, 1,2,4,5-tetrachlorobenzene, and the isomeric trichlorobenzenes [69,70]:

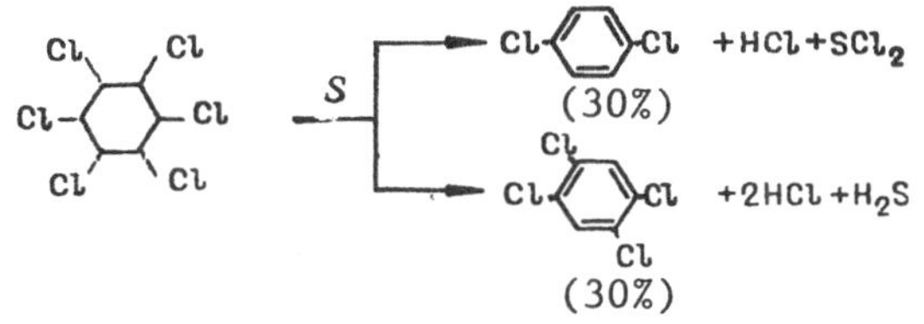

$$(4.66)$$

4.5. HALO-DERIVATIVES OF BENZENE, NAPHTHALENE, AND ANTHRACENE

Fluorobenzene does not react with sulfur at 250°C [71].

The reaction of sulfur with mono- or dichlorobenzenes at 250-350°C under pressure gives sulfur-containing polymers (40-80% S). The reaction is accompanied by the evolution of hydrogen chloride [72-77]. In the presence of excess sulfur (260°C), a higher-molecular-weight polymer insoluble in carbon disulfide is formed according to scheme (4.67) [74]:

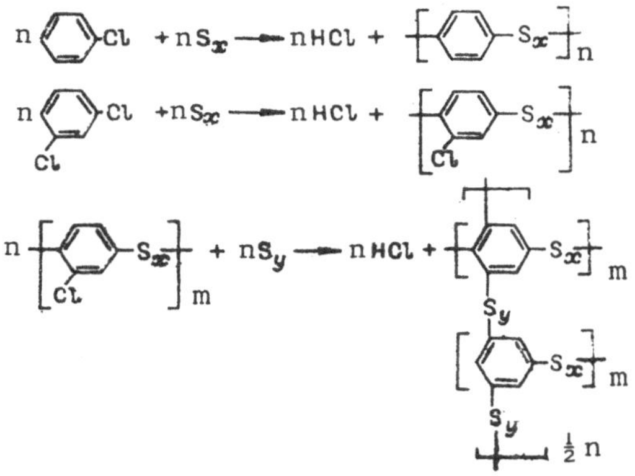

$$(4.67)$$

Heating sulfur with chlorobenzene, p-dichlorobenzene, 4-chlorobenzoic acid, 4-chlorophenol, 4-chloroaniline or 4-chlorotoluene in sealed ampoules in an atmosphere of nitrogen leads to cross-linked polyphenylene sulfides. A radical reaction mechanism has been discussed [78].

At 230-250°C, chlorobenzene reacts with sulfur according to the following scheme [79]:

$$4C_6H_5Cl \xrightarrow{S} C_6H_5S_nC_6H_5 + ClC_6H_4S_nC_6H_4Cl + S_mCl_2 + H_2S \qquad (4.68)$$
$$n \geqslant 2,\; m = 1,2$$

4-Chloronitrobenzene and its derivatives react with powdered sulfur in aqueous ethanol at 75°C to form the corresponding bis-nitrophenyl sulfides [80]:

$$O_2N-\underset{X}{\boxed{}}-Cl + S \longrightarrow O_2N-\underset{X}{\boxed{}}-S-\underset{X}{\boxed{}}-NO_2 \qquad (4.69)$$

X = H, CN, NO_2, SO_3NCH_3, SO_2NH_2

The reaction of sulfur with bromobenzene and bromo-derivatives of other aromatic hydrocarbons starts at 180°C and gives the corresponding diaryl polysulfides [79]. On heating bromobenzene with sulfur at 230°C under pressure both diphenyl sulfide and polysulfides are formed [81]. 4-Substituted bromobenzenes react with sulfur in an analogous manner at 230–250°C [71]. Donor-type substituents considerably accelerate this reaction [71]:

$$\underline{p}\text{-}XC_6H_4Br + nS \rightarrow XC_6H_4S_{n-2}C_6H_4X + S_2Br_2 \qquad (4.70)$$
$$n \geqslant 2 \quad X = H, F, Cl, C_6H_5, OC_6H_5$$

Iodobenzene and its analogs react with sulfur above 180°C to form diphenyl di- and polysulfides [79]. An admixture of cuprous sulfide exhibits a favorable effect upon the reaction of iodobenzene with sulfur giving rise to a 44% yield of diphenyl sulfide.

The reaction of molten hexachlorobenzene with sulfur and sodium sulfide in acetone and under pressure for thirty minutes gives pentachlorothiophenol [82]. The latter is also obtained by heating (108°C) under pressure hexachlorobenzene with sulfur and sodium sulfide in methanol [83].

The products of the reaction of substituted pentachlorobenzenes with excess sulfur at 325–350°C in an atmosphere of argon are thianthrenes (18 and 19) and linear oligomers (20 and 21) [84,85]:

R = Cl, CN, NCO

Heating di-. tri- or pentachlorobiphenyls with excess elemental sulfur at 300°C leads to the complete replacement of chlorine by sulfur and the formation of cross-linked biphenylene polysulfides [86].

Aromatic chloro-derivatives such as polychlorobiphenyl, for example, when sequentially heated with sulfur at 150, 260, 700. and 850°C, form activated charcoal (22% yield), noted for its high adsorptive capacity [87].

The reaction of a 1-naphthylamine nonachloro-derivative (22) with sulfur at 275-280°C affords a 79% yield of the azatrithiacyclopentaphenalene (23) [88]:

$$\underset{\underset{\displaystyle 22}{}}{} \quad \xrightarrow[-S_2Cl_2]{+\ S} \quad \underset{\underset{\displaystyle 23}{}}{} \tag{4.72}$$

The tetracyclic thiazine (25) is formed in 71% yield by treating the 1,5-diaminonaphthalene decachloro-derivative (24) with sulfur at 250-270°C [89]:

$$\underset{\underset{\displaystyle 24}{}}{} \quad \xrightarrow[-S_2Cl_2]{+\ S} \quad \underset{\underset{\displaystyle 25}{}}{} \tag{4.73}$$

Pentachlorophenyl isocyanate reacts with sulfur to give a thianthrene derivative [90]:

$$\xrightarrow{S} \tag{4.74}$$

On mixing chloronaphthalene with molten sulfur a porcelain-like substance which adheres well to metallic, wooden, and paper surfaces is obtained [92].

The reaction of 1,8-dichloro-, 9,10-dichloro-, 2,9,10-trichloro- and 1,3,9,10 -tetrachloroanthracene with excess sulfur has been patented as a method for the preparation of sulfur-containing vat dyes [93].

When heated with sulfur (290–295°C), tetrachloro-p-benzoquinone is converted to the corresponding thiophene derivatives (26–31) [93]:

27 X = S; X' = bond; 28 X = bond; X' = S

$$(4.75)$$

29 X = X^2 = bond; X' = X^3 = S;
30 X = X^3 = S; X' = X^2 = bond;
31 X = X^3 = bond; X' = X^2 = S

The reaction of tetrahalophthalic anhydrides with sulfur at 340° in the atmosphere of an inert gas leads to the formation of the corresponding tetrahaloidbenzodithiinetetraphthalic dianhydrides [94]:

$$(4.76)$$

R = Br, Cl

Tetrachloropyrimidine reacts with a 3-molar excess of sulfur at 300°C in a nickel autoclave to give the tetraazathianthrene (32) or its isomer (33) [95,96]. Under similar conditions, compounds (34 and 35) are formed from tetrachloropyrazine:

$$(4.77)$$

32   33

$$(4.78)$$

34   35

Pentafluoroiodobenzene when heated with sulfur at 230°C gives decafluorodiphenyl sulfide. Under analogous conditions, 1,2,3,4-tetrafluoro-5,6-diiodobenzene yields octafluorothianthrene [97,98]:

$$2 \; \text{(F)}_I \; + S \longrightarrow \text{(F)} - S - \text{(F)} \; + I_2 \tag{4.79}$$

$$2 \; \text{(F)}_I^I \; + 2S \longrightarrow \text{(F)} \langle \rangle \text{(F)} \; + 2I_2$$

At 320°C, octafluoro-2,2'-diiodobiphenyl reacts with sulfur to give octafluorodibenzothiophene [98]:

$$\text{(F)}_I \text{(F)}_I \xrightarrow{\; S \;} \text{(F)} \langle_S\rangle \text{(F)} \; + I_2 \tag{4.80}$$

The reaction of sulfur with 2,2'-dihalooctafluorodiphenyl sulfide leads to octafluorothianthrene [99]:

$$\text{(F)}_X^S \text{(F)}_X \; + S \longrightarrow \text{(F)} \langle_S^S \rangle \text{(F)} \; + X_2 \tag{4.81}$$

$$X = Br, \; I$$

1,2,4,5-Tetrafluorobenzene and its derivatives react with sulfur in the presence of antimony pentafluoride to give the corresponding bis-tetrafluoroaryl sulfides [100,101]:

$$X \underset{F \; F}{\overset{F \; F}{\text{(}}} H \xrightarrow[\text{SbF}_5]{S} X \underset{F \; F}{\overset{F \; F}{\text{(}}} -S- \underset{F \; F}{\overset{F \; F}{\text{(}}} X \; + H_2S \tag{4.82}$$

$$X = H, \; F, \; Br, \; SCH_3$$

1,2,4,5-Tetrafluorobenzene itself reacts with excess sulfur and forms poly(tetrafluorophenylene) sulfide:

$$\underset{F \; F}{\overset{F \; F}{\text{(}}} + 2n \; S \xrightarrow{\text{SbF}_5} \left[\text{(F)} -S- \text{(F)} -S \right]_n + H_2S \tag{4.83}$$

Reaction (4.82) is likely to involve the intermediate formation of a product formed by the reaction of sulfur with antimony pentafluoride $[S_n^{2+}(SbF_6)_2^-$ (n = 2,4,8,16) followed by an electrophilic attack of this species upon the fluorinated benzene molecule:

$$C_6F_5H \xrightarrow[\text{SbF}_5]{S} C_6F_5S_n^+ - \left[\begin{array}{l} \xrightarrow{\text{SbF}_5} C_6F_5S^+ \xrightarrow{C_6F_5H} \\ \xrightarrow{C_6F_5H} C_6F_5S_n C_6F_5 \xrightarrow{C_6F_5H} \end{array} \right] \rightarrow C_6F_5SC_6F_5 \tag{4.84}$$

When hexafluorobenzene is heated with sulfur a mixture of decafluorodiphenyl sulfide and disulfide is obtained [98]:

$$\langle F \rangle + nS \longrightarrow \langle F \rangle - S_n - \langle F \rangle + S_2F_2 \qquad (4.85)$$

$$n = 1,2$$

In the presence of sulfuric acid, the reaction of sulfur with fluorobenzenes (160–170°C) proceeds in a different direction [102, 103]:

$$X = F, \ Cl$$

4.6.   2-HALOTOLUENES

The reaction of sulfur with 2-bromotoluene, preferably in the presence of copper(II) sulfide, gives 1,2-benzodithiol-3-thione and a [1]benzothieno[3,2-b]-(1)benzothiophene [99,104]. The reaction of sulfur with α-derivatives of 2-bromotoluene is similar:

$$ (4.87) $$

$$ (4.88) $$

$$R = R' = R'' = H, \quad R = R'' = H, \quad R' = CH_3;$$
$$R = R' = H, \quad R'' = CH_3; \quad R = R'' = CH_3, \quad R' = H;$$
$$R = Br, \quad R' = CH_3, \quad R'' = H$$

The corresponding disulfides are the intermediate products in reactions (4.87) and (4.88). In the sulfuration of α-derivatives of 2-bromotoluene, i.e., 1-bromo-2-bromomethylbenzene and 2,2'-dibromodibenzyl disulfide, 1,2-benzodithiol-3-thione is also formed in 14 and 21% yield, respectively [104]:

$$\text{(4.89)}$$

The reaction of sulfur with 2-iodotoluene gives only traces of 1,2-benzodithiol-3-thione. 2-Chlorotoluene and 2-fluorotoluene do not react appreciably with sulfur at their respective boiling points [104]. Heating 2-chlorobenzyl chloride with sulfur and dimethyl-formamide gives 1,2-benzodithiol-3-thione [105].

2,6-Dichlorobenzylidene chloride reacts with sulfur and ammonia with the formation of 4-chlorobenzisothiazole according to the scheme [106]:

$$\text{(4.90)}$$

The interaction of polyhalomesitylene derivatives with sulfur carried out under similar conditions gives the compound below [105]:

$$\text{(4.91)}$$

R = Br, $N(CH_3)_2$, $N(C_2H_5)_2$, piperidine, SH

4.7. ARYL SUBSTITUTED HALOALKANES

4.7.1. Aryl Substituted Halomethanes

Studies of the solubility curves of sulfur in benzyl chloride have shown that, besides being soluble, sulfur also reacts with the solvent at elevated temperatures [107,108]. When benzyl chloride is boiled with sulfur hydrogen sulfide is evolved [109]. The interaction of benzyl chloride with sulfur at 200-240°C proceeds with the evolution of hydrogen chloride and hydrogen sulfide and gives a high yield of tetraphenylthiophene [110,111]. This reaction represents the simplest and most convenient method for the synthesis of tetraphenylthiophene:

$$4 C_6H_5CH_2Cl + 3S \longrightarrow C_6H_5 \text{-thiophene-} C_6H_5 + 2H_2S + 4HCl \qquad \text{(4.92)}$$

The reaction of benzyl chloride with sulfur and sodium methoxide yields the corresponding salt of dithiophenylacetic acid [112]:

$$\text{C}_6\text{H}_5\text{CH}_2\text{Cl} + \text{S} \xrightarrow{\text{MeONa}} \text{C}_6\text{H}_5\text{CH}_2\text{-}\underset{\underset{\text{S}}{\|}}{\text{C}}\text{-S-Na} \tag{4.93}$$

The reaction of 2-chloro-5-nitrobenzyl chloride with sulfur at 50-60°C in the presence of triethylamine for 21 h gives 5-nitro-1,-2-benzodithiol-3-thione in 89% yield [113]:

$$\tag{4.94}$$

Chlorodiphenylmethane readily undergoes dehydrochlorination on heating with sulfur and affords tetraphenylethylene in about 70% yield [114]:

$$2\,(\text{C}_6\text{H}_5)_2\,\text{CHCl} \xrightarrow{\text{S}} (\text{C}_6\text{H}_5)_2\,\text{C}=\text{C}\,(\text{C}_6\text{H}_5)_2 + 2\text{HCl} \tag{4.95}$$

The reaction of sulfur with benzylidene chloride at 240°C gives a 52% yield of 3-chloro-2-phenylbenzo[b]thiophene as the reaction product [115]:

$$2\,\text{C}_6\text{H}_5\text{CH.Cl}_2 + \text{S} \longrightarrow \quad + 3\,\text{HCl} \tag{4.96}$$

Under more drastic conditions (300°C), benzylidene chloride reacts with sulfur to give a 60% yield of [1]benzothieno[3,2-b]-[1]benzothiophene [116,117]:

$$2\,\text{C}_6\text{H}_5\text{CHCl}_2 + 2\text{S} \longrightarrow \quad + 4\,\text{HCl} \tag{4.97}$$

$$\underset{\widetilde{37}}{}$$

It is likely that the reaction occurs via the intermediate formation of α,β-dichlorostilbene which upon reaction with sulfur is easily converted (260-280°C) to (37) [118,119]. When a reaction is carried out between sulfur and an equimolar mixture of benzyl chloride and benzylidene chloride, the only isolated sulfuration product is tetraphenylthiophene (51% yield) [110].

Sulfur reacts with a large excess of benzotrichloride at 200-233°C to give tetrachloro-1,2-diphenylethane as the only reaction product [118]:

$$2\text{C}_6\text{H}_5\text{CCl}_3 + 2\text{S} \longrightarrow \text{C}_6\text{H}_5\text{CCl}_2\text{CCl}_2\text{C}_6\text{H}_5 + \text{S}_2\text{Cl}_2 \tag{4.98}$$

At higher temperatures (250-300°C), the reaction between these two reagents leads to (37) and 3-chloro-2-phenylbenzo[b]thiophene [118]:

$$2C_6H_5CCl_3 \xrightarrow[-S_2Cl_2]{+S} C_6H_5CCl=CClC_6H_5 \xrightarrow{+S}_{-HCl}$$

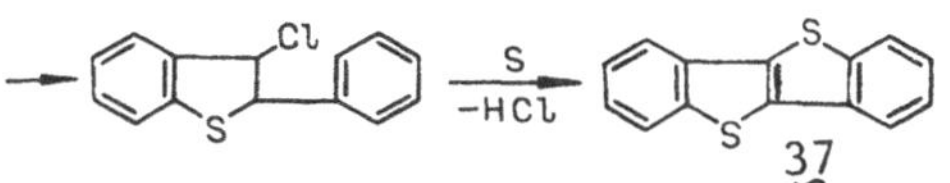

(4.99)

However, according to other data [114,117], carefully purified benzotrichloride and sulfur do not react at 240°C. This indicates that reactions (4.98) and (4.99) are probably initiated by traces of benzylidene chloride usually present in benzotrichloride. An equimolar mixture of benzotrichloride and benzyl chloride reacts with sulfur at 250-280°C to give a 38% yield of (37) [115].

Dichlorodiphenylmethane does not react appreciably with sulfur at 240°C. The reaction of sulfur with chlorotriphenylmethane gives triphenylmethane [117].

Arylbromomethanes react with sulfur in a different direction to that of the analogous chloro-derivatives. Benzyl bromide on heating with sulfur at 220°C gives 2-phenylbenzo[b]thiophene [114, 120:

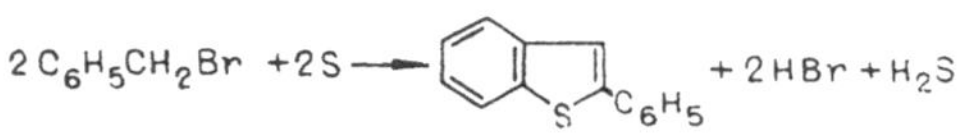

(>40%)

(4.100)

The yield of tetraphenylthiophene, which is also formed [cf. reaction (4.92)], does not exceed 10%, stilbene being a possible intermediate. The sulfuration of stilbene in the presence of hydrogen bromide at 220-240°C gives a 30% yield of 2-phenylbenzo[b]thiophene. The difference in the course of the sulfuration reactions of the bromo-derivatives when compared with the chloro-compounds is due to the specific catalytic effect of hydrogen bromide formed during the reaction:

$$2C_6H_5CH_2Br \rightarrow C_6H_5CHBrCH_2C_6H_5 \underset{-HBr}{\overset{HBr}{\rightleftharpoons}} C_6H_5CH=CHC_6H_5 \qquad (4.101)$$

An analogous closure of the benzo[b]thiophene ring is observed in the reaction of 4-chloro- and 4-bromobenzyl bromide with sulfur at 180-190°C giving 6-chloro- or 6-bromo-2-(4-chlorophenyl)benzo-[b]thiophene [114,120] respectively:

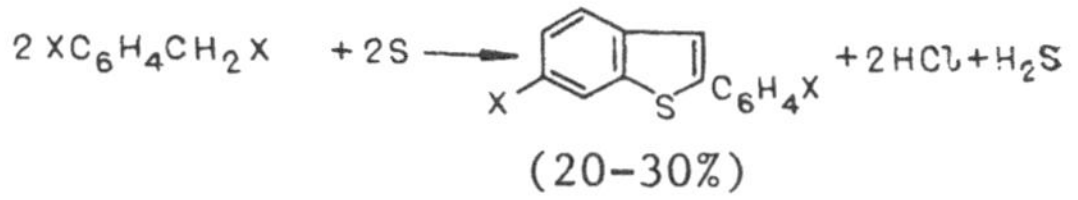

(20-30%)

(4.102)

X=Cl, Br

At higher temperatures, 4,4'-dichlorostilbene is the reaction product from the interaction of sulfur with 4-chlorobenzyl bromide. 3-Chlorobenzyl bromide reacts with sulfur at 190–210°C to give only a 6–7% yield of 3,3'-dichlorostilbene [114,120]:

$$2\ Cl\text{—}C_6H_4\text{—}CH_2Br \xrightarrow{S} Cl\text{—}C_6H_4\text{—}CH=CH\text{—}C_6H_4\text{—}Cl\ +2HBr \qquad (4.103)$$

When 2-chlorobenzyl bromide or benzylidene bromide are heated with sulfur, a low yield of [1]benzothieno[3,2-b]-[1]benzothiophene is obtained [120]:

$$2\ \underset{Cl}{\overset{CH_2Br}{C_6H_4}} \xrightarrow[-HCl,-HBr]{+2S} \cdots \xleftarrow[-4HBr]{+2S} 2\,C_6H_5CH_2Br_2 \qquad (4.104)$$

(<10%)

The interaction of sulfur with 2- and 4-methylbenzyl bromide at 220°C leads to the formation of the corresponding dimethylstilbenes in low yield:

$$\underset{R}{C_6H_4}\text{—}CH_2Br \xrightarrow{S} \underset{R}{C_6H_4}\text{—}CH=CH\text{—}\underset{R}{C_6H_4}\ +\ 2\ HBr \qquad (4.105)$$

(< 10%)

R = 2- or 4-CH$_3$

The reaction of 2-, 3- or 4-nitrobenzyl halides with "activated" sulfur yields the corresponding aromatic thioamides [121]:

$$NO_2\text{—}C_6H_4\text{—}CH_2X \xrightarrow[DMSO]{S_8/HNRR',\ 20°C} NO_2\text{—}C_6H_4\text{—}\underset{NRR'}{\overset{S}{C}} \qquad (4.106)$$

α-Chlorobenzylthioesters react with sulfur and secondary amines in a dipolar aprotic solvent at room temperature to give aromatic thiocarboxamides [122]:

$$R\text{—}C_6H_4\text{—}CH_2SR' \xrightarrow{Cl} \left[R\text{—}C_6H_4\text{—}\underset{SR'}{\overset{Cl}{CH}} \right] \xrightarrow{S_8/DMPh,\ HNR''_2} R\text{—}C_6H_4\text{—}\underset{NR''}{\overset{S}{C}} \qquad (4.107)$$

α-Chloroesters prepared by chlorinating benzyl thioethers react <u>in situ</u> with elemental sulfur in the presence of triethylamine to form dithioesters of benzoic acid [123] (Table 4.1):

$$R\text{—}C_6H_4\text{—}CH_2SR' \longrightarrow \left[R\text{—}C_6H_4\text{—}\underset{Cl}{\overset{}{CHSR'}} \right] \xrightarrow[N(Et)_3]{S_8,\ DMF} R\text{—}C_6H_4\text{—}\underset{SR'}{\overset{S}{C}} \qquad (4.108)$$

Table 4.1.   The Formation of Dithioesters of Benzoic
             Acid

R	R'	Time, min.	T, °C	Yield, %
H	CH_3	186	20	40
H	$C_6H_5CH_2$	240	20	24
4-CH_3O	4'-$CH_3OC_6H_4CH_2$	360	60	20
4-Cl	4'-$ClC_6H_4CH_2$	10	20	28
4-Br	4'-$BrC_6H_4CH_2$	10	0	40
4-NO_2	4'-$NO_2C_6H_4CH_2$	30	-20	9

4.7.2.   Side-Chain Halo-Derivatives of Ethylbenzene and Styrene

The isomeric side-chain monochloro- and monobromo-derivatives
of phenylethane (ethylbenzene), as well as the corresponding 1,2-
dihalo-1-phenylethanes, react with sulfur at 200-230°C to give 2,4-
diphenylthiophene [124,125]:

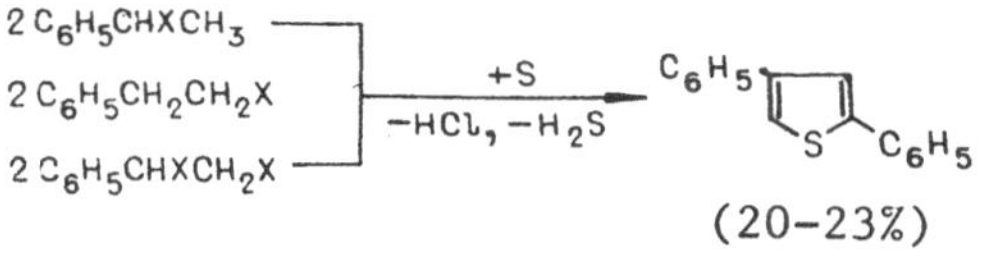

$$
\begin{array}{l}
2\,C_6H_5CHXCH_3 \\
2\,C_6H_5CH_2CH_2X \\
2\,C_6H_5CHXCH_2X
\end{array}
\xrightarrow[-HCl,\,-H_2S]{+S}
\underset{(20-23\%)}{C_6H_5\!\!\overset{}{\underset{S}{\bigcirc}}\!\!C_6H_5}
\qquad (4.109)
$$

$$X = Cl,\ Br$$

The intermediate in these reactions seems to be styrene or α-halostyrene whose interaction with sulfur also yields 2,4-diphenylthio-
phene [126,127]:

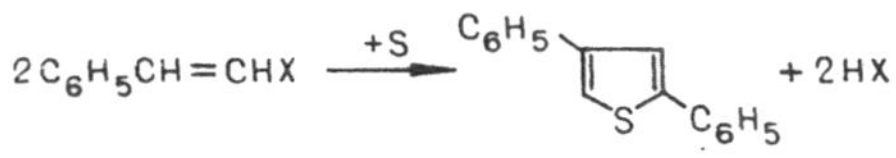

$$
2\,C_6H_5CH=CHX \xrightarrow{+S} C_6H_5\!\!\overset{}{\underset{S}{\bigcirc}}\!\!C_6H_5 + 2HX
\qquad (4.110)
$$

$$X = Cl,\ Br$$

1,1-Dihalo-1-phenylethanes when heated with sulfur readily
undergo dehydrohalogenation to α-halostyrenes whose reaction with
sulfur takes place at 220°C to give a mixture of 2,5- and 2,4-
diphenylthiophene in the ratio 3:1 (X = Cl) or 3:2 (X = Br):

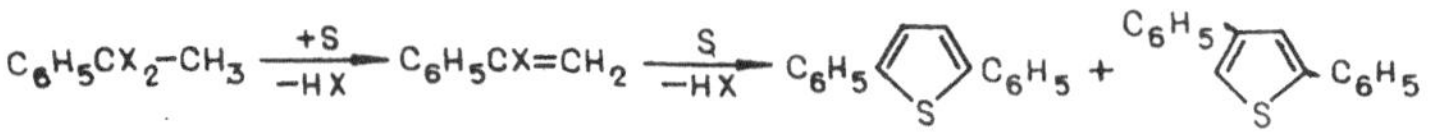

$$
C_6H_5CX_2-CH_3 \xrightarrow[-HX]{+S} C_6H_5CX=CH_2 \xrightarrow[-HX]{S} C_6H_5\!\!\overset{}{\underset{S}{\bigcirc}}\!\!C_6H_5 + \overset{C_6H_5}{\underset{S}{\bigcirc}}\!\!C_6H_5
\qquad (4.111)
$$

$$X = Cl,\ Br$$

Side-chain trihalo-substituted ethylbenzenes react with sulfur with the formation of the corresponding benzo[b]thiophene derivatives. 1,1,2-Trichloro-1-phenylethane reacts with sulfur at 225°C to form 2,3-dichlorobenzo[b]thiophene [125,128]. The interaction of sulfur with the hypothetical intermediate in reaction (4.112), i.e., α,β-dichlorostyrene, also gives 2,3-dichlorobenzo[b]thiophene (35% yield):

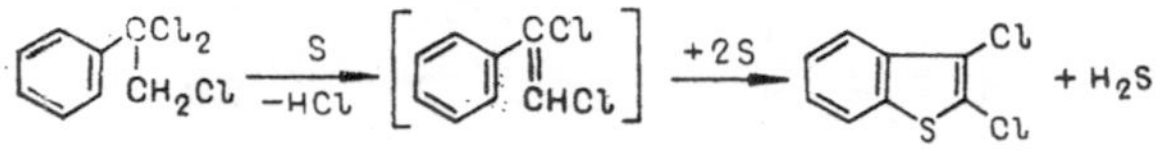

$$\text{(4.112)}$$

(28%)

1,2,2-Trichloro-1-phenylethane and β,β-dichlorostyrene give only tars on treatment with sulfur (230-240°C).

The reaction of sulfur with 1,1,2,2- or 1,2,2,2-tetrachloro-1-phenylethane at 230-240°C gives a high yield of 2,3-dichloro-benzo[b]thiophene [125,128]:

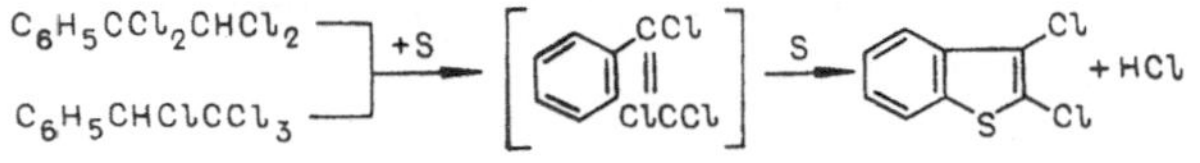

$$\text{(4.113)}$$

(60-70%)

The intermediate product of reaction (4.113), i.e., α,β,β-trichloro-styrene, on treatment with sulfur (225-240°C) also gives 2,3-di-chlorobenzo[b]thiophene (65% yield) [118]. The latter compound is also formed in the reaction of pentachloro-1-phenylethane with sulfur (230°C), but in low yield:

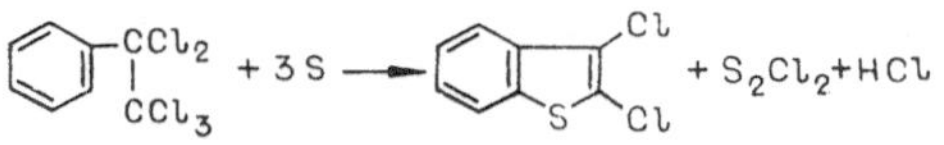

$$\text{(4.114)}$$

(16%)

The closure of the benzo[b]thiophene ring in reactions (4.113) and (4.114) is due to both dehydrohalogenation and dehalogenation involving sulfur (elimination of HCl and S_2Cl_2).

Side-chain β-bromo-substituted arylalkanes react with sulfur in a different way to the corresponding chloro-derivatives. 2-Bromo-1,2-dichloro-1-phenylethane and 1-bromo-1,2-dichloro-1-phenylethane react with sulfur at 220-240°C to form di-[1]-benzothieno-[2,3-b; 3',2'-e]-[1,4]dithiine [125]:

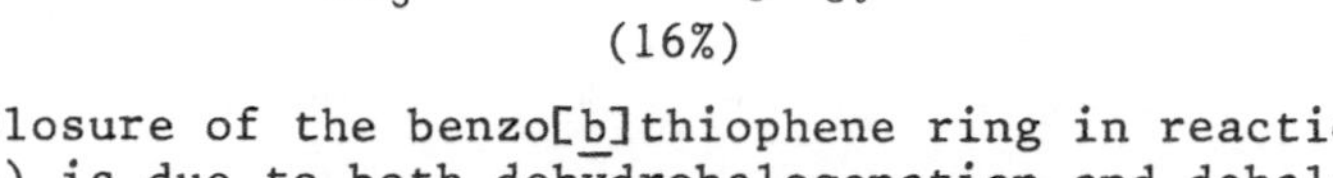

$$\text{(4.115)}$$

(35-52%)

The same compound was obtained in a 23% yield by treating sulfur
with β-bromo-α-chlorostyrene at 200–220°C, and also from 2,3-di-
chlorobenzo[b]thiophene and sulfur in the presence of hydrobromic
acid [125]:

$$(4.116)$$

$$2\,C_6H_5CCl_2CH_2Cl + 4\,S$$

Thus, the main process occurring in the reactions of sulfur
with side-chain halo-derivatives of ethylbenzene containing at least
one hydrogen atom in the side chain is dehydrohalogenation with the
formation of the corresponding styrene.

However, pentachloroethylbenzene is converted to α,β,β-trichloro-
styrene and disulfur dichloride:

$$C_6H_5CCl_2\,CCl_3 + 2S \longrightarrow C_6H_5CCl \!=\! CCl_2 + S_2Cl_2 \qquad (4.117)$$

When the intermediate products of the dehydrohalogenation are
styrene or β-halostyrene, they react further with sulfur and are
converted to 2,4-diphenylthiophene; cf. reactions (4.109) and (4.110).
When α-halostyrene is formed, it is converted mainly to 2,5-diphenyl-
thiophene and, to a lesser extent, to 2,4-diphenylthiophene.

When α,β-dichlorostyrene or α,β,β-trichlorostyrene are the
dehydrochlorination (or dechlorination) products, they react with
sulfur according to schemes (4.112) or (4.113) and are converted to
2,3-dichlorobenzo[b]thiophenes. Similarly, 1,3,5-trichloro-2,4,6-
tris-(trichlorovinyl)benzene reacts with sulfur to give the corresponding condensed heterocyclic compound [90]:

$$(4.118)$$

4.7.3. α- or β-Halo-Derivatives of Di-, Tri-, and Tetraphenylethane

The reaction of elemental sulfur with the α- or β-halo-derivatives of 1,1-diphenylethane, 1,1,2-triphenylethane, and 1,1,2,2-
tetraphenylethane gives the corresponding sulfur-containing heterocycle [129,130]. 2-Chloro-1,1-diphenylethylene is an intermediate
product of the reaction which is then converted into 3-phenylbenzo-
[b]thiophene on treatment with sulfur at 240°C:

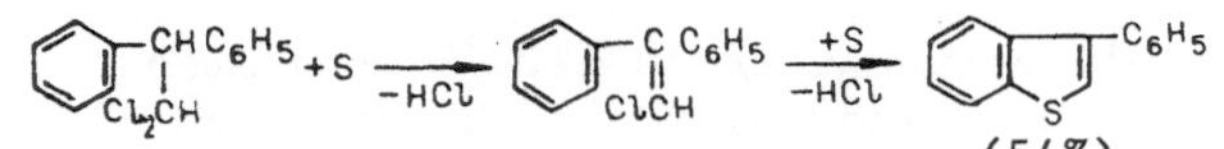

(4.119)

(54%)

The reaction of sulfur with 1,1,1-trihalo-2,2-diphenylethanes at 260-270°C gives a [1]benzothieno[2,3-b]-[1]benzothiophene (38) and its derivatives [129]:

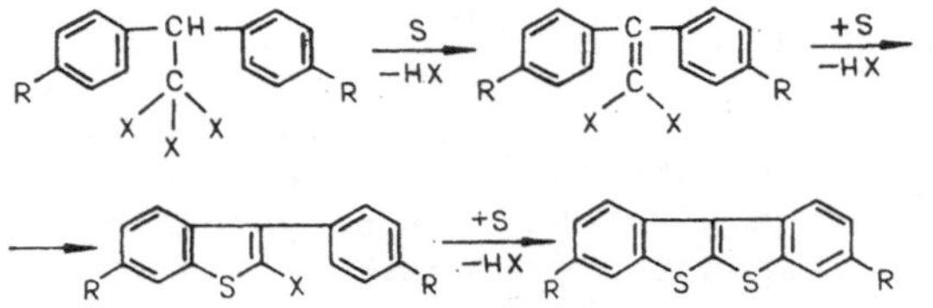

(4.120)

X = Cl, Br; R = H, Cl, Br

The mechanism of reaction (4.120) is confirmed by the formation of (38) in the reaction of sulfur with 1,1-dichloro-2,2-diphenyl-ethylene at 240-280°C [129]. When R = Cl, the intermediate reaction product in scheme (4.120) at 235-240°C is also 2,6-dichloro-3-(4-chlorophenyl)benzo[b]thiophene:

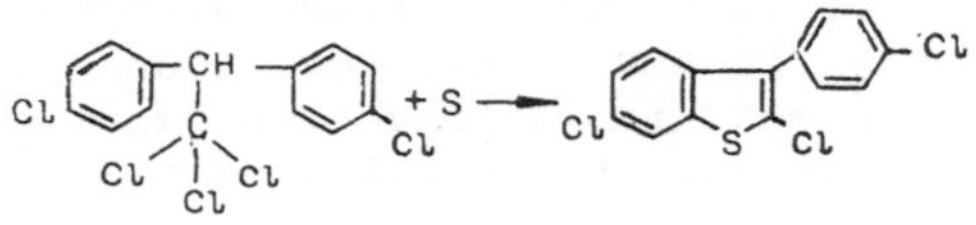

(4.121)

Tetrachloro-2,2-diphenylethane reacts with sulfur at 220-240°C, and also gives (38, R = H):

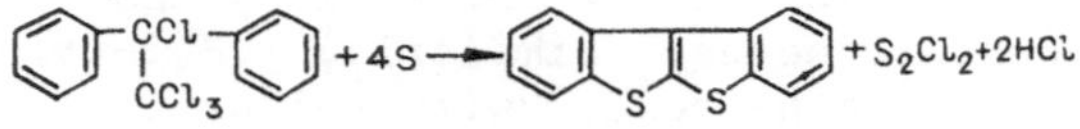

(4.122)

38 (41%)

1,2-Dichloro-1,1,2-triphenylethane reacts with sulfur at 230-240°C to give 2,3-diphenylbenzo[b]thiophene [131]. 2-Chloro-1,1,2-triphenylethylene formed as an intermediate, also gives 2,3-di-phenylbenzo[b]thiophene when heated with sulfur:

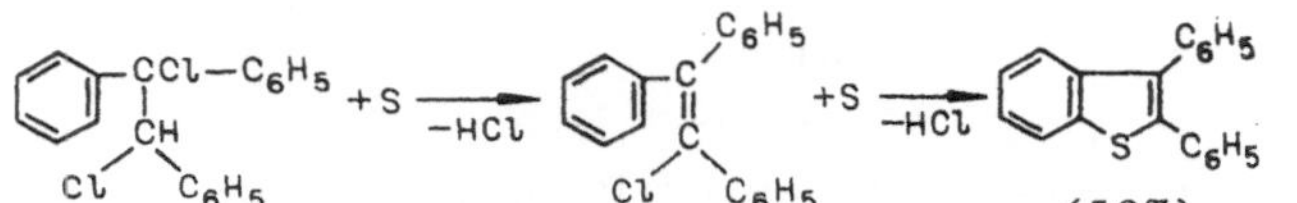

(4.123)

(52%)

The reaction of sulfur with 1,2-dichlorotetraphenylethane at 200-220°C leads to dechlorination to tetraphenylethylene [131]:

$$(C_6H_5)_2\, CClCCl\, (C_6H_5)_2 \xrightarrow{S} (C_6H_5)_2\, C{=}C\, (C_6H_5)_2 + Cl_2 \qquad (4.124)$$

(64%)

Thus, the reactions of sulfur with α- or β-halo-derivatives of 1,1-diphenylethane and 1,1,2-triphenylethane proceed via dehydrohalogenation of the aliphatic moiety of the molecule with the intermediate formation of a halo-substituted alkene which subsequently reacts with sulfur with the elimination of hydrogen halide (the hydrogen being eliminated is the <u>ortho</u> hydrogen in the aromatic ring) to afford a benzo[b]thiophene or [1]benzothieno[2,3-b]-[1]benzothiophene ring.

The reaction of sulfur with α- or β-halo-derivatives of 1,2-diphenylethane [132] gives the same sulfuration products as those obtained in the reactions of sulfur with benzyl and benzylidene halides. At 220-240°C, the reaction of sulfur with 1-chloro-1,2-diphenylethane and 1,2-dichloro-1,2-diphenylethane gives tetraphenylthiophene in $\sim$ 32 and 51% yields, respectively:

$$\left.\begin{array}{l} 2\,C_6H_5CHClCH_2C_6H_5 \\ \\ 2\,C_6H_5CHClCHClC_6H_5 \end{array}\right\} S \longrightarrow \underset{C_6H_5}{\overset{C_6H_5}{\bigotimes}}\underset{S}{\overset{C_6H_5}{}}C_6H_5 + 2\,HCl + 2\,H_2S \qquad (4.125)$$

2-Phenylbenzo[b]thiophene is the product obtained from the reaction between sulfur and 1,2-dibromo-1,2-diphenylethane (similar to benzyl bromide). Tetraphenylthiophene is also formed in $\sim$5% yield.

$$C_6H_5CHBrCHBrC_6H_5 + S \longrightarrow \underset{S}{\bigotimes}C_6H_5 + 2\,HBr \qquad (4.126)$$
$$(40\%)$$

The sulfuration of 1,2,2-trichloro-1,2-diphenylethane and tetrachloro-1,2-diphenylethane at 260°C gives (37), (similar to benzal chloride) [117,118].

The chloromethylated divinylbenzene-styrene polymer reacts with sulfur and sodium methoxide at 70°C to form the sodium salt of a sulfurated polymer containing 20% of sulfur. The latter displays cationic exchange properties and is selective towards Hg(II) ions [95].

4.7.4. α-, β-, and γ-Halo-Derivatives of Mono-, Di-, and Tetraphenylpropane

The sulfuration of α-, β-, or γ-mono-, di- and tri-halo-derivatives of 1-phenylpropane at 200-220°C gives an 11-18% yield of 5-phenyl-1,2-dithiol-3-thione. The highest yield is observed in the reaction between sulfur and 1-chloro-3-phenylpropane [133, 134]:

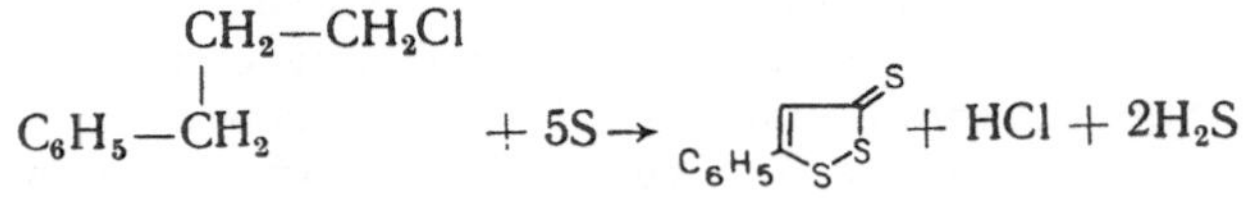

$$\underset{\text{C}_6\text{H}_5-\overset{\displaystyle |}{\text{CH}}_2}{\overset{\displaystyle \text{CH}_2-\text{CH}_2\text{Cl}}{}} \quad + 5\text{S} \rightarrow \quad + \text{HCl} + 2\text{H}_2\text{S} \qquad (4.127)$$

(18%)

The isomeric tetrachloro-1-phenylpropanes when heated with sulfur give 4-chloro-5-phenyl-1,2-dithiol-3-thione, the yield depending on the position of the chlorine atoms in the side chain. Thus, the yield varies from 5 to 35%. 2-Chloro-2-phenylpropane, 1,2-dichloro-2-phenylpropane, 1,1,2-trichloro-2-phenylpropane or their mixtures react with sulfur under analogous conditions to give 5-phenyl-1,2-dithiol-3-thione in approximately the same yield [133, 134]:

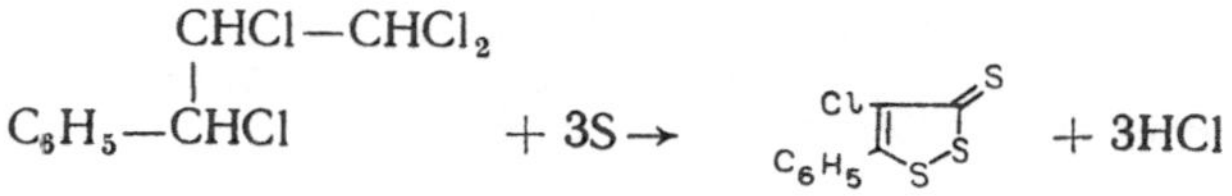

$$\underset{\text{C}_8\text{H}_5-\overset{\displaystyle |}{\text{CHCl}}}{\overset{\displaystyle \text{CHCl}-\text{CHCl}_2}{}} \quad + 3\text{S} \rightarrow \quad + 3\text{HCl} \qquad (4.128)$$

(35%)

The reaction of sulfur with 1,1,2,2-tetrachloro-2-phenylpropane and 1,1,1,2,3-pentachloro-2-phenylpropane gives 1H-[1]benzothieno-[2,3-c]-[1,2]-dithiol-1-thione [135,136]:

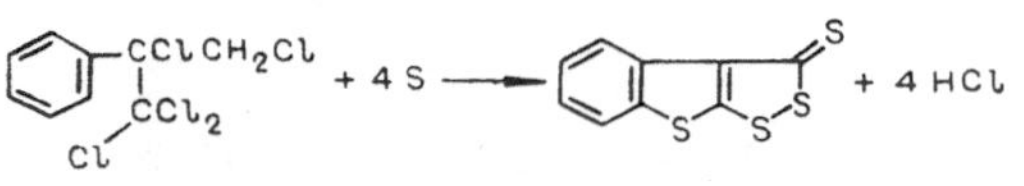

$$+ 4\text{S} \longrightarrow + 4\text{HCl} \qquad (4.129)$$

1-Chloro[1]benzothieno[2,3-c]-[1,2]-dithiolium chloride is the sulfuration product obtained from the reaction between 1,1,2,3,3-pentachloro-2-phenylpropane and sulfur:

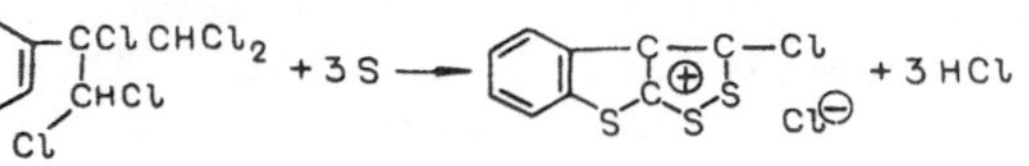

$$+ 3\text{S} \longrightarrow + 3\text{HCl} \qquad (4.130)$$

α-, β-, and γ-penta-, hexa-, and heptachloro-derivatives of 1-phenyl- and 2-phenylpropane when heated with sulfur give mainly tar-like products [137,138].

4.8. HALO-DERIVATIVES OF HETEROCYCLIC COMPOUNDS

The interaction of sulfur with halo-derivatives of hetero-aromatic compounds has not been sufficiently studied.

2-Iodothiophene reacts with sulfur and is converted to dithieno-[2,3-b;3',2'-e]-[1,4]-dithiine via the intermediate formation of bis-2-thienyl sulfide [79]:

$$2 \underset{S}{\langle} I + S \xrightarrow{-I_2} \underset{S}{\langle}\underset{S}{\rangle} \xrightarrow[-H_2S]{+S} \underset{S}{\langle\rangle} \qquad (4.131)$$

When 2-chloromethylthiophene is heated with sulfur at 185°C one obtains 1,2-di-(2-thienyl)ethylene [139]:

$$2 \underset{S}{\langle}-CH_2Cl_2 + S \longrightarrow \underset{S}{\langle}-CH=CH-\underset{S}{\rangle} + 2\,HCl \qquad (4.132)$$

However, tetra-(2-thienyl)thiophene is not formed. 3-Chloro-2-phenylbenzo[b]thiophene reacts with sulfur at 300°C to give [1]-benzothieno[3,2-b]-[1]benzothiophene [111,117]:

$$\underset{S}{\langle}\overset{Cl}{\underset{C_6H_5}{}} \xrightarrow{+S} \langle\rangle + HCl \qquad (4.133)$$

When sulfur reacts with 2,3-dichlorobenzo[b]thiophene in the presence of hydrobromic acid, di-[1]benzothieno[2,3-b;3',2'-e]-[1,4]-dithiine is formed [125]:

$$\underset{S}{\langle}\overset{Cl}{\underset{Cl}{}} + \overset{Cl}{\underset{Cl}{}\rangle}\underset{S}{} \xrightarrow{S,HBr} \langle\rangle \qquad (4.134)$$

2,6-Dichloro-3-(4-chlorophenyl)benzo[b]thiophene upon heating with sulfur gives the corresponding [1]benzothieno[2,3-b]-[1]benzothiophene [129]:

$$Cl\underset{S}{\langle}\overset{}{\underset{Cl}{}}\langle\rangle Cl \xrightarrow{+S} Cl\langle\rangle\underset{S\,S}{}\langle\rangle Cl \qquad (4.135)$$

The reactions of sulfur with the halo-derivatives of nitrogen-containing heterocycles are discussed in Section 7.11. The reactions of sulfur with the halo-derivatives of organo-metallic compounds are discussed in Chapter 8.

4.9. THE MECHANISM OF THE INTERACTION OF SULFUR WITH HYDROCARBONS AND THEIR HALOSUBSTITUTED DERIVATIVES

The data discussed above indicate that sulfur can act as a dehydrogenating, dehydrohalogenating, or dehalogenating agent in its reactions with organic halides.

A mechanism for the dehydrogenation of alkanes to alkenes by sulfur involving the formation of hydrogen sulfide was proposed in 1919 [140]. However, it is difficult to agree with the proposed mechanism, principally because atomic sulfur is not formed under the conditions studied by the authors (320–350°C). A free-radical mechanism for the reaction of sulfur with hydrocarbons has been invoked to explain the transformation of arylalkanes to arylalkenes [141]:

$$C_6H_5CH_2CH_2C_6H_5 + \cdot S \cdot \longrightarrow C_6H_5CH_2\overset{\cdot}{C}HC_6H_5 + HS \cdot \qquad (4.136)$$

$$C_6H_5CH_2\overset{\cdot}{C}HC_6H_5 + HS \cdot \longrightarrow C_6H_5CH = CHC_6H_5 + H_2S \qquad (4.137)$$

This scheme suffers from the same difficulty as the previous mechanism and cannot apply to the reactions of sulfur with organic halides because, for example, the conversion of aryl substituted haloalkanes to aryl substituted haloalkenes or arylalkenes takes place without the formation of hydrogen sulfide. Hydrogen sulfide is formed in only a few cases in which thiophene ring closure occurs, e.g., the reaction of sulfur with α-halo-derivatives of toluene, ethylbenzene, and 1,2-diphenylethane, or where a benzothiophene ring is formed, e.g., the reaction of sulfur with benzyl bromide and 1,1,2-trichloro-1-phenylethane.

As indicated above, cyclooctasulfur molecules undergo ring opening when heated above 180°C and form linear biradicals, $\cdot S_8 \cdot$, which can undergo both polymerization and depolymerization. The biradical character of the linear fragments, $\cdot S_8 \cdot$, is responsible for the similarity of their reactions to those of the carbenes [142-144]. Especially important are insertions into C-H and C-X (X = halogen) bonds [144].

It has been shown experimentally that insertion into C-H bonds can occur not only with atomic sulfur, but also with molecular sulfur, S_n. Thus, the formation of thiols is observed on heating sulfur with alkanes, benzene, and numerous olefins [145].

One has to assume that organic oligosulfanes, $R-S_nH$, are the primary products in the formation of thiols. These compounds are quite unstable, even at room temperature, and easily decompose according to the scheme [146]:

$$\overset{\diagup}{\underset{\diagdown}{}}C-H + \cdot S_n \cdot \rightarrow \overset{\diagup}{\underset{\diagdown}{}}C-S_n-H \rightarrow \overset{\diagup}{\underset{\diagdown}{}}C-SH + \cdot S \cdot_{n-1} \qquad (4.138)$$

The possibility of such a process is also indicated by the reaction of sulfur with phenols which proceeds as follows:

$$HOC_6H_5 + \cdot S_n \cdot \rightarrow HOC_6H_4S_nH \xrightarrow[-H_2S]{+C_6H_5OH} HOC_6H_4S_{n-1}C_6H_4OH \qquad (4.139)$$
$$n = 2-8$$

Elemental sulfur can also be inserted into C-X (X = halogen) bonds. Compounds of the $R-S_nX$ type are very unstable [147,148]. The insertion of linear molecules into C-H and C-X bonds, not necessarily a one-step process, is responsible, in our opinion, for the dehydrogenating, dehydrohalogenating, and dehalogenating effects of sulfur at elevated temperatures. These processes can be represented as follows:

1. Intermolecular dehydrogenation

$$\text{>C-H} + \cdot S_n \cdot \rightarrow \text{>C-}S_n\text{H} \xrightarrow{+\text{H-C<}} \text{>C}S_{n-1}\text{C<} + H_2S \qquad (4.140)$$

2. Intramolecular dehydrogenation

$$\underset{H\ \ H}{\text{>C-C<}} + \cdot S_n \cdot \rightarrow \underset{H\ \ \overset{|}{S_n}}{\text{>C-C<}} \rightarrow \text{>C=C<} + HS_nH \qquad (4.141)$$
$$\qquad\qquad\qquad\qquad\qquad\qquad H_2S + \cdot S\cdot_{n-1}$$

3. Carbene formation

$$-CH_2X + \cdot S_n \cdot \rightarrow -CH_2-S_nX \rightarrow -CH\!\!-\!\!S_{n-1} \rightarrow -CH: + \cdot S_n \cdot + HX \qquad (4.142)$$

4. Alkene formation

a) $\quad -CH: + -CH_2X \rightarrow -CH_2CHX- \xrightarrow{-HX} -CH=CH- \qquad (4.143)$

b) Intramolecular dehydrohalogenation

$$\underset{H\ \ X}{\text{>C-C<}} + \cdot S_n \cdot \rightarrow \underset{H\ \ S_n}{\text{>C-C<}} \rightarrow \text{>C=C<} + HS_nX \qquad (4.144)$$
$$\qquad\qquad\qquad\qquad\qquad\qquad HX + \cdot S\cdot_{n-1}$$

c) Intramolecular dehalogenation

$$\underset{X\ \ X}{\text{>C-C<}} + \cdot S_n \cdot \rightarrow \underset{X\ \ S_n}{\text{>C-C<}} \rightarrow \text{>C=C<} + XS_nX \qquad (4.145)$$
$$\qquad\qquad\qquad\qquad\qquad\qquad S_2X_2 + \cdot S\cdot_{n-2}$$

A scheme describing the intermolecular formation of sulfur-containing heteroarenes by the reaction of sulfur with arylalkanes and arylalkenes has been proposed [141]. Thus, for example, the formation of phenyl derivatives of thiophene from stilbene or styrene has been explained as follows:

$$RCH=CHR' \xrightarrow{\cdot S\cdot} RCH-\dot{C}HR' \xrightarrow{RCH=CHR'} R'CH-CHR \rightarrow$$
$$\qquad\qquad\qquad \underset{S\cdot}{|} \qquad\qquad\qquad \underset{RCH\ \ CHR'}{|\quad\ |}$$
$$\qquad\qquad\qquad\qquad\qquad\qquad\qquad S$$

$$\rightarrow \underset{RCH\ \ CHR'}{\overset{R'CH-CHR}{|\qquad|}} \xrightarrow{2\cdot S\cdot} \underset{RC\ \ CR'}{\overset{R'C-CR}{\|\quad\|}} + 2H_2S \qquad (4.146)$$
$$\qquad\quad S \qquad\qquad\qquad\quad S$$

$$R = C_6H_5,\ H;\ \ R' = C_6H_5$$

The addition of the sulfur biradical to the alkene gives an unstable adduct which undergoes cyclization as soon as it attains the necessary conformation for the formation of a stable ring [141].

The formation of thiophene-like sulfur heterocycles, benzo[b]-thiophene, [1]benzothieno[2,3-b]-[1]benzothiophene and [1]benzo-[3,2-b]-[1]benzothiophene from alkenes or halo-substituted alkenes, can occur via insertion of the sulfur biradical into the C-H or C-X (X = halogen) bond followed by cyclization. Cyclization can be either intermolecular or intramolecular. In the case of an inter-molecular reaction, the sulfur heterocycle is formed by the inter-action of sulfur with two alkene molecules, probably according to the following scheme [132]:

$$
\begin{array}{c}
-CH \\
\parallel \quad + \cdot S_n \cdot \rightarrow \\
-CH
\end{array}
\quad
\begin{array}{c}
-CS_nH \\
\parallel \\
-CH
\end{array}
\qquad (4.147)
$$

$$
\begin{array}{cc}
-CS_nH & HC- \\
\parallel \quad + \quad \parallel \\
-CH & HC-
\end{array}
\rightarrow
\left[
\begin{array}{c}
S_{n-1} \!\!\diagup\!\! \overset{S}{\diagdown} \!\! H \\
\overset{\diagdown}{X} \overset{}{C} \cdots C \\
\parallel \quad \parallel \\
HC- \; -CH
\end{array}
\right]
\rightarrow H_2S + \cdot S'_{n-1}
\qquad (4.148)
$$

$$
+
\begin{array}{cc}
-C-C- \\
\parallel \quad \parallel \\
HC - CH
\end{array}
\xrightarrow{+\cdot S_n}
\left[
\begin{array}{c}
-C-C- \\
\parallel \quad \parallel \\
-C \quad C \\
\diagup S' \diagdown H \\
S_{n-2} \text{---} SH
\end{array}
\right]
\rightarrow
\begin{array}{c}
-C-C- \\
\parallel \quad \parallel \\
-C \quad C- \\
\diagdown S \diagup
\end{array}
+ H_2S + \cdot S'_{n-2}
$$

$$
\begin{array}{c}
-CH \\
\parallel \quad + \cdot S_n \cdot \rightarrow \\
-CX
\end{array}
\quad
\begin{array}{c}
-CS_nH \\
\parallel \\
-CX
\end{array}
\qquad (4.149)
$$

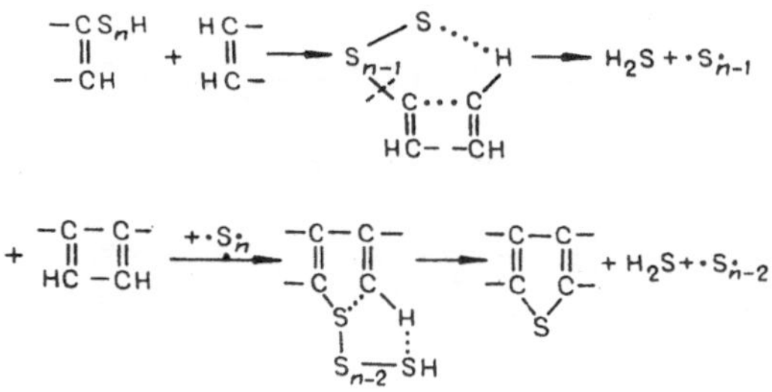

$$
\begin{array}{cc}
-CS_nH & XC- \\
\parallel \quad + \quad \parallel \\
-CX & HC-
\end{array}
\rightarrow
\left[
\begin{array}{c}
H \cdots \\
S_n \!\!\diagup\!\! \diagdown X \\
C \cdots C \\
\parallel \quad \parallel \\
XC- \; -CH
\end{array}
\right]
\rightarrow HX + \cdot S'_n
$$

$$
+
\begin{array}{cc}
-C - C- \\
\parallel \quad \parallel \\
-CX \quad HC-
\end{array}
\xrightarrow{+\cdot S\cdot}
\left[
\begin{array}{c}
-C-C- \\
\parallel \quad \parallel \\
C \quad C- \\
X \diagup \; S \\
H - S_{n-1}
\end{array}
\right]
\rightarrow
\begin{array}{c}
-C-C- \\
\parallel \quad \parallel \\
-C \quad C- \\
\diagdown S \diagup
\end{array}
+ HX + \cdot S'_{n-1}
\qquad (4.150)
$$

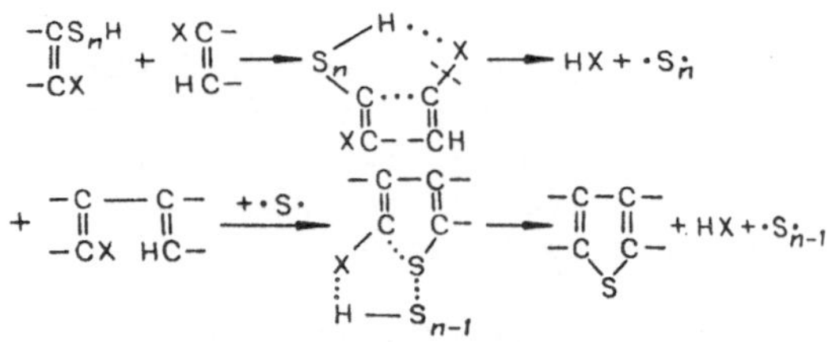

Both reactions, i.e., (4.148) and (4.150), occur via insertion of the $\cdot S_n \cdot$ biradical into the C-H bond. The insertion product so formed reacts with a second alkene and is converted to the corre-sponding diene, via the intermediate formation of a cyclic activated complex. The possibility of the intermediate formation of cyclic activated complexes in the reaction of sulfur with olefins was initially suggested by Syrkin (however, the complexes proposed by Syrkin had a different structure) [149]. The cyclization of a diene to thiophene occurs via a similar complex.

The mechanism of formation of 1,2-dithiol-3-thione from halo-arylpropanes can be illustrated by the following scheme [135]:

$$\begin{array}{ccc} \overset{\displaystyle >}{\underset{\displaystyle /}{C}}\!-\!\overset{\displaystyle <}{C} + \cdot S_n^{\cdot} & \longrightarrow & \overset{\displaystyle >}{\underset{\displaystyle /}{C}}\!-\!\overset{\displaystyle <}{C} & \longrightarrow & \overset{\displaystyle >}{}C\!=\!C\overset{\displaystyle <}{} + HX + \cdot S_n^{\cdot} \end{array} \tag{4.151}$$

The direction in which the insertion of the $\cdot S_n \cdot$ biradical takes place is determined by the relative reactivities of the C–H and C–X bonds.

$$\begin{array}{c} -\overset{|}{C}-CH_3 \\ \| \\ -\overset{|}{C}X \end{array} + \cdot S_n^{\cdot} \longrightarrow \longrightarrow \longrightarrow \begin{array}{c} -C-CH_2 \\ \| \ \ | \\ -C \ \ S \\ \diagdown\!S\!\diagup \end{array} + HX + \cdot S_{n-2}^{\cdot} \tag{4.152}$$

$$\begin{array}{c} -C-CH_2 \\ \| \ \ | \\ -C \ \ S \\ \diagdown\!S\!\diagup \end{array} + \cdot S_n \cdot \to \begin{array}{c} -C-C\!=\!S \\ \| \ \ | \\ C \ \ S \\ \diagdown\!S\!\diagup \end{array} + H_2S + \cdot S \cdot_{n-2} \tag{4.153}$$

$$\begin{array}{c} -C-CH_2X \\ \| \\ -CH \end{array} + \cdot S_n^{\cdot} \longrightarrow \longrightarrow \begin{array}{c} -C-CHX \\ \| \ \ | \\ -C \ \ S \\ \diagdown\!S\!\diagup \end{array} + H_2S + \cdot S_{n-3}^{\cdot} \tag{4.154}$$

$$\begin{array}{c} -C-CHX \\ \| \ \ | \\ -C \ \ S \\ \diagdown\!S\!\diagup \end{array} + \cdot S_n \cdot \to \begin{array}{c} -C-C\!=\!S \\ \| \ \ | \\ -C \ \ S \\ \diagdown\!S\!\diagup \end{array} + HX + \cdot S \cdot_{n-1} \tag{4.155}$$

$$\begin{array}{c} -C-CH_3 \\ \| \\ -CX \end{array} + \cdot S_n^{\cdot} \longrightarrow \longrightarrow \begin{array}{c} -C-CH_2 \\ \| \ \ | \\ -C \ \ S \\ \diagdown\!S\!\diagup \end{array} + HX + \cdot S_{n-2}^{\cdot} \tag{4.156}$$

$$\begin{array}{c} -C-CHX \\ \| \ \ | \\ -C \ \ S \\ \diagdown\!S\!\diagup \end{array} + \cdot S_n \cdot \to \begin{array}{c} -C-C\!=\!S \\ \| \ \ | \\ -C \ \ S \\ \diagdown\!S\!\diagup \end{array} + HX + \cdot S \cdot_{n-1} \tag{4.157}$$

The formation of thiirane, thiolane, and 1,4-dithiane rings, e.g., in the reaction of sulfur with tetrafluoroethylene, can be explained by primary addition of the $\cdot S_n \cdot$ biradical to the double bond [135]:

$$\overset{\displaystyle \diagdown}{\underset{\displaystyle \diagup}{C}}\!=\!C\overset{\displaystyle \diagup}{\underset{\displaystyle \diagdown}{}} + \cdot S_n \cdot \to \cdot S_n\!-\!\overset{|}{\underset{|}{C}}\!-\!\overset{|}{\underset{|}{C}}\cdot \to \overset{\displaystyle \diagdown}{\underset{\displaystyle \diagup}{C}}\!-\!C\overset{\displaystyle \diagup}{\underset{\displaystyle \diagdown}{}} + \cdot S \cdot_{n-1} \tag{4.158}$$

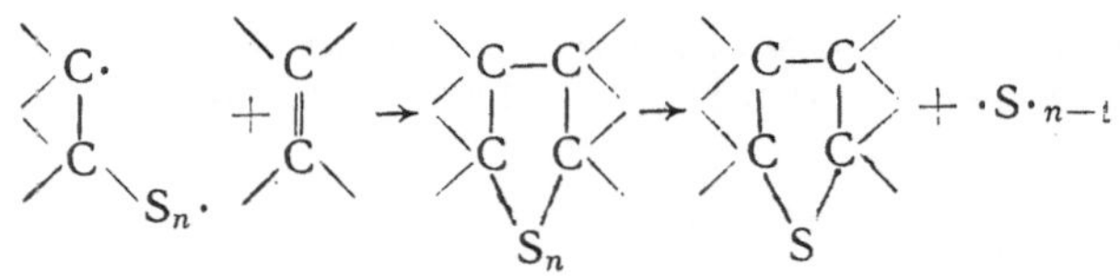

$$+ \cdot S \cdot_{n-t} \qquad (4.159)$$

$$+ 2 \cdot S \cdot_{n-t} \qquad (4.160)$$

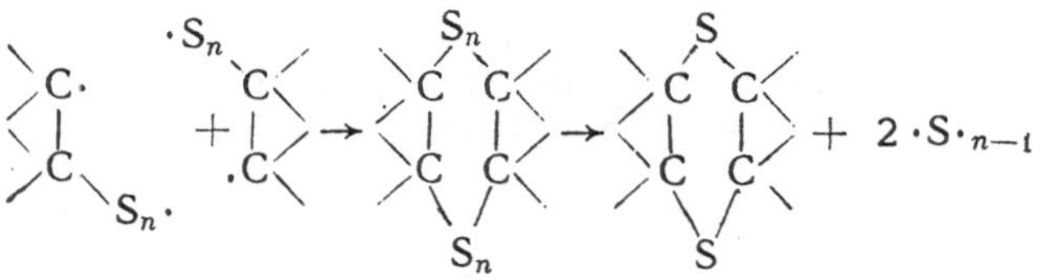

It is possible that the intramolecular ring closure in the formation of the thiophene system in the reaction of sulfur with alkenes and haloalkenes can also occur via scheme (4.147), by dehydrohalogenation or dehalogenation of the intermediate thiolanes. In agreement with this, intramolecular formation of the thiophene ring from dienes, e.g., in the reaction of sulfur with hexachloro-butadiene, can be represented as follows:

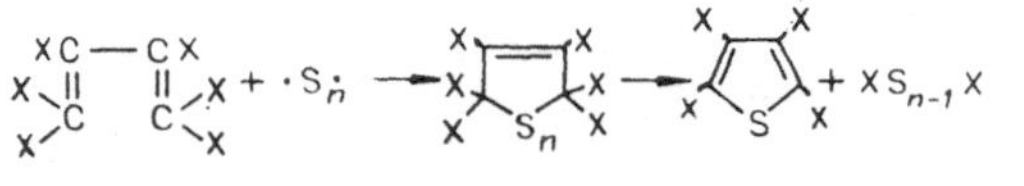

$$(4.161)$$

$$X = H, \; Cl, \; Br$$

The formation of thiophene derivatives, for example in the reaction of sulfur with nonterminal alkynes, may be demonstrated by the following scheme:

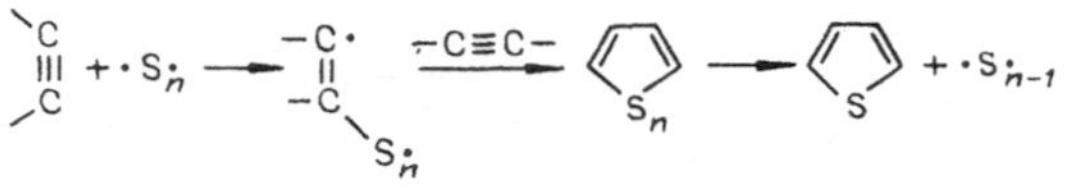

$$(4.162)$$

In the case of an intramolecular reaction, the mechanism of the formation of benzo[b]thiophenes can be illustrated as follows:

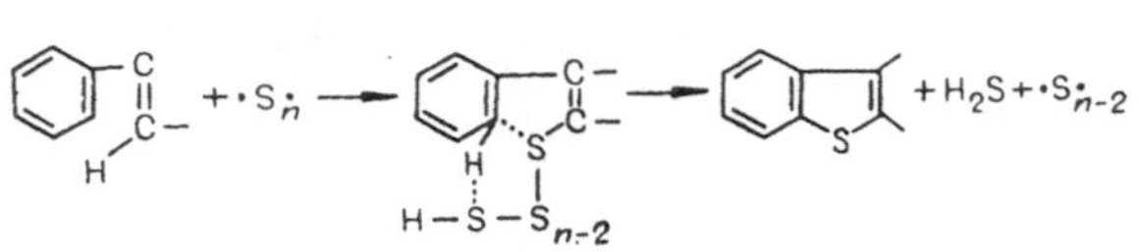

$$+ H_2S + \cdot S \cdot_{n-2} \qquad (4.163)$$

$$+ HX + \cdot S \cdot_{n-1} \qquad (4.164)$$

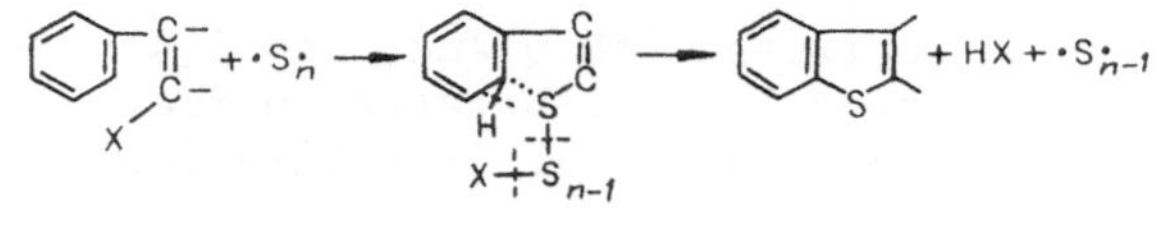

The reaction starts with the insertion of the $\cdot S_n \cdot$ biradical into a C–H or C–X bond at the unsaturated carbon atom and is completed by the formation and decomposition (in one step) of an intra-molecular complex. This also makes it possible to explain the subsequent formation of [1]benzothieno[2,3-b]-[1]benzothiophene from phenylhalo-substituted benzo[b]thiophenes.

All the previously discussed hypothetical schemes treat the reactions of sulfur with hydrocarbons and their halo-derivatives from just one point of view, i.e., insertion of the $\cdot S_n \cdot$ biradicals into C-H and C-X bonds or their addition to multiple bonds, and may thus account for the various reactions occurring between sulfur and organic compounds. Nevertheless, in many cases it is still not clear whether the primary reaction step is the insertion of a biradical into a C-H or C-X bond, or whether it is its addition to a multiple bond.

In spite of all the previous discussions, the experimental data available so far are not sufficient to reject views stating that the reaction of sulfur with arylaliphatic hydrocarbons and their halo-derivatives proceeds via the intermediate formation of free radicals of the benzylic type [133]. The mechanism of dehydrohalogenation and dehalogenation by sulfur can then be represented as follows:

Intermolecular dehydrohalogenation

$$
\begin{aligned}
-CH_2X + \cdot S_n \cdot &\longrightarrow -CH_2 \cdot + \cdot S_nX \\
2-CH_2 \cdot &\longrightarrow -CH_2-CH_2- \\
-CH_2-CH_2- + \cdot S_nX &\longrightarrow -CH_2\overset{\cdot}{C}H- + HS_nX \\
-CH_2\overset{\cdot}{C}H- + \cdot S_nX &\longrightarrow -CH=CH- + HS_nX \\
HS_nX &\longrightarrow HX + \cdot S_n \cdot
\end{aligned}
$$

(4.165)

Intramolecular dehydrohalogenation and dehalogenation

$$
\begin{aligned}
{>}CH-CX{<} + \cdot S_n \cdot &\to {>}CH-\overset{\cdot}{C}{<} + \cdot S_nX \\
{>}CH-\overset{\cdot}{C}{<} + \cdot S_nX &\to {>}C=C{<} + HS_nH \\
H-S_n-X &\to HX + \cdot S_n \cdot \\
{>}CX-CX{<} + \cdot S_n \cdot &\to {>}CX-\overset{\cdot}{C}{<} + \cdot S_nX \\
{>}CX-\overset{\cdot}{C}{<} + \cdot S_nX &\to {>}C=C{<} + XS_nX \\
XS_nX &\to S_2X_2 + S_{n-2}
\end{aligned}
$$

(4.166)

This mechanism, i.e., the free-radical mechanism for the reaction of sulfur with arylhaloalkanes, is favored by the formation of radicals in the photolysis of compounds of the series $C_6H_5CH_2X$, $C_6H_5CHX_2$, and $C_6H_5CX_3$ [150]:

$$
\begin{aligned}
C_6H_5CH_2X &\longrightarrow C_6H_5CH_2 \cdot + X \cdot \\
C_6H_5CHX_2 &\longrightarrow C_6H_5CHX \cdot + X \cdot
\end{aligned}
$$

(4.167)

At the same time, reactions (4.167) do not occur. The intermediate formation of radicals of the benzylic type also explains why the arylaliphatic compounds react with sulfur more readily than aliphatic compounds. Ring closures are also facilitated in benzylic systems.

$$C_6H_5CH_2X \longrightarrow C_6H_5CH: + HX$$
$$C_6H_5CH_2X \longrightarrow C_6H_5CHX \cdot + H \cdot \qquad (4.168)$$

Intermolecular closure of the thiophene ring is shown by the following scheme:

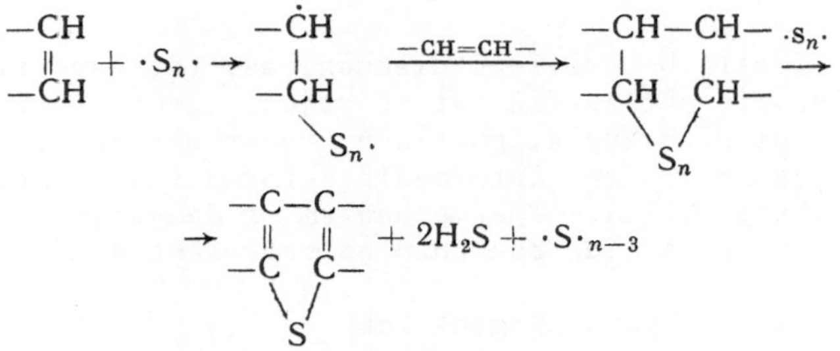

$$
\begin{array}{l}
-CH \\
\| \quad + \cdot S_n \cdot \rightarrow \\
-CH
\end{array}
\quad
\begin{array}{l}
-\overset{\cdot}{C}H \\
| \\
-CH \\
\quad \backslash S_n \cdot
\end{array}
\quad \xrightarrow{-CH=CH-} \quad
\begin{array}{l}
-CH-CH- \cdot S_n \cdot \\
| \qquad | \\
-CH \quad CH- \\
\quad \backslash S / \\
\quad \quad n
\end{array}
\longrightarrow
$$

$$
\rightarrow \quad
\begin{array}{l}
-C-C- \\
\| \quad \| \qquad + 2H_2S + \cdot S \cdot_{n-3} \\
-C \quad C- \\
\quad \backslash S /
\end{array}
\qquad (4.169)
$$

Intramolecular closure of the thiophene ring may be represented as follows:

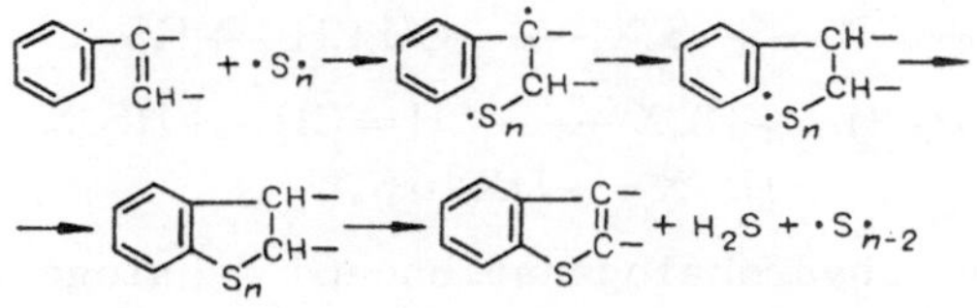

$$ \qquad (4.170) $$

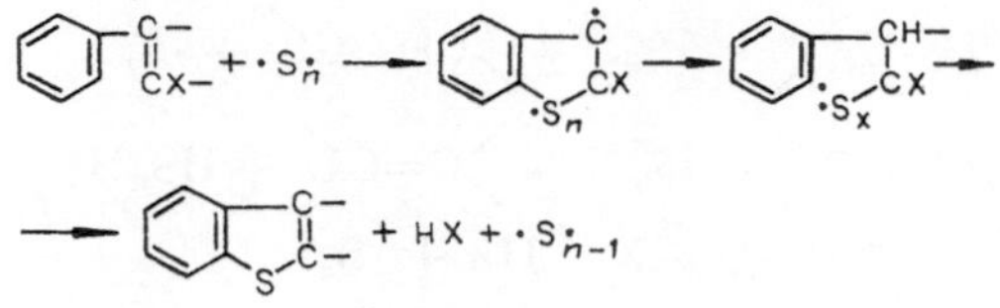

$$ \qquad (4.171) $$

In both cases of the formation of the benzo[b]thiophene ring, the reaction occurs via addition of the $\cdot S_n \cdot$ biradical to the terminal carbon atom, i.e., in the direction common to radical reactions, and subsequent rearrangement (with hydrogen atom transfer) of the benzylic-type radical formed (it seems that this reaction is catalyzed by hydrobromic acid with subsequent cyclization). The process is completed by either simultaneous or successive extrusion of excess sulfur, hydrogen sulfide or halogen hydracid. Thus, the thermal reactions of sulfur with hydrocarbons and their halo-derivatives clearly follow a radical mechanism.

REFERENCES

1. P. Klason, Ber., 20, 2376 (1887).
2. V. A. Gomin, Zh. Obshch. Khim., 6, 852 (1936).
3. R. de Fazi, Gazz. Chim. Ital., 54, 251 (1924).
4. S. R. Sterlin, L. S. Zhuravkova, and V. L. Dyatkin, USSR Patent 417,419 (1974).
5. A. von Bartal, Ber., 38, 3067 (1903).
6. D. M. Marquis, U.S. Patent 2,962,529 (1960); C. A., 55, 7285 (1961).
7. G. A. R. Brandt, H. T. Emeleus, and R. N. Haszeldine, J. Chem. Soc., 2198 (1952).
8. J. Baer, Brit. Patent 279,406 (1926); C. A., 22, 2856 (1929).
9. J. Baer, Swiss Patent 127,540 (1926), 132,505-8 (1926); C. A., 23, 5026 (1930).
10. W. J. Middleton, U.S. Patent 3,069,395 (1963).
11. R. E. Banks, G. M. Hasiman, R. N. Haszeldine, and A. Peppin, J. Chem. Soc., C 1171 (1966).
12. M. Elbanowski, Przem. Chem., 12, 95 (1971).
13. F. Boberg, Liebigs Ann. Chem., 679, 109 (1964).
14. M. Hauptschein and A. Grosse, J. Am. Chem. Soc., 73, 5461 (1951).
15. M. Hauptschein and R. E. Oesterling, U.S. Patent 3,256,328 (1966).
16. M. Hauptschein and R. E. Oesterling, U.S. Patent 3,209,036 (1965).
17. P. D. Faurote and J. G. O'Rear, J. Am. Chem. Soc., 78, 4999 (1956).
18. A. N. Akapyan, A. M. Saaklan, and G. A. Dzhauri, Arm. Khim. Zh., 21, 414 (1968).
19. G. von Dyke Tiers, J. Org. Chem., 26, 2538 (1961).
20. A. Jonczyk, Angew. Chem., Int. Ed. Engl., 18, 217 (1979).
21. R. Mayer, W. Thiel, and H. Viola, Z. Chem., 19, 56 (1979).
22. R. Mayer, W. Thiel, H. Viola, and E. A. Jauer, East Ger. Patent 133,663 (1979); C. A., 91, P74217 (1979).
23. W. Thiel, H. Viola, and R. Mayer, Z. Chem., 17, 366 (1977).
24. R. Mayer, H. Viola, and B. Hopf, Z. Chem., 18, 90 (1978).
25. M. Mayer, H. Viola, and M. Thiel, East Ger. Patent 127,428 (1977).
26. S. Suzuki, U.S. Patent, 3,929,896 (1975); C. A., 84, 105000 (1976).
27. M. Kunio and H. Hitoshi, Kobunshi Kagaku, 30, 622 (1973); C. A., 80, 71325 (1974).
28. L. G. Anello and R. F. Swenly, U.S. Patent 3,821,290 (1974); C. A., 81, 104751 (1974).
29. E. M. Lown, E. L. Dedio, O. P. Strausz, and H. E. Gunning, J. Am. Chem. Soc., 89, 1056 (1967).
30. R. H. Robson and Th. B. Doger, U.S. Patent 3,278,553 (1968); Ref. Zh. Khim., 24N332P (1968).
31. R. L. Hodgson and E. J. Smuthy, U.S. Patent 3,394,146 (1968).

32. H. J. Pettelkau and W. Göbel, Ger. Patent 2,645,920 (1978); Ref. Zh. Khim., 40309P (1979).

33. G. R. Schultze and F. Boberg, Ger. Patent 1,128,432 (1961); C. A., 12479 (1962); U.S. Patent 3,062,833 (1962); C. A., 58, 9082 (1963).

34. F. Boberg, Angew. Chem., 72, 629 (1960).

35. J. Park, S. Hopwood, and J. Lacher, J. Org. Chem., 23, 1169 (1958).

36. C. E. Aho, Fr. Patent 1,405,653 (1963); C. A., 65, PC5640g (1963).

37. E. Geering, J. Org. Chem., 24, 1128 (1959).

38. W. Middleton, E. Howard, and W. Sharkey, J. Org. Chem., 30, 1375 (1965).

39. O. M. Nefedow, Angew. Chem., 78, 1055 (1966).

40. A. C. Pierce, U.S. Patent 3,520,903 (1970).

41. K. V. Martin, J. Chem. Soc., 2944 (1964).

42. K. V. Martin, U.S. Patent 3,048,629 (1962); C. A., 58, 459 (1963).

43. W. R. Brasen, H. N. Cripps, C. G. Bottomley, H. W. Farlon, and C. G. Krespan, J. Org. Chem., 30, 4188 (1965).

44. C. G. Krespan and W. R. Brasen, J. Org. Chem., 27, 3995 (1962).

45. C. G. Krespan, U.S. Patent 3,119,836 (1963); C. A., 59, 50225, 3910p (1963).

46. D. M. Marquis, U.S. Patent 3,097,236 (1963); C. A., 59, 13825 (1963).

47. C. G. Krespan and C. M. Langkammer, J. Org. Chem., 27, 3584 (1962).

48. C. G. Krespan, J. Org. Chem., 27, 3588 (1962).

49. G. G. Belenkin, V. L. Kopaevich, L. S. German, and I. L. Knunyants, Dokl. Akad. Nauk SSSR, 201, 603 (1971).

50. T. Satokawa and Y. Ohsaka, Jpn. Patent 7,737,000 (1977).

51. C. G. Krespan, U.S. Patent 3,088,935 (1963).

52. C. G. Krespan, U.S. Patent 3,149,124 (1964).

53. H. Broun, J. Org. Chem., 22, 715 (1957).

54. H. G. Horn, Chem. Ztg., 95, 893 (1971).

55. S. P. Sterling, B. L. Dyatkin, and I. L. Knunyants, Izv. Akad. Nauk. SSSR, Ser. Khim., 2583 (1967).

56. B. L. Dyatkin, S. P. Sterling, L. G. Tsurkanova, and I. L. Knunyants, Dokl. Akad. Nauk SSSR, 183, 598 (1968).

57. B. L. Dyatkin, S. P. Sterling, L. G. Tsurkanova, B. I. Martynov, E. I. Misov, and I. L. Knunyants, Tetrahedron, 29, 2759 (1973).

58. C. G. Krespan, U.S. Patent 3,069,431 (1962).

59. C. J. Benning, U.S. Patent 2,968,659 (1961); C. A., 55, 13460 (1961).

60. G. Beck and H. Holdschmidt, Ger. Patent 2,213,865 (1973); C. A., 80, 27243 (1974).

61. G. Beck and H. Holdschmidt, Ger. Patent 2,331,795 (1975); C. A., 82, 170883 (1975).

62. G. Beck and H. Holdschmidt, Ger. Patent 2,214,610 (1973); Ref. Zh. Khim., 20N623 (1973).

62a. E. Degener, G. Beck, and H. Holdschmidt, Ger. Patent 2,231,698 (1974); C. A., 80, 82951 (1974).
63. C. G. Krespan, B. McKusick, and T. Cairrs, J. Am. Chem. Soc., 82, 1515 (1960).
64. H. A. Wiebe, S. Braslavsky, and J. Heickolen, Can. J. Chem., 50, 2721 (1972).
65. C. G. Krespan, U.S. Patent 3,052,691 (1962).
66. C. G. Krespan and D. C. Engbad, J. Org. Chem., 33, 1850 (1968).
67. R. D. Chambers, Partingstons International Symposium on Fluorine Chemistry, California (1973), p. 0-1.
68. O. P. Strausz, J. Font, E. L. Dedio, P. Kebarle, and H. E. Gunning, J. Am. Chem. Soc., 89, 4805 (1967).
69. F. Becke, Angew. Chem., 72, 867 (1960).
70. Ger. Patent 1,078,556 (1960); C. A., 55, 14378 (1961).
71. S. Oae and G. Zsuchida, Tetrahedron Lett., 1283 (1972).
72. Netherland Patent 6,501,993 (1964); C. A., 64, 6837 (1966).
73. B. Hortling and J. J. Lindberg, Chem. Scr., 2, 179 (1972); C. A., 78, 44048 (1973).
74. M. Schmidt, Inorg. Macromol. Rev., 1, 101 (1970).
75. A. D. Macullum, J. Org. Chem., 13, 154 (1948).
76. R. M. Lenz and W. K. Carrington, J. Polym. Sci., 43, 167 (1960).
77. R. M. Lenz and W. K. Carrington, J. Polym. Sci., 41, 333 (1959).
78. B. Hortling and I. I. Lindberg, Macromol. Chem., 179, 1707 (1978).
79. M. G. Voronkow and A. N. Perferkowitsch, Angew. Chem., 81, 257 (1969).
80. Z. Vrba and J. Kvapil, Czech. Patent 168,330 (1977); C. A., 88, 50483 (1978).
81. S. Oae and Y. Tsuchida, Jpn. Kokai, 74, No. 18 (1974); C. A., 80, 133021 (1974).
82. D. J. Harper, Brit. Patent 1,428,761 (1976); C. A., 85, 109793 (1976).
83. M. Blazejak and J. Hayden, Ger. Patent 1643320 (1975); Ref. Zh. Khim., 20N179P (1976).
84. G. Beck, H. Heitzer, and H. Holdschmidt, XXIV IUPAC Congress, Abstracts of Papers, Hamburg (1973), p. 448.
85. G. Beck and H. Holdschmidt, Ger. Patent 2,229,162 (1974); C. A., 80, 83004 (1974).
86. S. Oae, M. Kozuka, N. Furukuwa, Y. Tsuchida, and H. Yamamoto, Jpn. Kokai, 74, 127,954 (1954); C. A., 82, 155718 (1975).
87. H. Miyake, E. Takahashi, H. Kamiyama, K. Terada, Jpn. Kokai, 75, 62192 (1975); C. A., 83, 195901 (1975); ibid., 75, 36,393 (1975); C. A., 84, 107840 (1976).
88. G. Beck and H. Holdschmidt, Ger. Patent 2,224,835 (1973); C. A., 80, 83021 (1974).
89. G. Beck and H. Holdschmidt, Ger. Patent 2,224,746 (1973); C. A., 80, 480022 (1974).
90. G. Beck, H. Heitzer, and H. Holdschmidt, XXIV IUPAC Congress, Abstracts of Papers, Hamburg (1973), p. 448.

91. J. B. Rayman and H. Gibson, Brit. Patent 292,057 (1927); C. A.,
     23, 1229 (1930).

92. G. Kalischer, H. Salkowski, and F. Frister, Ger. Patent 480,377
     (1929); C. A., 23, 5046 (1929).

93. G. Beck, H. Holdschmidt, and K. Ley, Ger. Patent 2,224,836
     (1973); C. A., 80, 47965 (1974).

94. G. Beck and H. Holdschmidt, Ger. Patent 2,224,834 (1973);
     C. A., 80, 48009 (1974).

95. Y. Fujimoto and J. Masamura, Ger. Patent 2,325,647 (1973);
     C. A., 80, 96968 (1974).

96. G. Beck and R. Braden, Ger. Patent 2,229,163 (1974); C. A., 80,
     83076 (1974).

97. S. C. Cohen, M. L. N. Reddy, and A. G. Massey, Chem. Commun.,
     451 (1967).

98. C. S. Cohen and M. L. H. Reddy, J. Organomet. Chem., 11, 563
     (1968).

99. C. S. Cohen and A. G. Massey, J. Organomet. Chem., 12, 341
     (1968).

100. G. G. Jakobson, G. G. Furin, and T. V. Terent'eva, Izv. Akad.
     Nauk. SSSR, Ser. Khim., 2128 (1972).

101. G. G. Jakobson, G. G. Furin, and T. V. Terent'eva, Zh. Org.
     Khim., 10, 799 (1974).

102. E. K. Fields, U.S. Patent 3,449,119 (1970).

103. E. K. Fields, U.S. Patent 3,514,480 (1970).

104. M. G. Voronkov and L. M. Khokhlova, Zh. Org. Khim., 10, 811
     (1974).

105. L. P. Broun and M. Thompson, J. Chem. Soc., Perkin Trans. 1,
     863 (1974).

106. F. Becke and H. Hagen, Liebigs Ann. Chem., 729, 146 (1969).

107. G. G. Boguskii, Zh. Russ. Khim. O-va., 36, 1554 (1904).

108. J. S. Teletov and N. D. Pelik, Ukr. Khim. Zh., 4, 387, 457
     (1929).

109. H. R. Kruyt, Z. Phys. Chem., 65, 486 (1909).

110. M. G. Voronkov and V. E. Udre, USSR Patent 165,470 (1964);
     Byull. Izobret., No. 19, 17 (1964).

111. M. G. Voronkov and V. E. Udre, Khim. Geterotsikl. Soedin., 148
     (1965).

112. A. Aman, H. Koening, and H. Gierte, Ger. Patent 2,306,543
     (1974); C. A., 81, 136150 (1974).

113. H. Hagen and H. Fleig, Ger. Patent 2,460,783 (1976); C. A., 85,
     12899 (1976).

114. M. G. Voronkov and V. E. Udre, USSR Patent 203,697 (1967);
     Byull. Izobret., No. 21, 32 (1967).

115. M. G. Voronkov and V. E. Udre, USSR Patent 172,833 (1965);
     Byull. Izobret., No. 14, 33 (1965).

116. M. G. Voronkov and V. E. Udre, USSR Patent 169,127 (1964);
     Byull. Izobret., No. 6, 28 (1965).

117. M. G. Voronkov and V. E. Udre, Khim. Geterotsikl. Soedin., 683
     (1965).

118. E. J. Geering and G. Island, U.S. Patent 3,278,552 (1966).

119. M. G. Voronkov and V. E. Udre, USSR Patent 199,909 (1967);
      Byull. Izobret., No. 16, 29 (1967).
120. M. G. Voronkov and V. E. Udre, Khim. Geterotsikl. Soedin., 527
      (1966).
121. R. Mayer, H. Viola, J. Reichert, and W. Krause, J. Prakt. Chem.,
      320, 313 (1978).
122. R. Mayer, W. Thiel, and H. Viola, Z. Chem., 16, 395 (1976).
123. W. Thiel, H. Viola, and R. Mayer, Z. Chem., 17, 92 (1977).
124. M. G. Voronkov, V. E. Udre, and A. O. Taube, USSR Patent
      371,233 (1972); Byull. Izobret., No. 12, 74 (1973).
125. M. G. Voronkov, V. E. Udre, and E. P. Popova, Khim. Geterotsikl.
      Soedin., 1003 (1967).
126. J. Drake, Brit. Patent 2,538,722 (1951); C. A., 45, 4209 (1951).
127. H. Wegteche, Chem. Rev., 39, 219 (1946).
128. M. G. Voronkov, V. E. Udre, and E. P. Popova, USSR Patent
      274,118 (1970); Byull. Izobret., No. 21, 29 (1970).
129. M. G. Voronkov and V. E. Udre, Khim. Geterotsikl. Soedin., 43
      (1968).
130. M. G. Voronkov and V. E. Udre, USSR Patent 181,087 (1966);
      Byull. Izobret., No. 19, 29 (1966).
131. M. G. Voronkov and V. E. Udre, USSR Patent 230,184 (1968);
      Byull. Izobret., No. 34, 34 (1968).
132. M. G. Voronkov and V. E. Udre, Khim. Geterotsikl. Soedin., 457
      (1970).
133. M. G. Voronkov, T. V. Lapina, and E. P. Popova, Khim.
      Geterotsikl. Soedin., 633 (1967).
134. M. G. Voronkov and T. V. Lapina, USSR Patent 175,516 (1965);
      Byull. Izobret., No. 25, 24 (1965).
135. M. G. Voronkov, T. V. Lapina, and E. P. Popova, Khim.
      Geterotsikl. Soedin., 49 (1968).
136. M. G. Voronkov, T. V. Lapina, and E. P. Popova, USSR Patent
      213,898 (1968); Byull. Izobret., No. 11, 36 (1968).
137. M. G. Voronkov, V. E. Udre, and T. V. Lapina, Second Organic
      Sulphur Symposium, Groningen, Abstracts of Papers, 38 (1966).
138. M. G. Voronkov, Izv. Akad. Nauk. SSSr, Ser. Khim., No. 3, 33
      (1963).
139. M. G. Voronkov, A. N. Perferkovich, M. P. Gavar, and G. V.
      Ozolin', Khim. Geterotsikl. Soedin., 1183 (1970).
140. W. M. Bryce and C. Hinschelwood, J. Chem. Soc., 3379 (1949).
141. A. W. Horton, J. Org. Chem., 14, 761 (1949).
142. C. Walling, Free Radicals in Solutions, Wiley (1957).
143. A. M. Nefedov and M. N. Manakow, Angew. Chem., 78, 1039 (1966).
144. V. Kirmse, Carbene Chemistry, Academic Press, New York—London
      (1964).
145. W. A. Pryor, Mechanisms of Sulfur Reactions, McGraw-Hill,
      New York (1962).
146. L. Heitinger, Monatsh. Chem., 4, 163 (1883).
147. G. Schulz and H. Beyschlag, Ber., 42, 743 (1909).
148. B. Holenhery, Ber., 43, 220 (1910).
149. Ya. K. Sirkin, Izv. Akad. Nauk SSSR, Ser. Fiz., 38, 606 (1974).

150. G. Minkov, Frozen Free Radicals, Inostr. Lit., Moscow (1962), p. 173.
151. C. M. Hull, U.S. Patent 2,259,695 (1941); C. A., $\underline{36}$, 894 (1942).
152. A. P. H. Dupire, Fr. Patent 985,938 (1951); C. A., $\underline{49}$, 12768 (1955).
153. Brit. Patent 824,529 (1959); C. A., $\underline{54}$, 6116 (1960).
154. R. Delaplace, J. Pharm. Chim., $\underline{26}$, 139 (1922).
155. A. Efard, Ann. Chem. Phys., $\underline{2}$, 503 (1894).
156. J. H. Hildebrand and R. L. Scott, Solubility of Non-Electrolytes, 3d ed., New York (1950).
157. J. Timmermans, The Physico-Chemical Constants of Binary Systems in Concentrated Solutions, Vol. IV, New York (1960), p. 667.
158. M. Amadori, Gazz. Chim. Ital., $\underline{52}$, 1, 387 (1922).
159. J. A. Wilkinson, C. Nelson, and H. M. Wylde, J. Am. Chem. Soc., $\underline{42}$, 1377 (1920).
160. D. H. Wester and A. Bruins, Pharm. Weekblad, $\underline{51}$, 1443 (1914).
161. C. G. Gemmelaro, Bol. Minero, Soc. Nacl. Mineria (Chile), $\underline{52}$, No. 56, 484 (1940).
162. W. Jacek, Rocz. Chem., $\underline{6}$, 501 (1926).
163. K. A. Hofmann, H. Kirmreuther, and A. Thal, Ber., $\underline{43}$, 183 (1910).
164. H. Rheinholdt and K. Schneider, J. Prakt. Chem., $\underline{120}$, 238 (1929).

5
Organic Sulfur Compounds

5.1. THIOLS

The reaction of sulfur with aliphatic thiols is always accompanied by the evolution of hydrogen sulfide. Reactions of alkanethiols with sulfur are base-catalyzed and normally lead to the formation of a mixture of dialkyl polysulfides, R_2S_n ($n \geqslant 2$) [1-7]. The reaction takes place at 25-60°C in polar solvents and in the presence of aliphatic amines:

$$2RSH + (n—1)S \longrightarrow RS_nR + H_2S \tag{5.1}$$

A maximum yield of the disulfides ($n = 2$) from primary and secondary alkanethiols is observed at the molar ratio RSH:S = 2.5. When this ratio is increased, a mixture of polysulfides is formed. When heated with sulfur, tert-alkanethiols do not form any disulfides, presumably because of steric hindrance.

Under certain conditions, all alkanethiols can be converted into dialkyl trisulfides [5]. In the case of primary and secondary alkanethiols, an excess of the thiol and a high temperature prevent the formation of trisulfides because, under these conditions, trisulfides decompose with the formation of disulfides. The conversion of tert-alkanethiols to the corresponding dialkyl trisulfides is facilitated at elevated temperatures and by the presence of a polar solvent.

Tetrasulfides are the predominant reaction products when sulfur reacts with tert-alkanethiols at low temperatures (25°C) (nonpolar solvent, ratio RSH:S = 1) [5,8].

173

Dodecanethiol reacts with sulfur in the presence of the zinc salts of organic acids to give didodecyl disulfide and zinc sulfide [9]. The reaction of sulfur with mercaptobenzothiazole proceeds similarly [10].

The possibility of sulfur being inserted into the C–S bond is demonstrated by the exchange reaction between ^{35}S and 1,3-benzo-thiazole-2-thiol which, in principle, proceeds with the participation of the sulfur atom of the mercapto group [11]. 2-Mercapto-3-pentanone reacts with sulfur from –70 to +20°C with the formation of bis-3-pentanon-2-yl disulfide [12]:

$$CH_3CH_2COCHSH + S \longrightarrow$$
$$\overset{|}{C}H_3$$

$$\rightarrow CH_3CH_2COCH-S_2-CHCOCH_2CH_3 \hspace{2cm} (5.2)$$
$$\overset{|}{C}H_3 \hspace{0.5cm} \overset{|}{C}H_3$$

It is likely that the first step in reaction (5.1) is the interaction of the alkanethiol with a base (B:) [5]:

$$RSH+B:\rightleftarrows[RS^{\delta-}\dots H^{\delta+}\dots B]\rightleftarrows RS^-BH^+ \hspace{2cm} (5.3)$$

$$\begin{matrix}[RS^{\delta-}-H^{\delta+}-B]\\ RS^-BH^+\end{matrix} + S\overset{\diagup S\diagdown}{\underset{S}{\underset{\diagdown S}{\diagup}}}{\underset{\diagup}{\overset{\diagdown S}{S}}}{\underset{\diagdown S \diagdown}{\diagup}}S \rightarrow \begin{matrix}R-S_8-S^-BH^+\\ RSH \downarrow\uparrow\\ RS_9H + RS^-BH^+\end{matrix} \hspace{1cm} (5.4)$$

$$RS_9H+RS^-BH^+ \longrightarrow RS_nR+HS^-BH^+ \hspace{2cm} (5.5)$$
$$n=2-9$$

$$mRS_nR+2z[RS^-BH^+] \longrightarrow yRS_xR+zH_2S+2zB: \hspace{1cm} (5.6)$$
$$x = 2, 3, \text{ or } 4$$

Nucleophilic attack of the thiolate ion upon the sulfur molecule leads to the opening of the eight-membered ring and formation of a linear adduct RS_n-RH^+ which further reacts with the starting thiol and gives an alkyl hydropolysulfide. Subsequent interaction of the thiolate with the alkyl hydropolysulfide leads to the formation of a mixture of dialkyl polysulfides [reaction (5.5)]. The statistical probability of the interaction of thiolate ion with the alkyl polysulfides determines the nature of the final products in reaction (5.1) [5].

The reaction of sulfur with alkanethiols is used in practice to free oils from malodorous admixtures [13], to obtain oil additives [14], and to synthesize polymeric materials based on polythiols which are used for markings on highway surfaces [15].

Heating higher tert-alkanethiols with sulfur at 200-220°C gives
1,2-dithiol-3-thiones [2]. The intermediate products formed in the
reaction are the alkenes $(H_3C)_3C-CH_2-\underset{CH_3}{C}=CH_2$ and $(H_3C)_2C-CH=C(CH_3)_2$:

$$2 (CH_3)_3CCH_2C(CH_3)_2SH + 8S \longrightarrow (CH_3)_3CCH_2 \underset{S}{\overset{S}{\diagdown}} S + CH_3 \cdots + 2 H_2S \qquad (5.7)$$

$$(CH_3)_3CCH_2C(CH_3)(SH)CH_2C(CH_3)_3 + 4 S \longrightarrow (CH_3)_3C-CH_2 \cdots + H_2S \qquad (5.8)$$

3-Phenyl-2-propen-1-thiol is readily converted to 5-phenyl-1,-
2-dithiol-3-thione on treatment with sulfur [16]:

$$2 C_6H_5\overset{CHCH_2SH}{\underset{\|}{CH}} + 4 S \xrightarrow{50°C} C_6H_5 \cdots + H_2S \qquad (5.9)$$

Unsymmetric disulfides have been prepared by treating sodium
alkanethiolates with sulfur and alkyl halides in alcohol [17]:

$$C_6H_{13}SNa + S \rightarrow C_6H_{13}SSNa \xrightarrow{+ XR} C_6H_{13}SSR \qquad (5.10)$$

$$R = Et, Pr, Bu, iso-Bu, tert-Bu$$

The reaction of the disodium salt of dimercaptomaleonitrile with
sulfur and tetracyano-1,4-dithiine leads to the formation of the
sodium salt of 3,4-dimercaptoisothiazole-5-carbonitrile [18,19,20]:

$$NCC(SNa)=C(SNa)CN + S + \cdots \rightarrow \cdots \qquad (5.11)$$

Kinetic data for the reaction of various substituted thiophenols
with liquid sulfur at 130-160°C have been obtained and intermediates
of general formula RS_xH and HR_xH were identified. The reaction is
usually free radical involving an initiation period in which a
steady state concentration of sulfur radical species is established.
The reaction appears to be second order in mercaptan and third order
in sulfur [21]:

$$2XC_6H_4SH + S \longrightarrow (XC_6H_4)_2S_2 + H_2S \qquad (5.12)$$

$$X = 4-NH_2, 4-MeO, 4-Me, H, 4-F, 4-Cl, 4-Br, 4-NO_2$$

The joint effect of elemental sulfur and highly basic amines (piperidine or morpholine) on the indenethiol (1) in methanol, chloroform

or dimethylformamide in an inert atmosphere at $20°C$ affords 3-amino-2-phenylindene-1-thiones (3), probably via the 3-mercapto-2-phenylindene-1-thione (2) [22]:

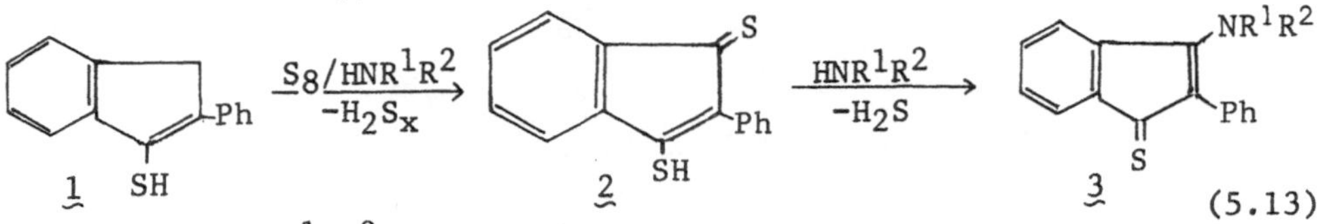

$$R^1, R^2 = (CH_2)_5(NH), \ (CH_2)_2O(CH_2)_2(NH)$$

(5.13)

The reaction of mercaptobenzothiazole or mercaptonaphthothiazole with sulfur and morpholine in the presence of oxidants yields the corresponding morpholyldithiotriazoles which may be used as rubber accelerators [23].

5.2. SULFIDES AND POLYSULFIDES

Dialkyl sulfides slowly react with elemental sulfur at $180°C$ and higher temperatures to give a mixture of the corresponding dialkyl polysulfides [24-27]. Dimethyl sulfide (or methanethiol) reacts with excess sulfur in the gas phase at $550-700°C$ for 5-30 seconds to give a 98% yield of carbon disulfide [28]:

$$RSR + (n-1)S \longrightarrow RS_nR \qquad (5.14)$$
$$R = alkyl, \ n \geqslant 2$$

Heating bis-3-methyl-1-butyl sulfide with excess sulfur at $200°C$ gives 4,5-dimethyl-1,2-dithiol-3-thione [29]:

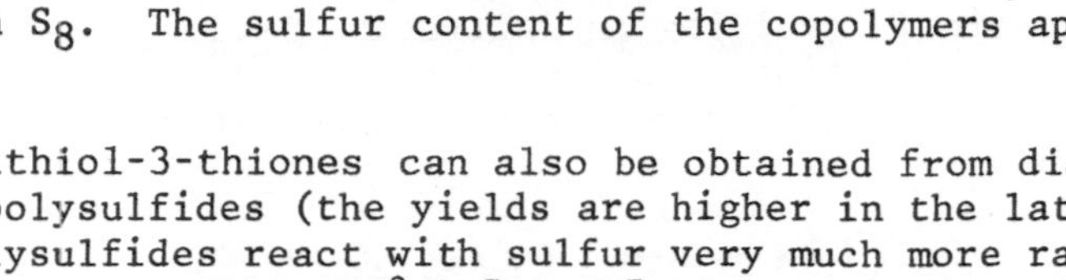

(5.15)

A high-molecular-weight $(5 \cdot 10^4)$ copolymer of elemental sulfur has been prepared by anionic copolymerization of 1,2-propylene sulfur with S_8. The sulfur content of the copolymers approached 90% [30].

1,2-Dithiol-3-thiones can also be obtained from dialkyl sulfides and polysulfides (the yields are higher in the latter case). Dialkyl polysulfides react with sulfur very much more rapidly at lower temperatures $(130-150°C)$ [31-35] with extension of the hydrocarbon chain, e.g.,

$$(CH_3)_3CS_2C(CH_3)_3 + S \longrightarrow (CH_3)_3CS_3C(CH_3)_3 \qquad (5.16)$$

$$C_2H_5S_4C_2H_5 + S \longrightarrow C_2H_5S_5C_2H_5 \qquad (5.17)$$

Or, according to a general scheme, reaction (5.18) is accelerated in the presence of Friedel-Crafts-type catalysts and takes place under even milder conditions [36]:

$$R-S_n-R' + mS \longrightarrow R-S_{n+m}-R' \qquad (5.18)$$
$$n=1-4, \quad m=1-15$$

Dimethyl disulfide and dimethyl trisulfide are converted to dimethyl hexasulfide under these conditions at 110-130°C. Diethyl sulfide is converted to diethyl pentasulfide.

Dibenzyl disulfide readily reacts with sulfur at 150°C in the absence of catalysts with the formation of dibenzyl tri-, tetra-, and pentasulfide [37-41]. The phase diagram of the system dibenzyl disulfide—sulfur has a maximum at 88.5% disulfide and 11.5% sulfur. This corresponds exactly to dibenzyl trisulfide. However, the temperature of this maximum (64.4-71.1°C) exceeds the melting point of dibenzyl trisulfide by 15-22°C [42]. The reaction products obtained from the interaction of sulfur with dibenzyl sulfide at 200°C are hydrogen sulfide (39.3%), H_2S_2, toluene, α-toluenethiol, and stilbene (41.5%) [43]. The mechanism of this reaction can be illustrated as follows:

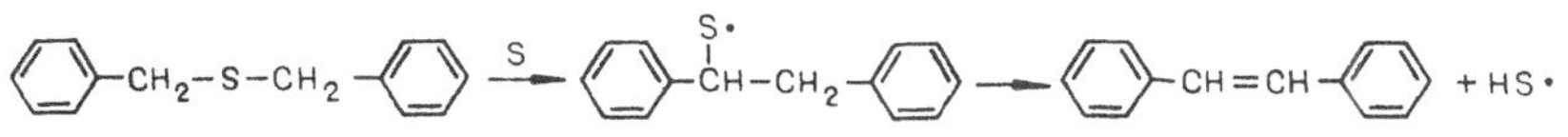

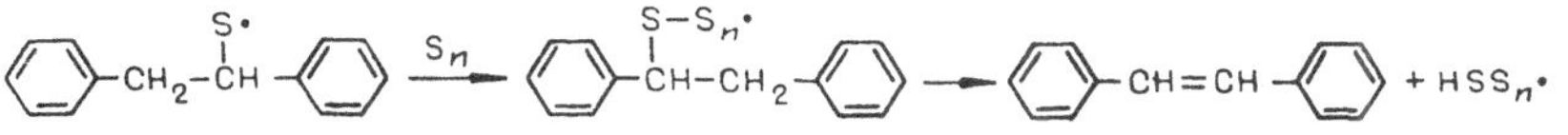

According to other data, the interaction of dibenzyl sulfide or dibenzyl disulfide with sulfur at 200-210°C leads to the formation of tetraphenylthiophene in relatively high yields [44]:

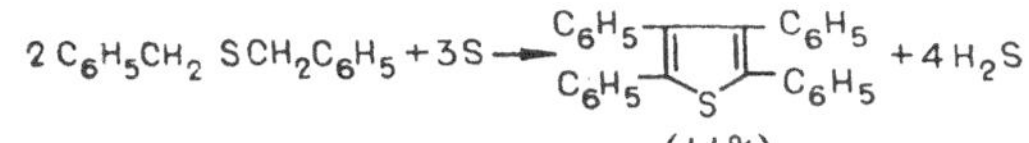

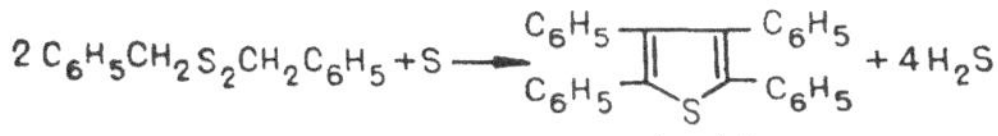

Stilbene is an intermediate product in reactions (5.21) and (5.22). It is obtained in a yield > 30% when it is continuously removed from the reaction mixture [37,44]:

$$C_6H_5CH_2S_nCH_2C_6H_5 + S \longrightarrow C_6H_5CH = CHC_6H_5$$

$$+ H_2S + (n-1)S \qquad\qquad n = 1, 2 \qquad\qquad (5.23)$$

Sulfur serves as a catalyst in this reaction; the mechanism may be illustrated as shown below:

$$C_6H_5CH_2SSCH_2C_6H_5 + \cdot S_n \cdot \longrightarrow C_6H_5CH_2S_{n+2}CH_2C_6H_5 \qquad (5.24)$$

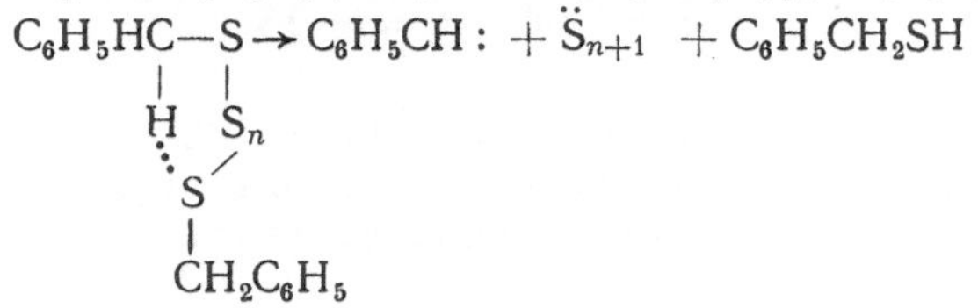

$$\underset{\substack{| \quad | \\ H \quad S_n \\ \ddots_{\diagdown} \\ S \\ | \\ CH_2C_6H_5}}{C_6H_5HC-S} \to C_6H_5CH: + \ddot{S}_{n+1} + C_6H_5CH_2SH \qquad (5.25)$$

$$2C_6H_5CH: \longrightarrow C_6H_5CH = CHC_6H_5 \qquad\qquad (5.26)$$

Under similar conditions, the interaction of sulfur with substituted dibenzyl sulfide and disulfide does not give the corresponding tetraphenylthiophenes but only substituted stilbenes (37-46% yield). The only exception is 2,2'-dimethyldibenzyl sulfide which gives a considerable amount of tar-like materials upon reaction with sulfur and thus only a 1-2% yield of 2,2'-dimethylstilbene is obtained [44]. Similar behavior is shown by bis-2-thienylmethyl sulfide which, on heating with sulfur at 185°C, is converted to 1,-2-bis(2-thienyl)ethylene [45]:

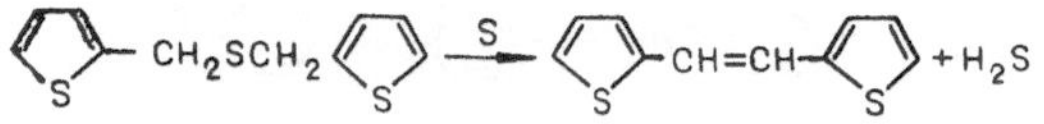

$$(5.27)$$

Diphenyl sulfide when heated with sulfur gives diphenyl disulfide (4) [46] or thianthrene (5) [47]:

$$C_6H_5SC_6H_5 + S \longrightarrow C_6H_5S_2C_6H_5 \qquad\qquad (5.28)$$
$$\underset{4}{}$$

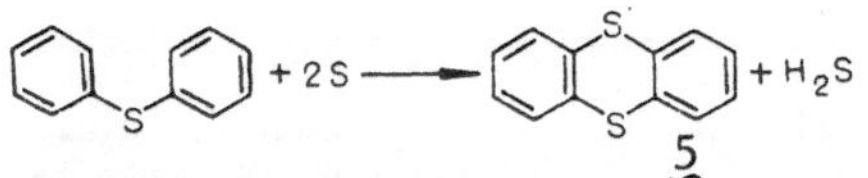

$$(5.29)$$

Halogenated diphenyl sulfides react with sulfur at 240-270°C in a different way to give halo-substituted benzenes as the principal reaction products, as well as halosulfides, disulfides, and polysulfides [48]:

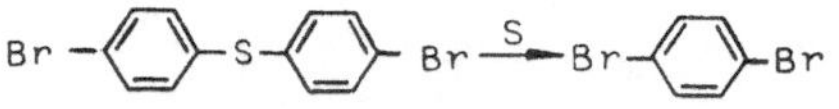

$$(5.30)$$

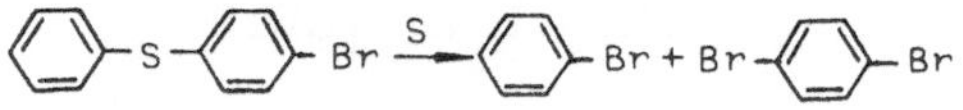

$$(5.31)$$

The reaction of sulfur with polyphenylene sulfides at 300°C considerably increases their molecular weight [49].

When diphenyl-N-p-tosylsulfimine and sulfur are refluxed in chlorobenzene diphenyl sulfide is formed [50]:

$$C_6H_5SC_6H_5 + S \rightarrow C_6H_5SC_6H_5 \qquad (5.32)$$
$$\downarrow$$
$$N\text{-}Ts \qquad\qquad (50\%)$$

Heating a diphenylsulfoximine with sulfur gives a diphenyl sulfoxide [50,51]:

$$\begin{array}{c} O \\ \uparrow \\ R\text{-}S\text{-}R' \\ \downarrow \\ NH \end{array} + S \xrightarrow{160°C} \begin{array}{c} R\text{-}S\text{-}R \\ \| \\ O \end{array} \qquad (5.33)$$

$$(86\%)$$

R = aryl, R' = alkyl

When dialkyl dithioacetals and trialkyl trithioorthoformates are heated with sulfur in the presence of Friedel-Crafts catalysts at 100-200°C, one to fifteen sulfur atoms can be incorporated into their molecules [52-54].

Allyl methyl sulfide and diallyl sulfide upon heating with sulfur in dimethyl sulfoxide (80-90°C) are converted into the corresponding disulfides. The reaction of allyl-1,1-d_2 methyl sulfide with sulfur gives allyl-3,3-d_2 methyl disulfide [55]. The allylic rearrangement is likely to occur according to the scheme:

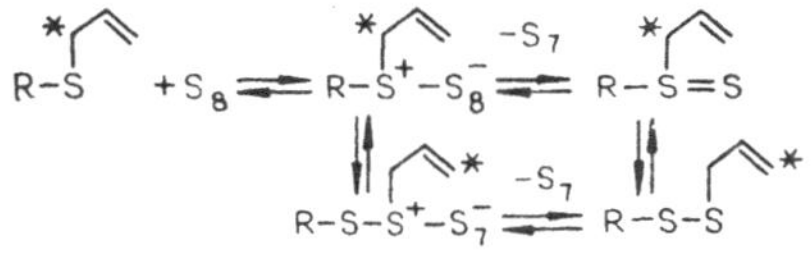

$$R\text{-}S + S_8 \rightleftharpoons R\text{-}S^+\text{-}S_8^- \xrightarrow{-S_7} R\text{-}S=S \qquad (5.34)$$

5.3. SULFIDES, SULFONES, SULFONIC ACIDS, AND THEIR DERIVATIVES

When aliphatic or aromatic sulfoxides and sulfones are heated to 250-360°C with sulfur they undergo reduction, e.g., [43,49,56-59]:

$$2(C_4H_9)_2SO + 2S \rightarrow (C_4H_9)_2S + (C_4H_9)_2S_2 + SO_2 \qquad (5.35)$$

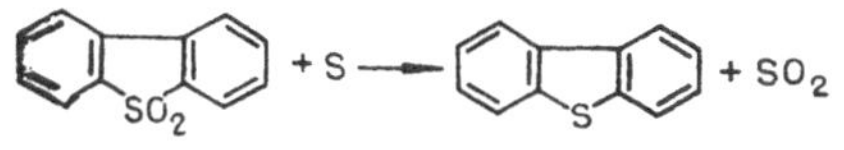

(5.36)

^{35}S-Thianthrene, ^{35}S-thioxanthene, and ^{35}S-dibenzothiophene S,S-dioxides upon heating with sulfur (320-346°C) exchange sulfur atoms to the extent of 80-90%. Presumably the exchange takes place according to the scheme [60]:

$$\text{(diagram)} \quad X=S, O, -, n=2 \tag{5.37}$$

The sulfoxides of the above-mentioned ^{35}S-compounds are reduced to the corresponding sulfides by sulfur at lower temperatures than the S,S-dioxides (250°C) and do not undergo an exchange reaction [60].

4,4'-Dichlorodiphenyl sulfoxide upon heating with sulfur at 240-270°C is converted to 1,4-dichlorobenzene [48]. Diethyl sulfoxide does not react with sulfur at 180°C [61]. The mechanisms of reduction of sulfoxides and sulfones by sulfur are different. The reaction of diphenyl sulfoxide with sulfur leads to a direct reduction of the sulfoxide group with cleavage of the S-O bond [58]:

$$\text{(diagram)} \tag{5.38}$$

The reduction of diphenyl sulfone is accompanied by cleavage of the C-S bond and 70-80% substitution of the sulfone group by sulfur [58]:

$$\text{(diagram)} \tag{5.39}$$

The first step of the reaction of sulfur with dibenzyl sulfoxide is dehydrogenation of the methylene group [43]:

$$\text{(diagram)} \tag{5.40}$$

The benzyl radical formed can disproportionate and/or react in several parallel directions. This explains the formation of products such as toluene, stilbene, benzaldehyde, and α-toluenethiol:

$$\text{(diagram)} \tag{5.41}$$

$$\text{(diagram)} \tag{5.42}$$

$$\text{(diagram)} \tag{5.43}$$

$$\text{(diagram)} \tag{5.44}$$

$$\text{(diagram)} \tag{5.45}$$

The reaction products obtained from the interaction of sulfur with
dibenzyl sulfone are hydrogen sulfide (32%), stilbene (68%), and a
small amount of water. Formation of sulfur dioxide is not observed.
Thus, the reactions of dibenzyl sulfide, sulfone, and sulfoxide with
sulfur are initiated by abstraction of a benzylic hydrogen atom by
a sulfur radical.

The abstraction of a hydrogen atom from the d-methylene group
of di-n-butyl sulfide, sulfoxide, and sulfone during the reaction
with sulfur is considerably more difficult than is the case with
the corresponding dibenzyl derivatives [43]. The abstraction of
hydrogen atoms from diphenyl sulfoxide and sulfone by sulfur does
not take place to any great extent within the temperature range
studied. The fate of the initially formed radical generated in the
reactions of sulfur with arylaliphatic and aliphatic sulfoxides and
sulfones depends on its stability and also on the relative ease of
cleavage of the C-C, C-S, C-O, and S-O bonds [43]. Diethyl ether
is the product from the reaction of diethyl sulfite with sulfur at
$200°C$. It is likely that it is formed by thermal decomposition of
the starting sulfite [61].

When sulfur is melted with sulfonic acids, derived from macromolecular compounds (resins, pitches), powdery products are obtained
which can be used as fungicides and insecticides [62].

5.4. THIOCARBONYL COMPOUNDS

The reaction of symmetric and unsymmetric thioketones, RCSR'
$(R,R' = CH_3, C_4H_9)$ with sulfur and carbon disulfide affords the
corresponding 1,2-dithiol-3-thione [63]:

$$-\overset{\overset{S}{\|}}{C}-CH_2 \rightleftarrows -\overset{\overset{SH}{|}}{C}=CH- \xrightarrow{\text{CS}_2, \text{ S}}_{\overline{\text{DMF},\text{Et}_2\text{N}}} \qquad + \qquad \qquad (5.46)$$

Heating a mixture of sulfur, morpholine, and thioacetophenone at
$145°C$ in the presence of lead oxide gives phenylthioacetomorpholide
[64]:

$$C_6H_5\underset{S}{\overset{\|}{C}}-CH_3 +S +HN\bigcirc O \longrightarrow C_6H_5CH_2\underset{S}{\overset{\|}{C}}-N\bigcirc O + H_2S \qquad (5.47)$$

1,3-Diphenyl-3-thioxo-1-propanone reacts with sulfur and morpholine at $145°C$ to give 1-benzoyl-2-phenylethane, β-benzylstyrene,
and thiobenzmorpholide [65]:

$$C_6H_5\underset{S}{\overset{\parallel}{C}}CH_2COC_6H_5 + S + HN\boxed{}O \longrightarrow C_6H_5COCH_2CH_2C_6H_5$$

$$+ C_6H_5CH=CHCH_2C_6H_5 + C_6H_5\underset{S}{\overset{\parallel}{C}}N\boxed{}O \qquad\qquad (5.48)$$

The reaction of sulfur with some 5-(1-alkenyl)-1,2-dithiol-3-thiones leads to the formation of thiophene derivatives [66]. The process takes place at 200°C in biphenyl, in the presence of N,N'-di-2-methylphenylguanidine:

$$(5.49)$$

R, R' = CH$_3$, C$_6$H$_5$; CH$_3$, 4-ClC$_6$H$_4$; H, C$_6$H$_5$; H, 4-ClC$_6$H$_4$

The interaction of sulfur with a 4-substituted-1,2-dithiol-3-thione in dimethylamine, dimethylformamide, or dimethyl sulfoxide gives the corresponding 5-mercapto-derivative [67]:

$$(5.50)$$

5.5. SULFUR-CONTAINING HETEROCYCLIC COMPOUNDS

The absolute rate constants for the reaction of sulfur S(^{3}P) with thiirane have been measured at 20°C [68].

Copolymerization of thiirane derivatives and sulfur under the action of UV-irradiation in the presence of organic derivatives of IA and IIB group metals yields soluble sulfur-containing polymers having polysulfide bonds in the macrochain [69]. A decrease in the reactivity of the thiiranes as well as the use of more active forms of elemental sulfur increases the possibility of copolymerization and formation of polysulfide bonds in the copolymers [70].

Tetrahydrothiophene is dehydrogenated by sulfur under pressure [71] to give thiophene and 3,3'-bithiophene:

$$(5.51)$$

Tetrahydrothiapyran on heating with sulfur yields 4H-thiapyran-4-thione [72]:

$$(5.52)$$

The reaction of trimeric thioaldehydes with sulfur at 50-200°C in aqueous media in the presence of acids or bases (pH 1.5-1.0) gives rubber-like products [73].

The reaction of thiachromane with sulfur affords thiothiacoumarin and small amounts of thiothiachromone [74]:

$$(5.53)$$

The reaction of sulfur with 2-(1-)alkenylthiophenes gives 4-(or 5-) thienyl-1,2-dithiol-3-thiones [75,76]:

$$(5.54)$$

$$(5.55)$$

$$(5.56)$$

$$(5.57)$$

$$(5.58)$$

Thieno[3,2-b]-[1]benzothiophenes are formed in the dehydrogenation of the corresponding dihydro-derivatives with sulfur [77, 78]. Sulfur does not react with benzo[b]thiophene at its boiling point, and may be used for its purification [79]:

$$(5.59)$$

The reaction of sulfur with benzo[b]thiophenes containing an α-hydroxy-α-methylbenzyl group at C-2 or C-3 involves cyclization to a thiophene ring [78]:

$$(5.60)$$

$$\text{(5.61)}$$

The reaction of 3-phenylbenzo[b]thiophene with sulfur at 150-200°C in the presence of aluminium chloride gives [1]benzothieno[3,2-b]-[1]benzothiophene [80]. This reaction presumably occurs with migration of the phenyl group from C-3 to C-2:

$$\text{(5.62)}$$

In the absence of aluminium chloride, the above-mentioned reactions (300-310°C) give the expected [1]benzothieno[2,3-b]-[1]benzothiophene:

$$\text{(5.63)}$$

2-Phenylbenzo[1,2-b;5,4-b']dithiophene reacts with sulfur at 300-310°C to form di[1]benzothieno[3,2-b;3',2'-f]-[1]benzothiophene [80]:

$$\text{(5.64)}$$

The sulfuration of 2,3'-bibenzo[b]thiophene in the presence of aluminium chloride gives two isomers, i.e., di[1]benzothieno-[2,3-b;2',3'-d]thiophene and di[1]benzothieno-[3,2-b;2',3'-e]-thiophene [80]. Similarly to reaction (5.62), the formation of the latter isomer is due to isomerization catalyzed by aluminium chloride:

$$\text{(5.65)}$$

^{35}S-Thianthrene and ^{35}S-thioxanthene exchange their sulfur atoms when heated with sulfur at 300°C (70-80% exchange) [60]. ^{35}S-Dibenzothiophene does not exchange sulfur under similar conditions [60].

For the reaction of sulfur with bis-2-thienylmethyl sulfide, see Section 5.2, reaction (5.27) [45].

The reactions of sulfur with 2-iodothiophene [81], 2-chloromethylthiophene [45], and chlorothianaphthenes [82-84] have been discussed in Section 4.8.

The iodine-catalyzed reaction of phenothiazine with sulfur affords phenothiazine-3-thiol isolated as the corresponding disulfide (oxidation by an air flow in boiling water) [84].

REFERENCES

1. G. Kvasnikov, A. Pfister, and P. Vekkiyuti, Fr. Patent 441,706 (1972); Byull. Izobr., No. 32, 150 (1974).
2. P. S. Landis and L. A. Hamilton, J. Org. Chem., 26, 274 (1961).
3. B. D. Vineyard, J. Org. Chem., 31, 2, 601 (1966).
4. B. D. Vineyard, Monsanto Tech. Rev., 12, 17 (1967).
5. B. D. Vineyard, J. Org. Chem., 32, 3833 (1967).
6. F. H. McMillan and J. A. King, J. Am. Chem. Soc., 70, 4144 (1948).
7. J. A. King and F. H. McMillan, J. Am. Chem. Soc., 68, 632 (1946).
8. A. P. Ko Zacik and H. Myers, Fr. Patent 1,527,462 (1968); C. A., 71, 2971g (1968).
9. C. M. Hull, R. S. Olsen, and W. G. Krance, Ind. Eng. Chem., 38, 1282 (1946).
10. J. Tsurugi and H. Fukuda, Rubber Chem. Technol., 33, 1, 217 (1960); C. A., 66, 14870g (1960).
11. T. Ko, Bull. Chem. Soc. Jpn., 43, 2626 (1970).
12. F. Asinger, M. Schaefer, and A. Saus, Monatsh. Chem., 96, 1265 (1965).
13. R. Eichstedt, Schriftner ver Wasser Boden, Luftys, Berlin-Dahlem (1971), p. 79; C. A., 76, 89661 (1972).
14. D. Harman and G. L. Perry, US Patent 2,562,144 (1951); C. A., 45, P9854c (1951).
15. US Patent 6,613,253 (1967).
16. E. E. Reid, Organic Chemistry of Bivalent Sulfur, Vol. III, Chem. Publ., New York (1960).
17. A. B. Kuliev, G. A. Zeinalova, M. S. Gasanov, and F. Yu. Aliev, Zh. Org. Khim., 14, 661 (1978).
18. E. J. du Pont de Nemours and Co., Neth. Appl. 7704108 (1977); C. A., 90, P137878 (1979).
19. S. A. Vladuchik, US Patent 4,066,656 (1978); C. A., 89, P44909 (1978).
20. E. J. du Pont de Nemours and Co., Belg. Patent 853,648 (1977); C. A., 89, P18357 (1978).
21. H. J. Langer and J. B. Hyne, Adv. Chem. Ser., 110 (Sulfur Rec. Trends), 113 (1972); C. A., 76, 152902c (1972).
22. V. A. Usov, K. A. Petriashvili, and M. G. Voronkov, Zh. Org. Khim., 16, 1550 (1980).
23. G. Gollmen, A. Friederich, and R. Schubart, Ger. Patent 2164480 (1973); Ref. Zh. Khim., 19N220 (1974).
24. H. Böttger, Liebigs Ann. Chem., 223, 335 (1884).
25. M. Müller, J. Prakt. Chem., 4, 39 (1871).
26. J. R. Van Wazer and D. Grant, Ann. Chem. Soc. Div. Polym.-Chem., Preprints, 5, 621 (1964).
27. D. Grant and J. R. Van Wazer, J. Am. Chem. Soc., 86, 3012 (1964).

28. B. Buathler and A. Combes, Ger. Patent 2,516,262 (1975); C. A., 84 (1976).
29. F. Wessely and A. Siegel, Monatsh. Chem., 82, 607 (1951).
30. S. Penczek, R. Slazak, and A. Duda, Nature (London), 273(5665), 738 (1978); C. A., 89, 215821s (1978).
31. P. Klässon, J. Prakt. Chem., 15, 193 (1887).
32. W. A. Schulze and W. Crouch, US Patent 2,529,355 (1950); C. A., 45, P2966gh (1950).
33. B. Hohmberg, Ann. Chem., 359, 81 (1908); Ber., 43, 220 (1910).
34. S. F. Birch, T. V. Cullum, and R. A. Dean, J. Inst. Petrol. (London), 39, 206 (1953); C. A., 48, 4428 (1954).
35. H. Krebs, Die Katalytische Aktivierung des Schwefels, Westdeutscher Verlag, Kölh und Opladen (1958).
36. J. D. Weleb, US Patent 3,075,019 (1963); C. A., 59, P449a (1963).
37. M. G. Voronkov, A. N. Perferkovich, V. A. Pestunovich, and V. É. Udre, Zh. Org. Khim., 3, 2211 (1967).
38. M. G. Voronkov, G. P. Sharonov, and V. V. Dolbin, Zh. Prikl. Khim. (Leningrad), 1562 (1961).
39. M. G. Voronkov, G. P. Sharonov, V. V. Dolbin, A. N. Perferkovich, and V. É. Udre, Russian Patent 165860 (1963); Byull. Izobr., No. 20 (1964).
40. V. V. Dolbin, Tr. Sverdlovsk. Sel'skokhoz. Inst., Vol. XIII, Sverdlovsk (1965).
41. M. G. Voronkov, G. P. Sharonov, and V. V. Dolbin, VIth Scientific Conf. Chem. Sulfur-Contg. Compds. in Petroleum and Petroleum Products, Izd. Akad. Nauk SSSR, Ufa (1961)
42. U. Minoura, Nippon Gomu Kyokaishi, 32, 184 (1959); C. A., 54, 8694 (1960).
43. W. Tagaki, S. Kiso, and S. Oae, Bull. Chem. Soc. Jpn., 38, 414 (1965).
44. M. G. Voronkov and A. N. Perferkovitch, Khim. Geterotsikl. Soedin., 51 (1971).
45. M. G. Voronkov, A. N. Perferkovich, M. P. Gavar, and G. V. Ozolin', Khim. Geterotsikl. Soedin., 1183 (1970).
46. F. Krafft and W. Vorster, Ber., 26, 2815 (1893).
47. F. Krafft and R. E. Lions, Ber., 29, 435 (1896).
48. J. H. Billman and G. Dongherty, J. Am. Chem. Soc., 61, 387 (1939).
49. G. C. Ray and D. A. Frey, US Patent 3,458,486 (1969); C. A., 71, 71393g (1963).
50. S. Oae, Y. Tsuchida, K. Tsujihara, and N. Furukawa, Bull. Chem. Soc. Jpn., 45, 2856 (1972).
51. S. Oae, Y. Tsuchida, and N. Furukawa, Bull. Chem. Soc. Jpn., 46, 650 (1973).
52. J. D. Webb, US Patent 2,966,521 (1960); C. A., 55, P8291d (1960).
53. J. D. Webb, US Patent 2,966,522 (1960); C. A., 55, P8294i (1960).

54. J. D. Webb, US Patent 3,075,020 (1958); C. A., 59, P2722d (1958).

55. R. D. Baechler, J. P. Hummel, and K. Mislow, J. Am. Chem. Soc., 95, 4442 (1973).

56. J. Böeseken, Recl. Trav. Chim. Pays-Bas, 30, 137 (1911).

57. N. M. Cullinane and C. G. Davies, Recl. Trav. Chim. Pays-Bas, 55, 881 (1936).

58. S. Oae and S. Kamamura, Bull. Chem. Soc. Jpn., 36, 163 (1963).

59. S. Kiso and S. Oae, Bull. Chem. Soc. Jpn., 40, 1722 (1967).

60. S. Oae, S. Makino, and Y. Tsuchida, Bull. Chem. Soc. Jpn., 46, 650 (1973).

61. H. Prinz, Ann. Chem., 223, 371 (1884).

62. A. Steindorf and K. Pfall, Ger. Patent 396,129 (1924); Chem. Zentralbl., II, 1019 (1924).

63. R. Couturier, D. Paquer, and A. Vibet, Bull. Soc. Chim. France, 1670 (1975).

64. G. Purrello, Gazz. Chim. Ital., 97, 539 (1967).

65. G. Purrello, Gazz. Chim. Ital., 95, 1072 (1965).

66. J. Brelivet, P. Appion, and J. Teste, C. R. Acad. Sci., C, 265, 1016 (1967).

67. J. P. Brown, J. Chem. Soc., C 1077 (1968).

68. R. B. Klemm and D. D. Davis, Int. J. Chem. Kinet., 5, 149 (1973).

69. A. D. Aliev, I. P. Solomatina, Zh. Zhumabaev, S. L. Alieva, and B. A. Krentsel, Russian Patent 516711 (1976); C. A., 85, 78664 (1976).

70. A. D. Aliev, Zh. Zhumabaev, A. Yu. Koshevnik, S. L. Alieva, and B. A. Krentsel, Azerb. Khim. Zh., No. 3, 77 (1980).

71. W. Friedmann, J. Inst. Petrol. (London), 37, 239 (1951).

72. R. Mayer and P. Fischer, Chem. Ber., 95, 1307 (1962).

73. C. A. Coltd, C. A. Curtis, and D. S. Stephens, Brit. Patent 574,270 (1945); C. A., 42, P7565g (1945).

74. R. Mayer and H. Damme, Z. Chem., 5, 150 (1965).

75. J. Teste and N. Lozac'h, Bull. Soc. Chim. France, 492 (1954).

76. J. Teste and N. Lozac'h, Bull. Soc. Chim. France, 137 (1955).

77. W. S. Parham and B. Gadsby, J. Org. Chem., 25, 234 (1960).

78. L. Szperl, Rocz. Chem., 18, 804 (1938); C. A., 33, 6304 (1939).

79. C. Hansen, J. Am. Chem. Soc., 69, 2908 (1947).

80. T. Sunivasa Murthy, L. J. Pandya, and B. D. Tubak, J. Sci. Ind. Res., 20B, 169 (1961).

81. M. G. Voronkov, and A. N. Perferkowitsch, Angew. Chem., 81, 257 (1969).

82. M. G. Voronkov and V. É. Udre, Russian Patent 169,127 (1964); Byull. Izobr., No. 6, 28 (1965).

83. N. G. Voronkov, V. É. Udre, and E. P. Popova, Khim. Geterotsikl. Soedin., 1003 (1967).

84. M. G. Voronkov and V. É. Udre, Russian Patent 199,909 (1967); Byull. Izobr., No. 16, 29 (1967).

85. E. A. Nodiff, P. C. Taunk, A. Cantor, and J. M. Hulsizer, Chem. Ind., 653 (1976).

6
Oxygen-Containing Compounds

6.1. COMPOUNDS WITH A HYDROXYL GROUP

6.1.1. Aliphatic Alcohols

Methanol and ethanol react with sulfur in the presence of
catalysts or upon ultraviolet irradiation [1]. The interaction of
ethanol with sulfur in the presence of molybdenum disulfide (or of
a mixed catalyst, 27% WS_2 + 3% NiS on Al_2O_3) at 180°C under pressure
gives a 31-37% yield of ethanethiol [2,3]. Photochemical reactions
of sulfur with methanol lead to the formation of dimethyl sulfide
and traces of methane and ethane. The reaction is accelerated in the
presence of sensitizers such as benzene, naphthalene, or pyrene [4,5].
Methanol reacts with excess sulfur in the gas phase at 550°C and
contact time of five seconds to form carbonyl sulfide and carbon
disulfide (77% and 21% yields, respectively) [6].

Heating aliphatic alcohols (C_2-C_{10}) with sulfur and red phosphorus at 80-150°C leads to the formation of dialkyldithiophosphorus
acids contaminated with dialkylthiophosphorus acid, thioalkyl esters
of dialkylthionophosphorus acid and trialkyl thiophosphates [7]:

$$nP + mS + 1ROH \rightarrow (RO)_2\overset{\overset{S}{\|}}{P}OH + (RO)_2\overset{\overset{S}{\|}}{P}SH + (RO)_2\overset{\overset{S}{\|}}{P}SR + (RO_3)P = S + H_2S$$

$$(6.1)$$

The reaction of higher alcohols (C_7-C_{10}) with sulfur and red
phosphorus carried out at 160-230°C involves decomposition of the
reaction mixture to form the corresponding α-alkenes, alkanethiols,
and dialkyl sulfides [7].

The reaction of 3-aminopropanol with sulfur and red phosphorus starts at 70°C self-heating to 180°C. This results in the replacement of oxygen atoms by sulfur atoms with the formation of a salt-like adduct in 46.6% yield [8]:

$$2H_2NCH_2CH_2OH + 2S + 2P \rightarrow NH=CHCH_2CH_2\overset{-}{S}\overset{+}{N}H_3CH_2CH_2CH_2SH$$

$$\downarrow + CH_3COCl \qquad\qquad (6.2)$$

$$CH_3CONHCH_2CH_2CH_2SCOCH_3$$

Treatment of the latter compound with acetyl chloride in the presence of triethylamine affords acetyl 3-(N-acetylamino)propyl sulfide.

Secondary alcohols undergo relatively ready dehydrogenation with sulfur to give the corresponding carbonyl compounds [9,10]. This reaction is used for the qualitative determination of secondary alcohols (blackening of lead(II) acetate by the hydrogen sulfide formed):

$$RR'CHOH + S \longrightarrow RR'CO + H_2S. \qquad\qquad (6.3)$$

The sulfuration of linalool gives a compound $C_{10}H_{18}OS_6$ of unknown structure [11]. According to other data, linalool, α-terpineol, and dihydro-α-terpineol react with sulfur to lose water and undergo cyclization to form p-cymene, dipentene, and an unknown terpene $C_{10}H_{16}$ or $C_{10}H_4$. Furthermore, a sulfide $C_{10}H_{18}S$ has been isolated; it seems that this sulfide is a mixture of two sulfur-containing compounds, one of which contains a thioether group whereas the other compound contains a thiocarbonyl group [12,13].

6.1.2. Arylaliphatic Alcohols

Sulfur reacts with arylaliphatic alcohols on heating and the reaction yields dehydration and dehydrogenation products [14]. Thus, the reaction of sulfur with benzyl alcohol gives dibenzyl ether, benzaldehyde, and stilbene:

$$2C_6H_5CH_2OH \overset{+S}{\rightarrow} C_6H_5CH_2OCH_2C_6H_5$$

$$2C_6H_5CH_2OCH_2C_6H_5 \overset{+2S}{\longrightarrow} 2C_6H_5CHO + C_6H_5CH=CHC_6H_5 + 2H_2S \qquad (6.4)$$

When the isomeric xylenols are heated with sulfur under carbon dioxide, the products formed are methyl-substituted benzaldehydes and benzoic acids together with α,β-dimethylstilbene, but no dixylyl ether. The corresponding ether is formed only in the reaction of sulfur with nitrotoluyl alcohol [14].

Diphenylmethanol, on heating with sulfur (150°C), is almost completely converted into the corresponding ether [15]:

$$2(C_6H_5)_2 CHOH \overset{+S}{\rightarrow} [(C_6H_5)_2 CH]_2O + H_2O \qquad (6.5)$$

Heating 1-naphthylphenylmethanol with a small amount of sulfur leads predominantly to the corresponding ether and also gives small amounts of 1-naphthylphenyl ketone and 1-benzylnaphthalene [16]:

$$C_6H_5(1\text{-}C_{10}H_7)CHOH \overset{S}{\rightarrow} C_6H_5(1\text{-}C_{10}H_7)CHOCH(1\text{-}C_{10}H_7)C_6H_5$$
$$+ 1\text{-}C_{10}H_7COC_6H_5 + 1\text{-}C_{10}H_7CH_2C_6H_5 + H_2S \qquad (6.6)$$

The reaction of benzoin and sulfur at 230°C gives a quantitative yield of benzil [17]:

$$C_6H_5CH(OH)COC_6H_5 + S \longrightarrow C_6H_5COCOC_6H_5 + H_2S \qquad (6.7)$$

6.1.3. Polyhydroxy-Alcohols and Saccharides

1-Thioglyceric acid is obtained when anhydrous glycerol is heated with sulfur [18]:

$$HOCH_2CH(OH)CH_2OH + 2S \longrightarrow HOCH_2CH(OH)CSOH + H_2S \qquad (6.8)$$

The reaction of glycerol with sulfur at 290-300°C gives 2-propene-1-thiol and diallyl hexasulfide [19,20]:

$$3HOCH_2CH(OH)CH_2OH \overset{+S}{\rightarrow} CH_2=CH-CHSH + (CH_2=CH-CH_2)_2S_6 \qquad (6.9)$$

The interaction of glycerol with sulfur at 145-200°C under pressure yields thioacrolein [21]:

$$HOCH_2CH(OH)CH_2OH \overset{S}{\rightarrow} CH_2=CH-CHS \qquad (6.10)$$

In the presence of yeast, glycerol reacts with sulfur to give large amounts of hydrogen sulfide and carbon dioxide. It is likely that sulfur oxidizes glycerol to glyceraldehyde or methylglyoxal. These compounds then undergo further transformations according to the yeast fermentation scheme [22]:

$$HOCH_2CH(OH)CH_2OH \overset{S}{\rightarrow} HOCH_2CH(OH)CHO + H_2S \qquad (6.11)$$

Only a few saccharides react with sulfur upon heating. Thus levulose, in the presence of dextrose (glucose), reacts with sulfur when heated in glycerol in the presence of lead(II) acetate [23]. Two secondary hydroxyl groups are oxidized with the formation of an α,β-dioxo alcohol:

$$HOCH_2CH(OH)CH(OH)CH(OH)CH(OH)CH_2OH + S \longrightarrow$$
$$\longrightarrow HOCH_2CH(OH)CH(OH)COCOCH_2OH + H_2S \qquad (6.12)$$

A ketose reacts with sulfur in an analogous manner. Glucose and
sucrose do not react with sulfur and this may be utilized for the
separation of these sugars [23,24]. When sulfur is melted with
sucrose, large amounts of hydrogen sulfide are formed [25].

6.1.4. Cycloaliphatic Alcohols

When sulfur reacts with tertiary and secondary cycloaliphatic
alcohols (derivatives of cyclohexanol, cyclohexenols, and cyclo-
hexylcarbinols), their dehydration and dehydrogenation takes place
simultaneously and aromatic rings are formed [26,27]. Thus, 2-2'-
methylphenyl-3-methylcyclohexanol upon heating with sulfur gives
bi-(2-methylphenyl) in a smooth reaction [28]:

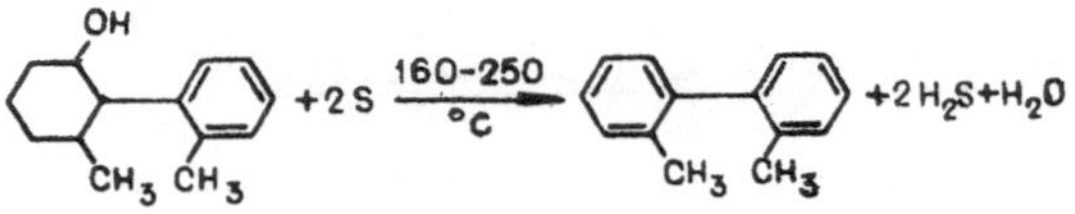

$$(6.13)$$

A further similar reaction is shown below [29]:

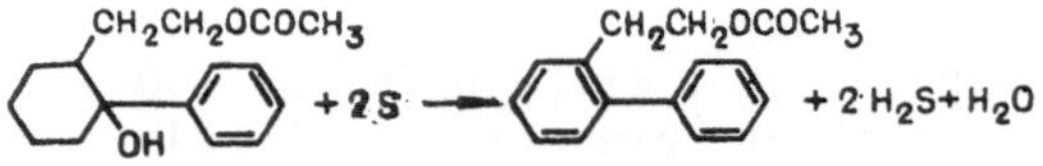

The interaction of sulfur with elemol ($\underline{1}$) which contains two
geminal groups in the cyclohexene ring (this makes the dehydrogenation to the corresponding aromatic derivative more difficult)
gives eudalene and a sulfuration product $C_{14}H_{18}S$ which has been
assigned a benzo[b]thiophene structure ($\underline{2}$) [30-32]:

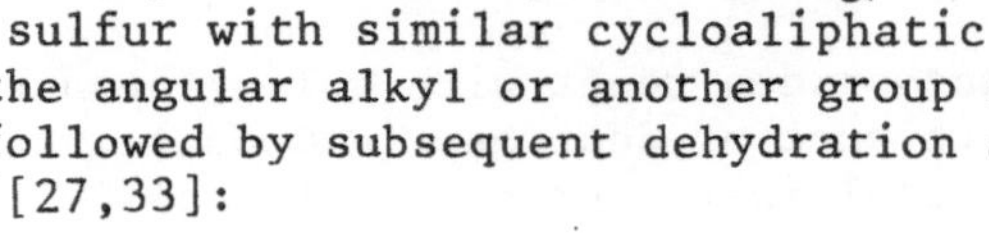

$$(6.14)$$

However, a more likely sulfuration product obtainable from elemol
is 4-isopropyl-1-methyl-2-3'-thienylbenzene ($\underline{3}$). In many cases,
the reaction of sulfur with similar cycloaliphatic alcohols leads
to cleavage of the angular alkyl or another group from the cycloaliphatic ring followed by subsequent dehydration and dehydrogenation, e.g., [27,33]:

$$(6.15)$$

eudalene

The dehydrogenation of abietinol (4) involves loss of water and a retropinacol rearrangement with the formation of retene (5) and homoretene (1-ethyl-7-isopropylphenanthrene) (6) [34-37]:

$$(6.16)$$

When sulfur reacts with secondary hydroaromatic alcohols, sometimes only dehydrogenation and cleavage of "excess" geminal and angular alkyl and carboxyl groups takes place, e.g., [38,39]:

ursolic acid   2,9-dimethyl-10-hydroxypicene   β-amyrin

$$(6.17)$$

The dehydrogenation of most triterpenes with sulfur gives hydroxygatalene (trimethylnaphthol) [40,41], whereas the dehydrogenation of the diterpenoid alcohol totarol gives 1-methylphenanthren-7-ol [42] and the dehydrogenation of octahydrophenanthren-9-ol gives 9-phenanthrol [43].

Lupeol and sulfur when refluxed in benzyl acetate give trithiolupeol $C_{30}H_{48}OS_3$ [44]. It is likely that the latter compound is a 3-thioxo-1,2-dithiol:

lupeol

$$(6.18)$$

The reaction of borneol with sulfur at 180-200°C gives the corresponding sulfide and disulfide [45,46].

6.1.5. Phenol and its Derivatives

The phenolic hydroxyl group is stable towards heating with sulfur up to 300°C. Phenol and its derivatives, as well as 1-naphthol, when heated with excess sulfur in the presence of basic catalysts (K_2CO_3, KOH, NaOH, $C_6H_5NH_2$, CH_3COOK) give sulfur-containing resins [47-54]. With an increasing ratio of sulfur to substrate there is an increase in the melting point and in the hardness of the resins, and their solubility in common organic solvents decreases. Sulfur-containing phenol-based resins are used as glues, antioxidizing agents, additives in adhesive formulas, in cold dyeing, and in the synthesis of individual benzenethiols [48,49,55-60].

The reaction of phenol with sulfur at 120-160°C in the presence of sodium hydroxide gives sodium polysulfide as an intermediate; sodium polysulfide then acts as a catalyst. The reaction is inhibited by the bis(hydroxyphenyl) disulfide formed. This auto-inhibition increases with increasing temperature and may be correlated with the concentration of the reagents [61]. Heating phenol with the stoichiometric amount of sulfur in a solution of sodium carbonate in aqueous glycerol at 120°C gives bis-4-hydroxyphenyl disulfide [62,63]:

$$HOC_6H_5 + 2S \rightarrow HO\text{——}S\text{-}S\text{——}OH + H_2S \qquad (6.19)$$

A similar reaction at 180°C leads to the formation of the corresponding sulfide in low yield. Similarly, mono- and disulfides may be obtained at higher temperatures from sulfur and substituted phenols containing OH, OR, COR, COOH, COOR, or SO_3H groups on the aromatic ring [62,63].

The reaction of equivalent amounts of sulfur and phenol in the presence of sodium hydroxide at 150°C leads predominantly to bis-(o-hydroxyphenyl) polysulfides [57-60]. The subsequent catalytic hydrogenation of the latter over cobalt sulfide gives a 65% yield of mercaptophenols [64].

The polycondensation of 1 mole of phenol and 0.5-3.0 mole of elemental sulfur in the presence of alkali catalysts in an aqueous medium at 140-145°C affords a sulfur-containing phenol resin useful for the synthesis of ionites [65].

Phenol present in large excess reacts with sulfur when heated at 140-180°C for 6-24 hours to give a quantitative yield of a mixture of isomeric bis(hydroxyphenyl) sulfides (2,2'-, 2,4'- and 4,4'-isomers) in a 45:45:10 ratio [66,67]:

$$2\underset{}{\bigcirc}^{OH} + 2S \xrightarrow{NaOH} \underset{}{\bigcirc}^{OH}-S-\underset{}{\bigcirc}^{OH} + H_2S \qquad (6.20)$$

The following mechanism for the reaction of phenol with sulfur is proposed [66]:

$$S^{2-}+2PhOH \longrightarrow 2PhO^{-}+H_2S \quad .$$

$$(6.21)$$

$$S_8 \longrightarrow S_7+H_2S \longrightarrow S_5+H_2S \longrightarrow S_3+H_2S \longrightarrow S_1+H_2S$$
$$S_7 \longrightarrow S_6+H_2S \longrightarrow S_4+H_2S \longrightarrow S_2+H_2S \longrightarrow S_1+\tfrac{1}{2}H_2S$$

A phenoxy nucleophile is generated in a limiting step in reaction (6.21). Phenols, being weak acids, are in equilibrium with hydrosulfide ions, and this increases the concentration of phenoxy nucleophiles:

$$ArOH+HS^{-} \rightleftharpoons ArO^{-}+H_2S$$

In the presence of catalytic amounts of alkali (NaOH) sterically hindered phenols (e.g., 2,6-disubstituted) do not react with sulfur. When stoichiometric amounts of a base (NaOH) are used, 2,6-dimethyl- or 2,6-diphenylphenol react with sulfur at 170°C to give a 30% yield of the corresponding bis(4-hydroxyphenyl) sulfides [68,69]:

$$(6.22)$$

$$R=CH_3, \quad C_6H_5$$

At a lower temperature (80°C) under analogous conditions and in the presence of stoichiometric amounts of potassium hydroxide, the corresponding bis(4-hydroxyphenyl) polysulfides are formed from 2,6-disubstituted phenols and sulfur [70]:

$$(6.23)$$

$$R=CH_3, \quad C_6H_5$$

When acrylonitrile is quickly added to a mixture of 2,6-di-
methylphenol and sulfur in the presence of sodium hydroxide in catalytic amounts at 80°C, bis(3,5-dimethyl-4-hydroxyphenyl) sulfide is
formed in 85–90% yield [68]:

$$\text{(structure)} + 3S + 2\,H_2C=CH-CN \xrightarrow[\text{NaOR}]{80°C} \text{(structure)} + S_2(CH_2CH_2CN)_2 \qquad (6.24)$$

Ethylene oxide acts according to scheme (6.24). The role of acrylonitrile or ethylene oxide in reaction (6.24) is to bind polysulfide
anions of sulfur, thus promoting selective formation of monosulfides
in the reactions of phenols with sulfur under mild conditions [68]:

$$(6.25)$$

The reaction of phenol with sulfur in the presence of bases
(NaOH) and propylene oxide at 120–155°C for 1.5–7 hours leads to
a mixture which consists of 2-2'-hydroxypropylthiophenol, 2-(2'-
hydroxypropyl)benzene and 1-(2'-hydroxypropoxy)-2-(2'-hydroxy-
propylthio)benzene [71]. The mixtures formed were recommended as
anti-oxidants. The reaction with 2-methylphenol proceeds in a
similar manner.

On heating a mixture of alkylphenol, sulfur and alkali hydroxide, a mixture is formed which is then subjected to a secondary
heating with 2 moles of alkali-earth hydroxide. The superbasic
sulfurated alkylphenolate obtained is used as a lubricant additive
[72,73].

The sulfuration of chlorophenols by a mixture of elemental sulfur and sulfur chloride in the presence of aluminium chloride with subsequent reduction of the intermediate in an alkaline medium gives chlorothiophenols in high yield [74].

The highest yield of the sulfuration product of 3-methylphenol is obtained after five hours at 170°C at the molar ratio of sulfur: 3-methylphenol:NaOH = 2:1:0.06 [75].

The reaction of cresols with sulfur affords the corresponding bis(cresyl) sulfides [75a]:

$$\text{HO} \underset{R^2}{\overset{R^1 \quad R^3}{\diagdown}} + S \rightarrow \text{HO} \underset{R^2}{\overset{R^1 \quad R^3}{\diagdown}} \!\!\!-\!S\!-\!\! \underset{R^2}{\overset{R^3 \quad R^1}{\diagdown}} \text{OH} \qquad (6.26)$$

$$R^1 = H, Br; \quad R^2 = H, Me, Br; \quad R^3 = H, Me$$

2,2',4-Trihydroxybenzophenone upon treatment with sulfur is converted to 5,5'-epidithio-2,2',4-trihydroxybenzophenone [76]:

$$(6.27)$$

Similarly, 2,4-dihydroxyphenyl 2-hydroxyphenyl sulfone, 3-aminophenyl 3-hydroxyphenyl sulfone, and bis(2-hydroxyphenyl)amine and phenolphthalein react with sulfur to give the corresponding epidisulfides [76]:

$$(6.28)$$

$$(6.29)$$

$$(6.30)$$

When eugenol and isoeugenol react with sulfur at 190–230°C, 5-(4-hydroxy-3-methoxyphenyl)-1,2-dithiol-3-thione is formed [77-83]:

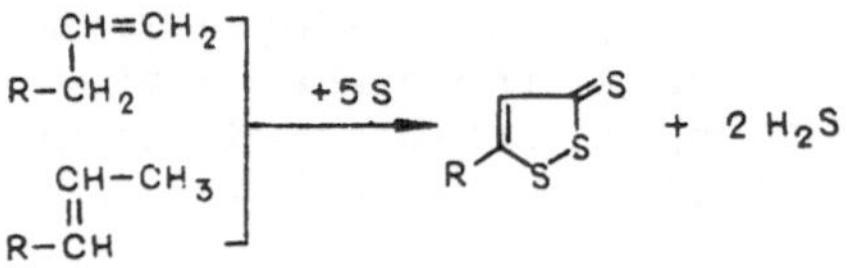

$$R = 4\text{-}OH,\ 3\text{-}OCH_3,\ C_6H_3$$

The sulfuration of 2-(2- and 1-propenyl)phenol gives thiocoumarin [81,82]:

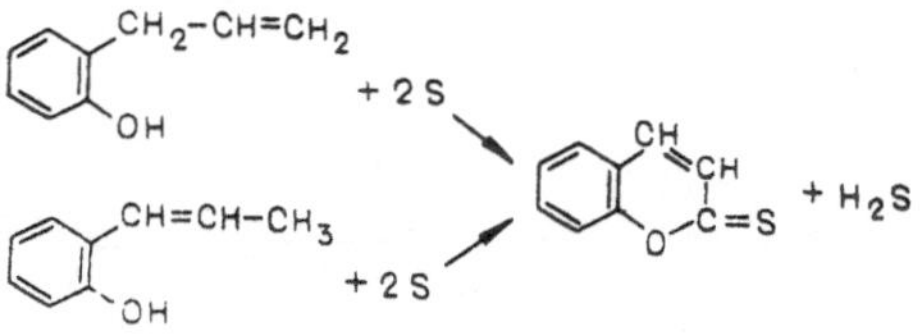

The interaction of polyhalophenols with sulfur in the presence of acids gives the corresponding bis(hydroxypolyhalophenyl) sulfides and oligosulfides [84]:

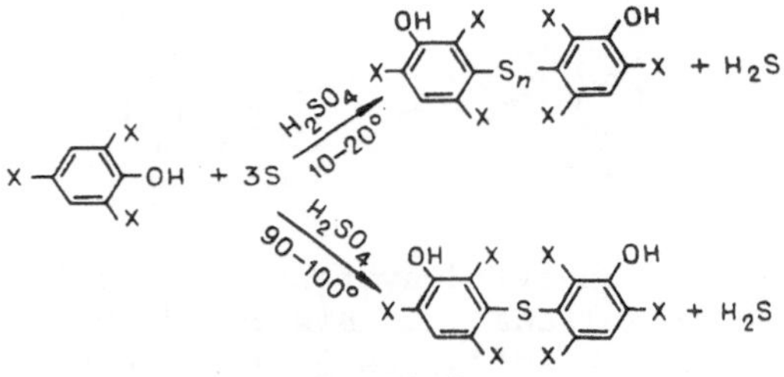

$$X = Cl,\ Br$$

Heating alkylphenols (C_{4-30}) with sulfur in the presence of other components yields additives for lubricants [85].

6.2. ETHERS

The reaction of sulfur with ethers has been studied in most detail in the case of diphenyl ether and its derivatives (the Ferrario reaction) [86]. Practically no reaction is observed when diphenyl ether and sulfur are heated to 160–170°C. However, in the presence of aluminium chloride the reaction starts readily at 70–90°C and gives a high yield of phenoxathiin [86-92]:

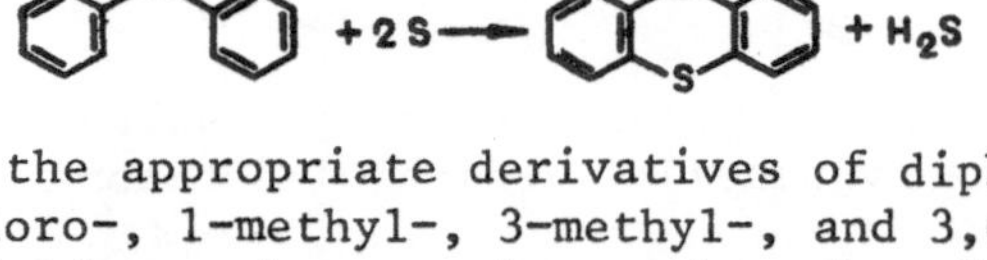

Similarly, the appropriate derivatives of diphenyl ether give 1-chloro-, 3-chloro-, 1-methyl-, 3-methyl-, and 3,6-dimethyl-phenoxathiin; yields are however lower than that for the unsubstituted phenoxathiin [88,91,93].

A process for the polymerization of vinyl ethers in the presence of elemental sulfur (1-5%) allowing high-molecular-weight
polymers to be prepared has been developed [94].

Alkyl benzyl ethers react with sulfur at 170-180°C to give
dehydrogenation and condensation products according to the scheme
[95]:

$$2C_6H_5CH_2OCH_2R + 2S \longrightarrow C_6H_5CHO + RCH=CHR + 2H_2S \qquad (6.35)$$
$$R = CH_3, \ C_2H_5$$

Dibenzyl ether and sulfur give tetraphenylthiophene on heating
[95]:

$$(6.36)$$

Oxirane and sulfur give a highly viscous oil containing 1.2%
sulfur upon heating at 170°C [96]. The reaction of oxiranes with
sulfur and hydrogen sulfide affords polythiodiglycols in quantitative yield [97]:

$$R-\overset{\displaystyle}{C}H-CH_2 + S + H_2S \xrightarrow{H_2O(NaSH)} RCH(OH)CH_2(S_n)CH_2CH(OH)R$$
$$R = H, \ CH_3, \ C_6H_5 \qquad (6.37)$$

Heating tetrahydrofuran with elemental sulfur and red phosphorus at 180-200°C under pressure gives a 45% yield of tetrahydrothiophene [98]. Furan dissolved in ammonia gives succinic acid
upon oxidation with sulfur. The reaction of 2-methyltetrahydrofuran
with sulfur at 200-250°C under pressure gives 2-methyltetrahydro-
thiophene and 2-methylthiophene [99]:

$$(6.38)$$

$$(52\%)$$

Under analogous conditions, 2-methylfuran gives 2-methylthiophene
[99].

The interaction of sulfur with dixanthylene (280°C) involves
cleavage of the double bond and formation of xanthione [100]:

$$(6.39)$$

The reaction of 4-alkoxy-, 4-aryloxy-, 4-aroyloxyisopropyl-benzenes, with sulfur in the presence of the mercury salt of acet-amide in boiling dichlorobenzene does not affect the ether function, and the corresponding 1,2-dithiol-3-thiones are then formed [101] (Table 1).

Sulfur reacts analogously with estragole, safrole, anethole [methyl-4-(1-propenyl)phenyl ether], eugenol and isoeugenol methyl ethers, isosafrole and 4-methoxyisopropenylbenzene at 190-230°C, with retention of the ether function [77-82,102-107] (Table 6.1). Furthermore, the reaction of sulfur with anethole yields also $C_{20}H_{20}O_2S$ (20% yield); probably a thiophene derivative, and a small amount of $C_{20}H_{19}O_2S_2$ which may well contain two condensed thiophene rings. In the case of trithioanethole, the sulfuration product of anethole, an incorrect structure was originally assigned [82,83, 102-107].

6.3. ALDEHYDES

When sulfur is heated with paraformaldehyde, rubber-like sub-stances are obtained. They are soluble in carbon disulfide and have been suggested as rubber vulcanization accelerators [108,109].

The interaction of sulfur with valeraldehyde(pentanal) at 250°C gives thiovaleraldehyde and valeric acid (pentanoic acid) [110]:

$$2C_4H_9CHO + S \longrightarrow C_4H_9CHS + C_4H_9COOH \tag{6.40}$$

Further sulfuration of thiovaleraldehyde gives a compound with the composition $C_5H_6S_3$ to which Barbaglia assigned the structure of the "trithiovaleraldehyde" [111,112]:

$$C_4H_9CHS + 4S \rightarrow \underset{\underset{S}{\diagup}\; \underset{S}{\diagup}\; \underset{H}{}}{CH_2 - CH - CH - CH - C} \overset{S}{\diagup} + 2H_2S \tag{6.41}$$

However, this compound is probably 5-ethyl-1,2-dithiol-3-thione:

$$\underset{C_2H_5 - CH_2}{\overset{CH_2 - CHS}{|}} + 4S \rightarrow C_2H_5 - \text{[ring]} S + 2H_2S \tag{6.42}$$

Acrolein undergoes rapid polymerization at room temperature in the presence of sulfur [113].

Table 6.1. The Reaction of Isopropyl- and Isopropenyl Benzenes Sulfur

Initial ethers	T, °C	Reaction Products (yield, %)	References
4-ROC$_6$H$_4$CH(CH$_3$)$_2$ R = alkyl, aryl, aroyl	boiling dichloro-benzene	4-ROC$_6$H$_4$ (thiophene-S structure)	101
CH=CH$_2$ | R-CH$_2$ R = 4-CH$_3$OC$_6$H$_4$, (benzodioxole structure) (aryl with OCH$_3$, OCH$_2$COOCH$_3$ substituents)	190-230	(thiophene-S structure, R, S)	77-82 102-107
CH-CH$_3$ || R-CH R = 4-CH$_3$OC$_6$H$_4$, 3,4-(CH$_3$O)$_2$C$_6$H$_3$, (benzodioxole structure) (aryl with OCH$_3$, OCH$_2$COOCH$_3$ substituents)	190-230	(thiophene-S structure, R, S)	77-82 102-107
R-C-CH$_3$ || CH$_2$ R = 4-CH$_3$OC$_6$H$_4$	200	(R, thiophene-S structure)	102-107
R-C-CH$_3$ || C$_6$H$_5$-CH cis- and trans-isomers R = CH$_3$OCO	200	(H$_5$C$_6$, R, S structure) (H$_3$C, H$_5$C$_6$, S, O structure)	81,32

When benzaldehyde is heated with sulfur at 180°C, stilbene is formed in addition to benzoic acid and trimeric thiobenzaldehyde, presumably according to the following scheme [113]:

$$2C_6H_5CHO + S \longrightarrow {}^1\!/_3(C_6H_5CHS)_3 + C_6H_5COOH$$

$$C_6H_5CHS + SHCC_6H_5 \xrightarrow{\Delta} C_6H_5CH{=}CHC_6H_5 + 2S$$

(6.43)

Cinnamaldehyde reacts with sulfur in dimethylformamide to give 5-phenyl-1,2-dithiol-3-thione, probably according to the following scheme [114]:

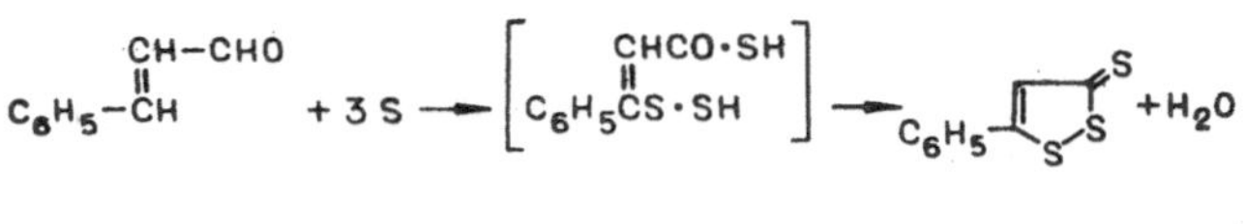

$$\underset{C_6H_5-CH}{\overset{CH-CHO}{\|}} + 3\,S \longrightarrow \left[\underset{C_6H_5CS\cdot SH}{\overset{CHCO\cdot SH}{\|}}\right] \longrightarrow \underset{C_6H_5}{\overset{}{}}\!\!\bigcirc\!\!\overset{S}{\underset{S}{}}\!\!S + H_2O \qquad (6.44)$$

6.4. THE REACTION OF SULFUR WITH ALDEHYDES IN THE PRESENCE OF AMMONIA OR AMINES

The reaction of aldehydes with sulfur in the presence of ammonia or amines gives derivatives of 1,3-thiazolidine (tetrahydro-1,3-thiazole), thiazoline(2,5-dihydro-1,3-thiazole), 1,3-thiazole, 1,4-thiazine, or the corresponding thiocarboxamides.

In the reaction of acetaldehyde with ammonia and sulfur the latter does not participate.

Propanal and butanal react with sulfur and ammonia to give a low yield of 2,5-dihydro-1,3-thiazole [115]:

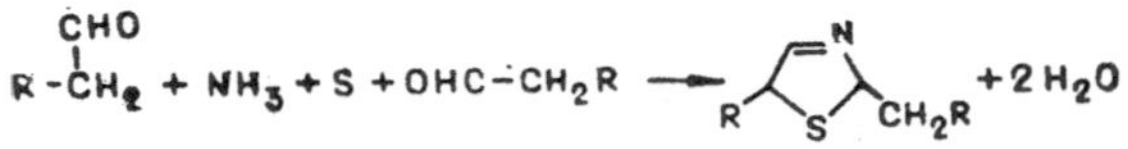

$$\underset{R-CH_2}{\overset{CHO}{|}} + NH_3 + S + OHC{-}CH_2R \longrightarrow \underset{R}{\overset{N}{\diagdown\!\!\diagup}}\underset{S}{\diagdown}CH_2R + 2H_2O \qquad (6.45)$$

$$(10\%)$$

$$R{=}CH_3,\ C_2H_5$$

The 2,5-dihydro-1,3-thiazoles formed are readily dehydrogenated by sulfur to give 1,3-thiazoles:

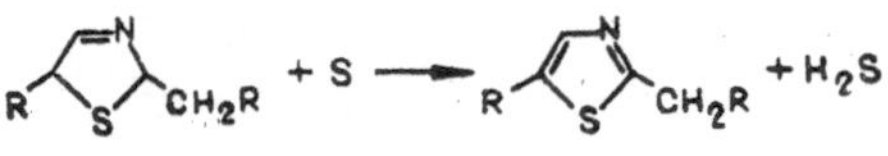

$$\underset{R}{\overset{N}{\diagdown\!\!\diagup}}\underset{S}{\diagdown}CH_2R + S \longrightarrow \underset{R}{\overset{N}{\diagdown\!\!\diagup}}\underset{S}{\diagdown}CH_2R + H_2S \qquad (6.46)$$

Propanal, butanal and pentanal react with ethylenimine (aziridine) and sulfur in the presence of dimethylformamide or anhydrous potassium carbonate to give a mixture of 2-alkyl-5,6-dihydro- 4H-1,-4-thiazines and 2-alkyltetrahydro-1,3-thiazines [116]. Under these conditions, acetaldehyde forms 2-methyltetrahydro-1,3-thiazole as the only product:

$$RCH_2CHO + S + HN\overset{\ }{\diagup} \longrightarrow \underset{R}{\overset{H}{\underset{\ }{\diagdown}}} S + \underset{RH_2C}{\overset{HN-}{\diagdown}} S + H_2O \qquad (6.47)$$

$$R = CH_3, \ C_2H_5, \ C_3H_7$$

Unlike organic amines, primary organosilicon amines (7) react with sulfur and formaldehyde in refluxing aqueous ethanol to form the corresponding substituted formamides (8) and thioureas (11) [117,118]. Shortening the reaction time from twelve to five hours made it possible to obtain, together with (8) and (11), 1,3,5-tris-(trialkylsilylpropyl)hexahydrotriazines (10). The latter seem to be intermediates in reaction (6.48) as they react with sulfur in an aqueous ethanol solution to form a mixture of (8) and (11) in a 1:1:6 ratio. The formation of triazines (10) indicates that one of the reaction stages involves the formation of an aldimine (9). Only the use of preparative gas chromatography allowed organo-silicon isothiocyanates (12) to be isolated from the reaction mixture.

$$R_3Si(CH_2)_nNH_2 \ + \ CH_2O \ \rightarrow \ \left[R_3Si(CH_2)_nNHCH_2OH \right]$$

$$\underset{7}{\ } \qquad\qquad\qquad\qquad\qquad\qquad\qquad\qquad (6.48)$$

$$(CH_2)_nSiR_3 \qquad -H_2O \ \bigg\Vert\! \ +H_2O$$

$$\overset{|}{N}$$

$$\underset{10}{\overset{S}{\overset{\frown}{-}} R_3Si(CH_2)_nN\underset{\ }{\diagdown} N(CH_2)_nSiR_3} \ \rightleftharpoons \ \left[R_3Si(CH_2)_nN=CH_2 \right] \qquad \overset{S}{-H_2S}$$

$$\underset{10}{\ } \qquad\qquad\qquad\qquad\qquad\qquad\qquad \underset{9}{\ }$$

$$\underset{H_2O}{\ } \rightarrow \left[R_3Si(CH_2)_nNH \right]_2 C{=}S \ + \ R_3Si(CH_2)_nNCS \qquad R_3Si(CH_2)_nNHCHO$$

$$\underset{11}{\ } \qquad\qquad\qquad \underset{12}{\ } \qquad\qquad\qquad\qquad \underset{8}{\ }$$

$$n = 3; \ R = CH_3, \ C_4H_9$$

The reaction of benzaldehyde with sulfur and secondary dialkyl-amines at 65–80°C gives N,N-dialkylthiobenzamides [119]. Substituted benzaldehydes react similarly. 4-Hydroxybenzaldehyde reacts with sulfur and morpholine in a similar manner [120].

$$2,4\text{-}R'R''C_6H_3CHO + S + HNR_2 \rightarrow 2,4\text{-}R'R''C_6H_3CSNR_2 + H_2O \qquad (6.49)$$

$$R = \text{alkyl}, \ R' = H, \ 4\text{-OH}, \ R'' = H, \ 2\text{-alkoxy}$$

The condensation of aromatic o-haloaldehydes with sulfur and ammonia leads to benzoisothiazoles (13), (14), and (15) [121–126]:

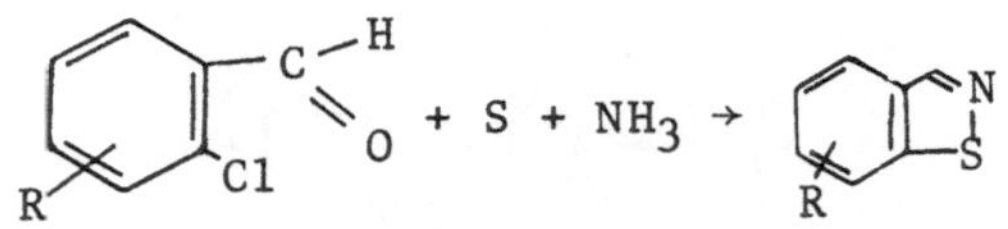

(6.50)

13

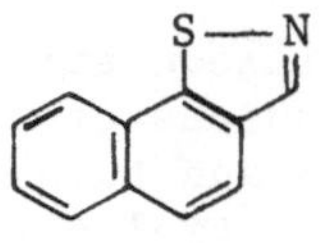

14 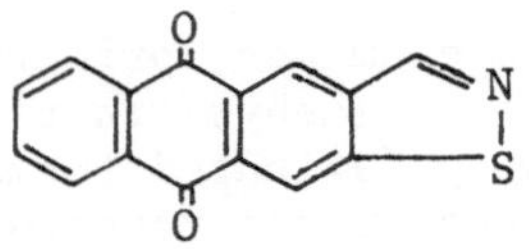   15

The interaction of 2-halobenzaldehydes with sulfur and 1,2-
or 1,3-diaminoalkanes yields 2-(2'-imidazolin-2'-yl)thiophenols (16)
[127]:

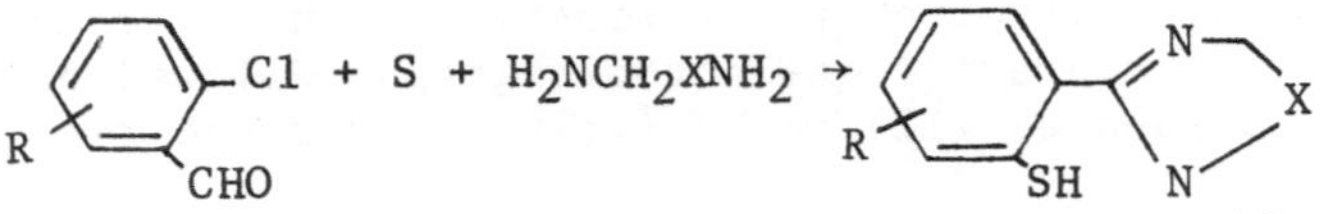

(6.51)

16

$$X = CH_2, \ CHMe, \ CH_2CH_2$$
$$R = H, \ 4\text{-}Cl, \ 4\text{-}Br, \ 4\text{-}NO_2, \ NH_2, \ CH_3COO, \ NH_2SO_2$$

The reaction of aromatic aldehydes with sulfur and α-amino-
benzyl cyanide leads to 2-substituted 5-alkylideneamino-4-phenyl-
1,3-thiazoles [128]:

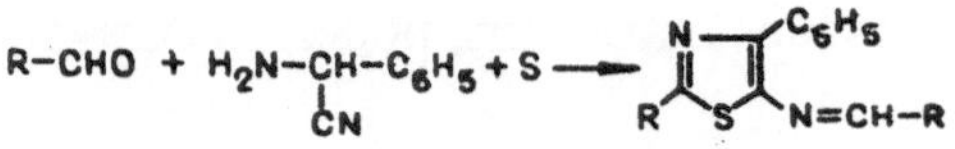

(6.52)

(20-70%)

$$R' = C_6H_5, \ 4\text{-}CH_3OC_6H_4,$$
$$4\text{-}CH_3C_6H_4, \ 4\text{-}NO_2C_6H_4, \ C_6H_4CH=CH, \ n\text{-}C_3H_7$$

6.5. KETONES

In contrast to aldehydes, aliphatic ketones do not react with
sulfur at 150°C [113]. The reaction of 1,3,4-triphenyl-1-butanone
with sulfur at 240-250°C gives 2,3,5-triphenylthiophene [129,130]:

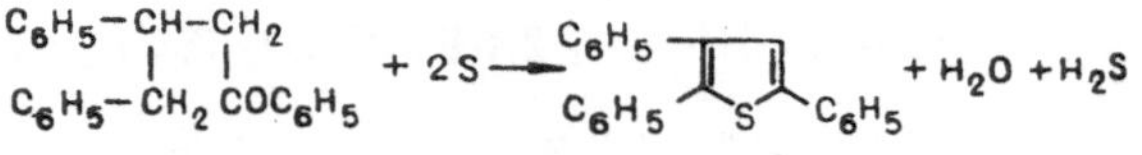

(6.53)

5,6-Dibenzoyl-1,3-diphenylisobenzothiophene (17) is formed by
the action of sulfur on 1,2-dibenzoyl-4,5-dibenzylbenzene (16a)
[131]:

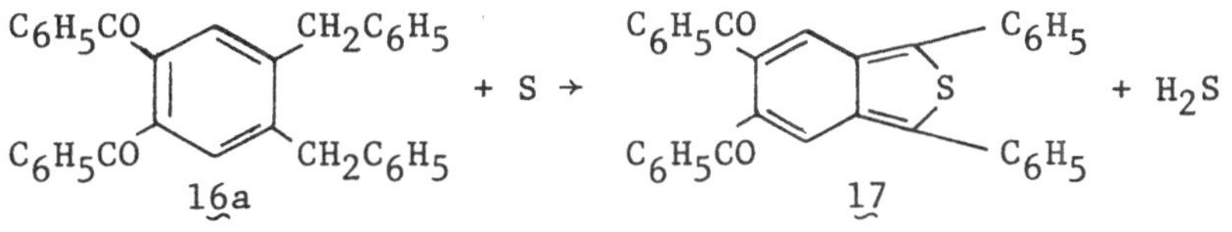

Heating aliphatic and arylaliphatic ketones with sulfur and phosphorus pentasulfide in biphenyl yields a 1,2-dithiol-3-thione [132-134]:

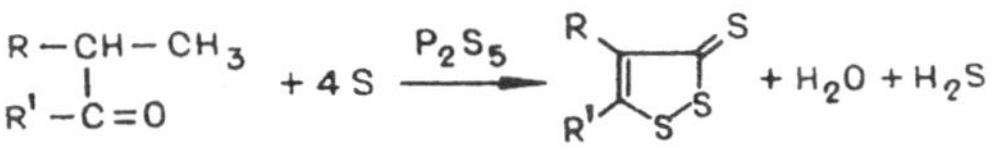

$$(20-35\%)$$

R, R′=H, CH$_3$; CH$_3$, CH$_3$; CH$_3$, C$_2$H$_5$; C$_6$H$_5$, CH$_3$; H, 4-CH$_3$C$_6$H$_4$;

H, 2,4-(CH$_3$)$_2$C$_6$H$_3$; H, 3,4-(CH$_3$O)$_2$C$_6$H$_3$; CH$_3$, C$_6$H$_5$; CH$_3$,

4-CH$_3$C$_6$H$_4$; CH$_3$, 4-CH$_3$OC$_6$H$_4$; CH$_3$, 2,4-(CH$_3$)$_2$C$_6$H$_3$; CH$_3$,

2,6-(CH$_3$)$_2$C$_6$H$_3$; CH$_3$, 3,4-(CH$_3$)$_2$C$_6$H$_3$; CH$_3$, 2,4-(CH$_3$O)$_2$C$_6$H$_3$;

CH$_3$, 2,6-(CH$_3$O)$_2$C$_6$H$_3$; CH$_3$, 3,4-(CH$_3$O)$_2$C$_6$H$_3$; CH$_3$, 2,4,6-

(CH$_3$)$_3$C$_6$H$_2$; C$_6$H$_5$, C$_6$H$_5$

The interaction of sulfur and phosphorus pentasulfide with halo-ketones gives 1,2-dithiol-3-thiones in much higher yield [135]:

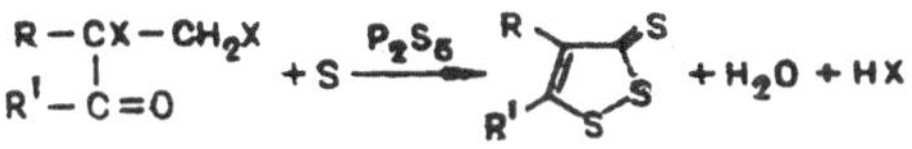

$$x=Cl, Br$$

1-Phenyl-2-propanone reacts with sulfur in boiling dimethylformamide to give 5-phenyl-1,2-dithiol-3-thione [114]:

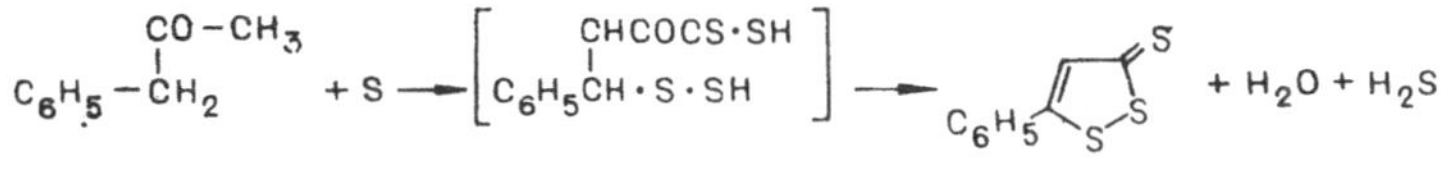

When sulfur is heated with unsaturated ketones, viscous sulfur-containing polymers are formed which inhibit the oxidation of lubricants [136].

The reaction of sulfur with α-methylchalcones gives 4-acyl-1,2-dithiol-3-thiones [137]:

$$R'-\langle\text{benzene}\rangle-CO-\underset{R''}{\underset{|}{C}}=CR''' + 4S \xrightarrow{210 \atop °C} R'-\langle\text{benzene}\rangle-CO-\langle\text{dithiole}\rangle + H_2S \qquad (6.57)$$

$$R'=H, Cl, CH_3O;$$
$$R''=H, CH_3O;$$
$$R'''=CH_3, C_6H_5, 4\text{-}CH_3OC_6H_4$$

The reaction of cinnamoyl chloride with sulfur gives 5-phenyl-1,2-dithiol-3-one [138,139]:

$$\underset{C_6H_5-\overset{\|}{C}H}{\overset{CH-COCl}{}} + 2S \longrightarrow \underset{C_6H_5}{\langle\text{dithiol-one}\rangle} + HCl \qquad (6.58)$$

(20%)

2-Methylcyclohexanone reacts with sulfur and phosphorus penta-sulfide to give a small amount of 1,2-benzodithiol-3-thione [132].

The sulfuration of 2-alkylidenecyclopentanones and 2-aryl-methylene-5-isopropylidenecyclopentanones at 200°C leads to the formation of the corresponding thiophene derivatives:

$$\langle\text{structure}\rangle \xrightarrow{S} \langle\text{structure}\rangle \qquad (6.59)$$

$$\langle\text{structure}\rangle \xrightarrow{S} \langle\text{structure}\rangle + \langle\text{structure}\rangle \qquad (6.60)$$

$$\langle\text{structure}\rangle \xrightarrow{S} \langle\text{structure}\rangle \qquad (6.61)$$

$$R-CH=\langle\text{structure}\rangle + 3S \xrightarrow{175-235°C} R-CH=\langle\text{structure}\rangle + R-CH_2-\langle\text{structure}\rangle + H_2S \qquad (6.62)$$

$$R=C_6H_5, \underline{o}\text{-}CH_3OC_6H_4, 4\text{-}CH_3OC_6H_4, 3,4\text{-}(CH_3O)_2C_6H_3,$$
$$\underline{o}\text{-}ClC_6H_4, 4\text{-}ClC_6H_4, 3,4\text{-}Cl_2C_6H_3, 4\text{-}CH_3C_6H_4$$

R may also be 2-furyl, 2-thienyl, 3-methyl-2-thienyl, 5-methyl-2-thienyl, 4,5-dimethyl-2-thienyl, and 5-ethyl-2-thienyl.

The reaction of 2-cinnamylidenecyclohexanones with sulfur gives the intramolecular 2-(5-phenyl-1,2-dithiolium-3-yl) phenoxide [142]. Compound (18) is also formed in the sulfuration of (5-phenyl-1,2-dithiol-3-ylidene)-2-cyclohexanone:

$$R = H, CH_3 \tag{6.63}$$

$$\tag{6.64}$$

Isophorone reacts with sulfur at 25°C in the presence of a base to give a purple-red liquid which has been assigned the unlikely structure of a thioozonide [143]:

$$\tag{6.65}$$

When isophorone is heated with sulfur to 200-300°C, the compounds formed are 5,5-dimethyl-3-thioxo-4,5,6,7-tetrahydro-1,2-benzodithiol-7-one (19), 7,7-dimethyl-3-thioxo-4,5,6,7-tetrahydro-1,2-dithiol-5-one (20), and 2,2,7,7-tetramethyl-1,2,3,4,6,7,8,9-octahydro[1]benzothieno[3,2-b]-[1]benzothiophene-4,9-dione (21) [144]:

$$\tag{6.66}$$

The yield of (21) increases with increasing temperature.

The reaction of dimedone with sulfur in benzene leads to the corresponding sulfide [145]:

$$\tag{6.67}$$

The sulfuration of carvone leads to the dehydrogenation of the aromatic ring with the formation of condensed thiophene and 3-thioxo-1,2-dithiol rings [146]:

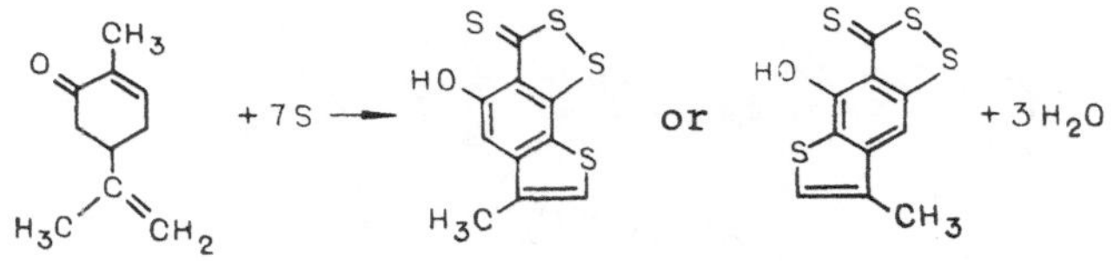

(6.68)

It is likely that the carbonyl group in the hydroaromatic ring is converted from its enol form into a phenolic function during dehydrogenation with sulfur. α-Tetralone is dehydrogenated by sulfur at 240°C to 1-naphthol [147] whereas 2,4,5,6,7,7a-hexahydrobenzo-[b]thiophen-4-one dehydrogenated at 247°C in diphenyl ether gives benzo[b]thiophene-4-ol [148]:

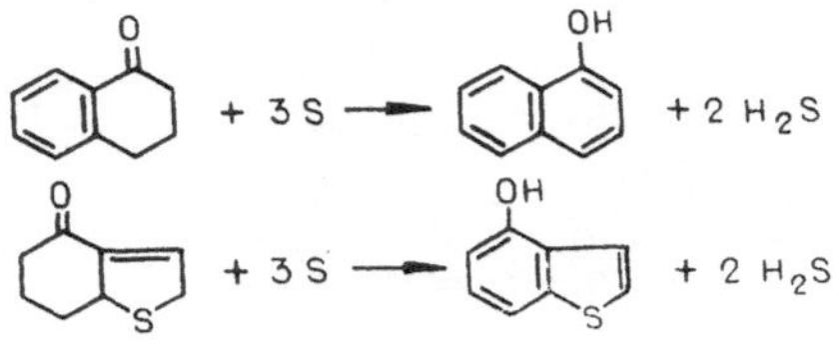

(6.69)

The carbonyl group remains intact in the reaction of exocyclic ketones of the hydroaromatic series with sulfur and only dehydrogenation of the hydroaromatic system takes place [149], e.g., Eq. (6.70):

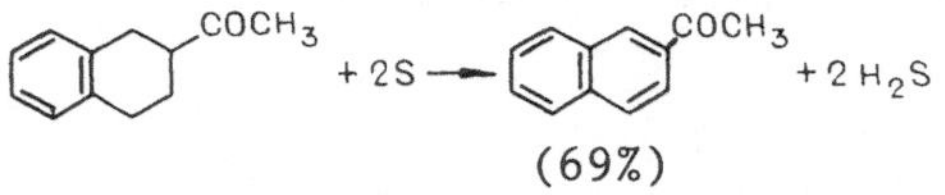

(6.70)

(69%)

4,5-Dibenzoyl-1,2-diphenyl-1-cyclohexene heated with sulfur gives the corresponding tetraphenylbenzo[c]thiophene [150]:

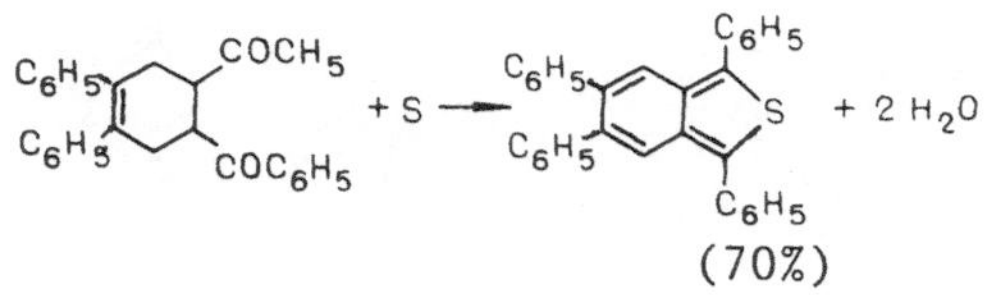

(6.71)

(70%)

5-Acetyl-1,3-cyclohexadiene and its homologs are easily dehydrogenated by sulfur to the corresponding arylaliphatic ketones [151]:

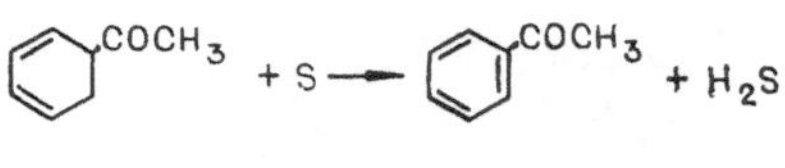

(6.72)

Acetophenone reacts with sulfur at 155-175°C to give 1,4-diphenylbutane-1,4-dione [152]:

$$2C_6H_5COCH_3 + S \longrightarrow C_6H_5COCH_2CH_2COC_6H_5 + H_2S \qquad (6.73)$$

1,4-Diphenylbutane-1,4-dione further reacts with sulfur at
increased temperatures to give 2,5-diphenylthiophene [153]. The
interaction of acetophenone with sulfur and phosphorus pentasulfide
at 200°C gives only 2,5-diphenylthiophene [132]:

$$\begin{array}{c} C_6H_5COCH_2 \\ | \\ C_6H_5COCH_2 \end{array} + S \longrightarrow \underset{C_6H_5 \ \ S \ \ C_6H_5}{\fbox{}} \ +2H_2O \tag{6.74}$$

The reaction of acetophenone with sulfur at 180-210°C gives
thioindigo as well, obviously due to a parallel reaction [153]:

$$\overset{CO}{\underset{}{\bigcirc}}\diagdown CH_3 + 6S \rightarrow \underset{S}{\overset{CO}{\bigcirc}}C=C\underset{CO}{\overset{S}{\bigcirc}} + 4H_2S \tag{6.75}$$

The latter product becomes the principal product in the reaction of
2-chloroacetophenone with sulfur in dimethylformamide [154]. In the
presence of aqueous sodium hydroxide, acetophenone is oxidized by
sulfur to benzoic acid [155]:

$$\underset{Cl}{\overset{CO-CH_3}{\bigcirc}} + S \xrightarrow{HCON(CH_3)_2} \underset{S}{\overset{O}{\bigcirc}}\diagdown\diagup\underset{O}{\overset{S}{\bigcirc}} + HCl + H_2S \tag{6.76}$$

When 4-methylphenyl phenyl ketone is heated with sulfur to
190-210°C for 135 hours, 1,2-bis(4-benzoylphenyl)ethane is obtained
[156]:

$$C_6H_5COC_6H_4CH_3 + CH_3C_6H_4COC_6H_5 + S \rightarrow \begin{array}{c} C_6H_5COC_6H_4CH_2 \\ | \\ C_6H_5COC_6H_4CH_2 \end{array} + H_2S \tag{6.77}$$

Under analogous conditions, 4-ethylbenzophenone gives 2,5-bis(4-
benzoylphenyl)thiophene [152]:

$$C_6H_5COC_6H_4\overset{CH_3}{\underset{}{CH_2}} + \overset{CH_3}{\underset{}{CH_2}}C_6H_4COC_6H_5 + S \longrightarrow$$

$$C_6H_5COC_6H\overline{\underset{S}{\fbox{}}}C_6H_4COC_6H_5 \ \ +4H_2S \tag{6.78}$$

The reaction of 1- and 2-acetylnaphthalene with sulfur at 230-
250°C gives thioindigoid dyes of the naphthalene series (1,2- and
2,1-naphthothioindigo, respectively) [153,157,158]:

$$\overset{CO}{\underset{CH_3}{\bigcirc\bigcirc}} +6S \longrightarrow \overset{CO}{\underset{S}{\bigcirc\bigcirc}}C=C\overset{CO}{\underset{}{\bigcirc\bigcirc}} + 4H_2S \tag{6.79}$$

$$\underset{CO}{\overset{}{\bigcirc\bigcirc}}\diagdown CH_3 +6S \longrightarrow \overset{S}{\underset{CO}{\bigcirc\bigcirc}}C=C\overset{CO}{\underset{S}{\bigcirc\bigcirc}} + 4H_2S \tag{6.80}$$

The interaction of sulfur with 5-acetyl-1,2,6,7,8,8a-hexahydro-acenaphthylene at 180°C leads to the dehydrogenation of both the six-membered and the five-membered ring with the formation of 5-acetylacenaphthylene [159]:

$$\text{(6.81)}$$

When sulfur reacts with α,β,γ,δ-unsaturated ketones of the aromatic series, thiophene derivatives and cyclic 1,2-dithiols are formed [160]. The yield of compounds (22) and (23) depends on the nature of the substituents R and R' and on the reaction temperature:

$$\text{(6.82)}$$

R, R′ = CH₃, C₆H₅, 4-CH₃OC₆H₄

2a,3,4,5-Tetrahydrocholanthren-1-one is dehydrogenated by sulfur at 220°C to give cholanthren-1-one [161]:

$$\text{(6.83)}$$

The sulfuration of 2-isopropylidene- and 2-benzylidene-1-indanone, as well as 2,3-dibenzyl-1-indenone, at 200°C gives the corresponding thiophene derivatives [141,162]:

$$\text{(6.84)}$$

R = C₆H₅,   o-CH₃OC₆H₄,   4-CH₃OC₆H₄,   3,4-(CH₃O)₂C₆H₃,
o-ClC₆H₄, 4-Cl, C₆H₄, 3,4-Cl₂C₆H₃, 4-CH₃C₆H₄

R may also be: 2-furyl, 2-thienyl, 3-methyl-2-thienyl, 5-methyl-2-thienyl, 4,5-dimethyl-2-thienyl, and 5-ethyl-2-thienyl.

Tetraphenyl- or tetra-4-methoxyphenylcyclopentadienone and its derivatives when melted with sulfur (270-350°C) give the corresponding tetra-arylthiophene [163,164]:

$$R = C_6H_5, \ 4\text{-}CH_3OC_6H_4 \tag{6.85}$$

When 2-methylanthraquinone is heated with sulfur, anthracene series dyes are formed, i.e., Cibanone Yellow at 280-290°C and Cibanone Orange at 320-336°C [165,166].

Tetrahydrothiophen-3-one and methyl cyanoacetate react with sulfur at 40°C in methanol in the presence of diethylamine to give the dihydrothieno[2,3-b]thiophene (25). The sulfide (24) was also isolated at room temperature or above 60°C [167]:

$$\tag{6.86}$$

Thiophenes (25a) have been prepared in 15-80% yield by treating aldehydes, ketones or their mercapto-derivatives with sulfur and nitriles [168]:

$$\tag{6.87}$$

$$R = H, \ CH_3, \ R^1 = H;$$
$$RR^1 = (CH_2)_4, \ R^2 = H, \ 3\text{-}Cl, \ 4\text{-}Cl, \ 2\text{-}NO_2, \ 4\text{-}CH_3, \ 3\text{-}NO_2$$

6.6. THE SIMULTANEOUS INTERACTION OF ELEMENTAL SULFUR AND AMMONIA WITH KETONES

Simultaneous action of sulfur and ammonia upon aliphatic and arylaliphatic ketones gives 2,5-dihydro-1,3-thiazoles in high yields [169](cf. Table 6.2). This reaction was discovered by Asinger and bears his name [170]:

$$\tag{6.88}$$

In the case of lower ketones (methyl ethyl ketone, diethyl ketone, and cyclohexanone) reaction (6.88) takes place at room temperature and normal pressure or even at 0°C. In the case of higher or branched ketones (dipropyl ketone, isopropyl methyl ketone, etc.), the reaction occurs only in boiling benzene and with removal of the water formed.

The yield of 2,2,4-trimethyl-2,5-dihydro-1,3-thiazole from acetone is very low (7%). Together with the latter compound, 2,2,-4,6,6-pentamethyl-2H-5,6-dihydro-1,3-thiazine (6%), mesityl oxide, triacetoneamine, and 2-mercapto-2-methylcyclopentanone are formed [171]:

$$3\,CH_3COCH_3 + S + NH_3 \longrightarrow \quad + \quad + \quad + H_2O \qquad (6.89)$$

Two possible mechanisms of reaction (6.88) can be proposed. It is likely that the Asinger reaction consists of two consecutive steps, the first being sulfuration with the formation of α-mercapto-ketones (elemental sulfur is activated in the presence of ammonia due to the formation of a small amount of ammonium polysulfides):

$$\underset{R'-CH}{\overset{R}{\underset{|}{\overset{|}{C}}-OH}} \rightleftarrows \underset{R'-CH_2}{\overset{R}{\underset{|}{\overset{|}{C}}=O}} + S \rightarrow \underset{R'CHSH}{\overset{R}{\underset{|}{\overset{|}{C}}=O}} \rightleftarrows \overset{R}{\underset{|}{\overset{|}{C}}-OH} \qquad (6.90)$$

It is not possible to isolate α-mercapto-ketones as intermediates, obviously because they react with ammonia and then decompose to the corresponding ketone and sulfur. Sulfuration occurs exclusively on the α-methylene group (not the α-methyl group) which explains the anomalous direction of reaction (6.88) with acetone:

$$\underset{R'-CHSH}{\overset{R-CO}{\underset{|}{|}}} \underset{NH_3}{\rightleftarrows} \underset{R'-CH_2}{\overset{R-CO}{\underset{|}{|}}} + S \qquad (6.91)$$

The second step of reaction (6.88) is the condensation of the α-mercapto-ketone with the starting ketone and ammonia:

$$\underset{R'-CH_2}{\overset{R}{\underset{|}{\overset{|}{C}}=O}} + \underset{HSCH-R'}{\overset{R}{\underset{|}{\overset{|}{C}}=O}} + NH_3 \longrightarrow \qquad (6.92)$$

Another possible path for the second step of the Asinger reaction (6.88) is condensation of the α-mercapto-ketone with the addition product of ammonia and the ketone [169].

Table 6.2. Yields of 2,5-Dihydro-1,3-thiazoles Obtained by the Reaction of Ketones, Sulfur, and Ammonia according to Reaction (6.88)

No.	R^1	R^2	R^3	R^4	R^5	R^6	Yield of Dihydro-thiazole, %	References
1	CH_3	H	H	H	CH_3	CH_3	7.5	171
2	CH_3	H	H	CH_3	CH_3	CH_3	85	172
3	CH_3	H	H	C_2H_5	H	C_3H_7	2–22	173,174
4	CH_3	H	H	C_2H_5	C_2H_5	CH_3	77–98	173,174
5	C_2H_5	H	H	CH_3	CH_3	C_2H_5	85	175
6	C_3H_7	H	H	C_2H_5	C_2H_5	C_3H_7	90	176
7	CH_3	H	CH_3	CH_3	CH_3	CH_3	42	177
8	CH_3	H	CH_3	CH_3	H	$CH(CH_3)_2$	20	177
9	C_2H_5	H	CH_3	CH_3	CH_3	C_2H_5	8	178
10	C_2H_5	H	CH_3	CH_3	CH_3	$CH(CH_3)_2$	92	178
11	CH_3	H	H	$CH(CH_3)_2$	H	$CH_2CH(CH_3)_2$	20–30	173,174
12	CH_3	H	H	$CH(CH_3)_2$	$CH(CH_3)_2$	CH_3	70–80	173,174
13	CH_3	H	CH_3	C_2H_5	H	$CH(CH_3)(C_2H_5)$	1–90	173
14	CH_3	H	CH_3	C_2H_5	$(CH_3)(C_2H_5)$	CH_3	10–99	173
15	CH_3	H	C_2H_5	C_2H_5	H	$CH(C_2H_5)_2$	5	173

Table 6.2 (continued)

No.	R^1	R^2	R^3	R^4	R^5	R^6	Yield of Dihydro- thiazole, %	References
16	CH_3	H	C_2H_5	C_2H_5	$(C_2H_5)_2$	CH_3	5	173
17	C_6H_{13}	H	H	$c\text{-}C_5H_{11}$	$c\text{-}C_5H_{11}$	C_6H_{13}	90	173
18	$c\text{-}C_5H_{11}$	H	H	H	$c\text{-}C_5H_{11}$	CH_3	50	179
19	$c\text{-}C_5H_{11}$	H	H	H	H	$c\text{-}C_5H_{11}$	50	179
20	CH_3	H	H	C_6H_5	C_7H_5	CH_3	90	180
21	CH_3	H	H	$-(CH_2)_2-COOX$	$-(CH_2)_2-COOX$	CH_3	75	181
				X = H, CH_3, C_2H_5, Na				
22	C_6H_5	H	H	CH_3	CH_3	C_6H_5	19-60	182
23	C_6H_5	H	H	C_2H_5	C_2H_5	C_6H_5	61	182
24	C_6H_5	H	CH_3	CH_3	$(CH_3)_2$	C_6H_5	31	182

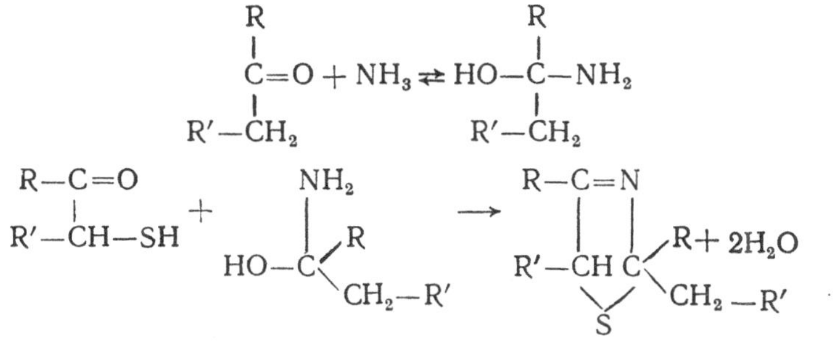

$$\begin{array}{c} R \\ | \\ C\!=\!O + NH_3 \rightleftharpoons HO\!-\!\overset{|}{C}\!-\!NH_2 \\ | \\ R'\!-\!CH_2 \qquad\qquad R'\!-\!CH_2 \end{array} \qquad (6.93)$$

$$\begin{array}{c} R\!-\!C\!=\!O \\ | \\ R'\!-\!CH\!-\!SH \end{array} + \begin{array}{c} NH_2 \\ | \quad R \\ HO\!-\!C \\ \quad CH_2\!-\!R' \end{array} \longrightarrow \begin{array}{c} R\!-\!C\!=\!N \\ | \qquad | \\ R'\!-\!CH \; C \diagdown R + 2H_2O \\ \diagdown \quad \diagup \; CH_2\!-\!R' \\ S \end{array}$$

The possibility of reactions (6.92) and (6.93) is confirmed by the
formation of 2,5-dihydro-1,3-thiazole in the condensation of α-
mercapto-ketones with aldehydes, ketones, and esters in the presence
of ammonia [169]. 2-Mercapto-1-tetralone reacts with sulfur and
ammonia and gives the corresponding 2,5-dihydro-1,3-thiazole derivative [183].

Similarly to acetone, acetophenone gives an anomalous reaction
with sulfur and ammonia [184]. In this reaction, 2-methyl-2,4-
diphenyl-2,5-dihydro-1,3-thiazole (26) is formed in low yield when
the reaction is carried out at room temperature. Also formed are
2-methyl-2,4-diphenyl-Δ^3-imidazoline-5-thione (27) and a very small
amount of 2,4-diphenylthiophene (28) [184,185]:

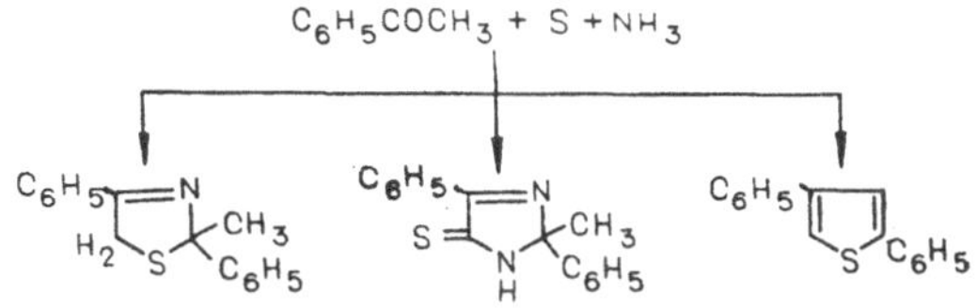

$$C_6H_5COCH_3 + S + NH_3$$

$$\begin{array}{ccc} 26 & 27 & 28 \end{array} \qquad (6.94)$$

When reaction (6.94) is carried out in boiling toluene with removal
of the water formed during the process (by distillation), the
products obtained are (26), a large amount of (28), and some 2,4-
diphenyl-1,3-thiazole. No (27) is formed. 2,4-Diphenylthiophene
(28) is formed in high yield on prolonged heating of (26) at 230-
250°C.

The yield of 2-methyl-2-phenyl-2,5-dihydroimidazole-5-thione
can be increased to 92% in the presence of excess sulfur and in the
presence of Schiff bases [186,187]. The following reaction mechanism may be postulated for scheme (6.94):

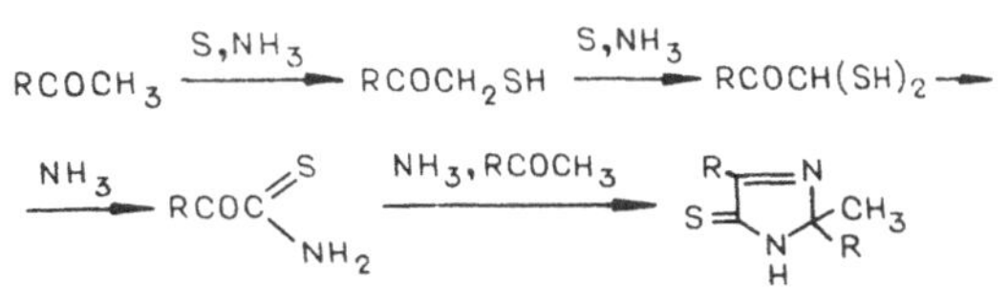

$$RCOCH_3 \xrightarrow{S,NH_3} RCOCH_2SH \xrightarrow{S,NH_3} RCOCH(SH)_2 \longrightarrow$$

$$\xrightarrow{NH_3} RCOC\diagup{\!\!\!S}_{\diagdown NH_2} \xrightarrow{NH_3, RCOCH_3} \begin{array}{c} R\!-\!\!=\!N \\ S\!=\!\!\diagdown_{N}\diagup\!\!\diagdown_{R}^{CH_3} \\ \;\;H \end{array} \qquad (6.95)$$

The reaction of sulfur and ammonia with cyclopentanone does not give the expected 2,5-dihydro-1,3-thiazole and the obtained products are dark, resinous, unidentified substances [188]. The reaction of sulfur and ammonia with cycloheptanone, cyclooctanone, cyclodecanone, 3,3,5,5-tetramethylcyclohexanone, ethyl 4-oxocyclohexanecarboxylate, tetrahydrothiapyran-4-one, and 1,4-dithia-6-cycloheptanone leads to the formation of the corresponding 2,5-dihydro-1,3-thiazoles in high yields [188-190]:

$$\text{(image)} + S + NH_3 \longrightarrow \text{(image)} + 2\ H_2O \qquad (6.96)$$

2-Hydroxyacetophenone reacts with sulfur and ammonia with the formation of the 5-thioxo-2,5-dihydroimidazole (29) [191]:

$$(6.97)$$

Upon prolonged heating (72 hours) the thioxo-group in (29) is then replaced by an imino-group with subsequent prototropic rearrangement of the C=N bond into the ring [191].

6.7. THE REACTION OF KETONES WITH SULFUR AND AMINES
(THE KINDLER REACTION)

The reaction of ketones with sulfur and primary amines proceeds at room temperature and atmospheric pressure and is catalyzed by methanol or acetic acid. 7-Alkylamino-7-aryl-8-thioxo-1,2,3,4,5,-6-hexathiocanes are the products in this reaction [192-195]:

$$\underline{p}\text{-}RC_6H_4\text{-}\underset{O}{\overset{}{C}}CH_3 + R'NH_2 + S_8 \longrightarrow \underline{p}\text{-}RC_6H_4\text{-}\underset{\underset{S_6}{\overset{}{\diagdown}}}{\overset{NH-R'}{\overset{|}{C}}}\text{-}C=S + H_2O + H_2S \qquad (6.98)$$

R=H, Cl, CH₃O.

R′=CH₃, C₂H₅, C₃H₇, iso -C₃H₇, C₆H₁₁, C₆H₅CH₂CH₂,

(C₂H₅)₂NCH₂CH₂, (CH₃)₂N(CH₂)₃,

(C₂H₅)₂N(CH₂)₃, CH₃OCH₂CH₂.

When R = $(C_2H_5)_2N(CH_2)_3$, reaction (6.98) is accompanied by a competing reaction:

$$C_6H_5COCH_3 + (C_2H_5)_2N(CH_2)_3NH_2 + S \rightarrow$$
$$\rightarrow (C_2H_5)_2N(CH_2)_3N = C(C_6H_5)CS_2H + H_2O + H_2S \qquad (6.99)$$
$$(51\%)$$

When R = CH$_3$CO, bis(7-alkylamino-8-thioxo-1,2,3,4,5,6-hexathiocan-7-yl)benzene is formed as a condensation product with the participation of both acetyl groups.

One of the modifications of the Willgerodt reaction is the condensation of ketones with sulfur and secondary amines, sometimes called the Kindler reaction [196,197].

The principal product of this reaction is the corresponding thiocarboxamide and. in some cases, other products are obtained some of which contain sulfur. Carboxylic acids obtained in the Kindler reaction are the result of the hydrolysis of the corresponding thiocarboxamides.

The reaction of acetophenone with sulfur and morpholine may be taken as an example. At 145°C it gives the morpholides of phenyl-thioacetic (30) and phenyl-1-thioglyoxylic acid (31) [198]:

$$C_6H_5COCH_3 + S + HN\underset{}{\bigcirc}O \rightarrow C_6H_5CH_2 - \underset{\underset{S}{\|}}{C} - N\underset{}{\bigcirc}O + C_6H_5CO - \underset{\underset{S}{\|}}{C} - N\underset{}{\bigcirc}O \qquad (6.100)$$
$$\underset{}{30}\ (21\%) \qquad\qquad \underset{}{31}\ (10.5\%)$$

N-Styrylmorpholine, which presumably is an intermediate in reaction (6.100), reacts with sulfur and morpholine to give (30). In the presence of lead oxide reaction (6.100) can be carried out at 85°C. In the presence of pyridine, acetophenone reacts with sulfur to give only the morpholide of phenylthioacetic acid.

The presence in the ketone of a methyl group or an alkyl chain at the carbonyl group is necessary for the formation of thiomor-pholides. In some cases, the thiomorpholides having a propyl or butyl chain in the molecule undergo cyclization to give dithiol-thiones or thiophenes [197-235] (Table 6.3).

The isomeric phenylbutanones upon heating with sulfur and morpholine (130-150°C) are converted to 2-morpholino-5-phenylthio-phene and the morpholide of 4-phenylthiobutanoic acid [210,211,218]. Substituted thiophenes are often the only products from the reaction of ketones with sulfur and morpholine:

$$\left.\begin{array}{l} C_6H_5CH_2CH_2COCH_3 \\[4pt] C_6H_5CH_2COCH_2CH_3 \\[4pt] C_6H_5COCH_2CH_2CH_3 \end{array}\right\} + S + HN\underset{}{\bigcirc}O \longrightarrow C_6H_5(CH_2)_3\overset{\overset{S}{\|}}{C}N\underset{}{\bigcirc}O + C_6H_5\text{—}\underset{S}{\square}\text{—}N\underset{}{\bigcirc}O \qquad (6.101)$$
$$(40\text{-}70\%) \qquad\qquad (15\text{-}20\%)$$

Table 6.3. Reactions of Ketones with Sulfur and Morpholine (the Kindler Reaction)

Initial Ketone	T, °C	Reaction Products (Yield, %)	Ref.
$C_6H_5COCH_3$	145	$C_6H_5CH_2\underset{\parallel}{C}-N\bigcirc O$ (21), $C_6H_5CO\underset{\parallel}{C}-N\bigcirc O$ (10.5)	197,198
$4-CH_3OC_6H_4COCH_3$	145	$4-CH_3OC_6H_4CH_2-\underset{\parallel}{C}-N\bigcirc O$ (96)	199
$4-C_2H_5OC_6H_4COCH_3$	145	$4-C_2H_5OC_6H_4CH_2-\underset{\parallel}{C}-N\bigcirc O$ (68)	200
$3-NO_2C_6H_4COCH_3$	130	$3-NO_2C_6H_4CH_2-\underset{\parallel}{C}-N\bigcirc O$ (69.5)	201
$3,4-Cl(OH)C_6H_3COCH_3$	140	$3,4-Cl(OH)C_6H_3-\underset{\parallel}{C}-N\bigcirc O$	202
$3,4-(OH)(COOH)C_6H_3COCH_3$	145	$3,4-(OH)(COOH)C_6H_3COCH_3$	203
$2,5-(CH_3O)_2C_6H_3COCH_3$	145	$2,5-(CH_3O)_2C_6H_3CH_2-\underset{\parallel}{C}-N\bigcirc O$	204
$1,3,5-C_6H_3(COCH_3)_3$	–	$1,3,5-C_6H_3(CH_3COOH)_3$ (95)	205

Initial Ketone	T, °C	Reaction Products (Yield, %)	Ref.
$4\text{-}C_6H_5COC_6H_4CH_2OC_6H_5$	150	$C_6H_5OCH_2C_6H_4CH_2CH_2\text{-}C\underset{\parallel S}{\text{-}}N{-}O\text{(morpholine)}$ (60)	206
$4\text{-}C_6H_5C_6H_4COCH_3$	150	$4\text{-}C_6H_5C_6H_4CH_2\text{-}C\underset{\parallel S}{\text{-}}N{-}O\text{(morpholine)}$ (95)	196
naphthalene $COCH_3$ / H_3CO	145	naphthalene $COOH$ / H_3CO	207
tetralin $COCH_3$	145	$COOH$	207
phenanthrene $COOH$	145	$COOH$	207
phenanthrene $COCH_3$	145	$COOH$	207
phenothiazine $COCH_3$, $R = H, CH_3$	145	phenothiazine $CH_2\text{-}C\underset{\parallel S}{\text{-}}N{-}O\text{(morpholine)}$, R	208

Table 6.3 (continued)

Initial Ketone	T, °C	Reaction Products (Yield, %)	Ref.
$C_6H_5COCH_2CH_2C_6H_5$	145	$C_6H_5CH=CHCH_2C_6H_5$, $C_6H_5C(=S)-N$(morpholino)	209
$C_6H_5CH_2COCH_2C_6H_5$			209
$4-CH_3OC_6H_4COCH_2CH_2C_6H_4OCH_3-4$	130	$4-CH_3OC_6H_4CH=CHCH_2C_6H_4OCH_3-4$, $4-CH_3OC_6H_4-C(=S)-N$(morpholino)	209
$4-CH_3OC_6H_4CH_2COCH_2C_6H_4OCH_3-4$			
$4-ClC_6H_4COCH_2CH_2C_6H_4OCH_3-4$	130	$4-ClC_6H_4CH=CHCH_2C_6H_4OCH_3$, $4-CH_3OC_6H_4-C(=S)-N$(morpholino),	209
$4-ClC_6H_4CH_2COCH_2C_6H_4OCH_3-4$		$4-ClC_6H_4-C(=S)-N$(morpholino)	
$C_6H_5CH_2CH_2COCH_3$	130	$C_6H_5(CH_2)_3-C(=S)-N$(morpholino) (48–72),	210,211
$C_6H_5CH_2COCH_2CH_3$		C_6H_5-(thiophene with N-morpholino) (13–20)	
$C_6H_5COCH_2CH_2CH_3$			
$C_6H_5COCH_2C_6H_5$	155	$C_6H_5CH=CHC_6H_5$ (44)	212
$C_6H_5COCH(C_6H_5)_2$	165	$(C_6H_5)_2C=CHC_6H_5$	212
$(C_6H_5)_2CO$	180	$(C_6H_5)_2CH_2$ (85), $(Ph_2CH)_2S_2$	213,214
$[4-(CH_3)_2NC_6H_4]_2CO$	180	$[4-(CH_3)_2NC_6H_4]_2C=S$ (5)	215

Initial Ketone	T, °C	Reaction Products (Yield, %)	Ref.
R = H, OCH$_3$	140	(30)	216
$C_6H_5COCH_2COCH_3$	145	(37)	217, 218
$C_6H_5COCOCH_3$	145	(5)	217
$C_6H_5COCOCH_2C_6H_5$	135	$C_6H_5COCH_2CH_2C_6H_5$ (20), $C_6H_5-\overset{\text{S}}{\underset{\|}{C}}-N\bigcirc O$ (5)	217
$C_6H_5COCH_2COC_6H_5$	145	$C_6H_5CH=CHCH_2C_6H_5$ (30), $C_6H_5COCH_2COCH_3$ (8), $C_6H_5CO-N\bigcirc O$ (5)	217
$C_6H_5COCOC_6H_5$	145	$C_6H_5COCH_2C_6H_5$ (50), $C_6H_5CH=CHC_6H_5$ (5), $C_6H_5COCH_2-\overset{\text{S}}{\underset{\|}{C}}-N\bigcirc O$, $C_6H_5COCH_2CH_2-\overset{\text{S}}{\underset{\|}{C}}-N\bigcirc O$	217
$C_6H_5COCH=CH_2$	160	(40)	219

Table 6.3 (continued)

Initial Ketone	T, °C	Reaction Products (Yield, %)	Ref.
4-$CH_3OC_6H_4COCH=CH_2$	160	4-$CH_3OC_6H_4COCH_2$-C(=S)-N(morpholine), 4-$CH_3OC_6H_4CH_2CH_2$-C(=S)-N(morpholine), 4-$CH_3OC_6H_4$ thiole-thione structure (50)	219
4-$ClC_6H_4COCH=CH_2$	160	4-$ClC_6H_4CH_2CH_2$-C(=S)-N(morpholine), 4-$ClC_6H_4COCH_2$-C(=S)-N(morpholine), 4-ClC_6H_4 thiole-thione structure (40)	219
$C_6H_5COCH=CHC_6H_5$	145	morpholino-thiophene C_6H_5 structure (10.6), C_6H_5-thiophene-C_6H_5 structure (8.1)	220
$C_6H_5CH=CHCH=CHCOC_6H_5$	135	C_6H_5,C_6H_5 dithiine structure, C_6H_5-thiophene-COC_6H_5 structure, C_6H_5-CHCOC$_6H_5$ structure	221
$C_6H_5CH=CHCOCH=CHC_6H_5$	135	C_6H_5,C_6H_5 dithiine structure (7), morpholino-thiophene C_6H_5 structure (6)	221

Initial Ketone	T, °C	Reaction Products (Yield, %)	Ref.
$C_6H_5C=CHCOCH_3$ (morpholino)	135	$C_6H_5-\overset{S}{\underset{\parallel}{C}}-N$(morpholino) (8), C_6H_5–thiophene–N(morpholine) (8.5), C_6H_5–thiophene=S morpholine (7.3)	218
$C_6H_5COCH_2CH_2N$(morpholine)	145	$C_6H_5CH_2CH_2-\overset{S}{\underset{\parallel}{C}}-N$(morpholine), $C_6H_5COCH_2-\overset{S}{\underset{\parallel}{C}}-N$(morpholine), thiophene=S	222
$4-RC_6H_4COCH_2CH_2-N$(morpholine); R = CH₃, OCH₃, Cl	145	$4-RC_6H_4CH_2CH_2-\overset{S}{\underset{\parallel}{C}}-N$(morpholine), $4-RC_6H_4COCH_2-\overset{S}{\underset{\parallel}{C}}-N$(morpholine), $4-RC_6H_4$–thiophene=S	222
$C_6H_5CH_2CH_2COCH_2-N$(morpholine)	150	C_6H_5–thiophene–N(morpholine) (25)	223
CH_3CO–benzofuran–C_2H_5	140	benzofuran(C_2H_5)–$CH_2-\overset{S}{\underset{\parallel}{C}}-N$(morpholine)	224

Table 6.3 (continued)

Initial Ketone	T, °C	Reaction Products (Yield, %)	Ref.
benzofuran with CH_3, $COCH_3$; R = H, Cl, F, CH_3, $(CH_3)_3C$, CH_3O, C_6H_5, $C_6H_5CH_2$, C_6H_{12}	140	benzofuran with CH_3, $CH_2{-}C({=}S){-}N$(morpholine), R	225
benzofuran COR; R = CH_3, C_2H_5	140	benzofuran $(CH_2)_nCOOH$ (20–30); n = 1,2	226
benzofuran COR, C_2H_5; R = CH_3, C_2H_5, $n{-}C_3H_7$	140	benzofuran $(CH_2)_nCOOH$, C_2H_5; n = 1,2,3	226
coumarin with CH_3, $COCH_3$, R; R = H, OCH_3	140	coumarin with CH_3, CH_2COOH, R	170
pyrazole with H_3C, $COCH_3$, CH_3, R	150	pyrazoline with H_3C, $CH_2{-}C({=}S){-}N$(morpholine), $NH_2^+Cl^-$, R; R = H, $COOC_2H_5$	227

Initial Ketone	T, °C	Reaction Products (Yield, %)	Ref.
	150		228
$C_6H_5-\underset{\underset{S}{\parallel}}{C}-CH_3$	145	$C_6H_5CH_2-\underset{\underset{S}{\parallel}}{C}-N$ (morpholine)	229
$C_6H_5-\underset{\underset{S}{\parallel}}{C}-CH_2COC_6H_5$	145	$C_6H_5COCH_2CH_2C_6H_5$, $C_6H_5CH=CHCH_2C_6H_5$, $C_6H_5-\underset{\underset{S}{\parallel}}{C}-N$ (morpholine)	230
		(70)	199
	150	(80)	231
	150	(63)	199,231

Table 6.3 (continued)

Initial Ketone	T, °C	Reaction Products (Yield, %)	Ref.
$COCH_3$ (pyridine)	150	morpholino-$N-C(=S)-CH_2-$ (pyridyl) (76.5)	231
(dimethylthiophene) $COCH_3$, CH_3	150	$CH_2-C(=S)-N$(morpholino), CH_3, H_3C-thiophene-S (66)	232
(dimethylthiophene) COC_2H_5, CH_3	150	$CH_2CH_2-C(=S)-N$(morpholino), CH_3, H_3C-thiophene-S (40)	232
$Cl_2CHCOCl$	in tri-chloro-ethylene	$RNR'-C(=S)-CONRR'$ NRR' = morpholino, piperidino, 1-pyrrollidinyl, piperazinyl	233
$RCOCH(Cl)R'$ R' = H, Cl	150	$RCO-C(=S)-NMe_2$ $R = Me_3C$, C_6H_5	234
$RCOCH_3$ $R = C_6H_5$; 2,4$(MeO)_2C_6H_3$	150	$C_6H_5-C(=S)-NH_2CH_2C_6H_5$, $RCH_2-C(=S)-NHC_6H_5$	235

Morpholinomethyl 2-phenylethyl ketone upon interaction with sulfur and morpholine gives only 2-morpholino-5-phenylthiophene [223]:

$$C_6H_5CH_2CH_2COCH_2N\bigcirc O + S + HN\bigcirc O \longrightarrow C_6H_5\overset{\parallel}{\underset{S}{\diagup}}N\bigcirc O \qquad (6.102)$$

(25%)

Table 6.2 shows the products from the Kindler reaction for various ketones.

Physical and analytical data for thioamides prepared from piperazines, sulfur and ketones or aldehydes of the aromatic and aliphatic series are reported in [236].

The reaction of aliphatic ketones ($H_5C_2COC_2H_5$, $H_3CCOCH(CH_3)_2$, $H_3CCOC_2H_5$, $H_5C_2COCH(CH_3)_2$) as well as alicyclic ketones (cyclopentanone, cyclohexanone, cycloheptanone, cyclooctanone) with elemental sulfur and aziridine occurs readily at room temperature and gives 5,6-dihydro-4H-1,4-thiazines (32) and 2,2-dialkyltetrahydro-1,3-thiazoles (33) [237-240]. The ratio of the reaction products (32):(33) in reaction (6.103) decreases with increasing temperature:

$$\begin{array}{l} R'-C=O \\ \qquad | \\ R''-CH_2 \end{array} + S + HN\big] \rightarrow \quad \underset{R''\diagup S}{\overset{R'}{\diagup}}\overset{H}{\underset{N}{|}} + \quad \underset{R''CH_2}{\overset{HN-}{\underset{R'}{|}}S} + H_2O$$

32   33   (6.103)

The principal product in the reaction of isopropyl phenyl ketone with sulfur and aziridine is 2-isopropyl-2-phenyltetrahydro-1,3-thiazole. 5,6-Dihydro-4H-1,4-thiazine is also formed in low yield [239].

It seems that the mechanism of the reaction of ketones with sulfur and aziridine is similar to that of the Asinger reaction. Aziridine acts as a sulfuration catalyst and participates in the cyclization. The by-products in reaction (6.103) are tetrahydro-1,3-thiazoles which may be formed during the interaction of the starting ketone with hydrogen sulfide formed in the dehydrogenation of aziridine with sulfur [240]. However, this possibility has not been proven experimentally and is thus quite speculative:

$$(6.104)$$

6.8. THE INTERACTION OF SULFUR AND AMMONIA OR AMINES WITH CARBON MONOXIDE

One of the methods for the synthesis of urea and its derivatives is based on the reaction of sulfur and ammonia (or primary amines) with carbon monoxide [241,242]:

$$2RNH_2 + CO + S \rightarrow RNH-\underset{\underset{O}{\|}}{C}-NHR + H_2S$$

$$(6.105)$$

$$R = H, \text{ alkyl}$$

The yield of urea derivatives varies within a wide range (1-92%) and depends on the structure of the starting amine, the temperature, the duration of the reaction, the pressure, and the ratio of the reagents:

$$CO + S \rightarrow COS \xrightarrow{2NH_3} H_2NCONH_4$$
$$\downarrow S$$
$$H_2NCONH_2 \xleftarrow{NH_3} HNCO + NH_4SH \rightarrow H_2S + NH_3$$

$$(6.106)$$

The interaction of carbon monoxide with sulfur, secondary amines, and methyl iodide gives the corresponding S-methyl N,N-dialkylthio-carbamates [243]. Reaction (6.107) is carried out at 120°C under 300 atmospheres pressure in tetrahydrofuran. Methyl iodide is then introduced into the reaction mixture at 0-10°C and atmospheric pressure:

$$2R_2NH + CO + S \xrightarrow{130°C} \left[R_2N-\underset{\underset{\ominus}{\|}}{\overset{O}{C}}-\overset{\oplus}{S}H_2NR_2 \right] \xrightarrow[0-10°C]{CH_3I} R_2N-\underset{\underset{O}{\|}}{C}-S-CH_3$$

$$(6.107)$$

$$R = C_4H_9, (CH_3)_2CH$$

6.9. CARBOXYLIC ACIDS

When carboxylic acids react with sulfur, the carboxyl group remains intact in most cases; however, decarboxylation is sometimes

observed. The formation of hydrogen sulfide in the reaction of
sulfur with saturated and hydroaromatic acids is due to the dehydrogenation of their side chains and rings.

6.9.1. Alkane-, Arylalkane- and Cycloalkanecarboxylic Acids

The sulfuration of aliphatic and resinous unsaturated carboxylic
acids at 120-160°C was studied in 1892. The reaction leads to the
so-called "thioacids" [244].

Butanoic acid and other alkanoic acids refluxed with sulfur give
thiophene which is considerably contaminated with carbon disulfide
[245].

The reaction of sulfur with stearic acid and the hydrogenated
acids of castor oil at 150-300°C leads to the evolution of hydrogen
sulfide and the formation of so-called "sulfurous balsams" [246].

The interaction of sulfur with phenylacetic acid at 220-260°C
gives tetraphenylthiophene [247]:

$$4\,C_6H_5CH_2COOH + S \longrightarrow \underset{C_6H_5\,\,\,S\,\,\,C_6H_5}{\overset{C_6H_5\quad C_6H_5}{\bigcirc}} + H_2S + CO_2 \qquad (6.108)$$

The dehydrogenating properties of sulfur are used to obtain
polycyclic aromatic acids from the corresponding hydroaromatic
derivatives. Thus, tetrahydronaphthalene- and tetrahydrophenan-
threnecarboxylic acids are dehydrogenated by heating with sulfur
and give the corresponding aromatic acids [248-257]:

$$\qquad (6.109)$$

The reaction of sulfur with 2-(1,2,3,4-tetrahydronaphthyl)-3-
propanoic acid affords 2-naphthyl-3-propanoic acid [258]. 1,7-Di-
methylnaphthalenecarboxylic acid is formed on heating the corresponding tetralincarboxylic acid with sulfur [259,260]:

$$\qquad (6.110)$$

The dehydrogenation of abietic and pimaric acids with sulfur
takes place with decarboxylation and the formation of retene [261-
265] and pimanthrene [266,267], respectively. The interaction of
sulfur with abietic acid at 177°C affords lubricants containing
5-15% of sulfur [268]:

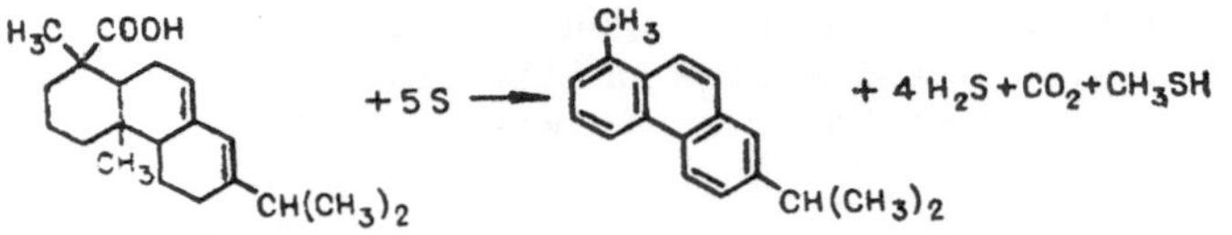

abietic acid   retene   (6.111)

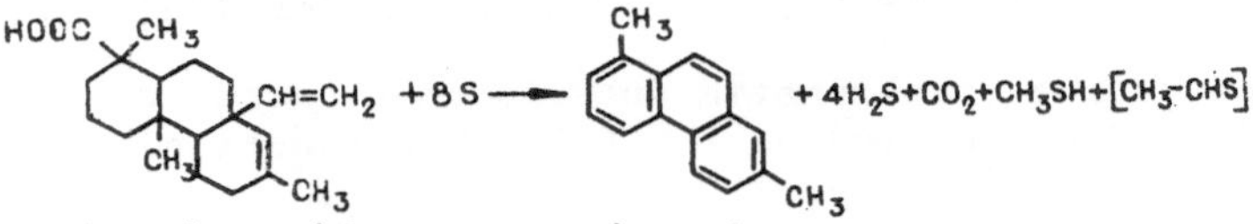

pimaric acid   pimanthrene   (6.112)

2-Benzyl-1,2,3,6-tetrahydrobenzoic acids react with sulfur and
phosphorus(V) oxide at 300-310°C to give alkyl-substituted anthra-
cenes [269]:

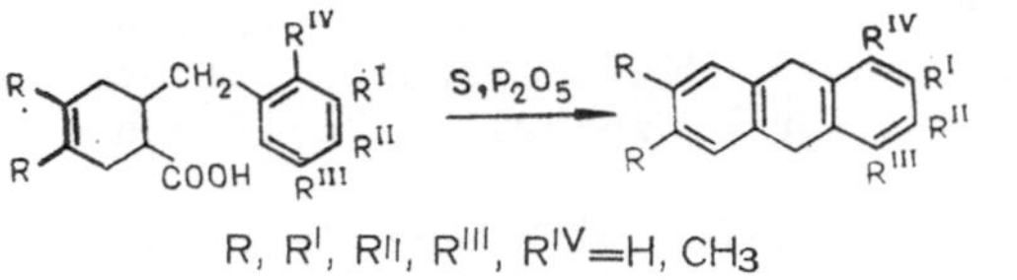

$$R, R^I, R^{II}, R^{III}, R^{IV} = H, CH_3$$

(6.113)

2-(4-Bromobenzyl)-1,2,3,6-tetrahydrobenzoic acids when heated
with sulfur and phosphorus(V) oxide undergo resinification. However,
in toluene they are converted to 2-bromo-9-(4-methylphenyl)anthracene
and in m-xylene to 6-bromo-2,3-dimethyl-10-(4-methylphenyl)anthracene
[270]:

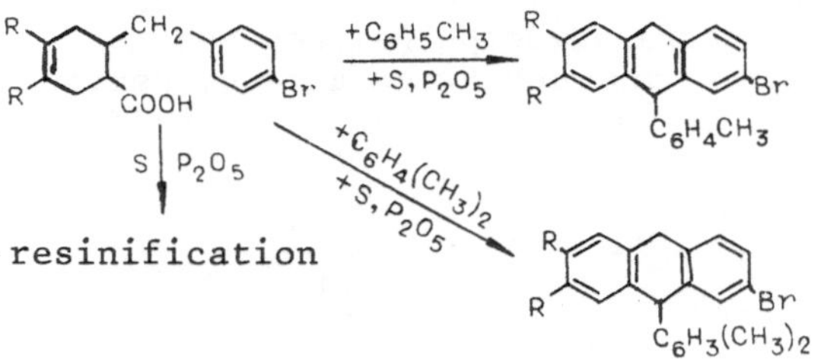

(6.114)

6.9.2. Unsaturated Acids

Crotonic acid when heated with sulfur gives thiophene and carbon
disulfide [245].

The interaction of sulfur with oleic acid or with linolenic acid
at 150-300°C gives a "sulfurous balsam" [246,271-281].

The rate of the reaction depends on the temperature, the amount
of sulfur used, and the nature of the acid. It has been stated that
in the first step of the reaction there is an addition of sulfur to
the double bonds of the unsaturated acids with the formation of
sulfide polymers of the type [276]:

$$CH_3(CH_2)_n-\underset{\diagdown S \diagup}{CH-CH}-(CH_3)_4COOH \qquad (6.115)$$

Further resinification leads to the formation of disulfide bridges.

Heating oleic acid with sulfur at about 140°C produces an adduct (addition of sulfur to the double bond) which on saponification gives a salt of a sulfur-containing acid [277]. Other higher unsaturated acids behave similarly. This reaction serves as the basis for the manufacture of "sulfur soaps" [244,278-280]:

$$C_{18}H_{34}O_2 + S \rightarrow C_{18}H_{34}SO_2 \xrightarrow{NaOH} C_{18}H_{33}SO_2Na + H_2O \qquad (6.116)$$

Oleic and stearic acid undergo sulfuration in the presence of sulfur and ^{35}S-labeled thiuram sulfides at 150-180°C at the expense of the thiuram sulfur [281]. When oleic acid is heated with 1% sulfur in the presence of water at 180°C or in its absence at 220°C in an atmosphere of carbon dioxide, elaidic acid is formed in 50% yield [282,283].

When oleic acid is distilled with sulfur (50% of the weight of the acid), margaric acid is obtained [284]. A crystalline sulfuration product has been obtained from the reaction of elaidic acid with sulfur. This product was not isolated in a pure state and is assumed to be a dithiane-elaidic acid [272,285].

The isomerization of maleic acid into fumaric acid is accelerated in the presence of sulfur [286].

When cinnamic acid is heated with sulfur, 2,5-diphenylthiophene, carbon dioxide, and hydrogen sulfide are formed [285,287-290]. Only carbon dioxide and styrene are formed using a small amount of sulfur at 180°C. In the presence of excess sulfur, cinnamic acid is rapidly decarboxylated to styrene at 180-300°C. Styrene then further reacts with sulfur and gives 2,5-diphenylthiophene. In a solvent (carbon disulfide), the yield of 2,5-diphenylthiophene is considerably increased:

$$C_6H_5CH=CHCOOH \xrightarrow{+S} C_6H_5CH=CH_2 + CO_2$$

$$2C_6H_5CH=CH_2 + 2S \longrightarrow C_6H_5\underset{S}{\boxed{}}C_6H_5 + H_2S \qquad (6.117)$$

Acrylic acid derivatives on heating with sulfur and phosphorus-(V) sulfide give the corresponding 3-thioxo-1,2-dithiol [291]. The yields of the 1,2-dithiols are higher when R' = C_6H_5, and R = alkyl or phenyl:

$$R-\underset{R'-CH}{\overset{CCOOH}{\underset{\parallel}{|}}} + 4\,S \longrightarrow \underset{R'}{\overset{R}{\bigsqcup}}{\overset{S}{\underset{S}{\diagdown}}}S + 2\,H_2O + H_2S \qquad (6.118)$$

$$(1.5\text{--}50\%)$$

$$R' = CH_3,\ C_6H_5;\ R = H,\ CH_3,\ C_6H_5,\ 4\text{-}CH_3OC_6H_4$$

6.9.3. Aromatic Carboxylic Acids

The interaction of 3- and 4-methylbenzoic acids with sulfur at 270–275°C lead to the corresponding stilbenedicarboxylic acids [292,293]:

$$(6.119)$$

R = 3- or 4-CH$_3$

If the hydrogen sulfide formed in the reaction is not removed, 3,3'- or 4,4'-(1,2-ethanediyl)dibenzoic acids and tetra-(3- or 4-carboxyphenyl)thiophenes are also formed [294].

6.10. ESTERS

6.10.1. Esters of Saturated Carboxylic Acids

When beef fat or other animal fats are melted with sulfur, a very large amount of hydrogen sulfide is formed. This reaction has been used as a method for the preparation of hydrogen sulfide [295-297].

Dehydrogenation with sulfur has been used as the basis for the synthesis of various esters of polycyclic aromatic acids [298,299]:

$$(6.120)$$

The dehydrogenation of the cyclic lactone of 3-(2-hydroxyphenyl)-propanoic acid with sulfur gives coumarin [299]:

$$(6.121)$$

6.10.2. Esters of Unsaturated Carboxylic Acids

The reaction of sulfur with esters of the simple aliphatic unsaturated acids was first studied by Michael in 1895 [300]. Michael postulated the possibility of the addition of one sulfur atom to the double bond with the formation of a thiirane ring. According to Michael, ethyl fumarate does not react with sulfur at 160-190°C but reacts at 205-210°C, and after twenty hours a compound is obtained which was assigned the structure of the corresponding thiirane:

$$S + \begin{array}{l} HC-COOC_2H_5 \\ \| \\ HC-COOC_2H_5 \end{array} \rightarrow S\begin{array}{l} CH-COOC_2H_5 \\ | \\ CH-COOC_2H_5 \end{array} \qquad (6.122)$$

The reaction of methyl crotonate with sulfur was thought to proceed in an analogous manner (48 hours) [300]:

$$S + \begin{array}{l} HC-COOCH_3 \\ \| \\ HC-CH_3 \end{array} \rightarrow S\begin{array}{l} CH-COOCH_3 \\ | \\ CH-CH_3 \end{array} \qquad (6.123)$$

Because both compounds isolated by Michael are thermally stable, not prone to polymerization, and undergo saponification with alcoholic base without any destruction of the sulfur bridge, the structures given above are unlikely [301,302]. The compounds are probably 1,4-dithiane derivatives:

$$\begin{array}{ccc} & S & \\ & / \ \backslash & \\ C_2H_5OOC-CH & CH-COOC_2H_5 \\ | & | \\ C_2H_5OOC-CH & CH-COOC_2H_5 \\ & \backslash \ / & \\ & S & \end{array} \qquad \begin{array}{ccc} & S & \\ & / \ \backslash & \\ C_2H_5OOC-CH & CH-CH_3 \\ | & | \\ CH_3-CH & CH-COOCH_3 \\ & \backslash \ / & \\ & S & \end{array}$$

$$\begin{array}{ccc} & S & \\ & / \ \backslash & \\ CH_3-CH & CH-CH_3 \\ | & | \\ C_2H_5OOC-CH & CH-COOC_2H_5 \\ & \backslash \ / & \\ & S & \end{array} \qquad (6.124)$$

The possibility of the formation of 1,4-dithiane derivatives in the reaction of unsaturated compounds with sulfur has been mentioned by a number of authors [277,303,304].

According to more recent data, sulfur reacts with esters of unsaturated acids at relatively low temperatures (up to 150°C) with the formation of thiophene derivatives [295,301]. Thus, the interaction of methyl acrylate with sulfur at 150-200°C gives a mixture of the methyl esters of 2,4- and 2,5-thiophenedicarboxylic acid. A small amount of methyl dithiopropanoate is also formed:

$$\underset{\substack{| \\ HCH}}{\overset{\substack{CH_3OCOCH \\ \|}}{}} + 4S + \underset{\substack{| \\ HCH}}{\overset{\substack{HCCOOCH_3 \\ \|}}{}} \longrightarrow CH_3OOC \underset{S}{\overset{}{\bigcirc}} COOCH_3$$

$$+ \underset{\substack{S}}{\overset{\substack{CH_3OOC}}{\bigcirc}} COOCH_3 + 2 H_2S \qquad (6.125)$$

The reaction of sulfur with methyl and ethyl crotonate at 150°C leads to the formation of a tetrahydrothiophene derivative, which readily loses sulfur, and a disulfide [301]:

$$2CH_3CH=CHCOOCH_3 \overset{2S}{\rightarrow} \underset{\substack{| \\ CH_3CH-CHCOOCH_3}}{\overset{CH_3CH-CHCOOCH_3}{}} \overset{S}{\underset{\searrow S \rightarrow S}{}} \overset{-S}{\rightarrow}$$

$$\rightarrow \underset{\substack{| \\ CH_3CH-CH-COOCH_3}}{\overset{CH_3CH-CH-COOCH_3}{}} \overset{}{\underset{S}{\searrow}} \qquad (6.126)$$

Dimethyl maleate and dimethyl fumarate react with sulfur with the formation of tetramethyl thiophenetetracarboxylate:

$$\underset{\substack{CH_3OCOCH \\ \| \\ CH_3OCOCH}}{} + 3S + \underset{\substack{HCCOOCH_3 \\ \| \\ HCCOOCH_3}}{} \longrightarrow \underset{\substack{CH_3OOC \\ CH_3OOC}}{\overset{}{\bigcirc_S}} \underset{\substack{COOCH_3 \\ COOCH_3}}{} + 2H_2S \qquad (6.127)$$

The interaction of acrylic acid derivatives (34) with sulfur, hydrogen sulfide and ammonium sulfide at 0-5°C in dichloroethane gives the corresponding polythiodipropanoic acid derivatives. Upon treatment of the reaction mixture with cyanide ions or sodium sulfite, tri- and tetrasulfides are converted to dithio analogs (35) [305]:

$$CH_2=CRCO_2X + S + H_2S \longrightarrow (CH_2CHRX)_2S_2$$
$$34 \qquad\qquad\qquad 35$$
$$R = H, \ X = CO_2CH_3, \ CN, \ CONH_2;$$
$$R = CH_3, \ X = CO_2CH_3$$

The interaction of naturally occurring esters of unsaturated fatty acids containing 16-24 carbon atoms, as well as the interaction of aliphatic alcohols, glycols, and glycerol with sulfur, leads to highly viscous sulfuration products which are used as additives in lubricants suitable for high pressures and drastic friction conditions [296,297,306-309].

The reaction of diethyl acetylenedicarboxylate with sulfur at 150°C gives tetraethyl thiophenetetracarboxylate [300]:

$$\underset{\substack{C_2H_5OCOC \\ \||| \\ C_2H_5OCOC}}{} + S + \underset{\substack{CCOOC_2H_5 \\ \||| \\ CCOOC_2H_5}}{} \longrightarrow \underset{\substack{C_2H_5OCO \\ C_2H_5OCO}}{\overset{}{\bigcirc_S}} \underset{\substack{COOC_2H_5 \\ COOC_2H_5}}{} \qquad (6.128)$$

Ethyl cinnamate reacts with sulfur at 250°C, but in a different manner to the acid. Its reaction with sulfur yields 5-phenyl-1,2-dithiol-3-one and ethanol [310-312]:

$$
\underset{C_3H_5-CH}{\overset{CHCOOC_2H_5}{\|}} + 2S \longrightarrow \underset{C_6H_5}{\text{[ring]}}{\overset{O}{}} + C_2H_5OH \qquad (6.129)
$$

Hydrocinnamic acid and diphenylthiophene are by-products of this reaction.

The sulfuration of methyl α-methylcinnamate gives 4-acetoxy-5-phenyl-1,2-dithiol-3-thione and 4-methyl-5-phenyl-1,2-dithiol-3-one [81,82] (Table 1):

$$
\underset{C_6H_5CH}{\overset{CH_3OCO-C-CH_3}{\|}} + 4S \longrightarrow \underset{C_6H_5}{\overset{CH_3OCO}{\text{[ring]}}}{\overset{S}{}} + H_2S \qquad (6.130)
$$

$$
\underset{C_6H_5CH}{\overset{CH_3-C-COOCH_3}{\|}} + 2S \longrightarrow \underset{C_6H_5}{\overset{CH_3}{\text{[ring]}}}{\overset{O}{}} + CH_3OH \qquad (6.131)
$$

The reaction of diisooctyl maleate with a stoichiometric amount of sulfur at 180-190°C for four hours leads to a compound which has been used as a rubber plasticizer [313]:

$$
S \underset{CO_2C_8H_{17}-iso}{\overset{CO_2C_8H_{17}-iso}{\text{[ring]}}} \qquad (6.132)
$$

On heating at 290°C in the presence of sulfur, ethyl eleostearate undergoes cyclization to form the ethyl ester of (6-butyl-cyclohexadien-2,4-yl)-1-octanoic acid [314]. Under the influence of sulfur (0.2%) the radical isomerization of the trans- to cis-C_{11}-C_{12} bond occurs and is followed by an electrocyclic reaction of the isomerization product.

Heating coumarin for 48 hours with excess sulfur at 220-230°C gives a condensed heterocyclic compound. Three isomeric structures are possible [315]:

$$\qquad (6.133)$$

The reaction of 8-acetyl-4-methylcoumarin derivatives with sulfur and morpholine yields the corresponding 8-carboxymethyl-4-methylcoumarins [316]:

$$R = R, \; OCH_3 \tag{6.134}$$

The interaction of 4-methylcoumarin-3-carbonitrile with sulfur and ammonia yields the 4H-thieno[3,4-c]-[1]benzopyran-3-amine (36) and its 1,1'-disulfide (38) [191]:

$$\tag{6.135}$$

The thiol (37) was not isolated.

6.10.3. Fatty Oils

The reaction of sulfur with the glycerides of higher unsaturated acids which represent the major components of plant oils and of most animal fats has been investigated by numerous researchers. The vulcanization of unsaturated fatty oils is of practical interest [317,318]. This reaction leads to plastics [318-326], rubber substitutes (brown factice) [327,328], film-forming materials (drying oils [329]) [319,329-335,339], plasticizers [290,336], de-emulsifiers [337,343], sulfur soaps [244,279,280,338], mercapto-acids, etc. Drying (or semidrying) oils when heated with sulfur (4-6%) at 120-170°C give a crackless putty [340].

Whereas sufficiently detailed literature is available on the technical use of sulfurated oils [318,336], their structure has not been extensively investigated.

The sulfuration products of oleic acid and of almond, walnut, castor, olive, rapeseed, poppyseed, and other higher unsaturated oils have found use as medicated sulfurous balsams and have been used for this purpose for a long time [341]. Their properties and recipes for their preparation are described in ancient collections of medicinal products. At the present time, sulfurous balsams are practically out of use.

The first studies of the sulfuration products of fatty oils and the processes occurring during their preparation were initiated

by Scheele in the 18th century. Scheele suggested the reaction of
sulfur with olive oil as a method for the preparation of hydrogen
sulfide [342]. Somewhat later, Reimsch [341] proposed the reaction
of sulfur with fatty oils for the same purpose. Reimsch studied the
reaction of linseed oil with sulfur at 230°C and was the first to
notice the formation of carbon disulfide in this reaction [341].
During the first half of the 19th century, several serious studies
(for their time) appeared in the literature and were devoted to the
sulfuration of fatty oils [271,272,341,344]. It was noted at that
time that the reaction of sulfur with oils only starts at 150°C and
then at 160°C hydrogen sulfide is formed. In the case of fifteen
fatty oils, conditions were studied at which oils react with sulfur,
and the amount of sulfur bonded in oils at 100-160°C was determined.
The characteristics of a sulfuration process of fatty oils was given.

One of the noteworthy studies was Anderson's work (1847) [344]
on the reaction of sulfur with oleic acid and with linseed and
almond oil. He concluded that the reaction of sulfur with fatty
oils leads to the replacement of hydrogen by sulfur in the molecule
of the fatty acid and not in the glycerol molecule.

In 1855, Gentele [345] described the sulfuration of linseed
oil. He used the sulfurous balsam formed in this reaction in a
formula for china gilding.

The amount of sulfur bound by linseed oil at 25-160°C was
studied by Pohl in 1870 [346]. He observed that treatment of the
oil below 140°C leads to practically no changes. The chemical
interaction of the oil with sulfur starts at 160°C.

The possibility of the formation of sulfuration products with
unsaturated acids and oils without the simultaneous evolution of
hydrogen sulfide (below 160°C) was not mentioned up to the end of
the 19th century. Such sulfuration products were simply considered
as solutions of sulfur in oils.

Thus, in heating sulfur with train oil and with lanolin, only
two reaction steps were established, i.e., the dissolution of sulfur
and its chemical bonding starting at 240°C and accompanied by the
evolution of hydrogen sulfide [280].

The conclusion that there is no reaction between rapeseed oil
and sulfur at 130-135°C was reached on the basis of the observation
that the saponification of the reaction mixture gave only a fatty
acid which contained practically no sulfur [303].

Riedel's patent indicates for the first time the possibility
of the addition of sulfur to oils at 120-160°C [244]. Low-temperature saponification of the sulfuration products obtained at 120-
160°C ("thiofats") gives salts of "thioacids," i.e., sulfur salts.

Saponification carried out at higher temperatures causes elimination of sulfur.

Heating unsaturated oils with sulfur gives a rubber substitute (brown factice) [347]. Subsequently it was shown that the sulfuration of unsaturated oils involves a step which takes place below the temperature at which the evolution of hydrogen sulfide begins. This step can be viewed as an addition of sulfur [277].

The sulfuration of fats is more complex than that of unsaturated acids, because it is accompanied by changes in their structure [277]. In connection with this result, it was concluded that, depending on the temperature, sulfur can be bonded to unsaturated oils and acids in different ways. This then leads to the formation of sulfuration products containing the same amount of sulfur, but with different properties [277,348]:

$$
R=S \qquad
\begin{array}{c}
S \\
/\backslash \\
R \quad R \\
\backslash / \\
S
\end{array}
\qquad
S\begin{array}{c}
\diagup R-S-R\diagdown \\
\diagdown R-S-R\diagup
\end{array}S
\qquad (6.136)
$$

R = unsaturated molecule

It was established that most of the sulfur in brown factice is bound by addition in the form of 1,4-dithiane derivatives [302]:

$$
\begin{array}{c}
R'-CH-CH-R'' \\
S \qquad\qquad S \\
R'-CH-CH-R''
\end{array}
\qquad (6.137)
$$

It was assumed that at low temperatures the addition of sulfur to unsaturated oils is also accompanied by substitution, but to a very small extent. Later it was shown [348] that the sulfuration of linseed oil at 160°C is a three-step process. The first step is the chemical reaction of sulfur with the oil which leads to the formation of a homogeneous substance. The second step is the accumulation of a polymer which is richer in sulfur than the remaining liquid, and which finally forms a gel which upon further heating yields a solid polymer.

In the first step of the sulfuration process no condensation of the glyceride molecules is observed and only the formation of sulfides takes place [349]. The second step of the sulfuration process is accompanied by a noticeable increase in the molecular weight indicating cross-linking of the glyceride molecules. The last step is a further polymerization via the glyceride moieties. The sulfuration of a triolein occurs similarly [317,318,350].

Hydrogen sulfide is not formed in the initial stage of the sulfuration of tung oil, which mainly consists of the glyceride of eleostearic acid containing three conjugated double bonds [351]. The amount of unsaturation of the oil decreases with an increasing amount of bound sulfur; however, the diene number remains unchanged until all the sulfur has reacted. It seems that the sulfur adds in a 1,6-manner, thus saturating one double bond, and steric factors make the remaining conjugated system less reactive. At higher temperatures, the diene number of tung oil can be lowered by heating with a small amount of sulfur. This prevents the oil from drying out [331,352].

In the case of the sulfuration of linseed oil, about two atoms of sulfur are bonded per each disappearing double bond. However, this reaction subsides rapidly [351].

The question concerning the structure of the sulfuration products of unsaturated acids and their esters remains unsolved. However, there are numerous data which support a 1,4-dithiane structure [277,347].

In this connection, it seems appropriate to mention the synthesis of thiiranes of unsaturated oils by the thiocyanate method and their further polymerization into factices [304,353,354]. When free acids are used instead of the unsaturated oils, crystalline compounds are obtained which have been assigned a 1,4-dithiane structure. These compounds were also obtained by treating unsaturated acids with disulfur dichloride, but not by direct sulfuration.

REFERENCES

1. H. Prinz, Liebigs Ann. Chem., 223, 355 (1884).
2. O. Weisser, J. Mostecký, and S. Landa, Brennst.-Chem., 44, 286 (1963).
3. S. Landa, O. Weisser, and J. Mostecký, Czech. Patent 108,422 (1961); C. A., 60, P9151g (1964).
4. J. Wojtczak and M. Elbanowski, Rocz. Chem., 40, 1963 (1966).
5. M. A. Elbanowski, Phosphorus Sulfur, 5, 111 (1978).
6. B. Buathier, A. Combes, F. Pierrot, and H. Guerpillon, Ger. Patent 2,540,128 (1976); C. A., 85, 23271 (1976).
7. F. H. Mazitova and V. K. Khairullin, Zh. Org. Khim., 50, 1718 (1980).
8. M. A. Vasyanina, Yu. A. Efremov, and V. K. Khairullin, Izv. Akad. Nauk SSSR, Ser. Khim., 1608 (1977).
9. F. Feigl, Spot Tests in Organic Analysis, Elsevier, New York (1960).
10. F. Feigl, V. Gentil, and C. Stark-Mayer, Microchim. Acta, 342 (1957).

11. H. Erdmann, Liebigs Ann. Chem., 362, 133 (1908).
12. A. Nakatasuchi, Chem. Ind. (Jpn.), 33, Suppl. Bind., 408 (1930);
     Chem. Zentralbl., 269 (1931).
13. A. Nakatasuchi, Chem. Ind. (Jpn.), 35, Suppl. Vol., 376, 377,
     372 (1932); Chem. Zentralbl., II, 3454 (1932).
14. C. Czped, Rocz. Chem., 6, 728 (1926); Chem. Zentralbl., I, 2985
     (1927).
15. L. Szperl and T. Wierusz-Kowalski, Chem. Polsk., 15, 28 (1918);
     C. A., 13, 2865 (1919).
16. L. Szperl, Rocz. Chem., 10, 252 (1930); Chem. Zentralbl., I,
     3436 (1930).
17. E. V. Tsmakhinskii and L. S. Malishevskaya, Dokl. Akad. Nauk
     SSSR, 17, 365 (1937); Zh. Org. Khim., 7, 2693 (1937); C. A.,
     32, 2926 (1938).
18. A. L. F. Bru, Fr. Patent 455,124 (1913); C. A., 8, 211 (1913).
19. C. H. Kantgen, Arch. Pharm. (Weinheim, Ger.), 28, 1 (1890).
20. C. H. Kantgen, Arch. Pharm. (Weinheim, Ger.), 28, 288 (1890);
     Ber., 23, 201R (1890).
21. J. Dekson, US Patent 2,067,261 (1937); C. A., 31, 1555 (1937).
22. H. Müller and L. Müller, Helv. Chim. Acta, 5, 628 (1922).
23. E. V. Tsmakhinskii, Dokl. Akad. Nauk SSSR, 17, 415; Zh. Org.
     Khim., 7, 2861 (1937); C. A., 32, 2513 (1938).
24. E. V. Zmachinskii, C. R. Acad. Sci. (Paris), 202, 668 (1936).
25. P. Mochalle, Ger. Patent 164,322 (1905); Chem. Zentralbl., II,
     1568 (1905).
26. L. Ruzicka, Forschr. Chem. Phys. Phys. Chem., 19A, No. 5,
     12 (1928).
27. L. Ruzicka, Angew. Chem., 51, 5 (1938).
28. W. Short, H. Stromberg, and A. E. Wiles, J. Chem. Soc., 319
     (1936).
29. N. Chaterjee, J. Indian Chem. Soc., 12, 591 (1935); C. A., 30,
     454 (1936).
30. L. Ruzicka and A. G. van Veen, Liebigs Ann. Chem., 476, 70
     (1929).
31. L. F. Fieser, J. Am. Chem. Soc., 55, 4977 (1933).
32. L. Ruzicka and M. Pfeiffer, Helv. Chim. Acta, 9, 841 (1926).
33. G. A. Kon and F. C. Ruzicka, J. Chem. Soc., 187 (1936).
34. L. Ruzicka and J. Meyer, Helv. Chim. Acta, 21, 565 (1938).
35. L. Ruzicka, H. Waldmann, P. J. Meyer, and H. Hösli, Helv. Chim.
     Acta, 16, 169 (1933).
36. L. Ruzicka, G. B. de Graaff, and H. Müller, Helv. Chim. Acta,
     15, 1300 (1932).
37. R. Haworth, J. Chem. Soc., 2717 (1932).
38. L. Ruzicka, M. Goldberg, and K. Hofmann, Helv. Chim. Acta,
     20, 325 (1937).
39. L. Ruzicka, H. Schellenberg, and M. Goldberg, Helv. Chim. Acta,
     20, 791 (1937).
40. L. Ruzicka, H. Bringger, R. Egli, L. Ekmann, M. Futter, and
     H. Hösli, Helv. Chim. Acta, 15, 431 (1932).
41. L. Ruzicka, H. Hösli, and K. Hofmann, Helv. Chim. Acta, 19, 370
     (1936).
42. W. Short and H. Stromberg, J. Chem. Soc., 516 (1937).

43. J. Cook, C. Hewett, and C. Lawrense, J. Chem. Soc., 71 (1936).

44. F. Biedeland, Arch. Pharm. (Weinheim, Ger.), 280, 304 (1942).

45. L. Haitinger, Monatsh. Chem., 4, 165 (1883).

46. C. Levevre and C. Desgrez, C. R. Acad. Sci. (Paris), 198, 1791 (1934).

47. Fr. Patent 537,207 (1922); Chem. Zentralbl., IV, 894 (1922).

48. Swiss Patent 104,015 (1924); Chem. Zentralbl., II, 1280 (1924).

49. Kh. Murav'ev, Zh. Khlopchatobumazhn. Promst., 6, No. 4, 32 (1936).

50. F. B. Downing, R. C. Cecrkson, and C. W. Hannum, US Patent 1,985,602 (1934); C. A., 29, 1101 (1935).

51. C. Ellis, US Patent 1,690,335 (1929); C. A., 23, 583 (1929).

52. C. Ellis, US Patent 1,636,596 (1924); C. A., 21, 2993 (1927).

53. A. E. Blumfeldt, Swiss Patent 92,408 (1921); Chem. Zentralbl., IV, 894, 1088 (1922).

54. A. E. Blumfeldt, Swiss Patent 94,801; 95,186; 95,328; 95,329; 95,187 (1922); Chem. Zentralbl., II, 411 (1923).

55. A. Giovanni, R. Russo, and F. Russo, US Patent 2,711,383 (1955); C. A., 49, P13683h (1955).

56. C. Ellis, US Patent 1,690,160 (1928); C. A., 23, 533 (1929).

57. E. J. Geering, US Patent 3,743,680 (1973); C. A., 79, 53014v (1973).

58. E. J. Geering and G. B. Stratton, US Patent 3,647,885 (1972); C. A., 76, 140178h (1972).

59. Fr. Patent 2,036,086 (1970); C. A., 75, P63390w (1971).

60. R. Mohlan, Chem. Ztg., 937 (1907).

61. M. Cherubim, Kautsch. Gummi, Kunstst., 19, 676 (1966); C. A., 67, 26146k (1967).

62. L. Haitinger, Monatsh. Chem., 4, 165 (1883).

63. C. Levevre and C. Desgrez, C. R. Acad. Sci. (Paris), 198, 1791 (1934).

64. E. J. Geering, US Patent 3,717,682 (1973); C. A., 79, 53014 (1973).

65. V. F. Martenkin, I. F. Popov, K. N. Abbasov, A. I. Kostina, and N. Ya Kucher, Russian Patent 390,120 (1973); Byull. Izobr., No. 30, 83 (1963).

66. A. J. Neale, P. J. S. Bain, and T. J. Rawlings, Tetrahedron, 25, 4583, 4593 (1969).

67. V. M. Kostyuchenko, T. G. Termigaliev, V. V. Bobilev, A. G. Rumyantsev, E. A. Yurkin, and A. F. Surkova, Neftepererab. Neftekhim., 15 (1976).

68. A. S. Hay and B. M. Boulette, J. Org. Chem., 41, 1710 (1976).

69. A. S. Hay, US Patent 3,953,519 (1976); C. A., 85, 159634 (1976).

70. T. Fujisawa, K. Hata, and T. Kojima, Synthesis, 5, 38 (1973).

71. A. S. Hay, US Patent 3,952,063 (1976); C. A., 85, 32618 (1976).

72. J. Crawford, US Patent 3,923,670 (1975); C. A., 84, 164418 (1976).

73. R. L. Sung, B. H. Zolski, and W. D. Foucher, Ger. Patent 2,535,472 (1976); C. A., 85, 80775 (1976).

74. V. G. Sharov, A. M. Potapov, A. Z. Vikulov, and T. V. Shutenova, Russian Patent 380,648 (1973); Byull. Izobr., 50(21), 86 (1973); C. A., 79, 78376 (1973).

75. I. I. Yukel'son, L. V. Fedotova, and V. V. Legacheva, Izv.
     Vyssh. Uchebn. Zaved., Khim. Khim. Tekhnol., 13, 1762 (1970).
75a. E. A. Hassan, A. M. Naser, and M. M. Wassel, Paintindia, 28, 13
     (1978); C. A., 90, 73352 (1979).
76. C. Levevre and C. Desgrez, C. R. Acad. Sci. (Paris), 198, 300
     (1934).
77. B. Böthier, Fr. Patent 871,802 (1941); C. A., 36, 52793 (1942);
     Chem. Zentralbl., II, 2205 (1942).
78. B. Böthier and A. Lüttringhaus, Liebigs Ann. Chem., 557, 89
     (1947).
79. B. Böthier, Offis. Pub. Board, Rept. 707 (1944).
80. A. Lüttringhaus, H. B. Köning, and B. Böttcher, Liebigs Ann.
     Chem., 560, 201 (1947).
81. N. Lozac'h and O. Gaudin, C. R. Acad. Sci. (Paris), 225, 1162
     (1947).
82. N. Lozac'h, Bull. Soc. Chim. France, 840 (1949).
83. N. Lozac'h, C. R. Acad. Sci. (Paris), 225, 686 (1947).
84. Brit. Patent 800,752 (1958); C. A., 53, P6160e (1959).
85. G. L. Weamer, Ger. Patent 1,280,645 (1969); C. A., 71, 21882v
     (1969).
86. C. L. Deasy, Chem. Rev., 32, 173 (1943).
87. E. Ferrario, Bull. Soc. Chim. France, 9, 536 (1911); Chem.
     Zentralbl., II, 214 (1911).
88. F. Ackermann, DRP Patent 234,743 (1910); Chem. Zentralbl., I,
     1468 (1911).
89. C. M. Suter, J. P. McKenzie, and C. E. Maxwell, J. Am. Chem.
     Soc., 58, 717 (1936).
90. C. M. Suter and C. E. Maxwell, Org. Synthesis, 18, 64 (1938).
91. C. M. Suter and F. O. Green, J. Am. Chem. Soc., 59, 2578 (1937).
92. G. M. Bennet, M. S. Lesvil, and E. E. Turrer, J. Chem. Soc.,
     444 (1937).
93. M. Tanotg, J. Pharm. Soc. Jpn., 58, 510 (1938); C. A., 32, 7467
     (1938).
94. B. A. Trofimov, L. Ya. Rappoport, L. V. Morozova, T. T. Minakova,
     G. N. Petrov, K. Yu. Salnis, A. Kh. Cherny, A. Ya. Pravenkaya,
     and R. S. Fainshtein, Zh. Prikl. Khim., 52, 1355 (1979).
95. L. Szperl, Congr. Nat. Quim. Purarpl., 9, IV, 233 (1934); Chem.
     Zentralbl., II, 1211 (1936).
96. D. Harman and Wm. E. Vaughan, US Patent 2,536,684 (1951);
     C. A., 45, P3157a (1951).
97. J. Bertozzi and R. Eugene, US Patent 3,778,478 (1973); C. A.,
     80, 59439 (1974).
98. V. K. Khairullin and M. A. Vasyanina, US Patent 3,933,856
     (1976); C. A., 84, 164597 (1976).
99. N. I. Shuikin and V. V. An, Izv. Akad. Nauk SSSR, Otd. Khim.
     Nauk, 2086 (1961).
100. A. Schoenberg, Ber., 58, 1793 (1925).
101. M. G. Voronkov and G. Lapina, Khim. Geterotsikl. Soedin., 5,
     592 (1970).

102. O. Gandin and R. Pottier, C. R. Acad. Sci. (Paris), 224, 479 (1947).
103. O. Gandin and N. Lozac'h, C. R. Acad. Sci. (Paris), 224, 577 (1947).
104. O. Gandin, Fr. Patent 941,543 (1949); C. A., 44, 9479 (1950).
105. J. Schmidt and A. Lespagnol, C. R. Acad. Sci. (Paris), 230, 551 (1950).
106. J. Schmidt and A. Lespagnol, C. R. Acad. Sci. (Paris), 230, 1771 (1950).
107. J. Schmidt and A. Lespagnol, Bull. Soc. Chim. France, 459 (1950).
108. J. Ban, Swiss Patent 132,222 (1924).
109. T. Kono, J. Soc. Rubber Ind. Jpn., 25, 33 (1952); C. A., 48, 11829b (1954).
110. G. A. Barbaglia, Ber., 13, 1574 (1880).
111. G. A. Barbaglia, Ber., 14, 2654 (1881).
112. G. A. Barbaglia and A. Marquardt, Gazz. Chim. Ital., 21, 195 (1891).
113. G. A. Barbaglia and A. Marquardt, Ber., 24, 1881 (1891).
114. L. P. Broun and M. Thompson, J. Chem. Soc., Perkin Trans., 863 (1974).
115. F. Asinger and M. Thiel, Angew. Chem., 70, 667 (1958).
116. F. Asinger, H. Offermans, D. Neuray, and P. Müller, Monatsh. Chem., 101, 1295 (1970).
117. M. G. Voronkov, N. N. Vlasova, A. E. Pestunovich, V. A. Pestunovich, and V. V. Keiko, Zh. Org. Khim., 66, 1657 (1976).
118. N. N. Vlasova, A. E. Pestunovich, V. A. Pestunovich, V. V. Keiko, T. K. Pogodaeva, and M. G. Voronkov, Izv. Akad. Nauk SSSR, Ser. Khim., 666 (1977).
119. H. Offermanns and F. Asinger, Austrian Patent 281,003 (1970); C. A., 73, 35079 (1970).
120. R. Gompper and R. R. Schmidt, Chem. Ber., 98, 1385 (1965).
121. H. Hagen and H. Fleig, Ger. Patent 2,503,699 (1976); C. A., 85, 177401 (1976).
122. H. Adolphi, H. Fleig, and H. Hagen, Ger. Offen. 2,626,967 (1978); C. A., P136608d (1978).
123. BASF A.-G., Belg. Patent 844,659 (1977); C. A., 88, P17341 (1978).
124. BASF A.-G., Neth. Appl. 7,608,382 (1978); C. A., 89, P24293 (1978).
125. J. Market and H. Hagen, Liebigs Ann. Chem., 768 (1980).
126. H. Hagen and D. Lenke, Ger. Offen., 2,734,882 (1979).
127. H. Hagen and H. Fleig, Liebigs Ann. Chem., 1994 (1975).
128. K. Gewald, H. Schonnelder, and U. Harn, J. Prakt. Chem., 316, 299 (1974).
129. M. G. Voronkov, V. É. Udre, and A. O. Taube, Russian Patent 371,233 (1972); Byull. Izobr., No. 12 (1973).
130. M. G. Voronkov, V. É. Udre, and A. O. Taube, IV Symposium on Organic Sulphur, Venezia-fondazione Glorgio cini, 15-20 Giugno, 1970.
131. L. Lepage and Y. Lepage, C. R. Acad. Sci. (Paris), Ser. C, 278, 541 (1974).

132. L. Legrand and N. Lozac'h, Bull. Soc. Chim. France, 1130 (1956).
133. L. Legrand and N. Lozac'h, Bull. Soc. Chim. France, 953 (1958).
134. N. Lozac'h and M. Denis, Bull. Soc. Chim. France, 1016 (1953).
135. L. Legrand and N. Lozac'h, Bull. Soc. Chim. France, 1686 (1959).
136. F. P. Otto, US Patent 2,520,101 (1948); C. A., $\underline{44}$, P10314c (1950).
137. J. Teste and R. Lefeuwre, Bull. Soc. Chim. France, 4027 (1966).
138. F. Boberg, Liebigs Ann. Chem., $\underline{678}$, 67 (1964).
139. F. Boberg, Angew. Chem., $\underline{72}$, 629 (1960).
140. Y. Poirier, L. Legrand, and N. Lozac'h, Bull. Soc. Chim. France, 1054 (1966).
141. Y. Poirier and N. Lozac'h, Bull. Soc. Chim. France, 1058 (1966).
142. R. Pinel, Y. Molleir, and N. Lozac'h, C. R. Acad. Sci. (Paris), $\underline{260}$, 5065 (1965).
143. O. Hupport, US Patent 2,397,355 (1946); C. A., $\underline{40}$, 3774 (1946).
144. M. Ebel and N. Lozac'h, Bull. Soc. Chim. France, 161 (1963).
145. N. Kajoba, Suom. Kemi, $\underline{138}$, 20 (1940).
146. J. Schmidt and A. Lespagnol, Bull. Soc. Chim. France, 459 (1950)
147. G. Darzens and G. Levy, C. R. Acad. Sci. (Paris), $\underline{194}$, 181 (1932).
148. R. P. Napfer, H. A. Kaufman, and P. R. Driscoll, J. Heterocycl. Chem., $\underline{7}$, 393 (1970).
149. A. Barbot, Bull. Soc. Chim. France, 161 (1963).
150. C. Allen and F. Gates, J. Am. Chem. Soc., $\underline{65}$, 1283 (1943).
151. A. A. Petrov, Dokl. Akad. Nauk SSSR, $\underline{53}$, 531 (1946).
152. T. W. Jezierski, Rocz. Chem., $\underline{10}$, 397 (1930).
153. T. W. Jezierski, Rocz. Chem., $\underline{14}$, 216 (1934).
154. J. P. Brown and M. Thompson, J. Chem. Soc., Perkin Trans. I, 863 (1974).
155. G. Toland, D. L. Hagmann, J. B. Wilkes, and F. J. Brutschy, Am. Chem. Soc. Div. Petrol. Chem. Preprints, $\underline{3}$, No. 2, B85 (1958); C. A., $\underline{55}$, 22082a (1961).
156. T. W. Jezierski, Rocz. Chem., $\underline{6}$, 738 (1926).
157. K. Dzienvonski, C. Z. Baraniecki, and L. Sternbauh, Bull. Sut. Acad. Polon., A 198 (1930); Chem. Zentralbl., II, 3408 (1930).
158. H. Vlajer, Chem. Ztg., No. 82 (1931).
159. J. V. Braun, E. Hahn, and J. Seeman, Ber., $\underline{55}$, 1687 (1922).
160. G. Pfister-Guillouzo and N. Lozac'h, Bull. Soc. Chim. France, 153 (1963).
161. W. S. Bachmann, J. Org. Chem., $\underline{3}$, 434 (1938).
162. Y. Poirier and N. Lozac'h, Bull. Soc. Chim. France, 1062 (1966).
163. W. Dilthey, W. Höschen, and H. Dierichs, Ber., $\underline{68B}$, 1159 (1935).
164. W. Dilthey, L. Graef, H. Dierichs, and W. Josten, J. Prakt. Chem., $\underline{151}$, 185 (1938).
165. Ger. Patent 209,233 (1908), Fr., $\underline{9}$, 815; Chem. Zentralbl., I, 1627 (1909)
166. Ger. Patent 209,232 (1909), Fr., $\underline{9}$, 814; Chem. Zentralbl., II, 627 (1909).
167. K. Gewald and J. Schael, J. Prakt. Chem., $\underline{315}$, 39 (1973).
168. C. Corral, R. Madronero, and N. Ulecia, Afinidad, $\underline{35}$, 129 (1978); C. A., $\underline{89}$, 129477 (1978).

169. F. Asinger and M. Thiel, Angew. Chem., 70, 667 (1958).

170. F. Asinger and H. Offermans, Angew. Chem., Int. Ed. Engl., 6, 907 (1967); C. A., 68, 29625 (1967).

171. M. Thiel and F. Asinger, Liebigs Ann. Chem., 610, 17 (1957).

172. F. Asinger, M. Thiel, and J. Kalsendorf, Liebigs Ann. Chem., 610, 25 (1957).

173. F. Asinger, W. Schäfer, G. Herkelmann, H. Römgens, B. D. Reintges, G. Scharein, and A. Wegerhoff, Liebigs Ann. Chem., 672, 156 (1964).

174. F. Asinger, W. Schäfer, B. D. Reintges, A. Wegerhoff, and G. Scharein, Liebigs Ann. Chem., 683, 121 (1965).

175. F. Asinger, M. Thiel, and E. Pallag, Liebigs Ann. Chem., 602, 37 (1957).

176. F. Asinger, M. Thiel, H. Kaltwasser, and G. Reckling, East Ger. Patent 14,350 (1958); C. A., 53, P10002c (1959).

177. M. Thiel, F. Asinger, and G. Reckling, Liebigs Ann. Chem., 611, 131 (1958).

178. F. Asinger, M. Thiel, and H. Sedlak, Liebigs Ann. Chem., 634, 164 (1959).

179. F. Asinger, M. Thiel, G. Peschel, and K. H. Meinicke, Liebigs Ann. Chem., 619, 145 (1958).

180. M. Thiel, F. Asinger, and M. Fedtke, Liebigs Ann. Chem., 615, 77 (1958).

181. F. Asinger, M. Thiel, H. Sedlak, O. Hampel, and R. Sowada, Liebigs Ann. Chem., 615, 84 (1958).

182. W. Schäfer and H. Triem, Monatsh. Chem., 97, 1510 (1966).

183. F. Asinger, H. Driem, and Sin-Gun An, Liebigs Ann. Chem., 643, 186 (1961).

184. F. Asinger, M. Thiel, P. Püchel, F. Haaf, and W. Schäfer, Liebigs Ann. Chem., 660, 85 (1962).

185. F. Asinger, W. Schäfer, and F. Haaf, Liebigs Ann. Chem., 672, 134 (1964).

186. F. Asinger, W. Schäfer, and A. Saus, Monatsh. Chem., 96, 1278 (1965).

187. F. Asinger, A. Saus, H. Offermanns, and H. D. Hahn, Liebigs Ann. Chem., 691, 92 (1966).

188. F. Asinger, M. Thiel, H. Isbeck, K. H. Gröbe, H. Grundman, and S. Tränker, Liebigs Ann. Chem., 634, 144 (1960).

189. F. Asinger, M. Thiel, and H. Kaltwasser, Liebigs Ann. Chem., 606, 67 (1957).

190. F. Asinger, W. Schäfer, M. Baumann, and H. Römgeng, Liebigs Ann. Chem., 672, 103 (1964).

191. W. Ried and E. Nyiondi-Bonguen, Liebigs Ann. Chem., 134 (1973).

192. F. Asinger and K. Halckur, Monatsh. Chem., 95, 24 (1964).

193. F. Asinger, H. W. Becker, W. Schäfer, and A. Saus, Monatsh. Chem., 97, 301 (1966).

194. F. Asinger, D. Neuray, E. Witte, and J. Hartig, Monatsh. Chem., 103, 1661 (1972).

195. F. Asinger, W. Schäfer, and H. W. Becker, Angew. Chem., 77, 41 (1965).

196. K. Kindler, Arch. Pharm., 265, 389 (1927).
197. E. Schwenk and E. Bloch, J. Am. Chem. Soc., 64, 3051 (1942).
198. G. Purrello, Gazz. Chim. Ital., 97, 539 (1967).
199. E. Schwenk and D. Papa, J. Org. Chem., 11, 798 (1946).
200. E. Francais, Fr. Patent 1,409,441 (1965); C. A., 64, 2019 (1966).
201. B. Singh, Synth. Commun., 8, 275 (1978); C. A., 89, 42686 (1978).
202. M. Makosza, H. Kycia, H. Jaroslwska-Sorochtej, and A. Zielinska, Pol. Patent 76,627 (1977); C. A., 90, P87062 (1979).
203. A. P. McFadden, R. C. Allen, and T. B. K. Lee, US Patent 4,118,401; C. A., 90, 54845 (1979).
204. O. P. Vig, S. D. Sharma, N. K. Verma, and V. K. Handa, Indian J. Chem. Sect. B, 15B, 988 (1977).
205. M. S. Newman and H. S. Lowrie, J. Am. Chem. Soc., 76, 6196 (1954).
206. H. Oelschlaeger, J. M. Iglesias, G. Goetze, and W. Schatton, Arzneim.-Forsch., 27, 1625 (1977); C.A., 88, 37719 (1978).
207. R. G. Jones, O. F. Soper, O. K. Behrens, and J. W. Corse, J. Am. Chem. Soc., 70, 2843 (1948).
208. B. Zupancic, Ger. Offen. 2,702,714 (1978); C. A., 89, 180029 (1978).
209. A. Compagnini and G. Purrello, Gazz. Chim. Ital., 95, 676 (1965)
210. F. Asinger and A. Mayer, Angew. Chem., 77, 812 (1965).
211. F. Asinger, A. Saus, and H. Mayer, Monatsh. Chem., 98, 825 (1967).
212. G. Purrello, Atti Accad. Gioenia Sci. Nat. Catania, 13, 45 (1963); C. A., 59, 7396 (1963).
213. J. Staněk, Chem. Listy, 45, 224 (1951); C. A., 46, 2528d (1952).
214. R. C. Moreau and N. Biju-Duval, Bull. Soc. Chim. France, 1527 (1958).
215. R. C. Moreau, Bull. Soc. Chim. France, 7532 (1958).
216. W. G. Dauben, R. P. Ciula, and J. B. Rogan, J. Org. Chem., 22, 362 (1957).
217. F. Bottino and G. Purrello, Gazz. Chim. Ital., 95, 1062 (1965).
218. G. Purrello, Gazz. Chim. Ital., 97, 549 (1967).
219. F. Bottino and G. Purrello, Gazz. Chim. Ital., 95, 693 (1965).
220. G. Purrello, Gazz. Chim. Ital., 97, 557 (1967).
221. G. Purrello, M. Piatteli, and A. Lo Vullo, Boll. Sed. Accad. Gioenia Sci. Nat. Catania, 4, 9, 33 (1967); C. A., 70, 38867 (1967).
222. G. Purrello, Gazz. Chim. Ital., 95, 1078 (1965).
223. G. Purrello, Gazz. Chim. Ital., 95, 699 (1965).
224. Jpn. Kokai 77,59,150 (1977); C. A., 88, P6721 (1978).
225. S. Yoshina, T. Kameyma, Y. Oiji, and A. Kiyohara, Ger. Offen. 2,631,189 (1977).
226. E. Bisagni and R. Royer, Bull. Soc. Chim. France, 86 (1962).
227. E. G. Brein and I. L. Finar, J. Chem. Soc., 2356 (1957).
228. D. L. Turner, J. Am. Chem. Soc., 70, 3961 (1948).
229. G. Purrello, Gazz. Chim. Ital., 97, 539 (1967).
230. G. Purrello, Gazz. Chim. Ital., 95, 1072 (1965).
231. R. L. Malan and P. M. Dean, J. Am. Chem. Soc., 69, 1797 (1947).
232. E. V. Brown and J. A. Blanchette, J. Am. Chem. Soc., 72, 3414 (1950).

233. S. L. Suminov and T. Voronina, Russian Patent 603,642 (1978); C. A., 89, 24365 (1978).

234. W. Kraemer, W. Draber, H. Timmler, and H. Foerster, Ger. Patent 2,460,909 (1976); C. A., 85, 94102 (1976).

235. E. P. Nakova, O. N. Tolkachev, and R. P. Evstigneeva, Zh. Org. Khim., 11, 2585 (1976).

236. C. B. Pollard and J. C. Braun, J. Am. Chem. Soc., 77, 6685 (1955).

237. F. Asinger, H. Offermans, W. Puerschel, K. H. Lim, and D. Neuray, Monatsh. Chem., 99, 2090 (1968).

238. F. Asinger, A. Saus, H. Offermans, D. Neuray, and K. H. Lim, Monatsh. Chem., 102, 321 (1971).

239. F. Asinger, H. Offermans, K. H. Lim, and D. Neuray, Monatsh. Chem., 101, 1281 (1970).

240. F. Asinger, H. Offermans, W. Puerschel, K. H. Lim, and D. Neuray, Monatsh. Chem., 99, 1695 (1968).

241. R. A. Franz, F. Applegath, F. V. Morriss, and F. Baiocchi, J. Org. Chem., 26, 3306 (1961).

242. R. A. Franz, F. Applegath, F. V. Morriss, F. Baiocchi, and C. Bolze, J. Org. Chem., 26, 3309 (1961).

243. D. W. Grisley, Jr., and J. A. Stephens, J. Org. Chem., 26, 3568 (1961).

244. J. D. Riedel, Ger. Patent 71,190 (1892); Ber., 26, 1025R (1893).

245. E. Seelig, Organische Reaktionen und Reagentien, Stuttgart (1899), p. 797.

246. J. Zajic, J. Zalud, and O. Kopecka, Sb. Vys. Sk. Chem. Technol. Praze, Potravinarska Technol., 8, 331 (1964); C. A., 66, 18415 (1967).

247. J. H. Ziegler, Ber., 23, 2471 (1890).

248. G. Darzens and A. Levy, C. R. Acad. Sci. (Paris), 194, 2056 (1932).

249. L. F. Fieser and E. B. Hershberg, J. Am. Chem. Soc., 57, 1508 (1935).

250. G. Darzens, C. R. Acad. Sci. (Paris), 183, 748 (1926).

251. L. F. Fieser and E. B. Hershberg, J. Am. Chem. Soc., 57, 1851 (1935).

252. G. Darzens, C. R. Acad. Sci. (Paris), 190, 1562 (1930).

253. G. Darzens and A. Levy, C. R. Acad. Sci. (Paris), 199, 1131 (1934).

254. G. Darzens and A. Levy, C. R. Acad. Sci. (Paris), 199, 1426 (1934).

255. G. Darzens and A. Levy, C. R. Acad. Sci. (Paris), 200, 2187 (1935).

256. G. Darzens and A. Levy, C. R. Acad. Sci. (Paris), 202, 427 (1936).

257. G. Darzens and A. Levy, C. R. Acad. Sci. (Paris), 194, 2056 (1932).

258. G. Darzens and A. Levy, C. R. Acad. Sci. (Paris), 201, 907 (1935).

259. G. Darzens, C. R. Acad. Sci. (Paris), 190, 1562 (1930).

260. G. Darzens and A. Heinz, C. R. Acad. Sci. (Paris), <u>184</u>, 33 (1926).
261. H. Berger, J. Prakt. Chem., 121, 133, 331 (1932).
262. L. Ruzicka and J. Meyer, Helv. Chim. Acta, <u>5</u>, 521 (1922).
263. A. I. Virtanen, Liebigs Ann. Chem., <u>424</u>, 150 (1921).
264. T. H. Easterfield and G. Bagley, J. Chem. Soc., <u>85</u>, 1238 (1904)
265. P. Levy, Z. Anorg. Chem., <u>81</u>, 145 (1913).
266. J. R. Hosking and W. T. McFadven, Chem. Ind. (London), <u>53</u>, 195 (1934).
267. L. Ruzicka and F. Balas, Helv. Chim. Acta, <u>6</u>, 677 (1923); <u>7</u>, 275 (1924).
268. Fr. Patent 828,934 (1938); C. A., <u>32</u>, P77178 (1938).
269. V. P. Skvarchenko, V. A. Puchnova, and P. Ya. Levina, Dokl. Akad. Nauk SSSR, <u>145</u>, 831 (1962).
270. V. P. Skvarchenko, V. A. Puchnova, and P. Ya. Levina, Zh. Org. Khim., <u>34</u>, 2210 (1964).
271. W. Radig, H. Harff, G. Ulex, and F. Schoy, Pharm. Z Bl., 815 (1833).
272. W. Radig, H. Harff, and G. Ulex, Arch. Pharm., <u>11</u>, 15 (1835); Chem. Zentralbl., <u>607</u>, 623 (1835).
273. R. Bebedikt and F. Ulza, Monatsh. Chem., <u>1</u>, 208 (1887); Chem. Zentralbl., 882 (1887).
274. A. H. Clark, C. Falconer Flint, and D. L. McGee, Chem. Ind. (London), 729 (1966).
275. A. Steigman, J. Soc. Chem. Ind., <u>64</u>, 119 (1945).
276. R. Salchow, Kautschuk, <u>14</u>, 12 (1938); Chem. Zentralbl., I, 1388 (1938).
277. J. Attschul, Angew. Chem., <u>8</u>, 535 (1895).
278. C. Bechert, Pharm. Ztg., <u>39</u>, 823 (1894); Apoth. Ztg., 858 (1894); Chem. Zentralbl., I, 113 (1895).
279. F. Hager, Pharm. Ztg., <u>39</u>, 351 (1893).
280. A. Seibers, Ger. Patent 56,065; 56,941 (1896); Ber., <u>24</u>, 510 998R (1891).
281. J. T. Tarasenko, Zh. Prikl. Khim. (Leningrad), <u>33</u>, 1203 (1960).
282. G. Rankoff, Ber., <u>62</u>, 2712 (1929).
283. G. Rankoff, Ber., <u>64</u>, 619 (1931).
284. Th. Anderson, Leibigs Ann. Chem., <u>63</u>, 370 (1874).
285. G. Rankoff and A. Popov, Annu. Univ. Sofia, Khim. Fakult., Livre 2, <u>45</u>, 127 (1948-1949); C. A., <u>45</u>, 1995b,d (1951).
286. H. Frendlich and G. Schikozz, Kolloid-Chem. Beihefte, <u>23</u>, 1 (1926); Chem. Zentralbl., I, 3130 (1926).
287. E. Baumann and E. Fromm, Ber., <u>28</u>, 890 (1895).
288. E. Fromm, P. Fante, and E. Liebsohn, Liebigs Ann. Chem., <u>457</u>, 267 (1927).
289. G. Rankov and A. Popov, Annu. Univ. Sofia, Khim. Fakult., Livre 2, <u>45</u>, 135 (1949).
290. K. Ellis, The Chemistry of Petroleum Derivatives, Vol. 1, The Chemical Catalog Co., Inc., New York (1934).
291. L. Jirousek and L. Stárka, Coll. Czech. Chem. Commun., <u>24</u>, 1982 (1959).

292. Wm. G. Toland, US Patent 2,677,703 (1954); C. A., _49_, P187c (1955).

293. Wm. G. Toland, Jr., J. B. Wilkes, and F. J. Brutschy, J. Am. Chem. Soc., _75_, 2263 (1953).

294. Wm. G. Toland, Jr., and J. B. Wilkes, J. Am. Chem. Soc., _76_, 307 (1954).

295. H. Hopff and J. von der Crone, Chimia (Switz.), _13_, 107 (1959).

296. J. B. Stucker, US Patent 2,541,789 (1951); C. A., 45; P4441b (1951).

297. K. C. Roberts, US Patent 2,549,525 (1951); C. A., _45_, P5921c (1951).

298. G. Darzens and A. Levy, C. R. Acad. Sci. (Paris), _203_, 669 (1936).

299. M. Meyer, R. Beer, and G. Lersch, Monatsh. Chem., _34_, 1665 (1913).

300. A. Michael, Ber., _28_, 1633 (1895).

301. T. W. Jauregg and B. E. Hackley, Jr., J. Org. Chem., _16_, 399 (1951).

302. B. D. Luft, The Chemistry of Rubber, Moscow (1930), p. 94.

303. R. Henrignes, Angew. Chem., _8_, 691 (1895).

304. H. P. Kaufmann, E. Gindsberg, W. Röttig, and R. Salchow, Ber., _70_, 2519 (1937).

305. R. L. Keener and H. R. Raterink, US Patent 3,769,315 (1973); C. A., _81_, 151578 (1974).

306. J. B. Stucker, E. T. Fronczak, and G. Wolfram, US Patent 2,483,600 (1949); C. A., _44_, P2748g (1950).

307. M. Degroote and B. Keiser, US Patent 2,372,366 (1945); C. A., _42_, 3936 (1948).

308. R. Miller, H. K. Latourette, and E. J. Rich, Jr., Brit. Patent 914,661 (1963); US Patent (1960); C. A., _58_, P10086b (1963).

309. A. Hribesch, W. Franke, W. Wolf, and F. Jantsch, Ger. Patent 932,571 (1955); C. A., _52_, P17693i (1958).

310. E. Baumann and E. Fromm, Ber., _30_, 110 (1897).

311. P. Friedlaender and S. Kielbasinski, Ber., _45_, 3389 (1912).

312. F. Bager, Ger. Patent 87,931 (1895); Ber., _29_, R745 (1896).

313. L. Tokarzewski, Pol. Patent 73,284 (1974); C. A., _84_, 18941 (1976).

314. T. Nagao, T. Yöichiro, and M. Noboru, Yukagaku, J. Jpn. Oil Chem. Soc., _27_, 83 (1978); Ref. Zh. Khim., 6Zh144 (1979).

315. L. Szperl and A. Chmielnicka, Rocz. Chem., _16_, 101 (1936).

316. K. H. Boltze and H. D. Dell, Angew. Chem., Int. Ed. Engl., _5_, 125 (1966); C. A., _64_, 12632h (1966).

317. P. Stamberger, Recl. Trav. Chim. Pays-Bas, _46_, 834 (1927).

318. P. Stamberger, Recl. Trav. Chim. Pays-Bas, _47_, 972 (1928).

319. L. Bäarnhielm and A. Jerrander, Ger. Patent 77,810 (1893); Ber., _28_, 1662 (1895).

320. A. Baer, Austral. Patent 58,735; Ger. Patent 253,277 (1912); Tsit. po N 293, O. Kaum, 85.

321. W. R. Brixoy, US Patent 714,858; 728,851; Tsit. po N 293, O. Kaum, 84.

322. C. F. Chosa, US Patent 1,671,229; 1,671,230 (1928); Chem. Zentralbl., II, _721_, 834 (1928).

323. M. Melamid, Ger. Patent 386,062 (1922), 434,143 (1923);
      Brit. C. A., B302 (1924); B495 (1927).
324. W. Nanfeldt, US Patent 1,959,686 (1934); C. A., $\underline{28}$, 4551 (1934)
325. N. Reif, Ger. Patent 288,532 (1915).
326. J. Wolff, Ger. Patent 20,483 (1882); Ber., $\underline{15}$, 3106R (1882).
327. H. Bagen, Ger. Patent 288,968 (1914); C. A., $\underline{10}$, 2542 (1916).
328. Brit. Patent 9,985 (1913); C. A., $\underline{8}$, 3507 (1914).
329. J. F. Morozov, Zh. Prikl. Khim., $\underline{4}$, 117 (1931).
330. G. Balbi, Olii minerali olii egrassi, Coloril Vermico, $\underline{22}$, 31
      (1942); Chem. Zentralbl., II, 1417 (1943).
331. E. Beringer and W. Zimmer, US Patent 1,103,473 (1914); C. A.,
      $\underline{8}$, 3127 (1914).
332. F. Crane, Brit. Patent 3,345 (1891); Chem. Ind. (London), $\underline{11}$,
      446 (1892).
333. H. A. Gardner, US Patent 1,992,570 (1935); Chem. Zentralbl.,
      II, 3709 (1935).
334. H. A. Gardner and G. G. Sward, Am. Paint and Varnish Mfr.
      Assoc., $\underline{388}$, 218 (1931); C. A., $\underline{25}$, 4418 (1931).
335. Brit. Patent 338,604 (1930); Chem. Zentralbl., I, 1976 (1931).
336. O. Kaum, Handbook on Synthetic Grease, ONTI (1934).
337. M. De Groote and B. Keiser, US Patent 2,372,366 (1945); C. A.,
      $\underline{39}$, 3684 (1945).
338. G. Bernstein, Z. Koll., $\underline{12}$, 193, 273 (1913).
339. H. A. Gardner and G. G. Sward, Am. Paint and Varnish Mfr.
      Assoc., $\underline{410}$, 129 (1932); Brit. C. A., B77 (1933).
340. F. Becher and C. Becher, East Ger. Patent 36,090 (1962);
      C. A., $\underline{63}$, P13579b (1965).
341. H. Reimsch, J. Prakt. Chem., $\underline{13}$, 136 (1838).
342. A. Romwalter, Math. Arz. Ungar. Akad. Wiss., $\underline{61}$, 122 (1942);
      C. A., $\underline{39}$, 3251 (1945).
343. P. Dekker, India Rubber, $\underline{92}$, 13 (1936).
344. T. Anderson, Liebigs Ann. Chem., $\underline{63}$, 370 (1847).
345. J. G. Gentele, Dingl. Pol. J., $\underline{137}$, 273 (1885).
346. J. J. Pohl, Dingl. Pol. J., $\underline{197}$, 508 (1870); Chem. Zentralbl.,
      677 (1870).
347. R. Henriques, Chem. Ztg., $\underline{17}$, 634, 707 (1893).
348. G. S. Whitby and H. D. Chataway, Chem. Ind., $\underline{45}$, 115T (1926).
349. J. S. Long, C. A. Knanss, and J. G. Smull, Ind. Eng. Chem.,
      1962 (1927).
350. B. C. Knight and P. Stamberger, J. Chem. Soc., 2791 (1928).
351. E. A. Hauser and M. C. Sze, J. Phys. Chem., $\underline{46}$, 118 (1942).
352. H. A. Gardner, US Patent 1,986,571 (1935); C. A., $\underline{29}$, 1270
      (1935).
353. J. Kombos, A. Kombos, and E. E. Engelke, Brit. Patent 265,994
      (1927); Chem. Zentralbl., I, 2949 (1927).
354. R. Salchow, Kautschuk, $\underline{14}$, 12 (1938).

7
Nitrogen-Containing Compounds

7.1. ALIPHATIC AND ARYLALIPHATIC AMINES

7.1.1. Primary Amines

As a rule, amines are efficient activators of sulfur in its reactions. Furthermore, amines themselves can react with sulfur and these reactions often take a rather unexpected course. Depending on the reaction conditions and on the structure of the respective amine, the reaction of aliphatic amines with sulfur leads to hydropolysulfides of amines, imides of sulfur or the corresponding imines, as well as to thiocarboxamides.

Within the framework of an intense search for rubber vulcanizers, the reactions of primary alkylamines occurring in the presence of hydrogen sulfide acceptors (PbO, HgO, MgO) were studied a long time ago. These reactions lead to the formation of bis-alkylaminotrisulfanes ($\underline{1}$) according to the following scheme [1-4]:

$$RNH_2 + 4S \xrightarrow[-(H_2S)]{PbO} RNH-S_3-NHR \qquad (7.1)$$
$$\underset{\underline{1}}{}$$

On heating in an inert solvent or in pyridine, alkylamines react with sulfur to give the corresponding thiocarboxamides ($\underline{2}$) [5,6]:

$$RCH_2NH_2 + 2S \xrightarrow[-H_2S]{100-200°C} RC\overset{\displaystyle S}{\underset{\displaystyle NH_2}{\big<}} \qquad (7.2)$$
$$\underset{\underline{2}}{}$$

However, 2-propanamine is, under these conditions, converted to 2-propanimine ($\underline{3}$) [7]:

$$(CH_3)_2CHNH_2 + S \xrightarrow[-H_2S]{} CH_3-\underset{\underset{\underset{\underline{3}}{NH}}{\|}}{C}-CH_3 \qquad (7.3)$$

No well-defined products were isolated when n-butylamine was heated with sulfur [8]. In aqueous medium, n-butylamine reacts with sulfur to give a hydropolysulfide $C_4H_9NH_2 \cdot H_2S_m$, whereas the expected composition would have been $(C_4H_9NH_2)_2 \cdot H_2S_m$ by analogy with α,ω-diaminoalkanes (cf. Eq. 7.17). Also the thiosulfate of the expected structure (4) is formed at the same time [9]:

$$4RNH_2 + 2(m+1)S + H_2O \xrightarrow{100°C} 2RNH_2 \cdot H_2S_4 + \underset{\underline{4}}{(RNH_2)_2 \cdot H_2S_2O_3} \qquad (7.4)$$

A very detailed study has been carried out on the reaction of sulfur with benzylamine. It was shown [10] that benzylamine reacts with sulfur at 180°C to give thiobenzamide. It was later confirmed [5] that benzylidenimine is also formed in this reaction:

$$2C_6H_5CH_2NH_2 + 3S \xrightarrow[-2H_2S]{180°C} C_6H_5C\overset{S}{\underset{NH_2}{\diagdown}} + C_6H_5CH=NH \qquad (7.5)$$

Subsequently, the above reaction was carried out at 200°C with an equimolar mixture of the starting materials, and N-benzylthiobenzamide (5) was obtained in high yield [11,12]. The following scheme has been proposed for this transformation [12]:

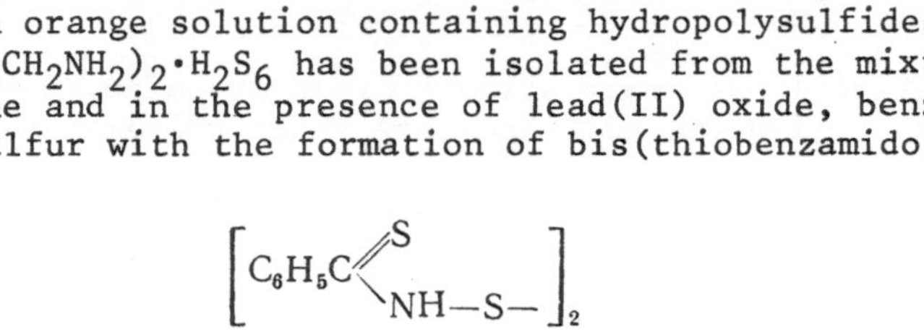

$$(7.6)$$
$$(7.7)$$

According to Levi [13], at low temperatures benzylamine and sulfur form an orange solution containing hydropolysulfides. The compound $(C_6H_5CH_2NH_2)_2 \cdot H_2S_6$ has been isolated from the mixture. At 20°C in benzene and in the presence of lead(II) oxide, benzylamine reacts with sulfur with the formation of bis(thiobenzamido) disulfide [2]:

$$\left[C_6H_5C\overset{S}{\underset{NH-S-}{\diagdown}} \right]_2$$

The reaction of sulfur with excess benzylamine at room temperature has been studied in detail [14,15]. It was shown that N-benzylidenebenzylamine (6), benzylammonium polysulfides (7), and ammonia are the final reaction products:

$$4C_6H_5CH_2NH_2 + nS \xrightarrow[-NH_3]{} C_6H_5CH=NCH_2C_6H_5 + (C_6H_5CH_2NH_2)_2 \cdot H_2S_n \quad (7.8)$$
$$n=6,7 \qquad\qquad\qquad\quad \underset{\sim}{6} \qquad\qquad \underset{\sim}{7}$$

It has been established that reaction (7.8) consists of the following steps [14]:

$$2C_6H_5CH_2NH_2 + (m+1)S \xrightleftharpoons[+H_2S]{-H_2S} C_6H_5CH_2NH-S_m-NHCH_2C_6H_5 \quad (7.9)$$

$$2C_6H_5CH_2NH_2 + (m-1)S + H_2S \rightarrow (C_6H_5CH_2NH_2)_2 \cdot H_2S_m \quad (7.10)$$

$$3C_6H_5CH_2NH-S_m-NHCH_2C_6H_5 \rightarrow C_6H_5CH=N-S_m-N=CHC_6H_5$$
$$+4C_6H_5CH_2NH_2 + 3(m-n)S \quad (7.11)$$

$$C_6H_5CH=N-S_m-N=CHC_6H_5 + 2C_6H_5CH_2NH_2 \rightleftharpoons 2C_6H_5CH=NH$$
$$+C_6H_5CH_2NH-S_m-NHCH_2C_6H_5 \quad (7.12)$$

$$C_6H_5CH=NH + C_6H_5CH_2NH_2 \rightarrow C_6H_5CH=NCH_2C_6H_5 + NH_3 \quad (7.13)$$

The principal products of the reaction of sulfur with α,ω-diaminoalkanes are their crystalline hydropolysulfides [8]:

$$H_2NCH_2CH_2NH_2 + S \xrightarrow{50°C} H_2NCH_2CH_2NH_2 \cdot H_2S_3 \quad (7.14)$$
$$(98\%)$$

$$H_2N(CH_2)_6NH_2 + S \xrightarrow{80°C} H_2N(CH_2)_6NH_2 \cdot H_2S_5 \quad (7.15)$$
$$(75\%)$$

It was not possible to isolate any well-defined products in the reaction of sulfur with 1,3-diaminopropane or diethylenetriamine, $H_2N(CH_2)_2NH(CH_2)_2NH_2$ [8]. Simultaneous interaction of sulfur and hydrogen sulfide with α,ω-diaminoalkanes in benzene, ethanol, or aqueous ethanol gives orange-red hydropolysulfides on heating [15–19]:

$$H_2N(CH_2)_nNH_2 + S + H_2S \rightarrow H_2N(CH_2)_nNH_2 \cdot H_2S_m \quad (7.16)$$
$$n=2-4; \qquad\qquad\qquad m=3,\ 5,\ 7$$

When the reaction of sulfur with α,ω-diaminoalkanes is carried out in the absence of hydrogen sulfide acceptors and without its subsequent introduction into the reaction zone, yellow or orange crystals of hydropolysulfides (8) are readily formed in aqueous solutions. At the same time, the stable crystalline thiosulfates of α,ω-diaminoalkanes (9) are also formed in good yields [9]:

$$3H_2N(CH_2)_nNH_2 + 2(m+1)S + 3H_2O \xrightarrow{100°C}$$
$$\rightarrow H_2N(CH_2)_nNH_2 \cdot H_2S_m + H_2N(CH_2)_nNH_2 \cdot H_2S_2O_3 \quad (7.17)$$
$$\underset{\sim}{8} \qquad\qquad\qquad\qquad \underset{\sim}{9}$$
$$n=2-6, \qquad\qquad m=3-7$$

In the above case, the formation of thiosulfates (9) is not due to atmospheric oxidation but is caused by the participation of hydroxide ions in the reaction [9]:

$$4S + OH^- \rightarrow S_2O_3^{2-} + 2S^{2-} + 3H_2O \qquad (7.18)$$

The hydropolysulfides of α,ω-diaminoalkanes are rather stable. The number of sulfur atoms in these compounds is basically determined by the length of the hydrocarbon chain and shows only a very limited dependence on the ratio of starting diamine to sulfur [8,9]. Thus, 1,2-diaminoethane always formed a hydrotrisulfide, whereas 1,4-diaminobutane formed a hydrosulfide with 6 or 7 sulfur atoms for the reagent ratios 1:2 and 1:4 respectively.

The reaction of diarylmethylamines with elemental sulfur gives diaryl thioketones when the reactants are heated [20,21]:

$$Ar_2CHNH_2 + S \xrightarrow[-NH_3]{140°C} Ar-\underset{\underset{S}{\|}}{C}-Ar \qquad (7.19)$$

Simultaneous action of sulfur and benzaldehyde upon 2-amino-cyclohexanol gives the N-thioacyl derivative of the above amine [22]:

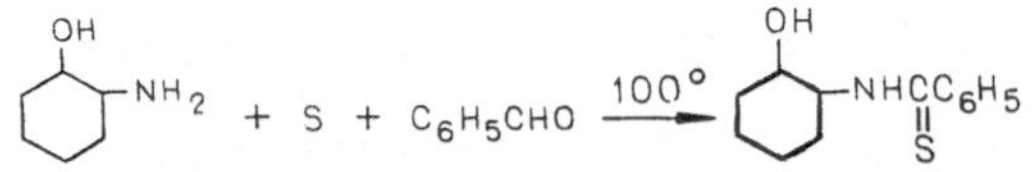

$$(7.20)$$

7.1.2. Secondary Amines

At 20-50°C, the lower dialkylamines and piperidine readily yield bis-dialkylaminopolysulfanes (10) in the presence of mercuric oxide or lead(II) oxide [13,23-26]. The following scheme has been proposed [25]:

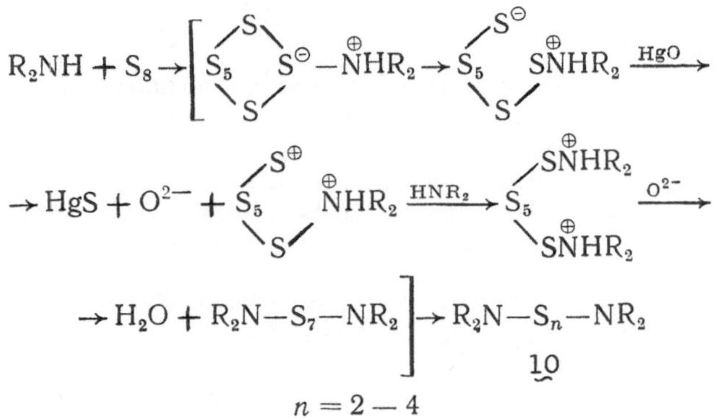

$$(7.21)$$

$$n = 2 - 4$$

In the absence of hydrogen sulfide acceptors the simultaneous action of sulfur and hydrogen sulfide upon secondary amines leads to hydropolysulfides [13,16,23,27,28]:

$$2R_2NH + nS + H_2S \xrightarrow{C_6H_6} (R_2NH)_2 \cdot H_2S_{n+1} \tag{7.22}$$

Di-n-butylamine and morpholine react with sulfur at 20-80°C with the formation of a complex mixture of products [8]. When morpholine is heated with sulfur at 158°C a 32% yield of oxalodimorpholide is obtained [171].

Dialkylamines containing a methylene group at the nitrogen atom give thiocarboxamides when heated with sulfur [6,29,30]:

$$2RCH_2NHR' + 2S \xrightarrow[\substack{-H_2S, \\ -RCH_2NH_2}]{100-200°C} RC\underset{NHR'}{\overset{S}{\diagup}} \tag{7.23}$$

Dibenzylamine heated with sulfur in the presence of catalysts is converted into N-benzylthiobenzamide [31]:

$$C_6H_5CH_2NHCH_2C_6H_5 + 2S \xrightarrow[-H_2S]{I_2} C_6H_5C\underset{NHCH_2C_6H_5}{\overset{S}{\diagup}} \tag{7.24}$$

The reaction of N-benzylaniline with sulfur at 220°C gives thiobenzanilide (11), whereas 2-phenylbenzothiazole (12) is obtained at higher temperatures [10]:

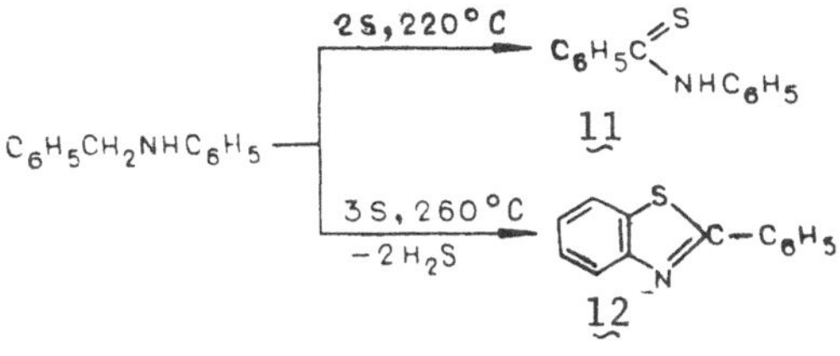

$$C_6H_5C\underset{NHC_6H_5}{\overset{S}{\diagup}} \tag{7.25}$$

11

(7.26)

12

Acetonitrile and 2-methyl-1,3-thiazole are the products obtained from the reaction of diethylamine with sulfur [32]:

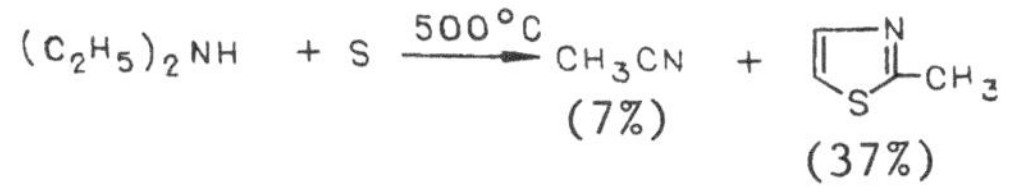

$$(C_2H_5)_2NH + S \xrightarrow{500°C} CH_3CN + \underset{(37\%)}{\text{(thiazole)}} \tag{7.27}$$

$$\underset{(7\%)}{}$$

The reaction of sulfur with piperidine has received special attention because the products are used as accelerators of rubber vulcanization. Piperidine reacts with sulfur in benzene in the absence of a catalyst, giving N-piperidinohydrosulfanes (13) [33-35]. Insertion of 2, 4, 5, or more sulfur atoms into the N-H bond is not observed. Compounds (13) decompose above 10°C with the formation of hydrogen sulfide and sulfur. They also undergo decomposition in the presence of inorganic acids [33]:

$$C_5H_{10} + nS \longrightarrow C_5H_{10}NS_nH \tag{7.28}$$

13

Heating piperidine with sulfur in benzene in the presence of an equimolar amount of lead(II) oxide leads to the formation of di-piperidinotri- and tetrasulfanes [24,36,37]:

$$C_5H_{10}NH + S + PbO \xrightarrow[-PbS]{} \begin{array}{c} \xrightarrow{40°C} C_5H_{10}N-S_3-NC_5H_{10} \\ \xrightarrow{60°C} C_5H_{10}N-S_4-NC_5H_{10} \end{array}$$

(7.29)

(7.30)

The interaction of pyrrolidine with sulfur leads to the polymerization products of pyrrole via the intermediate formation of the mercapto-derivatives of pyrrolidine and pyrroline in successive steps [38]. 3-Phenylpyrrolidine reacts similarly, whereas 2-phenyl-pyrrolidine is readily dehydrogenated with sulfur to give 2-phenyl-pyrrole [38]:

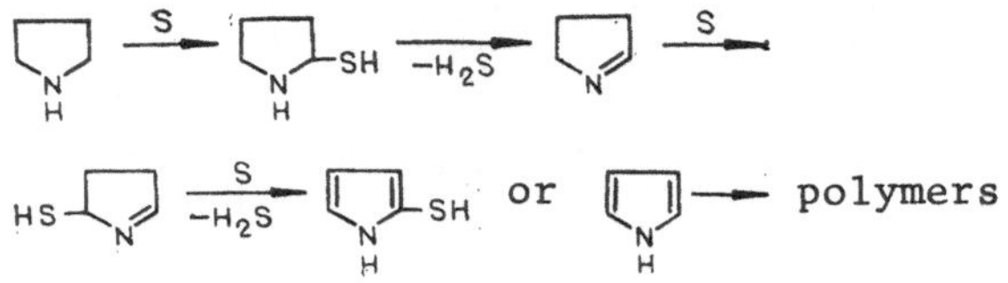

(7.31)

7.1.3. Tertiary Amines

Tertiary aliphatic amines do not react with sulfur at low temperatures [8,39-42]. Trialkylamines react with sulfur at 140-200°C to give N,N-dialkylthiocarboxamides [5,6,29]:

$$RNR'_2 + S \rightarrow RC\begin{array}{c} \diagup S \\ \diagdown NR'_2 \end{array}$$

(7.32)

R = alkyl, cycloalkyl, arylalkyl

Heating triethanolamine with sulfur at 80-100°C (sometimes in the presence of hydrogen sulfide) gives polysulfides containing 7-9% of catalytically active sulfur and useful in the vulcanization of rubber [3,4,13,43].

Simultaneous interaction of sulfur and hydrogen sulfide with lower trialkylamines yields the corresponding hydropolysulfides [16]:

$$2R_3N + nS + H_2S \xrightarrow{C_6H_6} (R_3N)_2 \cdot H_2S_{n+1}$$

(7.33)

R = CH_3, C_2H_5, $\underline{n}$-C_3H_7; n = 6, 8

Heating lithium dimethylamide with sulfur in benzene or tetra-hydrofuran gives a mixture of bis-dimethylamino sulfides and bis-dimethylamino disulfides (cf. Chapter 8) [44]:

$$(CH_3)_2NLi + nS \xrightarrow{80°C} (CH_3)_2N-S_n-N(CH_3)_2$$

(7.34)

7.1.4. α,β-Unsaturated Amines (Enamines)

Enamines readily react with sulfur if they contain hydrogen atoms in the β-position. The formation of enethiols (14) is the initial stage of the reaction of sulfur with enamines [45]. Enethiols (14) are highly reactive and are converted into more stable compounds via subsequent transformations:

$$RC=CHR' \underset{S}{\rightleftarrows} RC=C\begin{smallmatrix}SH\\R'\end{smallmatrix} \rightleftarrows RCHC\begin{smallmatrix}S\\R'\end{smallmatrix} \qquad (7.35)$$

$$\underset{NR_2''}{|} \qquad \underset{NR_2''}{|} \qquad \underset{NR_2''}{|}$$

$$14$$

Aliphatic enamines react with sulfur in dimethylformamide at room temperature to give the corresponding thiocarboxamides [45,46], e.g.,

$$RC=CH_2 + S \xrightarrow{DMF} RCH_2C\begin{smallmatrix}S\\NR_2'\end{smallmatrix} \qquad (7.36)$$

$$\underset{NR_2'}{|} \qquad (\sim75\%)$$

$$RC=CHCH_3 + S \xrightarrow{DMF} R(CH_2)_2C\begin{smallmatrix}S\\NR_2'\end{smallmatrix} \qquad (7.37)$$

$$\underset{NR_2'}{|} \qquad (20\text{-}60\%)$$

The transformations (7.36) and (7.37) are modifications of the Willgerodt-Kindler reaction. Because of this, the corresponding ketones, sulfur, and secondary amines may be used as the starting materials (cf. Chapter 6).

1-Piperidino-2-methylpropene (15) reacts with sulfur in dimethylformamide at room temperature and gives resinous thiols (16) (ν_{SH} 2550 cm^{-1}, $\nu_{C=C}$ 1650 cm^{-1}) which are readily converted into the crystalline N-phenyldithiourethanes (17) by reaction with phenyl isothiocyanate [47]:

$$15 \qquad\qquad 16 \qquad\qquad 17 \qquad\qquad (7.38)$$

Heating 1-morpholino-1-cyclohexene (18) with sulfur and subsequent hydrolysis of the intermediate reaction product (19), without prior isolation, gives perhydrothianthrene-4a,9a-diol (20) [48]:

$$18 \qquad\qquad\qquad 19 \qquad\qquad 20 \qquad (7.39)$$

N-Styrylmorpholine (21) heated with sulfur gives three sulfuration products, i.e., a substituted thiophene (22), a dithiafulvene
(23), and a thiocarboxamide (24) [49]:

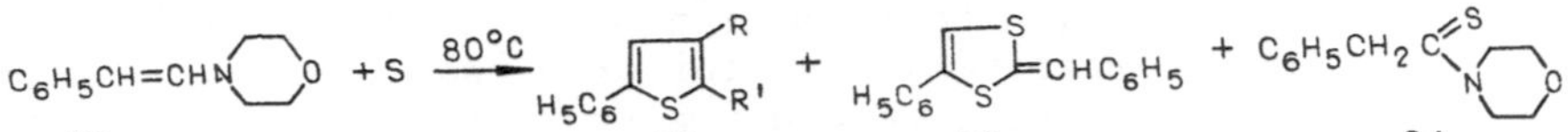

(7.40)

$$R=H, \; C_6H_5; \; R'=C_6H_5, \; H$$

When the ratio enamine:sulfur is 1:1, the yields are 18% (22),
0.08% (23), and 3% (24). At a ratio of 1:2, the yields are 9%, 7.5%,
and 52%, respectively. When the reaction is carried out in pyridine
at 80°C, the dithiafulvene (23) and the thiocarboxamide (24) are
obtained (at a 2:1 ratio, the yields are 11% (23) and 25% (24); at
a 1:1 ratio, they are 18% and 56%, respectively). No thiophene
derivative (22) is formed in this case.

Dithiafulvenes of the type (26) are formed in the reaction of
α-dialkylaminostyrenes (25) with sulfur [45]:

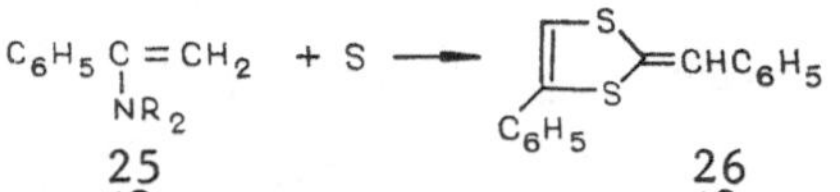

(7.41)

As a rule, aliphatic enamines readily undergo condensation with
the formation of 1-amino-1,4-dienes [50,51]. Thus, on heating
enamines with sulfur to facilitate the formation of 1-amino-1,4-
dienes, the sulfuration products of 1-amino-1,4-dienes are obtained,
i.e., the products are substituted thiophenes (27) [52,53]:

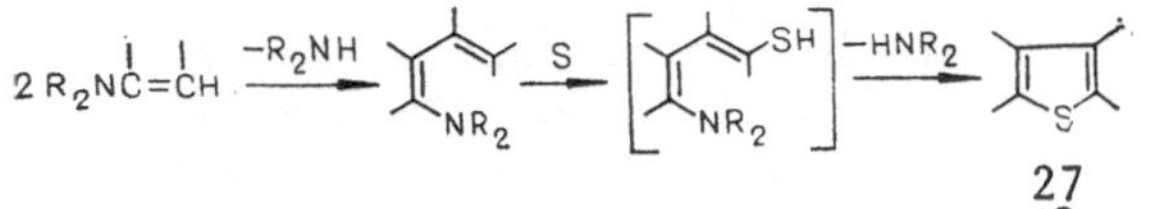

(7.42)

The reaction of enamines with a mixture of sulfur and carbon
disulfide has been extensively studied [45]. Alkyl-substituted
enamines readily react in dimethylformamide at 20°C to give a
3-thioxo-1,2-dithiol (28). In the case of enamines with aryl substituents at the α-carbon atom, the corresponding 2-thioxo-1,3-
dithiols (29) are obtained:

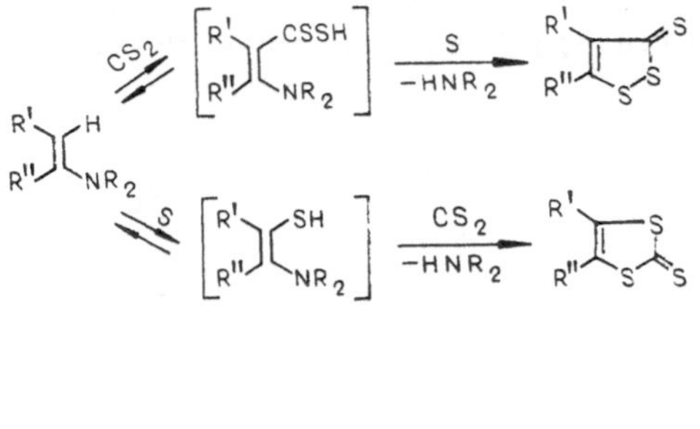

(7.43)

(7.44)

The difference in reactions (7.43) and (7.44) can be explained as being due to the different polarization of the double bonds in the starting enamines [45].

Also alicyclic enamines readily react with sulfur and, upon addition of "auxiliary" reagents (isothiocyanates, carbon disulfide, cyanamide), are converted to the corresponding substituted heterocyclic compounds, i.e., 2-thioxo-1,3-tetrahydrothiazoles (30), 3-thioxo-1,2-dithiols (31), and 1,3-thiazoles (32) [45,54,55]:

$$R'NCS \longrightarrow 30 \qquad (7.45)$$

$$CS_2, -HNR_2 \longrightarrow 31 \qquad (7.46)$$

$$H_2NCN, -HNR_2 \longrightarrow 32 \qquad (7.47)$$

2,2-Disubstituted-1-(N,N-dialkylamino)ethylenes (33) react with a mixture of sulfur and an isocyanate to form tetrahydro-1,3-thiazoles (34) [56]:

$$R_2NCH = CR'_2 + S + R''NCO \longrightarrow 34 \qquad (7.48)$$

$$33$$

The reaction of sulfur with enamines (33) and diisocyanates gives compounds (35), whereas the corresponding bis-enamines react with sulfur and phenyl isocyanate to give (36) [56]:

$$35 \qquad 36$$

$$Y = (CH_2)_6, \ C_6H_4$$

The interaction of aliphatic enamines with sulfur and ethyl cyanoacetate gives substituted aminothiophenes [57], e.g.,

$$X\!-\!N\!-\!CR'\!=\!CHR'' + S + H_2C \begin{array}{c} COOC_2H_5 \\ CN \end{array} \longrightarrow \qquad (7.49)$$

X = CH$_2$, O; R' = alkyl, cycloalkyl, COOC$_2$H$_5$; R'' = H, alkyl

1,1-Diaminoalkenes (37) react with sulfur to give high-melting betaines (38) rather than thiocarboxamides [58,59]:

$$CH_2=C\underset{NR_2}{\overset{NR_2}{\diagdown}} + S \longrightarrow \overset{S}{\underset{\ominus S}{\diagup}}C-C\underset{NR_2}{\overset{NR_2}{\diagdown}}\oplus \tag{7.50}$$

$$\underset{37}{} \qquad\qquad\qquad \underset{38}{}$$

The primary reaction products obtained from the reaction of sulfur with tris(phenylamino)ethylene (39) are the corresponding thiol (40), dithiooxanilide (41) and 1,2-bis(phenylamino)ethylene (42) [60]:

$$RCH=CR_2 + S \xrightarrow{150°} \overset{SH}{\underset{|}{RC}}=CR_2 + \overset{S\ \ S}{\underset{\|\ \ \|}{RC-CR}} + RCH=CHR \tag{7.51}$$

$$\underset{39}{} \qquad\qquad \underset{40}{} \qquad \underset{41}{} \qquad\quad \underset{42}{}$$

$$R=NHC_6H_5$$

The reaction of sulfur with enaminodithiocarboxylic acids (43) gives the corresponding 4,5,6,7-tetrahydrobenzo-1,2-dithiol-3-thione (44) [61]:

$$\tag{7.52}$$

Acyclic β-diketone enamines (β-aminovinylketones) do not react with sulfur under mild conditions [45], evidently, due to a lower reactivity of the C^2-H bond in the conjugated chain >N-C=CH-C=O which prevents the initial stage of the reaction with sulfur. Under analogous conditions, the C^2-H bond in molecules of alicyclic enaminoimines (β-aminovinylimines) also fails to react with sulfur [62]. N-Arylsubstituted 1-amino-3-imino-5,5-dimethyl-1-cyclo-hexenes (45), however, readily activate elemental sulfur analogously to strongly basic amines in dimethylformamide at 20°C and react with sulfur to form the corresponding N,N'-bis-organylaminopoly-sulfanes (46) and hexasulfides (47) of the initial enaminoimines in an inert atmosphere [62]. This reaction in the presence of hydrogen sulfide acceptors (PbO, HgO) leads to polysulfanes (46). When hydrogen sulfide is involved hexasulfides (47) are the main products. ortho-substituents (CH_3, Br) in the N-phenyl rings of enaminoimines (45) inhibit the reaction:

$$\tag{7.53}$$

$$R=C_6H_5, p\text{-}CH_3C_6H_4, p\text{-}BrC_6H_4, p\text{-}CH_3OC_6H_4$$

According to the data [62], the reaction in the presence of atmospheric oxygen affords compounds of type $(45) \cdot H_2S_6O_2$. There is some oblique evidence that these compounds, isolated for the first time, are salts of type (47) of the hypothetical dibasic acid $H_2S_6O_2$, a homolog of sulfoxylic acid H_2SO_2.

5,5-Dimethyl-3-phenylimino-1-piperidino-1-cyclohexene (48) fails to react under the conditions of the above reaction. However, in the presence of hydrogen sulfide, enaminoimine readily reacts with sulfur in dimethylformamide even at 20°C to form hexasulfide (49) and 5,5-dimethyl-3-phenylamino-2-cyclohexene-1-thione (50). Heating enaminoimines (45) or (48) with sulfur in dimethylformamide at 100-150°C leads to the evolution of hydrogen sulfide and the formation of a complex mixture of sulfurated products:

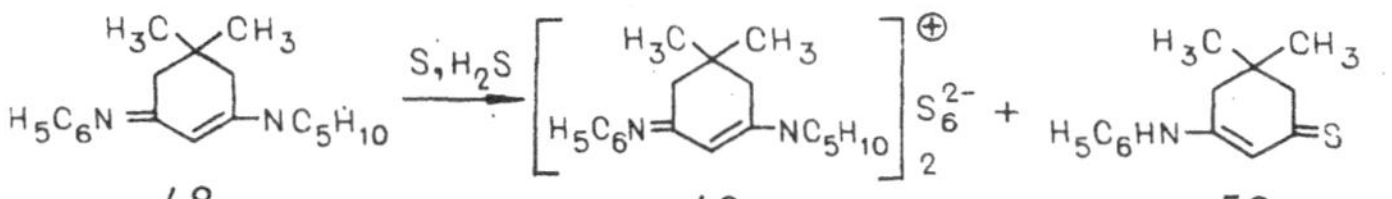

$$(7.54)$$

7.2. AROMATIC AMINES

7.2.1. Primary Amines

The reaction of sulfur with the simplest aromatic amine, aniline, has been studied in detail and has a long and confusing history [63-82]. For the first time the elucidation of the course and products of this reaction carried out at 120-140°C for 12-24 hours was achieved by Hinsberg [70]. He found that four crystalline compounds are present in the mixture, i.e., 4,4'-diaminodiphenyl sulfide (51) (the "März thioanilide" [64]), 2,4'-diaminodiphenyl sulfide (52), 2,4'-diaminodiphenyl disulfide (53), and diphenylamine. When aniline, its hydrochloride, and sulfur are melted at 190-200°C for two hours, a large quantity of phenothiazine (54) is formed [67]:

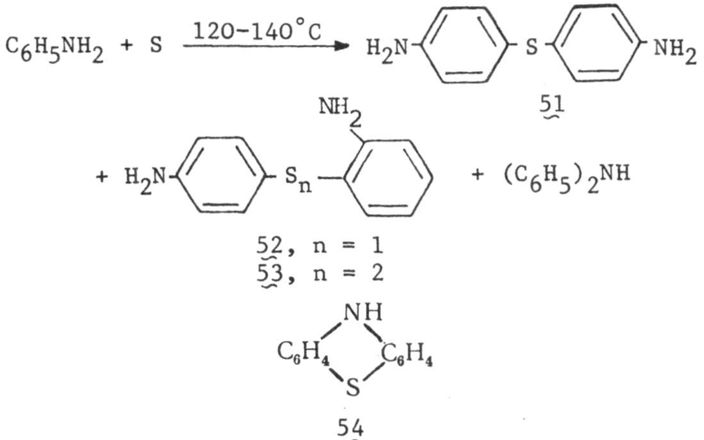

$$(7.55)$$

Subsequently it was shown that, when aniline is heated with sulfur in the presence of lead(II) oxide [71] or iodine [72], as well as when aniline hydrochloride is melted with sulfur in a 1:2 ratio [72], the stable 4,4'-diaminodiphenyl trisulfide (55) ("trithioaniline") is formed:

$$2C_6H_5NH_2 + 4S \xrightarrow{\text{PbO}}$$
$$\left.\begin{array}{c}\\ \\ 2C_6H_5NH_2 \cdot HCl + 4S \end{array}\right|_{-H_2S} \text{p-} H_2NC_6H_4S_3C_6H_4NH_2\text{-p} \quad (7.56)$$
$$\underset{55}{}$$

$$[C_6H_5NH-S_n-NHC_6H_5] \longrightarrow H_2NC_6H_4S_nC_6H_4NH_2 \quad (7.57)$$
$$\underset{56}{} \qquad n=1-3$$

It is assumed that diphenylaminosulfanes (56) are formed as intermediates in the reactions (7.55) and (7.56), etc., because N,N-disubstituted anilines do not react with sulfur under these conditions [1,71,73-75]. The monosulfide (56, n = 1) is the product obtained by cleavage of disulfane (56, n = 2) and not by direct interaction of aniline with sulfur [76].

Nitrogen-containing mercury derivatives are active catalysts in the reactions of primary aromatic amines with sulfur (e.g., mercuric derivatives of amides and imides, amines of mercuric salts) [77,78]. These catalysts make it possible to perform the reaction at room temperature or with moderate heating.

The first step in the reaction is the formation of sulfur diimides (57) [77]:

$$RNH_2 + S \xrightarrow[80°C]{\text{Hg(NHAC)}_2} RN=S \quad (7.58)$$
$$\underset{57}{}$$
$$R=C_6H_5, \text{p-}CH_3OC_6H_4, \text{m-}CH_3C_6H_4, 1\text{-}C_{10}H_7$$

The compounds (57) formed in the reaction lose sulfur in the presence of aluminium or zinc oxide. Because of this, it was not possible to isolate these compounds when the reaction mixture was worked up incorrectly [78] and what was obtained was a mixture of the corresponding azo-compounds (36-85%), benzidine, thiazine dyes (∿ 30%), and a small amount of azines. The corresponding hydrazo-compounds were postulated as intermediates in the reaction of sulfur with aniline [78].

The sulfuration of aniline and other arylamines with piperidinosulfenyl chloride gives the corresponding diarylaminosulfide (58) which upon heating gives the starting aniline and very reactive thionitrosoarenes (59) [79]:

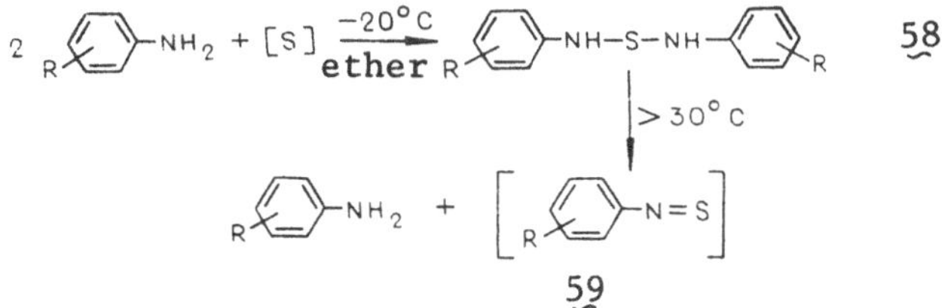

(7.59)

In the first papers devoted to the reaction of 4-methylaniline with sulfur in the presence of lead(II) oxide, it was pointed out that bis-2-amino-5-methylphenyl sulfide (60) ("thiotoluidine") is formed [64,80,81]. The reaction (7.60) in the absence of lead(II) oxide when carried out at 180-220°C gave a mixture of "thiotoluidine" (60), 2-(4-aminophenyl)-6-methylbenzo-1,3-thiazole (61) and "Primuline base" (62) [81-87]:

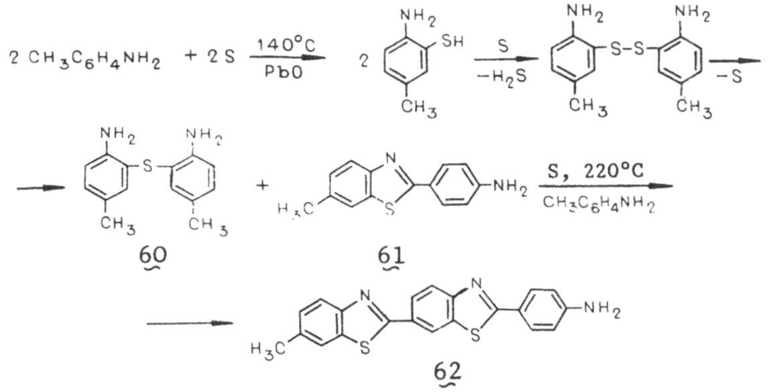

(7.60)

Reaction (7.60) is used industrially for the synthesis of certain sulfur-containing dyes (cf. Section 7.12). According to the data given in [1,88] heating 2-methylaniline with sulfur in the presence of lead(II) oxide gives bis-2-amino-5-methylphenyl trisulfide ("trithio-o-toluidine"), a rubber vulcanizer. When sulfur is heated with a mixture of toluidines or xylidines in the presence of certain organic bases (aniline, p-phenylenediamine, aminomethylpyridine), 1,3-thiazolic resins are obtained which are stable up to 700°C and used as cementing materials [89].

1-Naphthylamine reacts with sulfur at high temperatures with formation of bis-1-amino-2-naphthyl trisulfide (63) [77,90], which was erroneously assigned the structure of the corresponding disulfide [77] obtained previously under different conditions [91]:

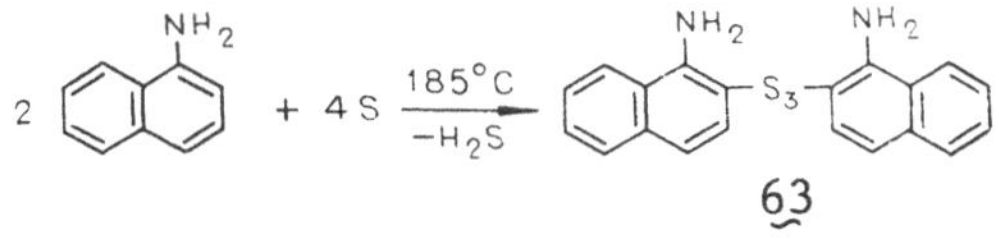

(7.61)

When a mixture of aniline, sulfur, and carbon disulfide is heated at 260°C for three hours, 1,3-benzothiazole-2-thiol is obtained in high yield [92]:

$$C_6H_5NH_2 + S + CS_2 \longrightarrow \text{(structure)} -SH \qquad (7.62)$$

Further search for the optimum conditions for the manufacture of accelerators for natural and synthetic rubber vulcanization, as well as herbicides, is being carried on. The reaction of aniline with sulfur and carbon disulfide is performed in a special tubular reactor at 240°C under pressure. The yield of benzothiazole-2-thiol amounts to 89% under these conditions [93]. A number of 2-mercapto-arylthiazoles are obtained in yields of 80% in the presence of special additives in the reaction mixture, e.g., cyanoamide, thiourea, its substitutes and derivatives, and ammonium rhodanide [94].

The reactions of aromatic diamines with sulfur are closely related to the methods of preparation of sulfur-containing dyes and are, therefore, discussed in Section 7.12.

7.2.2. Secondary Amines

N-Methylaniline refluxed with sulfur forms 1,3-benzothiazole and aniline [95,96]. When sulfur is heated with N-methylaniline at 240°C, N,N'-diphenylthiourea is obtained [97].

$$2 C_6H_5NHCH_3 + 3S \xrightarrow[-2H_2S]{} \text{(structure)} CH + C_6H_5NH_2 \qquad (7.63)$$

The reaction of diphenylamine with elemental sulfur has been studied in detail because this reaction forms the basis for the preparation of methylene blue. It was found in the last century [83,98] that when diphenylamine is melted with sulfur, phenothiazine ("thiodiphenylamine") is obtained:

$$(C_6H_5)_2NH + 2S \xrightarrow[-H_2S]{250-300°C} \text{(structure)} \qquad (7.64)$$

Various catalysts were later suggested ($AlCl_3$, $AlBr_3$, AlI_3, $FeCl_3$, $SbCl_3$, CuI_2, SI_2, I_2) to increase the rate of reaction (7.64), decrease its temperature, and increase the yield of phenothiazine up to 82-99% [99-107]. Derivatives of diphenylamine react with sulfur in a similar way [98,107-110] as well as those of phenyl-naphthylamine [111], dinaphthylamine [112], and 3-phenylaminoretene [113].

7.2.3. Tertiary Amines

When N,N-dimethylaniline is melted with sulfur, it gives mainly
1,3-benzothiazole [96,114]:

$$C_6H_5N(CH_3)_2 \; + \; S \xrightarrow[-3CS_2]{-3H_2S,} \underset{S}{\overset{N}{\bigcirc}}CH + \dot{C}_6H_5NHCH_3 + C_6H_5NH_2 \qquad (7.65)$$

The corresponding 10-(2-dialkylaminoalkyl)phenothiazines (64)
are the products of the reaction of N-(2-dialkylaminoalkyl)diphenyl-
amines with sulfur in the presence of iodine at 185°C [108]:

$$(C_6H_5)_2 N(CH_2)_n NR_2 \; + \; S \xrightarrow[-H_2S]{185°C} \underset{S}{\overset{(CH_2)_n NR_2}{\bigcirc\!\!\overset{N}{\bigcirc}}} \qquad (7.66)$$

$$R=CH_3, \; C_2H_5; \; n=2, 3 \qquad \underset{\sim}{64}$$

When N-methyl-2,2'-dinaphthylamine is heated with sulfur, the
corresponding 1,4-thiazine can be isolated [111].

7.3. IMINES AND OXIMES

Upon heating with elemental sulfur, diarylketimines form thioketones [21]:

$$Ar_2C=NH + S \xrightarrow{220°C} Ar_2C=S \qquad (7.67)$$

When arylmethylketimines are melted with sulfur, 2,4-diaryl-
thiophenes are formed [115]:

$$2 \; Ar-\underset{NAr'}{\overset{\|}{C}}-CH_3 \; + \; S \xrightarrow[-H_2NAr']{240°C} \underset{Ar}{\overset{Ar}{\underset{S}{\bigcirc}}} \qquad (7.68)$$

Alkylarylketimines react with sulfur in the presence of primary
aliphatic amines as catalysts and give the corresponding thiocarboxamides (65) as the final products [116]:

$$Ar-\underset{NR}{\overset{\|}{C}}-CH_3 \underset{amine}{\overset{S.20-60°C}{\rightleftharpoons}} Ar-\underset{NR}{\overset{\|}{C}}-C\overset{\diagup S}{\diagdown SH} \underset{+H_2NR}{\overset{-H_2NR}{\rightleftharpoons}} Ar-\underset{S}{\overset{\|}{C}}-C\overset{\diagup S}{\diagdown SH} \overset{+H_2NR'}{\rightleftharpoons}$$

$$\rightleftharpoons Ar-\underset{S}{\overset{\|}{C}}-C\overset{\diagup S}{\diagdown NHR'} + H_2S \rightleftharpoons Ar-\underset{SSH}{\overset{|}{C}H}-C\overset{\diagup S}{\diagdown NHR'} \xrightarrow[-S]{amine}$$

$$\longrightarrow \text{Ar}-\underset{\overset{|}{\text{SH}}}{\text{CH}}-\text{C}\underset{\text{NHR}'}{\overset{\text{S}}{\diagup}}\ \xrightarrow[-\text{s}]{\text{amine}}\ \text{Ar}-\text{CH}_2-\text{C}\underset{\text{NHR}'}{\overset{\text{S}}{\diagup}} \qquad (7.69)$$

$$\underset{65\ (82\%)}{}$$

When N-methylene-tert-octylamine is heated with sulfur in tert-octylamine at 155°C, tert-octyl isothiocyanate is formed [117]. Refluxing N-methylenecyclohexylamine with sulfur gives N,N'-dicyclohexylthiourea [118].

The reaction of N-methylethylidenimine with sulfur at high temperature leads to a mixture of acetonitrile and 2-methyl-1,3-thiazole [32,119]. When sulfur is heated with N-methylbenzylidenimine at 200°C, N-methylthiobenzamide is obtained [120]:

$$\text{CH}_3\text{CH}=\text{NCH}_2\text{CH}_3 + \text{S} \xrightarrow[-\text{H}_2\text{S}]{500°\text{C}} \underset{(11\%)}{\text{CH}_3\text{CN}} + \underset{(67\%)}{\overset{\text{N}}{\underset{\text{S}}{\diagdown}}\!\!-\!\text{CH}_3} \qquad (7.70)$$

An interesting method for the preparation of N,N'-disubstituted thioureas has been based on the reaction of formaldimines with sulfur and primary amines [121]:

$$\text{RN}=\text{CH}_2 + \text{S} + \text{R}'\text{NH}_2 \xrightarrow[-\text{H}_2\text{S}]{50-150°\text{C}} \text{RNH}-\underset{\overset{\|}{\text{S}}\ (\sim 80\%)}{\text{C}}-\text{NHR}' \qquad (7.71)$$

$$\text{R, R}' = \text{alkyl, cycloalkyl, arylalkyl, aryl}$$

The reaction of carbonimidic difluorides with sulfur in the presence of phosphorus pentafluoride gives the corresponding bis-trifluoromethylamino sulfide (66) [122]:

$$\text{RN}=\text{CF}_2 + \text{S} \xrightarrow{\text{PF}_5} \text{R}\underset{\overset{|}{\text{CF}_3}}{\overset{\overset{\text{CF}_3}{|}}{\text{NSNR}}} \qquad (7.72)$$

$$\underset{66}{}$$

The reaction of ketimines with a mixture of sulfur and carbon disulfide has been studied in detail [45,123]. Depending on the reaction conditions, the final reaction products are 1,2-dithiol-3-thiones (67) or 2,5-dihydro-1,2-thiazole-5-thiones (68), e.g.,

$$\text{R}'\text{CH}_2\overset{\overset{\text{R}''}{|}}{\text{C}}=\text{NR} \underset{\text{DMF}}{\rightleftharpoons} \text{R}'\text{CH}=\overset{\overset{\text{R}''}{|}}{\text{C}}-\text{NHR} \xrightarrow{\text{CS}_2} \text{R}'\overset{\overset{\text{HS}\diagdown\ \diagup\text{S}}{\text{C}}}{\underset{\overset{|}{\text{R}''}}{\text{C}}}=\text{CNHR} \qquad (7.73)$$

$$\xrightarrow[-\text{H}_2\text{NR}]{\text{S, 50°C}} \quad 67 \qquad \qquad \xrightarrow[-\text{H}_2\text{S}]{\text{S, 10-20°C}} \quad 68$$

The reaction of some alkyl(aryl) methyl ketoximes with sulfur at 200°C leads to the formation of the corresponding carboxylic acids in low yields [124].

7.4. AMIDES AND THIOCARBOXAMIDES

When acetanilide is heated with sulfur 2,2'-bibenzo-[d]-thiazole is obtained in low yield [125]:

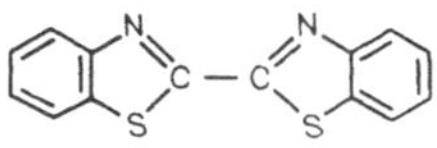

2-Substituted benzothiazoles are formed on melting sulfur with formanilide or benzanilide [125-127]. α- and β-Naphthylformamide, -acetamide, and -benzamide react in an analogous manner with the formation of substituted naphtho-[d]-thiazoles [127]:

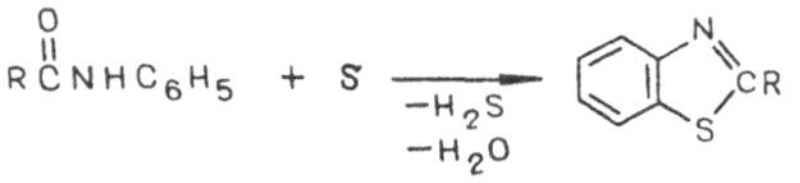

$$RCNHC_6H_5 + S \xrightarrow[-H_2O]{-H_2S} \text{[benzothiazole]} \qquad (7.74)$$

$$R = H, C_6H_5$$

At high temperatures, N-ethylthioacetamide reacts with sulfur to give 2-methyl-1,3-thiazole and acetonitrile [32]:

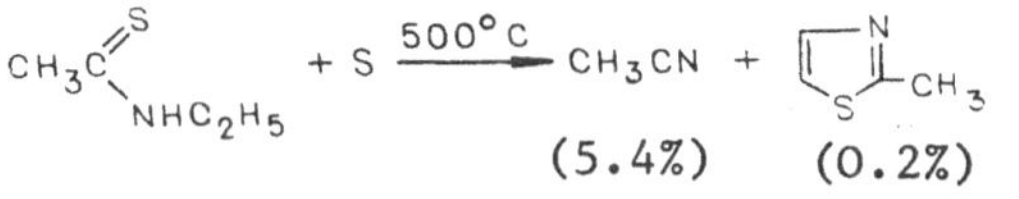

$$CH_3C(S)NHC_2H_5 + S \xrightarrow{500°C} CH_3CN + \text{[thiazole]CH_3} \qquad (7.75)$$
$$(5.4\%) \qquad (0.2\%)$$

The reaction of N,N-dimethyl-2-pyridinethiocarboxamide with sulfur leads to its partial demethylation [128]:

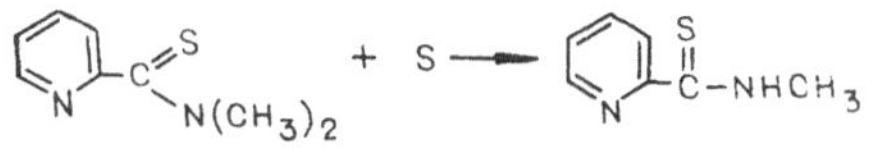

$$\text{[pyridine]}C(S)N(CH_3)_2 + S \longrightarrow \text{[pyridine]}C(S)-NHCH_3 \qquad (7.76)$$

When thioacetamide is heated with ^{35}S in toluene, there is an isotopic exchange and ^{35}S-thioacetamide is formed in high yield [129]. This reaction is first order with respect to sulfur:

$$CH_3C(S)NH_2 + {}^{35}S \rightarrow CH_3C({}^{35}S)NH_2 \qquad (7.77)$$
$$(96\%)$$

7.5. CARBAMIC ACID, THIOUREA, AND GUANIDINE DERIVATIVES

Ammonium phenyldithiocarbamate (69) heated with sulfur in boiling benzene gives N,N'-diphenylthiourea (70) [92]. At high temperatures, reaction (7.78) leads to the formation of 1,3-benzo-thiazole-2-thiol:

$$2C_6H_5NHC\overset{S}{\underset{SNH_4}{\diagdown}} + S \xrightarrow{80°C} C_6H_5NHCNHC_6H_5 \qquad (7.78)$$

$$\underset{\underset{69}{\sim}}{} \qquad\qquad\qquad \underset{S}{\overset{\|}{}}\ \underset{\underset{70}{\sim}}{}$$

It is assumed that its formation occurs via the intermediate formation of N,N'-diphenylthiourea (70) and phenyl isothiocyanate
according to the following scheme [92]:

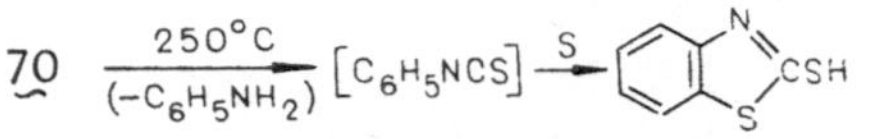

$$\underset{\sim}{70} \xrightarrow[(-C_6H_5NH_2)]{250°C} [C_6H_5NCS] \xrightarrow{S} \text{(benzothiazole)CSH} \qquad (7.79)$$

When a mixture of ammonium phenyldithiocarbamate and aniline
is heated with sulfur, 2-anilinobenzothiazole (71) is obtained in
good yield [92]:

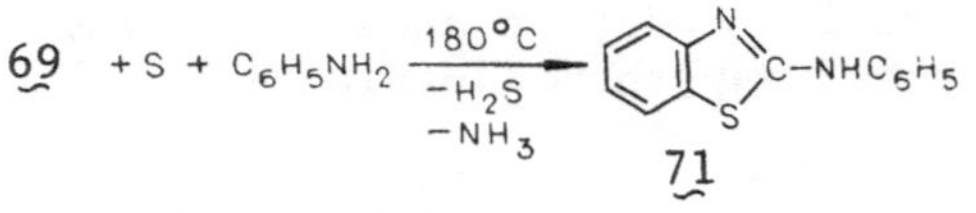

$$\underset{\sim}{69} + S + C_6H_5NH_2 \xrightarrow[\substack{-H_2S \\ -NH_3}]{180°C} \text{(benzothiazole)}C-NHC_6H_5 \qquad (7.80)$$

$$\underset{\underset{71}{\sim}}{}$$

When mono-, di-, or triphenylguanidine is melted with sulfur,
a mixture of sulfur-containing products is obtained from which 1,3-
benzothiazole-2-thiol can be isolated [130]. It is likely that the
course of this reaction is similar to that given in scheme (7.79).
When a mixture of sulfur, aniline, and mono- or diphenylguanidine
is heated, 2-anilinobenzothiazole (71) is obtained [130,131]. The
reaction of N,N'-diphenylguanidine with sulfur in boiling xylene
affords a mixture of products from which the linear polysulfides,
R-S$_n$-R (n = 2-5), of undetermined structure were isolated [132]:

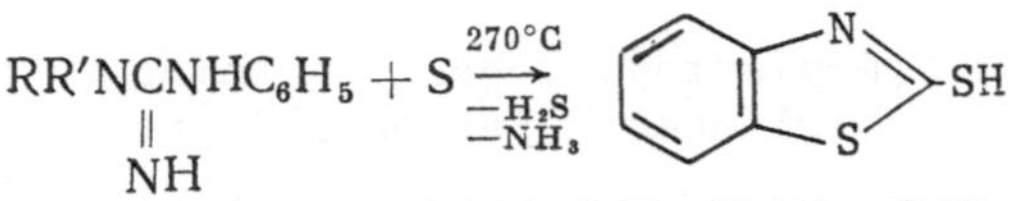

$$RR'NCNHC_6H_5 + S \xrightarrow[\substack{-H_2S \\ -NH_3}]{270°C} \text{(benzothiazole)}SH \qquad (7.81)$$

$$\underset{NH}{\overset{\|}{}}$$

$$R, R' = H;\ R = H,\ R' = C_6H_5;\ R, R' = C_6H_5$$

7.6. NITRILES

Aromatic nitriles react with sulfur in the presence of tertiary
amines as catalysts at low temperatures to give 1,2,3,4-trithiazoles
(72) [39], whereas 1,2,4-thiadiazoles (73) are obtained at higher
temperatures [133,134]:

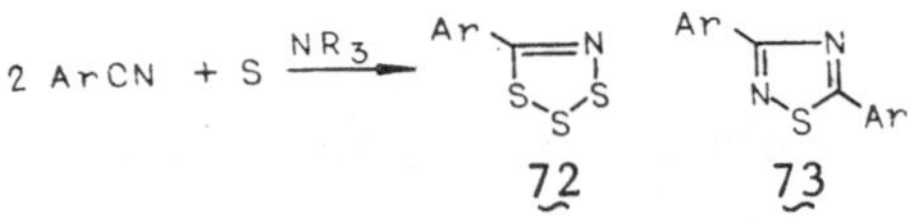

$$2\ ArCN + S \xrightarrow{NR_3} \underset{\underset{72}{\sim}}{\text{(trithiazole)}} \quad \underset{\underset{73}{\sim}}{\text{(thiadiazole)}} \qquad (7.82)$$

$$Ar = C_6H_5,\quad \underline{p}\text{-}CH_3C_6H_4,\ 2\text{-}C_{10}H_7$$

Nitriles containing an active methylene group give a specific reaction with sulfur in the presence of basic catalysts. Thus, malononitrile, for example, reacts with sulfur in the presence of dimethylformamide or triethylamine at room temperature to give 2,5-diaminothiophene-3,4-dicarbonitrile (74) and 3,5-diaminothiophene-2,4-dicarbonitrile (75) [135]:

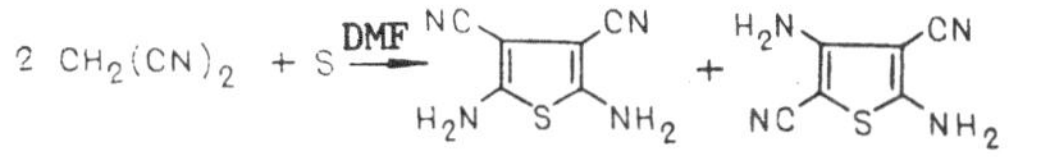

$$2 \ CH_2(CN)_2 \ + \ S \xrightarrow{DMF} \text{[structure 74]} + \text{[structure 75]} \tag{7.83}$$

74 (30%)   75 (20%)

The reaction of sulfur with C–H acids, such as the above nitriles, is one of the modifications of the Willgerodt reaction. The α-thiolation of the C–H acids is the primary reaction step [135, 136]. Thus, the formation of thiophene (74) can be explained by the following scheme [135]:

$$CH_2(CN)_2 + 2S \rightarrow (CN)_2C\begin{smallmatrix} SH \\ SH \end{smallmatrix} \underset{-H_2S}{\rightleftarrows} (CN)_2C=S \xrightarrow{CH_2(CN)_2}$$

$$\rightarrow (CN)_2CH-\underset{\underset{SH}{|}}{C}(CN)_2 \underset{-S}{\overset{NR_3}{\rightarrow}} (CN)_2CH-CH(CN_2) \overset{H_2S}{\rightarrow} 74 \tag{7.84}$$

Depending on the type of basic catalyst used in the reaction, ethyl cyanoacetate reacts with sulfur under analogous conditions to give diethyl 3,5-diaminothiophene-2,4-dicarboxylate (76), bis-(2-amino-3,4-diethoxycarbonyl-5-pyrryl) disulfide (77), diethyl 2-amino-5-mercaptothiophene-3,4-dicarboxylate (78), or ethyl 4-cyano-5-hydroxy-2-mercaptothiophene-3-carboxylate (79) [135]:

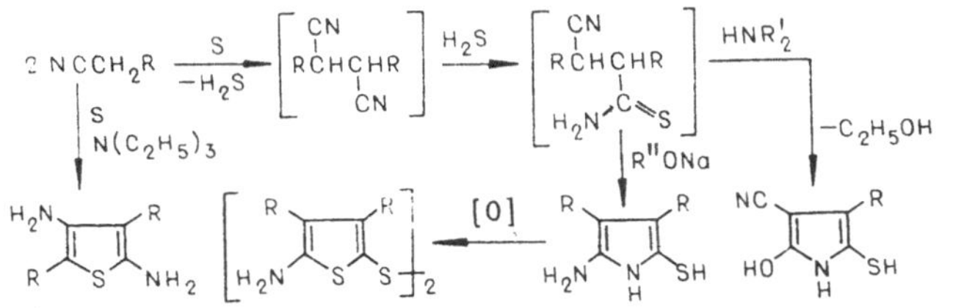

$$\tag{7.85}$$

76 (20%)   77 (26%)   78   79 (55–70%)

R = COOC$_2$H$_5$; R',R" = alkyl

2-Amino-5-mercaptopyrrole-3,4-dicarboxamide (80) may be obtained from cyanoacetamide in a similar way [135]:

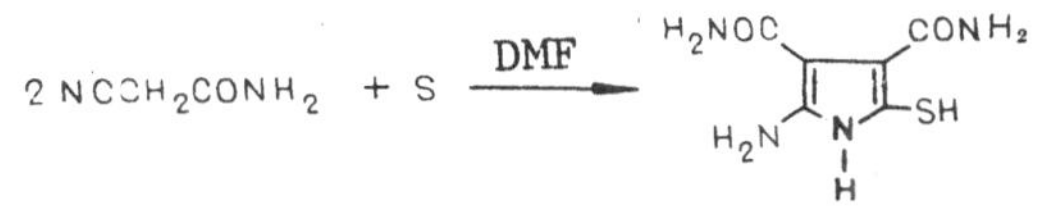

$$2 \ NCCH_2CONH_2 \ + \ S \xrightarrow{DMF} \text{[structure 80]} \tag{7.86}$$

80

When benzyl cyanide is heated with sulfur at approximately 175°C, α,β-dicyanostilbene (81) and 1,2-diphenylethane-1,2-dicarbonitrile (82) are formed. Prolonged heating at 190°C leads to the formation of a red polymer [137]:

$$4C_6H_5CH_2CN + S \xrightarrow[-H_2S]{175°C} \underset{\substack{| \\ CN \\ 81 \ (35\%)}}{\overset{\substack{CN \\ |}}{C_6H_5C=CC_6H_5}} + \underset{\substack{| \\ CN \\ 82 \ (1\%)}}{\overset{\substack{CN \\ |}}{C_6H_5CHCHC_6H_5}} \qquad (7.87)$$

The 2-cyanoisopropyl radical, formed from azoisobutyronitrile, reacts with sulfur to give the corresponding oligosulfides [138]:

$$AIBN \xrightarrow{80°C} 2\underset{\substack{| \\ CH_3}}{\overset{\substack{CH_3 \\ |}}{NCC}}\cdot \xrightarrow{nS} \underset{\substack{| \ \ | \\ CH_3 \ CH_3}}{\overset{\substack{CH_3 \ CH_3 \\ | \ \ |}}{NCC\text{-}S_n\text{-}CCN}} \qquad (7.88)$$
$$n = 1 - 3$$

α,β-Unsaturated nitriles react with sulfur in boiling methanol in the presence of di- or triethylamine and give the corresponding substituted thiophen-2-amines [45,134,139-142]. The reactions of type (7.89) can be directly carried out with a mixture of the corresponding ketones (β-diketones) with various nitriles containing an active methylene group in the presence of sulfur [139,143-148]:

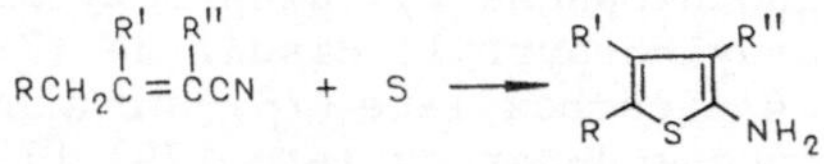

$$\underset{\substack{| \ \ | \\ RCH_2C=CCN}}{\overset{\substack{R' \ \ R'' \\ | \ \ |}}{}} + S \longrightarrow \underset{R \quad S \quad NH_2}{\overset{R' \qquad R''}{}} \qquad (7.89)$$

R, R' = H, alkyl, aryl; R' = CN, CONH₂, COOR (R = alkyl)

Many nitrogen- and sulfur-containing heterocyclic compounds are often synthesized by the reaction of sulfur with nitriles containing an activated methylene group. These reactions are carried out in the presence of carbon disulfide [45,149,150] or in the presence of organic isothiocyanates [151]. In the first case, the corresponding 1,2-dithiol-3-thiones (83) are obtained, whereas in the latter case 4-amino-2,3-dihydro-1,3-thiazole-2-thiones (84) are the reaction products. These reactions occur according to the following schemes:

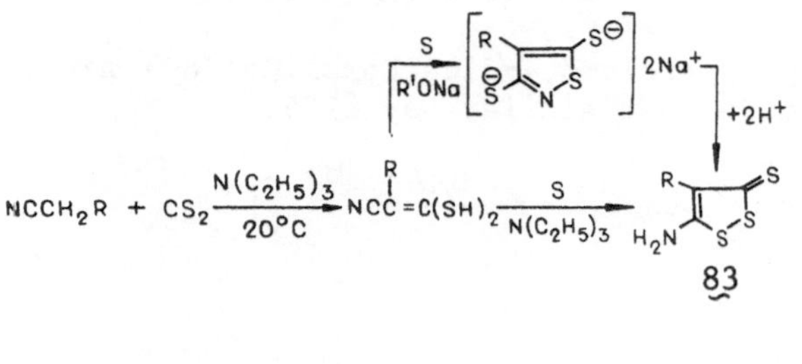

$$NCCH_2R + CS_2 \xrightarrow[20°C]{N(C_2H_5)_3} \underset{}{\overset{R \\ |}{NCC}}=C(SH)_2 \qquad (7.90)$$

83

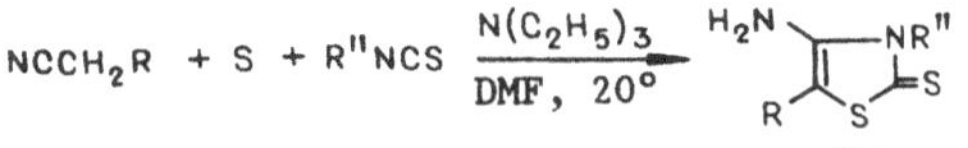

$$NCCH_2R \; + \; S \; + \; R''NCS \; \xrightarrow[\text{DMF, } 20°]{\text{N}(C_2H_5)_3} \; \underset{\underset{84}{}}{} \qquad (7.91)$$

R = CN, CONH$_2$, COOR (R = alkyl); R'' = alkyl, aryl

Reactions (7.89), (7.90), and (7.91) are modifications of the Asinger reaction and are performed in polar solvents in the presence of strongly basic secondary or tertiary amines as catalysts.

When benzoyl cyanide is heated with a mixture of sulfur and morpholine, the corresponding morpholide is obtained [152,153]:

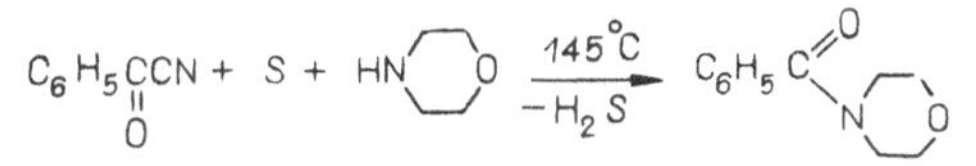

$$C_6H_5\underset{O}{\overset{}{C}}CN + \; S \; + \; HN\!\!\diagdown\!\!O \; \xrightarrow[-H_2S]{145°C} \; C_6H_5C\diagdown \qquad (7.92)$$

Under analogous conditions, benzyl cyanide is converted to thio-benzmorpholide (85) [153]. When phenyl isothiocyanate is used instead of morpholine, 4-amino-3,5-diphenyl-1,3-thiazole-2-thiones (86) are obtained [135]. One can assume that these reactions proceed according to the following scheme [135]:

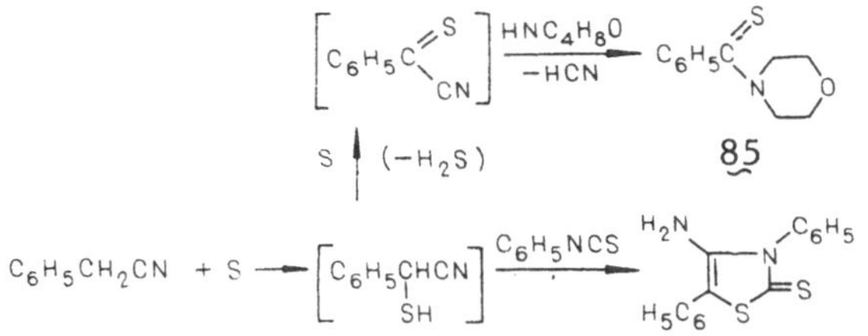

$$\left[C_6H_5C\diagup^{S}_{CN} \right] \; \xrightarrow[-HCN]{HNC_4H_8O} \; C_6H_5C \qquad (7.93)$$

$$\underset{85}{}$$

$$S \; \big\downarrow \; (-H_2S)$$

$$C_6H_5CH_2CN \; + \; S \; \longrightarrow \; \left[C_6H_5\underset{SH}{\overset{}{C}}HCN \right] \; \xrightarrow{C_6H_5NCS} \; \underset{86}{} \qquad (7.94)$$

Cyanophenylacetylene (87) and aryl cyanomethyl ketones (88) react with sulfur and morpholine with the formation of the corresponding isothiazoles (89) [153]:

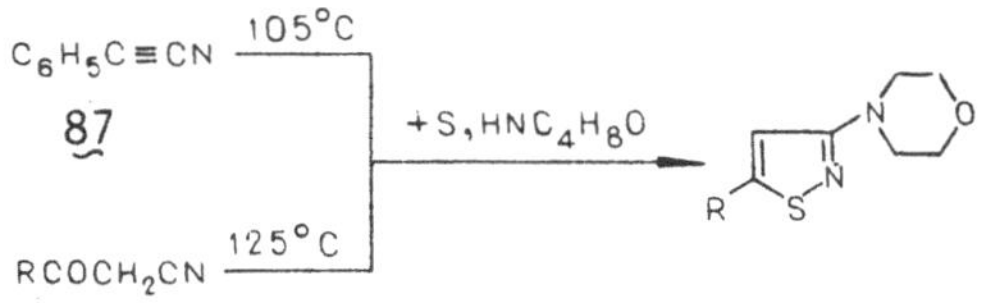

$$C_6H_5C\equiv CN \; \xrightarrow{105°C} \qquad (7.95)$$

$$\underset{87}{}$$

$$+S, HNC_4H_8O$$

$$RCOCH_2CN \; \xrightarrow{125°C} \qquad (7.96)$$

$$\underset{88}{} \qquad\qquad \underset{89}{}$$

R = C$_6$H$_5$, p-CH$_3$C$_6$H$_4$; p-CH$_3$OC$_6$H$_4$, p-ClC$_6$H$_4$

2-Imidazolines (90) are obtained when nitriles are treated with sulfur and vic-diaminoalkanes [70]:

$$RCN + \underset{\overset{|}{\underset{}{H_2NCHCH_2NH_2}}}{\overset{R'}{}} + S \xrightarrow{80-190°C} \underset{R'}{\overset{HN}{\underset{N}{\diagup}}}R \quad \underset{\widetilde{}}{90} \qquad (7.97)$$

$$R = Alkyl, \; aryl, \; 2-,3-, \; or \; 4-pyridyl; \; R' = H, \; CH_3$$

The reaction with sulfur may be used to introduce sulfur into the nitriles present in tall oil [154] and for the dehydrogenation of polyacrylonitrile [155].

The formation of S_2 and NCS molecules is observed in the flash photolysis of a mixture of cyanogen and carbonyl sulfide [156]. The authors of this paper assume that the primary reaction step is the formation of atomic sulfur in a singlet excited state (7.98) which can further react with the starting materials according to scheme (7.99):

$$COS \xrightarrow{h\nu} S(^1D) + CO \qquad (7.98)$$

$$2S(^1D) + C_2N_2 + 2COS \xrightarrow[-2CO]{} S_2 + NCS \qquad (7.99)$$

The reaction of phenylcyanamide with sulfur gives 1,3-benzo-thiazole-2-thiol [130]:

$$C_6H_5NHCN + S \longrightarrow \underset{S}{\overset{N}{\diagdown}}SH \qquad (7.100)$$

7.7. ISONITRILES AND ISOTHIOCYANATES

Ethyl and aryl isonitriles when heated with sulfur in ethanol or carbon disulfide give the corresponding isothiocyanates [157-160]:

$$RNC + S \xrightarrow{120°C} RNCS \qquad (7.101)$$

Aromatic isothiocyanates react with sulfur under more drastic conditions to form 2-mercaptoaryl isothiocyanates (91) which can further isomerize to 1,3-benzothiazole-2-thiols (92) [161]. 2-Naphthyl isothiocyanate gives an analogous reaction:

$$(7.102)$$

The reaction of alkyl and aryl isonitriles with sulfur and iodine pentafluoride in pyridine gives bis-trifluoromethylamino sulfides [122]:

$$RNC + S \xrightarrow{IF_5} \underset{\underset{CF_3}{|}}{\overset{\overset{CF_3}{|}}{RNSNR}} \qquad (7.103)$$

7.8. DIAZO COMPOUNDS AND AZIDES

Diaryldiazoalkanes refluxed with sulfur in toluene give tetra-arylthiiranes [162]. This reaction is considerably slower in diethyl ether [155]:

$$2Ar_2CN_2 + S \xrightarrow[-2N_2]{} \underset{S}{Ar_2C-CAr_2} \qquad (7.104)$$
$$(95\%)$$

Diazofluorene, diazoxanthene, and diazothioxanthene react in a similar manner [155]. The following mechanism has been proposed for these reactions [163]:

$$R_2CN_2 \xrightarrow[-N_2]{} R_2C: \xrightarrow{+S} [R_2C=S] \xrightarrow[-N_2]{R_2CN_2} \underset{S}{R_2C-CR_2} \qquad (7.105)$$

The intermediate formation of thioketones during the conversions of type (7.105) was first confirmed in the reaction of indene enaminodiazo-compounds (93) with sulfur which resulted in the isolation of corresponding 3-aminoindene-1-thiones (94) [62,164,165]:

$$(7.106)$$

Later it was established [166] that the reaction of the anthraquinone diazo-derivative (95) with sulfur was also completed at the thioketone (96) stage:

$$(7.107)$$

The reaction of diazoacetophenone and its derivatives with sulfur and secondary amines gives the corresponding amides of aryl-1-thioglyoxylic acids [167]:

$$RCCHN_2 + S + HN\underset{}{\bigcirc}X \xrightarrow{20°C} RCC'\overset{S}{\underset{O}{C}}N\underset{}{\bigcirc}X \qquad (7.108)$$

$$R = C_6H_5,\ 4\text{-}CH_3OC_6H_4,\ 4\text{-}NO_2C_6H_4 : X = CH_2, O$$

When alkoxycarbonyl azides are refluxed with sulfur in decalin, N-substituted cyclic sulfur imides are obtained [168]. In those cases in which a large excess of ethyl azidoformate is used in reaction (7.109), the presence of other sulfur imides among the reaction products has been detected by mass spectrometry: S_5R_3, S_4R_4, and S_3R_5 (R = $NCOOC_2H_5$):

$$N_3C\overset{O}{\underset{OR}{\diagdown}} + S \xrightarrow{126°C} S_7NCOOR \text{ and } S_6(NCOOC_2H_5)_2 \qquad (7.109)$$

$$R = C_2H_5,\ C_6H_5$$

7.9. N-HALOSUBSTITUTED DERIVATIVES

The reaction of sulfur with N,N-dichloroamides of arenesulfonic acids [169,170], N,N-dichlorourethanes [171], N,N-dichloroarylalkyl-amines, and N,N-dichloroacylamides [172] gives sulfur diimides (98). In the latter two cases iodine, iron(III) chloride, or aluminium chloride may be used as catalysts. It is likely that reaction (7.110) occurs via the corresponding imidosulfurous di-chlorides (97) which further react with a second molecule of the N-halo-derivative [172]:

$$RNCl_2 + S \xrightarrow{70°C} [RN=SCl_2] \xrightarrow[-2Cl_2]{RNCl_2} RN=S=NR \qquad (7.110)$$
$$\qquad\qquad\qquad 97 \qquad\qquad\qquad 98$$

$$R = ArSO_2,\ COO\text{-alkyl, alkyl}$$

α-Cyanoalkyl-N,N-dichloroamines (99) give only N-(1-cyano-alkyl)imidosulfurous dichlorides (100) upon reaction with sulfur [173]:

$$RR'C(CN)NCl_2 + S \rightarrow RR'C(CN)N=SCl_2 \qquad (7.111)$$
$$\quad 99 \qquad\qquad\qquad 100\ (50\text{-}77\%)$$

The reaction of sulfur with diethyl N-chlorocarbonimidate (101) and ethyl N-chlorobenzimidate (103) gives bis-ethoxycarbonyl-sulfur diimide (102) and dibenzoylsulfur diimide (104), respectively [174]:

$$2(C_2H_5O)_2C=NCl + S \xrightarrow[-2C_2H_5Cl]{70°C} (C_2H_5OCON=)_2S \qquad (7.112)$$
$$\quad 101 \qquad\qquad\qquad\qquad\qquad 102$$

$$2C_6H_5C(=NCl)OC_2H_5 + S \xrightarrow[-2C_2H_5Cl]{} (C_6H_5CON=)_2S \qquad (7.113)$$
$$\quad 103 \qquad\qquad\qquad\qquad\qquad 104$$

Compound (102) may also be obtained from the reaction of sulfur with N-chlorourethane (105) in the presence of pyridine [174]:

$$2C_2H_5OCONHCl + S + 2C_5H_5N \xrightarrow{80°C} 102 + 2C_5H_5N \cdot HCl \qquad (7.114)$$
$$105$$

At 20-25°C and in the presence of tetraethylammonium bromide, sulfur readily reacts with N-chlorobenzamide (106) in a ratio of 1:2 to form N-benzoyliminothionyl chloride (108) and benzamide [175]. The first stage of the reaction leads to N-benzoylamidosulfoxylic chloroanhydride (107) which further rapidly reacts with N-chlorobenzamide to give (108):

$$\underset{106}{C_6H_5CONHCl} + S \xrightarrow{20°C} \underset{107}{C_6H_5CONHSCl} \xrightarrow{C_6H_5CONHCl}$$

$$\rightarrow \underset{108}{C_6H_5CON=SCl_2} + C_6H_5CONH_2 \xrightarrow[-HCl]{50°C}$$

$$\rightarrow \underset{\substack{|\\ Cl \\ 109}}{C_6H_5CON=S-NHCOC_6H_5} \xrightarrow[-HCl]{80°C} \underset{110}{(C_6H_5CON=)_2S} \qquad (7.115)$$

When carried out at 40-50°C in the presence of iodine, the reaction yields N,N'-dibenzoylamidoiminosulfurous acid chloroanhydride (109) capable of converting to dibenzoylsulfur diimide (110) at a higher temperature [174,175]. N-Chlorobenzamide refluxed with sulfur in carbon tetrachloride in a 1:1 ratio gives, together with diimide (110), sulfoxylic acid N,N'-dibenzoyldiamide (111) in a 44% yield [175]. Compound (111) is formed due to the thermal decomposition of the chloroanhydride (107) which appears at the first stage of the reaction:

$$2C_6H_5CONHCl + S \xrightarrow{80°C} 110 + \underset{111}{(C_6H_5CONH)_2S} \qquad (7.116)$$

The reaction of N-chloroformimidoyl chlorides (112) with sulfur in the presence of chloride anions in catalytic amounts gives N-(chlorothio)formimidoyl chlorides (113) according to the following scheme [176]:

$$\underset{\substack{|\\ Cl \\ 112}}{RC=NCl} + Cl^- \underset{(-Cl_2)}{\longrightarrow} \left[\underset{\substack{|\\ Cl}}{RC=N^-} \right] \xrightarrow[(-Cl^-)]{+S,Cl_2} \underset{\substack{|\\ Cl \\ 113}}{RC=NSCl} \qquad (7.117)$$

N-Chloromorpholine (114) and N-chlorodiarylsulfonyl imides
(116) react with sulfur to give morpholine-N-sulfenyl chloride (115)
and sulfoxylic acid N,N,N',N'-tetrabenzenesulfonyl diamide (117),
respectively [174]:

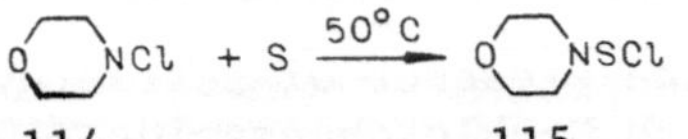

$$O\underset{114}{\underbrace{}}NCl + S \xrightarrow{50°C} O\underset{115}{\underbrace{}}NSCl \qquad (7.118)$$

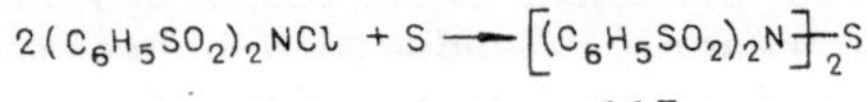

$$2(C_6H_5SO_2)_2NCl + S \longrightarrow \left[(C_6H_5SO_2)_2N\right]_2 S \qquad (7.119)$$
$$\underset{116}{} \qquad \underset{117}{}$$

N,N-Dichlorophosphoramides react vigorously with sulfur; no
reaction products, however, have yet been isolated [177].

7.10. NITRO-COMPOUNDS

The reactions of aromatic nitro-compounds with elemental sulfur
have been little studied. When nitrobenzene is refluxed with elemental sulfur, resinous products are obtained. Heating 1-nitro-
naphthalene with sulfur gives a mixture of colored identified
products and polymers, among which di-1-naphthylmono-, di-, and
trisulfides have been isolated [178]:

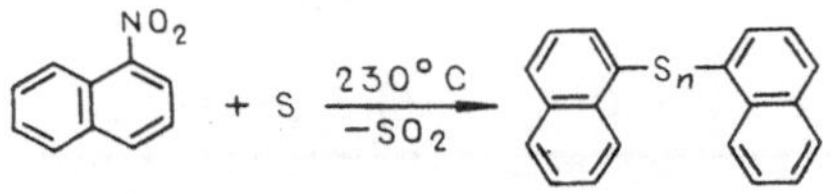

$$n = 1-3$$

The reaction of 2- or 4-nitrotoluene with sulfur in the presence
of sodium hydroxide leads to the formation of the corresponding
aminobenzaldehydes [179-181]:

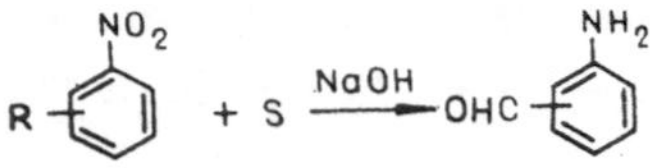

7.11. NITROGEN-CONTAINING HETEROCYCLIC COMPOUNDS

Elemental sulfur is often used for the dehydrogenation of
saturated nitrogen heterocycles some of which may also contain
sulfur:

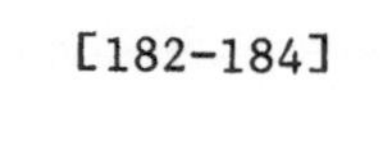

[182-184]

$$[185] \qquad (7.123)$$

$$[186] \qquad (7.124)$$

When N-alkyl-pyrrolidines and -piperidines are heated with sulfur, the reaction of the α-methylene group gives rise to the formation of the corresponding N-alkylthiolactams [38]:

$$\qquad (7.125)$$

$$(30-60\%)$$

$$R = CH_3, \ C_2H_5, \ C_4H_9; \ n = 1,2$$

The interaction of sulfur with 1-methyl-2-phenyl- or -2-(3-pyridyl)-pyrrolidines in toluene is accompanied by the dehydrogenation of the heterocycle and leads to the formation of the corresponding disulfides [38]. 1-Methyl-3-phenylpyrrolidine forms a solid polymer with sulfur at 105°C; this polymer is derived from the dehydrogenation product, i.e., from 1-methyl-3-phenylpyrrole [38]:

$$\qquad (7.126)$$

$$R = C_6H_5, \ 3\text{-}C_5H_4N$$

The reaction of nicotyrine[1-methyl-2-(3-pyridyl)pyrrole] with sulfur gives the cyclic bis-disulfide (118) [38,184]:

$$\qquad (7.127)$$

$$118$$

$$R = 3\text{-}C_5H_4N$$

The reaction of indole with sulfur has been studied under varying conditions [187-191]. At 180-190°C, indole undergoes dehydrogenation with sulfur and gives 3,3'-biindole (119) [188, 190]. Prolonged heating of indole with excess sulfur at 150-160°C gives a mixture of cyclic bis-disulfides (120) and (121) [189]. When this reaction at 190-200°C was studied, a compound was formed which was assigned structure (122) [188,189]. However, later the interaction of indole with sulfur was described by the following equations [191]:

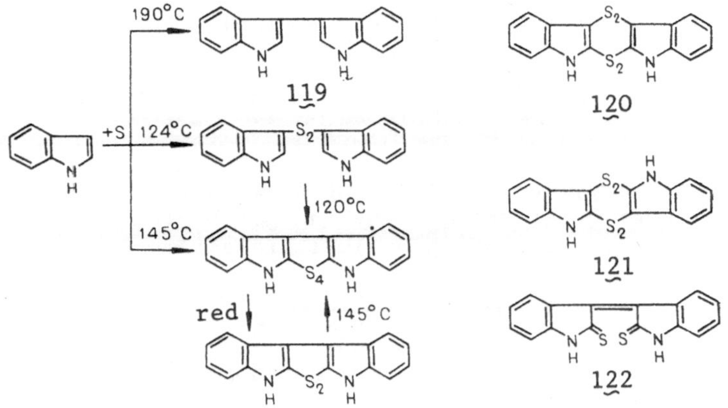

119   120   121   122

The reaction of 2- and 3-methylindole with sulfur gives crystalline, sulfur-containing compounds whose structure remains unknown [188].

Carbazole reacts with sulfur very slowly and under drastic conditions (240°C, 38 hours) with the formation of three sulfuration products, i.e., $C_{12}H_8NS$, $C_{24}H_{14}N_2S_5$, and $C_{24}H_{14}N_3S^-$ whose structure remains unknown. In addition to these three compounds, dicarbazolyl trisulfide (123) is also formed [192]. When the potassium salt of carbazole is heated with sulfur in benzene, the corresponding 9,9'-dicarbazolyl sulfide (124) is obtained [44]:

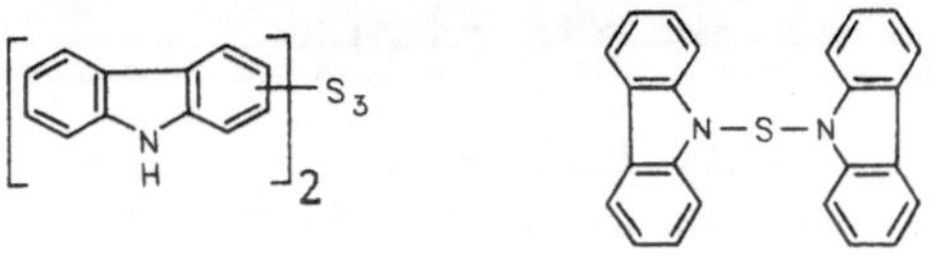

123   124

The reaction of pyridine with sulfur was described for the first time in 1914 [193]. However, only the formation of hydrogen sulfide was observed and no other individual reaction products were isolated. Later it was shown [194] that pyridine does not react with sulfur under these conditions and that the formation of hydrogen sulfide is due to the presence of picoline as an admixture.

4-Methylpyridine reacts with sulfur at 135-200°C with the formation of 1,2-bis-4-pyridylethane (125) and a small amount of 1,2,3-tris-4-pyridylpropane (126). When the reaction with sulfur is carried out at 290-330°C, tetrakis-4-pyridylthiophene (127) can be isolated as one of the reaction products. In the presence of sodium hydroxide at 135-200°C, 1,2-bis-4-pyridylethylene (128) is obtained as one of the reaction products, in addition to (125) and (126). An increase in the amount of sulfur used in the reaction does not show any effect on the yields of the products in reactions (7.128) to (7.130). However, increased reaction time and an increase in temperature lead to an increase in the yield of the thiophene (127) [195]:

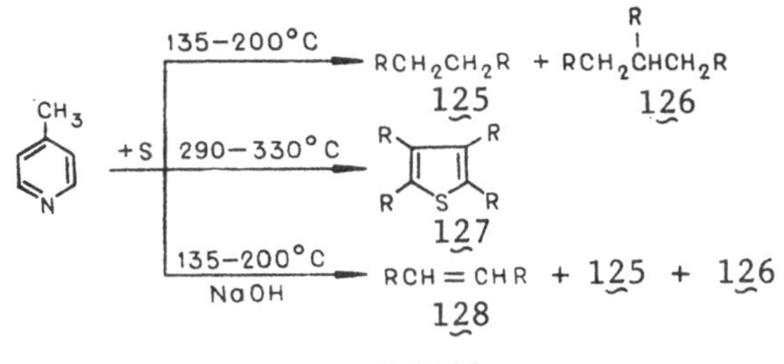

$$
\begin{array}{l}
\text{135--200°C} \quad RCH_2CH_2R \;+\; RCH_2\overset{\overset{R}{|}}{C}HCH_2R \qquad (7.128) \\
\qquad\qquad\qquad \underset{\widetilde{125}}{} \qquad\quad \underset{\widetilde{126}}{} \\
\text{290--330°C} \qquad \underset{\widetilde{127}}{\text{thiophene}} \qquad\qquad\qquad (7.129) \\
\underset{NaOH}{\text{135--200°C}} \quad RCH=CHR \;+\; \widetilde{125} \;+\; \widetilde{126} \qquad (7.130) \\
\qquad\qquad\qquad \underset{\widetilde{128}}{}
\end{array}
$$

$$R = 4\text{-}C_5H_4N$$

Prolonged heating of 2-methylpyridine with sulfur gives the complex disulfide (129) [194]. The addition of sodium hydroxide to the mixture accelerates the reaction [196]:

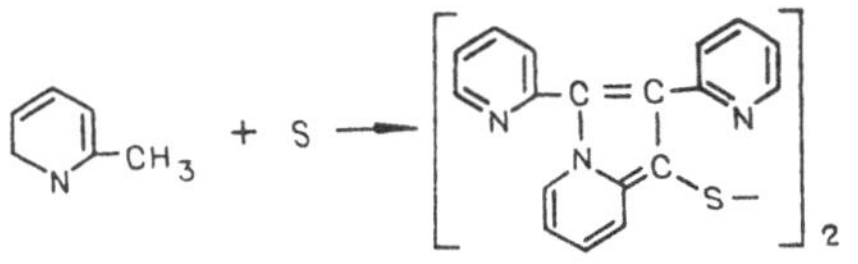

$$ \text{(7.131)} $$

$$\underset{\widetilde{129}}{}$$

The interaction of 2-,3-, and 4-methylpyridine and of 2- and 4-ethylpyridine with elemental sulfur in the presence of various amines has been studied in detail. The reaction is a modification of the Willgerodt-Kindler reaction (cf. Section 3.6) and is based on the transformation of the methyl group of the nitrogen hetero-cycle into a thioamide group. Thus, the reaction with morpholine, dimethylamine, or some primary amines occurs according to the following scheme [88,197-203]:

$$
RCH_3 + 3S + HNR'R'' \xrightarrow[-H_2S]{150-170°C} R-C\overset{\displaystyle S}{\underset{\displaystyle NR'R''}{<}} \qquad (7.132)
$$

$$R = 2\text{-}, \; 3\text{-}, \text{ or } 4\text{-}C_5H_4N; \; 4\text{-}CH_2C_5H_4N$$

2-Methylpyridine with sulfur and some aromatic and hetero-aromatic amines (aniline, 2- and 3-methylaniline, 4-methoxyaniline, pyridin-2-amine) forms the corresponding thioamides (130), whereas with other aromatic amines (2-methoxyaniline, 1-naphthylamine) it reacts under similar conditions to give the corresponding 2-substi-tuted 1,3-benzothiazoles (131) [204-209].

The yield of the benzothiazoles (131) increases with increasing temperature. The corresponding amidines (132) are by-products formed in the reaction of 2-methylpyridine with sulfur [209]:

$$RCH_3 + 3S + H_2NAr \xrightarrow[-H_2S]{160-180°C} RC\underset{\underset{\displaystyle S\downarrow 220°C}{NHAr}}{\overset{S}{\diagup}} \quad \underset{130}{}$$

$$RC\underset{NHAr}{\overset{NAr}{\diagup}} \quad + \quad RC\underset{\underset{131}{}}{\overset{N}{\diagdown}S} \tag{7.133}$$

$$\underset{132}{}$$

R = 2-, 4-C$_5$H$_4$N; 4-CH$_2$C$_5$H$_4$N, 4-CH$_2$CH$_2$C$_5$H$_4$N

   4-Methylpyridine, 4-ethylpyridine, and 4-propylpyridine react
similarly with sulfur and aromatic amines. At 160°C, the principal
reaction products are thiocarboxamides (130) [199,204,210,211],
whereas at 180-220°C the corresponding 1,3-benzothiazole (131) is
the main reaction product [199,201,209,211]. 3-Methylpyridine does
not give this reaction [199].

   Instead of using primary aromatic amines in the reaction of
sulfur with 2- and 4-methylpyridine, one can use the corresponding
nitro-compounds because they undergo reduction under the reaction
conditions employed in the sulfuration process [207,208,212].

   2-Methylpyridine reacts with sulfur and aromatic diamines at
140-160°C in several different ways. In the reaction with p-phenylenediamine [205] or 4-nitroaniline [208] it forms N,N'-(p-phenylene)-
bis-pyridine-2-thiocarboxamide (133) which is readily oxidized to
2,6-bis-2-pyridylbenzo[1,2-d; 4,5-d']thiazole (134) [205]:

$$RC\underset{NHC_6H_4NH}{\overset{S}{\diagup}}\underset{}{\overset{S}{\diagdown}}CR \rightleftharpoons \left[RC\underset{NC_6H_4N}{\overset{SH}{\diagup}}\underset{}{\overset{HS}{\diagdown}}CR \right] \xrightarrow[-H_2X]{[X]} RC\underset{N}{\overset{S}{\diagup}}\underset{S}{\overset{N}{\diagdown}}CR \tag{7.134}$$

$$\underset{133}{} \qquad\qquad \underset{134}{}$$

R = 2-C$_5$H$_4$N; X = O; S

   The reaction of 2-methylpyridine with sulfur in the presence
of benzidine or its analogs gives both the corresponding thiocarboxamides (135) and the bis-thiocarboxamides (136) and (137). The
latter are formed especially in the presence of excess 2-methyl-
pyridine and sulfur [205,213,214]:

$$H_2NC_6H_4-Y-C_6H_4NH\overset{S}{\overset{\|}{C}}R \xrightarrow{S,\ \overset{}{\underset{N}{\bigcirc}}-CH_3} R\overset{S}{\overset{\|}{C}}NHC_6H_4-Y-C_6H_4NH\overset{S}{\overset{\|}{C}}R$$

$$\underset{135}{} \qquad\qquad\qquad \underset{136}{} \tag{7.135}$$

$$\text{or} \quad R\overset{S}{\overset{\|}{C}}NHC_6H_4C_6H_4NH\overset{S}{\overset{\|}{C}}R \qquad \underset{137}{}$$

R = 2-C$_5$H$_4$N; Y = —, O, $-\overset{\|}{\underset{O}{C}}-$, $>$C(CH$_3$)$_2$, O=S=O

The reaction of sulfur, 2- or 4-methylpyridine, and hydrazine
gives the corresponding 2,5-dipyridyl-1,3,4-thiadiazoles [209] in
good yields, according to the scheme [191]:

$$2 \cdot RCH_3 + 6S + N_2H_4 \xrightarrow[-4H_2S]{140°C} \left[\begin{array}{c} SH \\ RC=N-N=CR \\ HS \end{array} \right] \xrightarrow{-H_2S} \underset{138}{\overset{N-N}{\underset{R}{\bigvee}}} R \qquad (7.136)$$

$$R = 2-, 4-C_5H_4N$$

The principal reaction products in the reaction of 2- or 4-
methylpyridine, 4-ethyl- or 4-propylpyridine with thiosemicarbazide
and sulfur are the corresponding thiosemicarbazones (139) [88,215]:

$$RCH_2R' + 2S + H_2NNHCNH_2 \xrightarrow{-2H_2S} RR'C = NNHCNH_2 \qquad (7.137)$$
$$\underset{S}{\|} \qquad\qquad \underset{139}{\overset{}{}} \underset{S}{\|}$$

$$R = 2-, 4-C_5H_4N; \ R' = H, CH_3, C_2H_5$$

Reaction (7.137) occurs with other monoacylhydrazines as well [88].
A very interesting method for the preparation of N-methyl-2-pyridine-
thiocarboxamide (140) is based on the reaction between 2-methyl-
pyridine, sulfur, and N-methylformamide [123]:

$$\text{[pyridine-CH}_3] + S + HCONHCH_3 \longrightarrow \text{[pyridine-}C\overset{S}{\underset{NHCH_3}{\diagdown}}] \qquad (7.138)$$
$$\underset{140}{} \ (50\%)$$

When a mixture of sulfur and 2-(β-anilinoethyl)pyridine (141)
is refluxed in pyridine, 2-pyridylthioacetanilides (142) are formed
in fair yield [215a]:

$$\text{[pyridine-CH}_2CH_2NH\text{-[}C_6H_4\text{]-}R] + S \longrightarrow \text{[pyridine-CH}_2C\overset{S}{\underset{NH\text{-[}C_6H_4\text{]-}R}{\diagdown}}] \qquad (7.139)$$
$$\underset{141}{} \qquad R = H, CH_3O, Cl, NO_2 \qquad \underset{142}{}$$

Dimethylpyridines (lutidines) react with sulfur and with aliphatic and aromatic amines in an unexpected way [198,204,214,216].
When 2,3-, 2,4-, or 2,6-dimethylpyridine or 5-ethyl-2-methylpyridine
react with sulfur and primary tert-alkyl amines (2:6:1), only the
2-methyl group reacts and the corresponding thiocarboxamides (143)
are obtained [198,204]. There have been attempts to explain the
impossibility of a simultaneous reaction on both methyl groups from
the point of view of the resonance structures of 2- and 4-methyl-
pyridines and their respective reactivities [198]:

$$R = 3\text{-}CH_3,\ 4\text{-}CH_3,\ 6\text{-}CH_3,\ 5\text{-}C_2H_5;$$
$$R' = \underline{tert}\text{-}C_4H_9,\ \underline{tert}\text{-}octyl,\ C_6H_5,\ 4\text{-}CH_3C_6H_4,\ 2\text{-}C_5H_4N$$

2,6-Dimethylpyridine reacts analogously with sulfur and aromatic or heterocyclic amines [204] or p-nitrotoluene (the latter is reduced by hydrogen sulfide to the corresponding amine) [216]. When the ratio 2,6-dimethylpyridine:sulfur:aromatic diamine is 4:10:1, the corresponding bis-thiocarboxamides (144) are the principal reaction products, whereas the corresponding polythiocarboxamides (145) are obtained when the ratio is 1:10:2 [214]:

$$R = H,\ CH_3,\ CH_3O$$

The reaction of 2,4-dimethylpyridine with sulfur and amines takes place with the participation of both methyl groups. The isomer with the thioamide group in the 2-position is usually formed in a higher yield than the 4-isomer. A large excess of 2,4-dimethylpyridine and sulfur with respect to the amine (or the nitro-compound) tends to increase the yield of substituted thiocarboxamides [198, 216]. Only the thioanilide of 2-methyl-4-pyridinecarboxylic acid is isolated in the reaction of 2,4-dimethylpyridine with sulfur and aniline, whereas the reaction with 4-nitrotoluene gives the corresponding 4-methylbenzenethiol. The reaction with pyridine-2-amine proceeds similarly [204]. All three possible thioamides are obtained in the reaction with 4-methylaniline [216], i.e., N-aryl-pyridinethiocarboxamides (146), (147), and (148):

$$146\ (40\%) \qquad 147\ (20\%) \qquad 148\ (10\%)$$

$$R = \underline{p}\text{-}CH_3C_6H_4$$

It is assumed that the interaction of 2,4-dimethylpyridine with sulfur and other amines takes place with the participation of both methyl groups as well, although it is difficult to isolate all the reaction products.

On heating pyridine, 2- or 4-methylpyridine N-oxides with sulfur at 150°C, oxygen is readily abstracted to give pyridine and

methylpyridine, respectively [217]. A mixture of 2-methylpyridine
N-oxide, sulfur, and aniline forms, under the same conditions, N-
phenylpyridinethiocarboxamide (40%), 2-(2-pyridyl)benzo-1,3-thiazole
(30%), 2-methylpyridine, and a salt $C_6H_5NH_2,5H_2SO_4$ [217].

The reaction of sulfur with quinoline, in the presence of
copper or zinc powder or sulfuric acid as the catalysts at 170-220°C,
leads to a mixture of various sulfuration products, the principal
product being 1,4-dithiano[2,3-b;6,5-b']diquinoline (149) [218-221]:

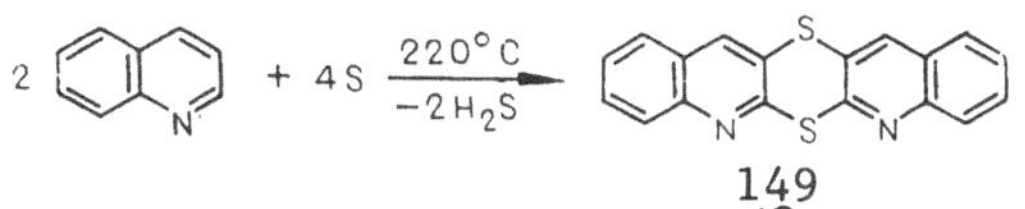

$$2 \quad \text{[structure]} \quad + 4S \xrightarrow[-2H_2S]{220°C} \text{[structure]} \tag{7.142}$$

149

2-Methylquinoline (quinaldine) reacts with sulfur and primary
aromatic amines to give the corresponding thiocarboxamides (150)
and 2-(2-quinolyl)-1,3-benzothiazoles (151) [222]:

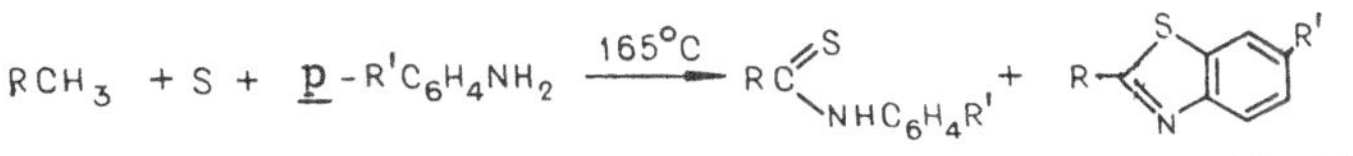

$$RCH_3 + S + \underline{p}\text{-}R'C_6H_4NH_2 \xrightarrow{165°C} RC \overset{S}{\underset{NHC_6H_4R'}{\diagdown}} + \text{[structure]} \tag{7.143}$$

150 (50-76%) 151 (3-14%)

$$R = \text{[structure]} \quad ; \quad R' = H, \text{ alkyl, alkoxyl}$$

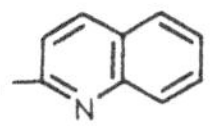

The reaction of sulfur with quinaldine in the presence of sodium
hydroxide gives 1,2-bis-2-quinolylethane (152) and a small amount
of a compound containing nitrogen and sulfur of unknown structure
[223,224]:

$$2RCH_3 + S \xrightarrow{NaOH} RCH_2CH_2R \tag{7.144}$$

152

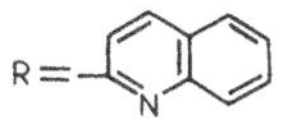

$$R = \text{[structure]}$$

The reaction of 2,2,4-trimethyl-1,2-dihydroquinoline (153) with
sulfur in dimethylformamide leads to the formation of the 1,2-
dithiol-3-thione (154) [225]:

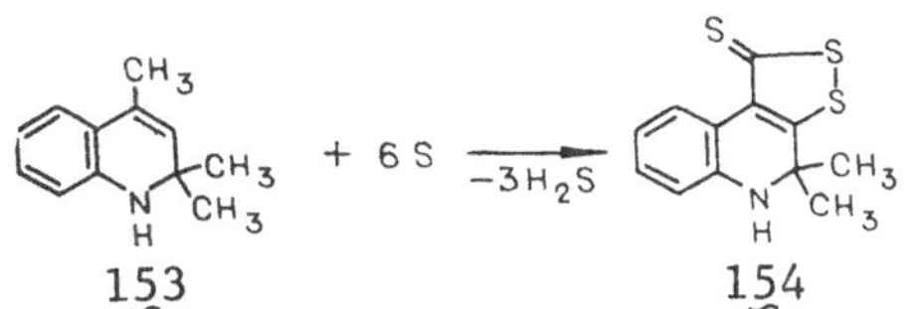

$$\text{[structure]} \quad + 6S \xrightarrow{-3H_2S} \text{[structure]} \tag{7.145}$$

153   154

When acridine is melted with sulfur at 190-200°C, 9-thioacridone
is obtained [218]:

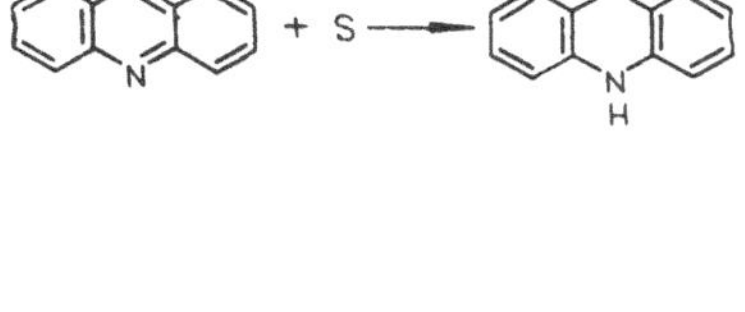

$$\text{[structure]} \quad + S \longrightarrow \text{[structure]} \tag{7.146}$$

When sulfur is melted with nitrogen-containing heterocyclic
ethylene derivatives, cleavage of the exocyclic double bond occurs
and formation of the corresponding thiocarbonyl compounds (155)
[226] and (156) [227] takes place:

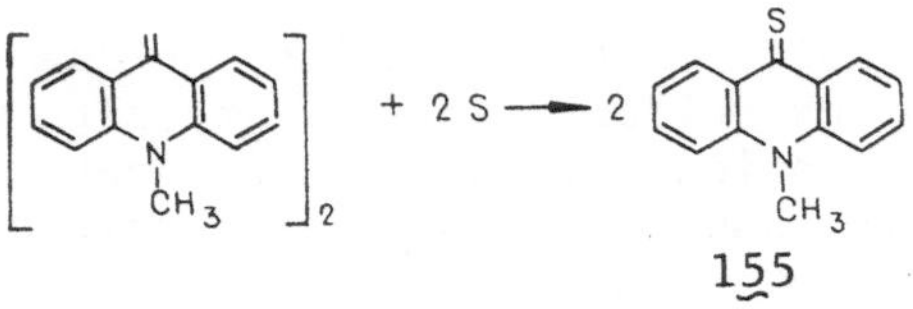

(7.147)

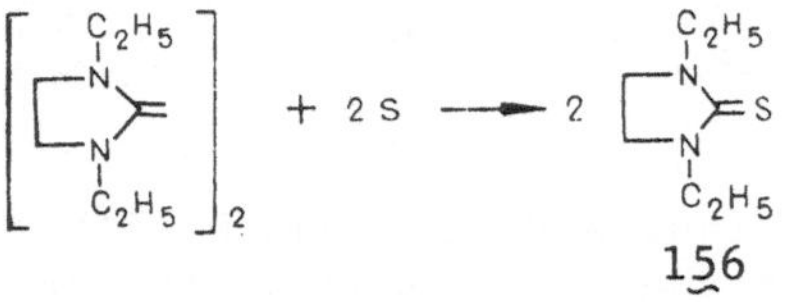

155

(7.148)

156

From the theoretical point of view, the reaction of a quaternary
salt of acridine with aniline in the presence of sulfur is of
interest. This reaction leads to 9-4'-aminophenylacridine (158)
[228]. The reaction is a two-step process. Because of the increased
electron density in the 4-position of aniline and decreased electron
density in the 9-position of the acridine derivatives, an intermolecular intermediate complex (157) between aniline and the acri-
dinium salt is formed in the first step. This complex then undergoes dehydrogenation in the presence of sulfur in the second step:

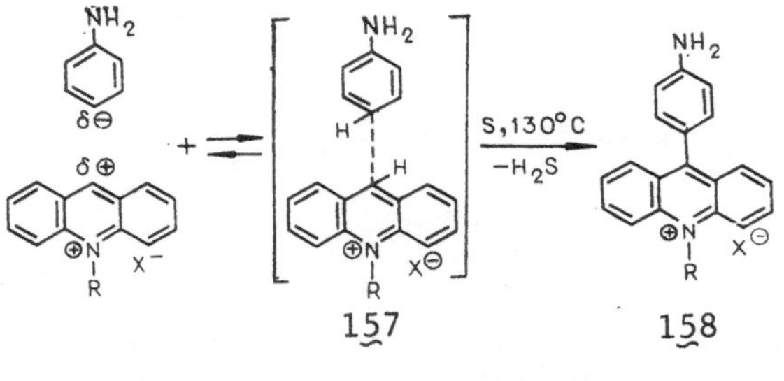

(7.149)

157   158

2-Methyl-1,3-thiazole reacts with sulfur and amines (aniline,
4-methoxyaniline, 4-ethoxyaniline, pyridin-2-amine) and forms the
corresponding 1,3-thiazole-2-thiocarboxamide in low yield [229],
since the process is accompanied by the destruction of the thiazole
ring. Below 130°C, polymers are obtained which have a very low
solubility. Some 1,3-thiazole derivatives undergo isotopic exchange
with ^{35}S [230]:

(7.150)

The reaction of 2-methyl-1,3-benzothiazole with sulfur and aromatic amines gives the corresponding 1,3-benzothiazole-2-thiocarboxanilide (159) and 2,2'-bi-(1,3-benzothiazole) (160) [88,202, 231]:

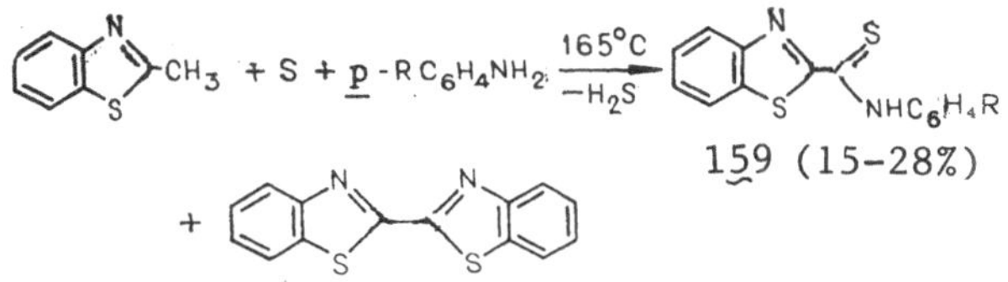

159 (15-28%)

160 (14-28%)

(7.151)

R = alkyl, alkoxyl

2-Ethyl- and 2-methyl-1,3-benzothiazole react with sulfur and thiosemicarbazide under more drastic conditions than in reaction (7.137) and give the corresponding hydrazone (161) and azine (162), respectively [88]:

$$RCH{=}NNHC({=}S)R \qquad RC(CH_3){=}N{-}N{=}C(CH_3)R$$

161 162

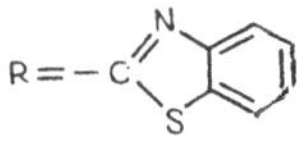

The reaction of sulfur with 3-methyl-2,3-dihydro-1,3-benzothiazole [232] or 1,3-dimethyl-2,3-dihydrobenzimidazole [233,234] gives the corresponding 1,3-benzothiazole and benziminazole-2-thione (163) and (164) under mild conditions:

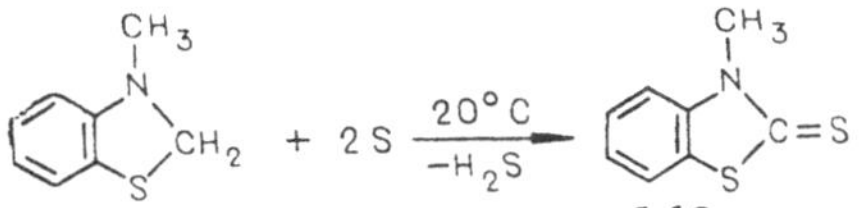

163

(7.152)

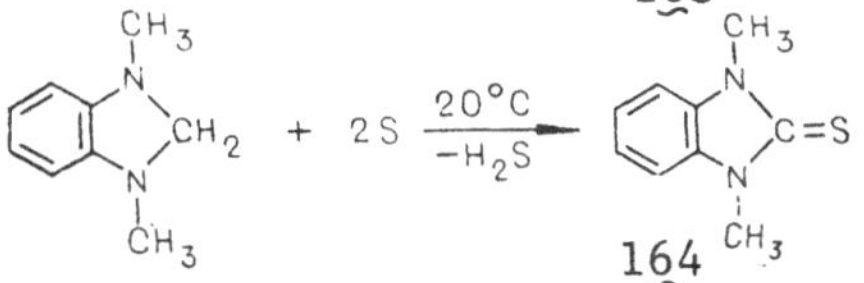

164

(7.153)

1-Methylbenzimidazole when melted with sulfur gives a high yield of 1-methylbenzimidazole-2-thiol [235]:

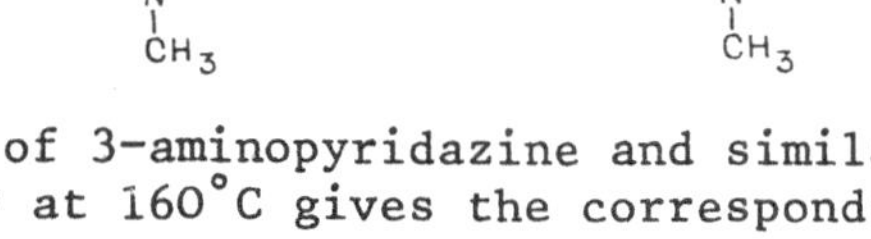

(7.154)

The reaction of 3-aminopyridazine and similar nitrogen heterocycles with sulfur at 160°C gives the corresponding heteroarenethiols [203], e.g.,

The reaction of methylpyrazine with sulfur and dimethylformamide in the presence of iodine gives N,N-dimethylpyrazinethiocarboxamide (165) [236]:

$$\text{(7.155)}$$

Substituted 6-methyl-2-oxo- (or 2-thioxo)-1,2,3,4-tetrahydropyrimidines (166) react with sulfur when refluxed in dimethylformamide or dimethyl sulfoxide to give the corresponding 1,2-dithiol-3-thiones (167) [237,238]:

$$\text{(7.156)}$$

X=O, S

N,N-Dimethylaniline when melted with 1,4-quinoxaline hydrochloride in the presence of sulfur in an inert atmosphere gives products arising from the replacement of a hydrogen atom in quinoxaline, i.e., 3-(4-N,N-dimethylaminophenyl)-1,2-dihydroquinoxaline-2-thione (168) and 3-(4-N,N-dimethylaminophenyl)quinoxaline (169). The following conversion scheme has been postulated:

$$\text{(7.157)}$$

When N-substituted 1,3-benzo[d]thiazine-8-thiones (170) are
melted with sulfur, the reaction occurs at the methylene group and
the corresponding dithiones (171) are formed in high yields [240]:

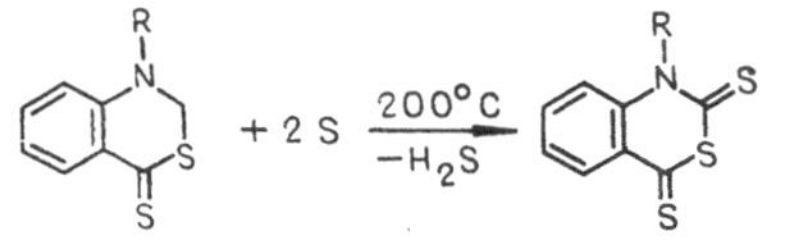

$$\qquad\qquad\qquad\qquad (7.158)$$

170   171 (90%)

Heating methylene-bis-4-morpholine with sulfur or morpholino-
methanethiol with sulfur and morpholine leads to the formation of
morpholinium morpholinodithiocarboxylate (172) [241]:

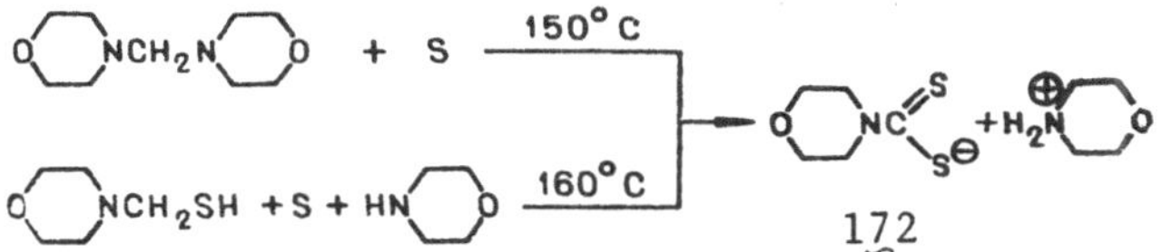

$$\qquad\qquad\qquad\qquad (7.159)$$

172

The reaction of 1,3,5-trimethylhexahydrotriazine (173) [241]
with sulfur at 150°C gives N,N'-dimethylthiourea [117]:

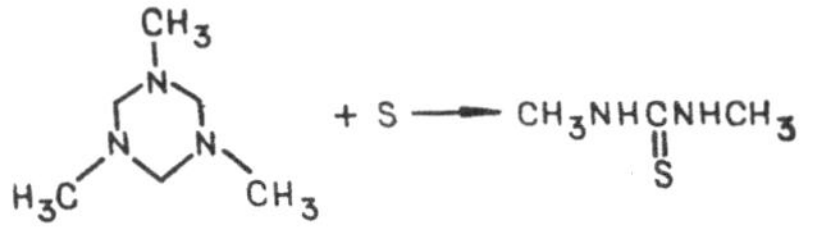

$$+ S \longrightarrow CH_3NHCNHCH_3 \qquad\qquad (7.160)$$

173

Heating hexamethylenetetramine (urotropine) with sulfur at
165°C gives hydrogen sulfide and leads to the formation of a mixture
of sulfuration products, some of which are soluble in water and give
an orange-red precipitate, $Pb_2C_2N_2S_3$, with lead acetate [242].

Noteworthy is a method of sulfuration of some five-membered
nitrogen heterocyclic salts by sulfur in the presence of sodium
hydride in dimethylformamide. Thus, pyrazolium-substituted tosylates
(174) convert for example readily under the same conditions to
corresponding 3-pyrazoline-5-thiones (175) [243]. 1,2,3-Triazolium-
[244], 1,2,4-triazolium- [245,246], imidazolium- [247] and thiazo-
lium-substituted [248] tosylates are sulfurated in a similar way:

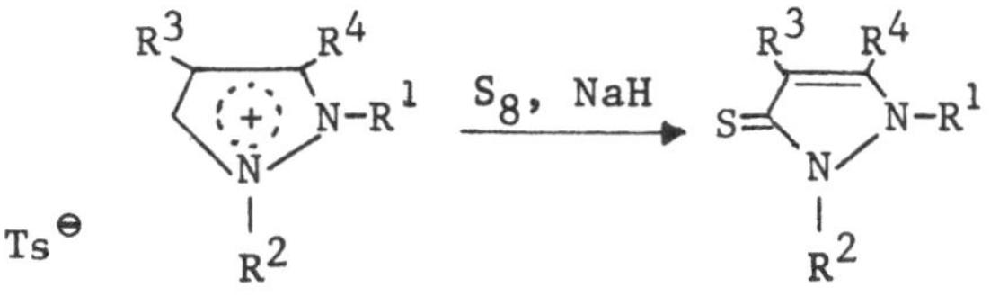

$$\xrightarrow{S_8, \text{ NaH}} \qquad\qquad (7.161)$$

174   175

A process for the preparation of ammonium dithiocarbamates of the type $RR'NC(S)S^{\ominus} \cdot H_2N^{\oplus}RR'$, based on the reaction of heterocyclic diamines with sulfur at 80-100°C has been patented [249].

The reactions of alkaloids (atropine, morphine, strychnine, scopolamine) with elemental sulfur in boiling xylene afford a mixture of products containing from one to three sulfur atoms [250].

7.12. SULFUR DYES BASED ON THE REACTION OF ORGANIC COMPOUNDS WITH SULFUR

Sulfur dyes are dyes capable of dyeing fibers in sodium sulfite solution directly, without preliminary mordanting. They are an intermediate group of dyes between substantive (direct) dyes and vat dyes.

Sulfur dyes are formed on heating certain groups of nitrogen-containing organic compounds, especially amino- and nitro-derivatives in the aromatic series, with sulfur, sodium sulfite, or sodium polysulfides [86,251-262].

Originally, yellow and brown sulfur dyes (Laval's paste) were obtained in 1873 by treating sawdust, bran, horns, and other organic waste with sulfur and sodium polysulfide at 300°C. The higher the temperature of the melt, the darker the dye.

A systematic study of sulfur dyes was initiated in 1893, when Vidal started to use p-phenylenediamine, p-aminophenol, nitrochloro-benzenes, and other nitrogen-containing aromatic compounds and melted them with sulfur and alkali metal sulfides. Later on, many interesting sulfur dyes were discovered which had some practical importance and some of which are still important today (sulfur black). Dyes which are today known under the general name of "sulfur" dyes are also marketed under other names, such as catigenic, cryogenic, thiogenic, immedial, etc. dyes.

The structure of sulfur dyes has not yet been sufficiently elucidated. However, at least some points have been clarified during the past thirty years.

Sulfur dyes are resistant with respect to rubbing, washing (laundering), and the action of light and can be of different colors, with the exception of pink, ruby-red, red, and violet. As far as their brilliance and hue purity are concerned, they usually do not compare well with other types of dyes. Sulfur dyes are widely used for dyeing fabrics and yarns obtained from plant fibers. One of the important reasons for their use is that they can be manufactured very inexpensively. However, fibers dyed with sulfur dyes often lose their mechanical strength, because the colloidal sulfur remaining in the fibers is gradually oxidized by atmospheric oxygen to give sulfurous acid which attacks cellulose.

At the present time, sulfur dyes are losing their importance and their production is decreasing in many countries because of the development of highly competitive dyes of other types.

7.12.1. Synthesis of Sulfur Dyes

Sulfur dyes are obtained by the sulfuration of nitrogen-containing organic compounds with sulfur or with mono- or polysulfides of alkali metals at high temperatures. Two techniques are used, i.e., baking and boiling [86,251-262].

Baking is carried out at 200-300°C in the absence of solvents. This technique is used to obtain yellow, orange, and brown dyes. The tone of the dyes depends on the temperature and on the duration of baking. More drastic conditions lead to darker dyes.

Boiling is performed in an aqueous or ethanolic medium at 100-150°C, with sodium polysulfides as the principal sulfuration agents. This method is used to obtain green, blue, and black dyes.

Small admixtures of organic or inorganic substances are often used in both of the above procedures. Thus, for example, addition of phenol, glycerol, or naphthalene makes it possible in many cases to decrease the temperature of the process and to obtain dyes of higher quality. The introduction of cupric sulfate gives a greenish tint to the blue sulfur dyes and a reddish tint to the brown sulfur dyes.

The isolation technique used to obtain sulfur dyes from the reaction mixture after baking or boiling is identical. The melt after baking or the precipitate after boiling is dissolved in a solution of sodium sulfite, the solution filtered in order to remove all the mechanical and insoluble admixtures, and the dye precipitated by salting out, acidification, or oxidation with atmospheric oxygen. The isolated dye is then worked up in an appropriate manner.

7.12.2. Properties of Sulfur Dyes

As a rule, sulfur dyes are insoluble in water, acids, and in most bases. They are amorphous powders, soluble in sodium sulfide solutions with the formation of lightly colored leuco-compounds which can be converted into the corresponding insoluble form by oxidation with atmospheric oxygen. The leuco-compounds of sulfur dyes can be used as direct dyes for cellulose, are well adsorbed by cellulose fibers, and washing or oxidation converts them into the

corresponding insoluble dyes on the fibers. Fibers of animal origin
(wool, silk) are normally not dyed with sulfur dyes because the
strongly alkaline solutions of their leuco-compounds damage these
fibers. As a whole, the general formula of a sulfur dye can be
expressed as R-S-S-R. Thus, the main reaction which takes place
during dyeing can be shown by the following equation:

$$2RSH \underset{2H^+}{\overset{[O]}{\rightleftharpoons}} RSSR$$

leuco-compound   dye

The solubility of the leuco-compounds in an alkaline solution
is due to the reaction

$$RSH + NaOH \rightarrow RSNa + H_2O$$

All the reactions which take place during dyeing with a sulfur
dye can be expressed by the following scheme:

$$RSSR \xrightarrow{+2H^+} 2\ RSH \xrightarrow[-H_2O]{+2NaOH} 2\ RSNa \xrightarrow[-2NaOH]{+2\ H_2O} 2\ RSH \xrightarrow[-H_2O]{+\ O} RSSR$$

7.12.3. Structure of Sulfur Dyes

The complexity of the sulfuration process leading to sulfur
dyes and their amorphous and polymeric nature has so far prevented
the elucidation of their structure. It has been possible to establish, however, that the black, blue, and green sulfur dyes are
basically 1,4-thiazine (176) derivatives. The reddish-brown dyes
contain 1,4-diazine rings (178) [251,253,262]:

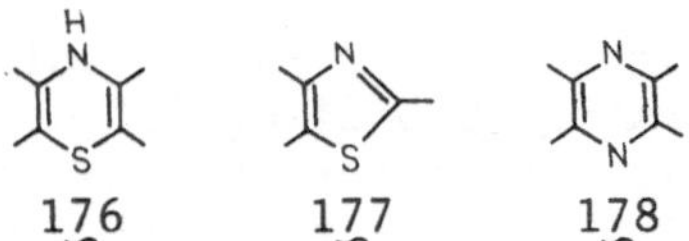

176   177   178

The molecules of sulfur dyes possess thiol, sulfide, di- and
polysulfide, sulfoxide, thianthrene, and other sulfur-containing
groups. The formation of these groups has been established in
numerous cases in which various individual aromatic amines and
diamines were used to synthesize the respective sulfur dyes.

Thus, 2,4-diaminotoluene melted with sulfur gives initially a
labile polysulfide (179) which is subsequently converted to a more
stable disulfide. The latter can be converted into the monosulfide
on prolonged heating or at high temperatures [263,264]. m-Phenylene-
diamine, 1,2,4-triaminobenzene, triaminobiphenyl [263], and 2,4-
diaminoacetanilide [265] react in a similar way:

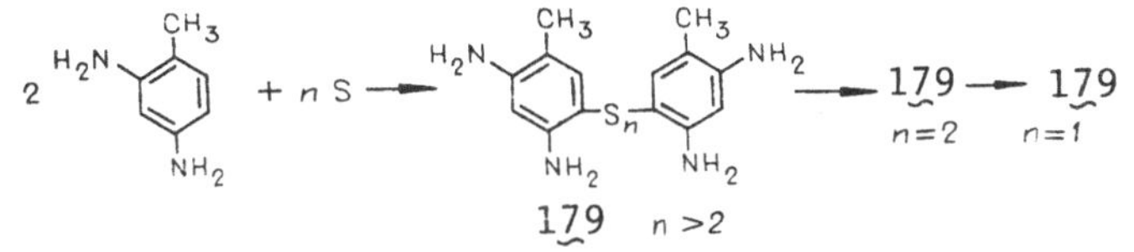

$$(7.162)$$

The mechanism of the reaction of sulfur and p-phenylenediamine in tetralin is expressed by the following scheme [266]:

$$H_2N-C_6H_4-NH_2 + S_8 \longrightarrow H_2N-C_6H_4-N^{\oplus}H_2-S-S_6-S^{\ominus} \rightarrow$$

$$H_2N-C_6H_4-N^{\oplus}H_2-S_x^{\bullet} + {}^{\bullet}S_y^{\ominus}; \qquad (7.163)$$

$$2\ H_2N-C_6H_4-N^{\oplus}H_2-S_x^{\bullet} \xrightarrow[-2H^{\oplus}]{} H_2N-C_6H_4-NH-S_x-S_x-NH-C_6H_4-NH_2 \xrightarrow{+S_8} \quad \text{etc.}$$

The formation of thiazine rings by the reaction of sulfur with p-phenylenediamine in 1-methy-2-pyrrolidone can occur according to the scheme [266, cf. 77,78,98,99,267,268]:

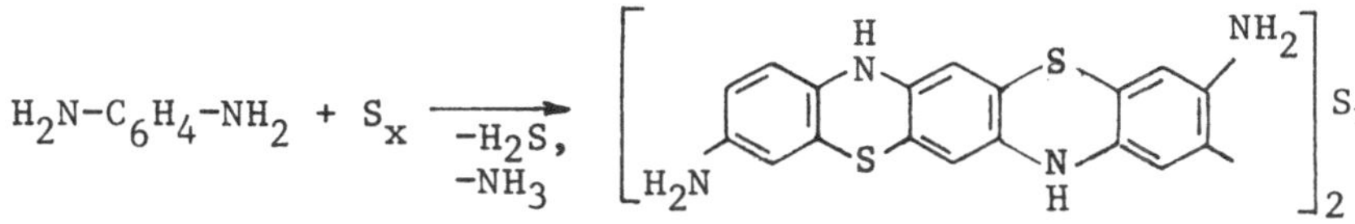

$$(7.164)$$

Processes leading to the formation of the thiazine ring in the preparation of black sulfur dyes from 4-aminophenol have been described [264]:

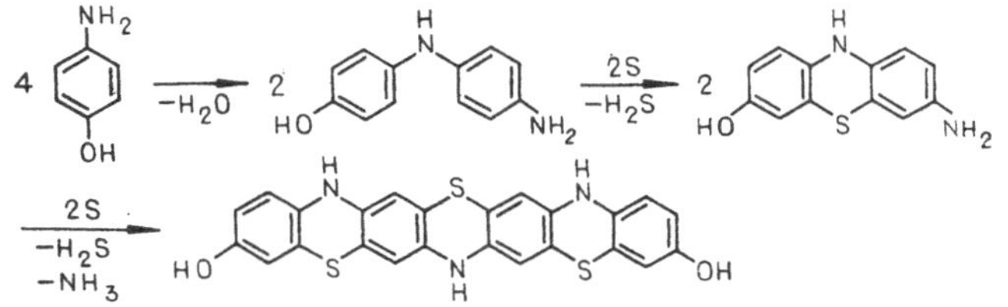

$$(7.165)$$

The formation of yellow, orange, and brown sulfur dyes has been studied much less from the mechanistic point of view; however, it seems that these dyes contain a thiazole ring (177). The thiazole ring is formed in the sulfuration of aromatic N-acylamines (cf. Section 7.4) or alkyl-substituted anilines (cf. Section 7.2.1).

The so-called "Primuline bases" obtained according to reaction (7.60) is not a single compound. It is likely that the main component in this substance is compound (62) (see p. 263).

The sulfuration of the "Primuline base" (62) [scheme (7.60)] in position 5 of the benzo-1,3-thiazole ring leads to a substantive yellow dye "Primuline" [84]. The base (62) and its sulfuration products have been used as the diazo components in the synthesis of some yellow substantive dyes (Erika, Geranin, Dianil Yellow) which are only rarely used today.

There are no data which make it possible to establish a relationship between the structure and the color of sulfur dyes and no rational chemical classification of sulfur dyes is available. As a rule, sulfur dyes are classified according to their color.

7.12.4. Individual Sulfur Dyes

Yellow sulfur dyes are normally obtained by baking aromatic amines with sulfur at 200-220°C. The positions of the substituents and their nature have to be such that the formation of a thiazole ring is possible [263,264,269,279]. The most frequently used amines are 2,4-diaminotoluene and its N-acyl derivatives, and nitro-derivatives of 2- and 4-methylaniline which may be reduced to the corresponding amines. As a rule, the yellow sulfur dyes exhibit poor stability with respect to light and washing.

Orange sulfur dyes are obtained by the prolonged baking above 220°C of the same reagents as the yellow dyes. Thus, Sulfur Orange is obtained by baking toluene-2,6-diamine with sulfur. It is assumed that the orange sulfur dyes also contain thiazole rings. The darker color of orange dyes as compared with that of the yellow dyes is presumably caused by a more complex structure of the orange dyes. They are more stable with respect to washing and to the action of light than the yellow dyes.

Brown sulfur dyes are obtained from nitro and other derivatives of benzene, biphenyl, naphthalene, and anthracene. As far as the reaction conditions used in their preparation are concerned, they are in between the conditions used for the yellow and orange dyes on one side, and the conditions used for the blue and black dyes on the other side. The brown dyes can be obtained both by baking or by boiling.

One of the dyes normally used is Sulfur Brown Ge obtained by the reduction of 2,4-dinitrotoluene with Na_2S_8 followed by baking the reaction mixture at 230°C. The dye Sulfur Brown is obtained from a mixture of 1,5- and 1,8-dinitronaphthalene under the same conditions used for the preparation of Sulfur Brown Ge.

$$(7.166)$$

The khaki dyes used for dyeing clothing also belong to this group. The demands in this case are very high: the dyes have to be stable with respect to light, washing, dust, and the action of sweat. One of the khaki dyes can be obtained by melting crude anthracene with sodium hydroxide and sulfur. However, in most cases the khaki dyes are mixtures obtained from the brown and black sulfur dyes.

Reddish and Violet Sulfur Dyes. So far it has been impossible to obtain sulfur dyes which are red or violet. The dyes which are manufactured as red or violet dyes possess just the appropriate tone but they are not actually red or violet. These dyes are closely related to the brown dyes and can be obtained from the same starting materials under somewhat different reaction conditions. The reddish tone can also be obtained by adding various admixtures to the melt, e.g., cupric sulfate. Good dyes may be obtained from starting materials which can form a 1,4-diazine ring during the reaction or which already possess this ring. The addition of cupric salts enhances the reddish color of the dyes in both cases. As an example of the first type of reagent, one can mention 4-amino-N-(2,4-dinitrophenyl)phenol which can also be used for the preparation of a yellow-brown sulfur dye:

$$(7.167)$$

As an example of the second type of reagent, let us mention the diazine which may be obtained by the oxidation of an equimolar mixture of 4-aminophenol and toluene-2,4-diamine. Prolonged heating (118°C) with Na_2S_4 in the presence of cupric sulfate gives the dye called "Sulfur Brown K." On further prolonged heating of the above azine with sodium polysulfide, which contains a larger amount of sulfur, Sulfur Bordeaux is obtained:

$$\longrightarrow \text{Sulfur Brown K} \qquad (7.168)$$

Sulfur dyes possessing a violet tone are formed during the sulfuration of 1,4-diazine derivatives at higher temperatures or by sulfuration of especially chosen intermediates.

Blue sulfur dyes, together with the black sulfur dyes, are most important. As a rule, they are obtained by boiling indoanilines with sodium polysulfides. These dyes contain a 1,4-thiazine ring.

The first blue sulfur dye was Sulfur Brilliant Blue K (Immedial Bright Blue) obtained by boiling the leuco-compound of indoaniline (obtained from N,N-dimethyl-p-phenylenediamine and phenol) with Na_2S_3 at 110-112°C. This dye, giving cotton a bright blue color, has been assigned the following structure:

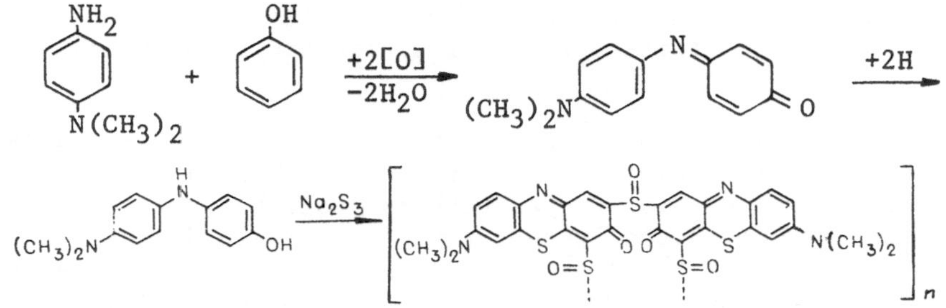

$$\text{Sulfur Brilliant Blue K} \qquad (7.169)$$

An interesting dye is Sulfur Carbazole Blue which is obtained by boiling indoaniline, based on carbazole and nitrosophenol, with sodium polysulfide in ethanol. In contrast to various other sulfur dyes, the above dye is not reduced by sodium sulfide and the leuco-compound is obtained by treating it with sodium hydrosulfite in an alkaline medium, similarly to the case of vat dyes. Sulfur Carbazole Blue is similar to indigo with respect to its stability toward light and chlorine, and also has a similar color. The dye has been assigned the following structure:

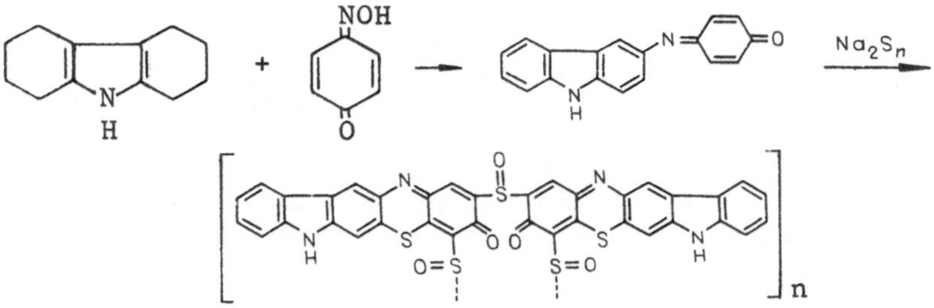

$$(7.170)$$

Sulfur Carbazole Blue

The formulae given above do not possess a disulfide group (-S-S-) which, upon the reduction of the dye, could give a mercapto-group (-SH) enabling the leuco-compounds to be soluble in alkaline solution. It is likely that the absence of the disulfide group is compensated by the presence of a sulfoxide group (-S-) which can also be reduced to a thiol.   ‖
                                                                      O

Sulfur Blue K, which is somewhat reddish, is also of interest. It may be obtained by boiling the leuco-compound of indoaniline (from nitrosophenol and 2-methylaniline) with Na_2S_5 at 110°C. If the same starting materials are boiled at 115-116°C for 15 hours, the dye Sulfur Dark Blue is obtained.

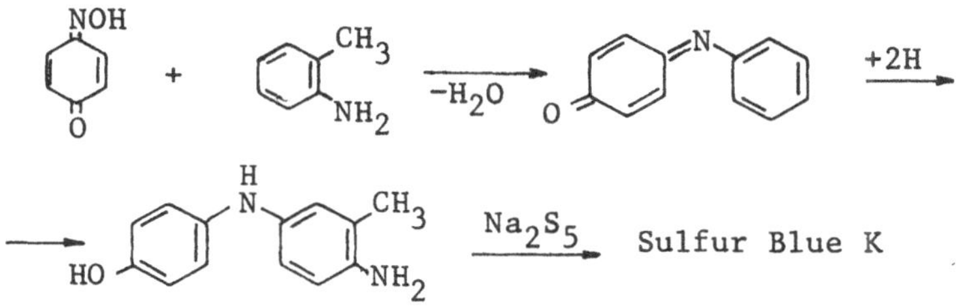

Sulfur Blue K (7.171)

Green sulfur dyes have been studied to a very limited extent, although some of them are important from the practical point of view. They are obtained by boiling indoanilines with Na_2S_n in the presence of cupric salts. These dyes contain 1,4-thiazine rings and it is likely that copper is complexed in the dyes in a 1:1 ratio.

An important dye of this type is Sulfur Bright Green B, obtained by boiling indoaniline (from nitrosophenol and 1-naphthylamine) with Na_2S_5 in the presence of cupric sulfate.

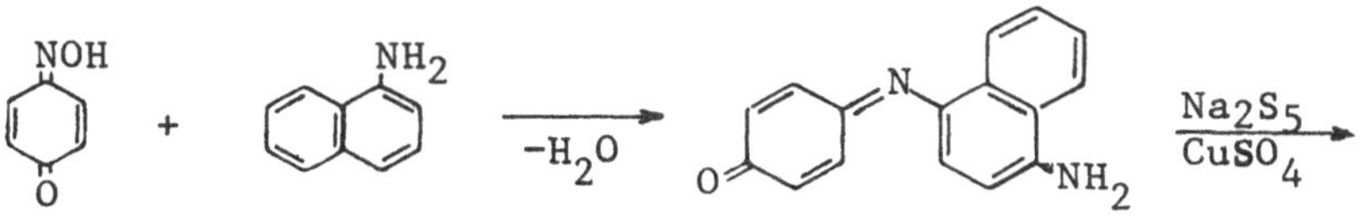

→ Sulfur Bright Green B (7.172)

Another purely green and bright dye is Sulfur Bright Green Ge which is obtained by boiling indoaniline (from nitrosophenol and 8-anilino-1-naphthalenesulfonic acid) with Na_2S_n at 102-103°C in the presence of cupric sulfate:

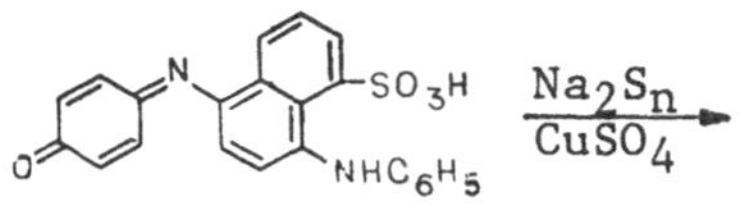

Sulfur Bright Green Ge (7.173)

Green dyes obtained by sulfuration of copper phthalocyanine possess unsurpassed stability with respect to light and chlorine among all the sulfur dyes. One of these dyes is Thinol Ultra Green BS. The dye Sulfur Bright Green Phthalocyanine manufactured in the U.S.S.R. possesses similar characteristics.

Black sulfur dyes are one of the most important groups of sulfur dyes. The most significant of these dyes is Sulfur Black obtained by boiling 2,4-dinitrochlorobenzene with Na_2S_4 at 110°C. This dye is widely used because it possesses valuable color characteristics and the starting materials are cheap and readily available. The leuco-compound of this dye is formed according to the following scheme:

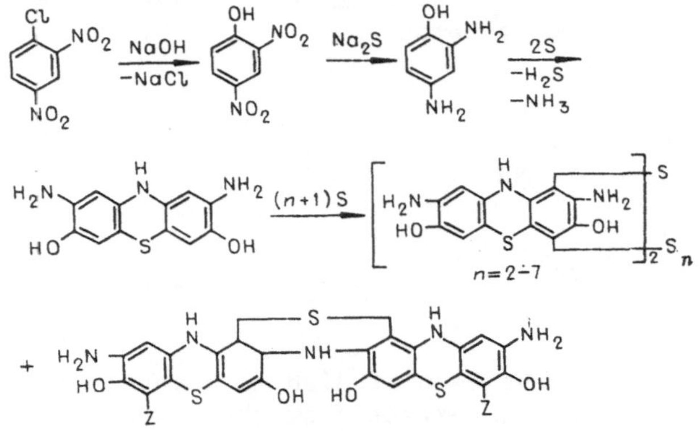

$$(7.174)$$

Z = sulfur-containing group

It is likely that the above formula is not the correct structural formula for Sulfur Black and it seems that this dye may be actually a mixture of several different compounds.

A few sulfur dyes have recently enriched the assortment of sulfur dyes. Thus, bordeaux, violet, and blue dyes have been synthesized from 1,2,4-trichloro-7-nitro-3H-phenothiazin-3-one (180):

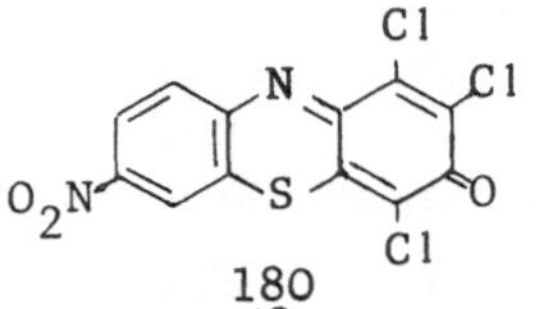

180

Thiosols. At the present time, one of the groups of dyes available on the market are water-soluble sulfur dyes which are called thiosols [260]. They are thioglycol or bisulfite derivatives of sulfur dyes (Dark Blue Thiosol K, Bright Green Thiosol Ge, Thiosols BS). The thioglycol derivatives are used to dye viscose in bulk, whereas the bisulfite derivatives are used to dye the viscose staple and cotton-paper fabrics. Some thiosols are good printing dyes for cotton-paper fabrics. Synthetic fibers dyed in bulk with thiosols are resistant to all types of extraneous effects.

7.12.5. Some Sulfur-containing Dyes

There are many dyes which are not sulfur dyes, but which may be obtained by various reactions of nitrogen-containing organic compounds with elemental sulfur.

Lauth's Violet is obtained by melting sulfur with p-phenylene-diamine [98,270,271] or 4-nitroaniline [272] (150-180°C) followed by oxidation of the obtained leuco-derivative with the formation of dyeing ions (181). This dye may also be obtained by treating p-phenylenediamine with hydrogen sulfide in the presence of iron (III) chloride [273].

leuco-compound   leuco-compound   181
(base)   (salt)   dyeing ion

Methylene Blue is obtained in a similar manner to Lauth's Violet from N,N-dimethyl-p-phenylenediamine [98,271,274-277]. However, the technical preparation of Methylene Blue utilizes thiosulfuric acid instead of elemental sulfur or hydrogen sulfide. The leuco-compound is then oxidized with potassium bichromate.

Methylene Blue

Thiazine dyes such as Lauth's Violet or Methylene Blue may also be obtained by the reaction of 4-chloronitrobenzene with Na_2S [275] or by the reaction of monochloroanilines with sulfur [278]. Lauth's Violet is practically not used anymore. Methylene Blue is used to dye wool and cotton; it is also used in bacteriology to dye certain pathogenic organisms, in medicine for dyeing some parts of living tissues (peripheral nerves), and as a disinfectant for some skin diseases and for skin and mucous membranes.

Auramine 00 is a yellow dye of the diphenylmethane type and is formed by the reaction of sulfur with N,N,N',N'-tetramethyldiamino-diphenylmethane in the presence of ammonium chloride, sodium chloride and gaseous ammonia [86,279]. Auramine is used to dye paper, silk, and leather:

$$RCH_2R + 2S \xrightarrow[-H_2S]{200°C} R\overset{\overset{\displaystyle S}{\|}}{C}R \xrightarrow[-H_2S]{NH_4Cl} R\overset{\overset{\displaystyle \overset{\oplus}{N}H_2}{\|}}{C}R \cdot Cl^{\ominus} \quad (7.175)$$

Auramine

$$R = 4 \text{-} (CH_3)_2NC_6H_4$$

A vat dye of the anthraquinone series, Cibanone Brown B, may be obtained by melting sulfur and 1-amino-(or -nitro)-2-methylanthra-quinone at 240-280°C [279]. The following method for the intro-duction of hydroxyl and amino groups is used in the synthesis of anthraquinone dyes [280,281]:

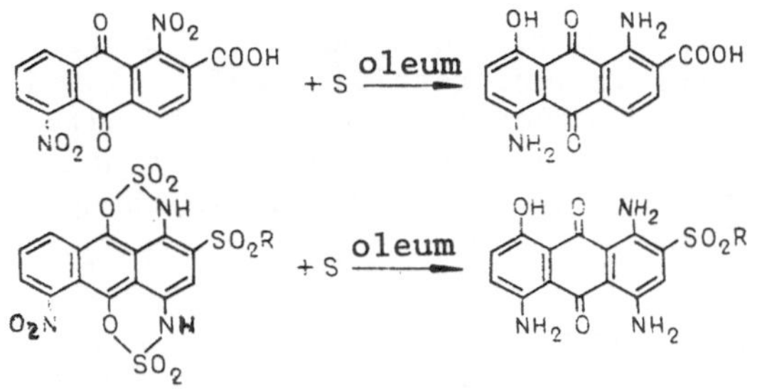

(7.176)

(7.177)

Indigo Blue may be obtained in good yield by dehydrogenation of 2-aminoacetophenone with sulfur in quinoline [282,283].

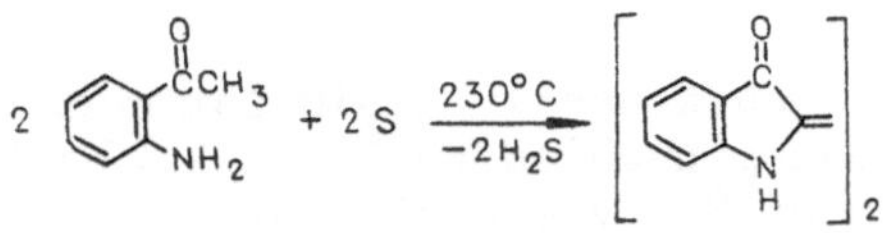

(7.178)

Other efforts to synthesize and utilize the "classical" sulfur-containing dyes (thiazine, thiazole, thioindigoid dyes, etc.) have practically stopped at the present time because all the possible ways of obtaining new and useful products have been exhausted [284].

REFERENCES

1. I. Ostromyslenskii, Zh. Russ. Fiz.-Khim. Ova., 47, 1885 (1915); Chem. Zentralbl., I, 911 (1916).
2. T. G. Levi, Gazz. Chim. Ital., 61, 294 (1931).
3. Ger. Patent 711,007 (1941); C. A., 37, 4077 (1943).
4. Ger. Patent 731,504 (1943); C. A., 38, 551 (1944).
5. K. Kindler, Liebigs Ann. Chem., 431, 187 (1923).
6. Ch. L. Levesque, US Patent 2,560,296 (1951); C. A., 46, 3558b (1952).
7. I. J. Damico, J. Org. Chem., 26, 3436 (1961).
8. K. Mori and Y. Nakamura, J. Org. Chem., 36, 3041 (1971).
9. M. G. Voronkov and A. Ya. Legzdyn', Zh. Org. Khim., 3, 465 (1967).
10. O. Wallach, Liebigs Ann. Chem., 259, 300 (1890).
11. T. G. Levi, Gazz. Chim. Ital., 59, 548 (1929).
12. F. H. McMillan, J. Am. Chem. Soc., 70, 868 (1948).
13. T. G. Levi, Gazz. Chim. Ital., 60, 975 (1930).
14. F. P. Olsen and Y. Sasaki, J. Am. Chem. Soc., 92, 3812 (1970).
15. Y. Sasaki and F. P. Olsen, Can. J. Chem., 49, 238 (1971).
16. H. Krebs, E. F. Weber, and H. Balters, Z. Anorg. Allg. Chem., 275, 147 (1954).
17. H. Krebs and K. Müller, Z. Anorg. Allg. Chem., 281, 187 (1955).
18. F. Feher and M. Baudler, Z. Anorg. Allg. Chem., 258, 132 (1949).
19. G. Peyronel, D. De Filippo, and G. Marcotrigiano, Gazz. Chim. Ital. 91, 1190 (1961).
20. A. Schönberg, O. Schütz, and S. Nickel, Ber., 61, 1375 (1928).

21. R. Möhlau, M. Heinze, and R. Zimmermann, Ber., 35, 375 (1902).
22. T. Tagushi and M. Kojima, J. Am. Chem. Soc., 78, 1464 (1956).
23. US Patent 1,719,920 (1929); C. A., 23, 4376 (1929).
24. T. G. Levi, Gazz. Chim. Ital., 61, 286 (1931).
25. H. Jenne and M. Becke-Goehring, Chem. Ber., 91, 1950 (1958).
26. M. Becke-Goehring, Angew. Chem., 73, 589 (1961).
27. H. Krebs, Die Katalytische Aktivierung des Schwefels, West-
    deutscher Verlag, Köln und Opladen (1958).
28. H. Krebs, Gummi Asbest, 8, 68 (1955); C. A., 49, 7283c (1955).
29. C. G. Moore and R. W. Saville, J. Chem. Soc., 2082 (1954);
    Rubber Chem. Technol., 28, 92 (1955).
30. B. A. Dogadkin and V. A. Shershnev, Usp. Khim., 30, 1013 (1961).
31. S. Tanaka, K. Hashimoto, and H. Watanabe, J. Pharm. Soc. Jpn.,
    93, 991 (1973).
32. N. Colebourne, Chem. Commun., 453 (1965).
33. A. S. Sultanov, A. S. Safaev, A. Kadyrov, and M. G. Solov'eva,
    Catalytic Treatment of Hydrocarbon Materials, TsNIITE Nefte-
    khimiya, Moscow (1968), Vol. 2, p. 122.
34. A. S. Sultanov, A. S. Safaev, and A. Kadyrov, Uzb. Khim. Zh.,
    No. 5, 94 (1970).
35. A. S. Safaev, A. K. Kadyrov, and M. A. Safaev, Uzb. Khim. Zh.,
    No. 4, 76 (1976).
36. N. V. Vasil'eva, G. K. Stergiu, Ya. V. Nadol'skii, and G. A.
    Kostyushko, Uzb. Khim. Zh., No. 6, 79 (1962).
37. K. Sharada and A. R. V. Musthy, Z. Anorg. Allg. Chem., 304, 344
    (1960).
38. S. Wawzonek and G. R. Hansen, J. Org. Chem., 31, 3580 (1966).
39. R. Mayer, Z. Chem., 13, 321 (1973).
40. R. E. Davis and H. F. Nakshbendi, J. Am. Chem. Soc., 84, 2085
    (1962).
41. P. D. Bartlett, A. K. Colter, R. E. Davis, and W. R. Roderick,
    J. Am. Chem. Soc., 83, 109 (1961).
42. P. D. Bartlett, E. Cox, and R. E. Davis, J. Am. Chem. Soc., 83,
    103 (1961).
43. B. L. Babitskii and L. E. Vinitskii, Kauch. Rezina, 20, 30
    (1961); C. A., 55, 26495d (1961).
44. P. Longi, R. Mantanga, and R. Mazzocchi, Chim. Ind. (Milan), 47,
    480 (1965); C. A., 63, 6893 (1965).
45. R. Mayer and K. Gewald, Angew. Chem., 79, 298 (1967).
46. R. Mayer and I. Wehl, Angew. Chem., 76, 861 (1964).
47. A. G. Makhsumov, I. T. Turdimukhamedova, and A. Safaev, Dokl.
    Akad. Nauk Uzb. SSR, 28(4), 44 (1971).
48. Y. Nomura and Y. Takeuchi, Bull. Chem. Soc. Jpn., 36, 1044
    (1963).
49. G. Gurrello and M. Piatteli, Boll. Sedute Accad. Gioenia Sci.
    Natur. Catania, 9(1), 54 (1967); C. A., 70, 3984 (1969).
50. C. Mannich and E. Kniss, Ber., 74, 1629 (1941).
51. W. Langenbeck, O. Godde, L. Weschky, and R. Schaller, Ber., 75,
    232 (1942).
52. F. Asinger and A. Mayer, Angew. Chem., 77, 812 (1965).
53. T. Bacchetti, A. Alemagna, and B. Danieli, Tetrahedron Lett.,
    2001 (1965).

54. H. Hartmann and R. Mayer, Z. Chem., 5, 152 (1965).
55. K. Gewald, H. Spies, and R. Mayer, J. Prakt. Chem., 312, 776 (1971).
56. K. Ley and R. Nast, Angew. Chem., 77, 544 (1965).
57. V. I. Shvedov and A. N. Grinev, Zh. Org. Khim., 1, 2228 (1965).
58. D. H. Klemens, A. J. Bell, and J. L. O'Brian, Tetrahedron Lett., 3257 (1965).
59. H. E. Winberg and D. D. Coffman, J. Am. Chem. Soc., 87, 2776 (1965).
60. Rin-Nosuke Shibata, Tech. Rep. Tohoku Imp. Univ., 8, 27 (1929); C. A., 24, 834 (1930).
61. R. Mayer and J. Jentzsch, J. Prakt. Chem., 23, 113 (1964).
62. V. A. Usov, L. V. Timokhina, and M. G. Voronkov, Zh. Org. Khim., 13, 2324 (1977); C. A., 88, 89234 (1978).
63. O. Meister, Ber., 2, 341 (1869).
64. V. Merz and W. Weith, Jahresber. Chem., 760 (1870); 711 (1871); Ber., 3, 978 (1870); 4, 384 (1871).
65. F. Krafft, Ber., 7, 384, 1164 (1874).
66. E. B. Schmidt, Ber., 9, 1050 (1876); 11, 1168 (1878).
67. K. A. Hoffman, Ber., 27, 2807, 3320 (1894).
68. R. Nietzki and H. Bothof, Ber., 27, 3261 (1894); 29, 2774 (1896).
69. F. Kehrmann and E. Bauer, Ber., 29, 2362 (1896).
70. O. Hinsberg, Ber., 38, 1131 (1905); 39, 2427 (1906).
71. H. H. Hodgson, J. Chem. Soc., 101, 1693 (1912).
72. H. H. Hodgson and A. G. Dix, J. Chem. Soc., 105, 952 (1914).
73. H. H. Hodgson, J. Soc. Dyers Colour, 40, 330 (1924); Chem. Zentralbl., I, 1398 (1925).
74. H. H. Hodgson, J. Chem. Soc., 125, 1855 (1924).
75. M. L. Moore and T. B. Johnson, J. Am. Chem. Soc., 51, 1287 (1935).
76. G. Schultz and H. Beyschlag, Ber., 42, 753 (1909).
77. A. P. Bindra, J. A. Elix, and C. C. Morris, Aust. J. Chem., 22, 2483 (1969).
78. P. S. Pishchemuka, Zh. Obshch. Khim., 10, 305 (1940); 21, 1689 (1951).
79. P. Tavs, Angew. Chem., 78, 1057 (1966).
80. J. Truhlar, Ber., 20, 664 (1887).
81. M. T. Bogert and M. R. Mandelbaum, J. Am. Chem. Soc., 45, 3045 (1923).
82. L. Gattermann, Ber., 22, 422 (1889).
83. R. Anschütz and G. Schultz, Ber., 22, 580 (1889).
84. A. G. Green, Ber., 22, 968 (1889).
85. P. Jacobson, Ber., 22, 330 (1889).
86. H. E. Fierz-David and L. Blangey, Grundlegende Operationen der Farbenchemie, Springer-Verlag, Wien (1952).
87. R. Wegler, E. Kühle, and W. Schäfer, Angew. Chem., 70, 351 (1958).
88. H. H. Hodgson and A. V. France, J. Chem. Soc., 1987 (1932).
89. A. A. Morton, J. B. Littlefield, and W. D. Mecum, US Patent 3,047,543 (1962); C. A., 57, 12722 (1962).
90. A. F. Cockerill, N. J. A. Gutteridge, D. M. Rackham, and C. W. Smith, Tetrahedron Lett., 3059 (1972).

91. P. T. Paul and L. B. Tewksbury, US Patent 2,435,508 (1948); C. A., 42, 5050 (1948).
92. E. Azzalin, Gazz. Chim. Ital., 55, 895 (1925).
93. Sumimoto Chemical Co., Ltd., Swiss Patent 562,234 (1975); C. A., 84, 13646f (1975).
94. Philips Petroleum Co., US Patent 3,946,030 (1976); C. A., 85, 21399n (1976).
95. A. Hantzsch and J. H. Weber, Ber., 20, 3119 (1887).
96. R. Möhlau and C. W. Krohn, Ber., 21, 59 (1888).
97. Y. Kawaoka, J. Soc. Chem. Ind. Jpn., 43, No. 2, Suppl. Binding, 53, 151 (1940); C. A., 34, 6131, 6487 (1940).
98. A. Bernthsen, Liebigs Ann. Chem., 230, 73 (1885); Ber., 16, 2896 (1883).
99. R. Möhlau, Chem. Ztg., 31, 937 (1907).
100. F. Ackermann, Ger. Patent 222,879 (1909); C. A., 4, 2882 (1910).
101. F. Ackermann, Ger. Patent 224,348 (1909); C. A., 5, 210 (1911).
102. Agfa, Fr. Patent 410,382 (1909).
103. Ger. Patent 237,771 (1911); Chem. Zentralbl., II, 1080 (1911).
104. F. Kehrmann and J. H. Dardel, Ber., 55, 2346 (1922).
105. E. P. Belokrinitskii, Zh. Prikl. Khim., 14, 187 (1941); C. A., 36, 1609 (1942).
106. D. E. Lenkov, P. Petrov, and V. Tsochev, God. Khim.-Tekhnol. Inst., 11(1), 35 (1964); C. A., 65, 13696 (1966).
107. S. P. Massie, Chem. Rev., 54, 797 (1954).
108. Brit. Patent 673,005 (1952); C. A., 47, 6990d (1953).
109. N. P. Thu-Cucy, Bun-Hoi, and N. D. Xuong, J. Chem. Soc., C, 87 (1966).
110. A. Bernthsen, Ber., 17, 2860 (1884).
111. O. Kym, Ber., 23, 2458 (1890).
112. Ch. Ris, Ber., 19, 2240 (1886).
113. K. J. Karrman, Sven. Kem. Tidskr., 58, 92 (1946); C. A., 40, 6072 (1946).
114. R. Möhlau and V. Klopfer, Ber., 31, 3164 (1898).
115. M. T. Bogert and P. P. Herrera, J. Am. Chem. Soc., 45, 238 (1923).
116. F. Asinger, A. Saus, H. Offermans, and F. A. Dagga, Liebigs Ann. Chem., 723, 119 (1969).
117. R. C. Mansfield, J. Org. Chem., 24, 1357 (1959).
118. W. Scott and G. W. Watt, J. Org. Chem., 2, 148 (1937).
119. N. Colebourne, R. G. Foster, and E. Robson, J. Chem. Soc., C, 685 (1967).
120. B. Böttcher and F. Bauer, Liebigs Ann. Chem., 568, 218 (1950).
121. Badische Anilin- und Soda-Fabrik. A.-G. Fr. Patent 1,585,353 (1970); C. A., 74, 41958 (1971).
122. T. E. Stevens, J. Org. Chem., 26, 3451 (1961).
123. Farbenfabriken Bayer, Ger. Patent 1,149,356 (1963).
124. J. Stanek, Coll. Czech. Chem. Commun., 12, 671 (1947); C. A., 42, 58761 (1948).
125. A. W. Hoffmann, Ber., 12, 2359 (1879); 13, 1223 (1880).
126. P. Jacobson, Ber., 19, 1067 (1886).
127. A. W. Hoffmann, Ber., 20, 1798 (1887).

128. T. Taguchi and K. Yoshihara, Chem. Pharm. Bull. (Tokyo), <u>11</u>, 430 (1963).
129. C. Skarez Contreras, Ann. Quim., <u>64</u>, 819 (1968); C. A., <u>70</u>, 56896 (1969).
130. G. Bruni and T. G. Levi, Gazz. Chim. Ital., <u>54</u>, 398, 402 (1924).
131. G. Bruni and T. G. Levi, Atti R. Accad. Lincei, Roma /5/, <u>32</u>, II, 313 (1924); Chem. Zentralbl., I, 2366 (1924).
132. B. A. Dogadkin and N. N. Pavlov, Dokl. Akad. Nauk SSSR, <u>138</u>, 1111 (1969).
133. W. Mack, Angew. Chem., <u>79</u>, 1106 (1967).
134. M. G. Toland, J. Org. Chem., <u>27</u>, 869 (1962).
135. K. Gewald, M. Kleinert, B. Thiele, and M. Hentschel, J. Prakt. Chem., <u>314</u>, 303 (1972).
136. J. P. Brown and M. Thompson, J. Chem. Soc., Perkin Trans. I, 863 (1974).
137. R. Mayer and St. Scheithauer, J. Prakt. Chem., <u>21</u>, 214 (1963).
138. D. J. Relyla, P. O. Tawney, and A. R. Williams, J. Org. Chem., <u>27</u>, 1078 (1962).
139. K. Gewald, E. Schinke, and H. Böttcher, Chem. Ber., <u>99</u>, 94 (1966).
140. V. I. Shvedov, V. K. Ryzhkova, and A. N. Grinev, Khim. Geterotsikl. Soedin., 239 (1967).
141. J. F. Tinney and D. Parkis, US Patent 3,558,606 (1971); C. A., <u>74</u>, 141622 (1971).
142. V. P. Arya and P. Chates, Indian J. Chem., <u>9</u>, 1209 (1971).
143. K. Gewald and E. Schinke, Chem. Ber., <u>99</u>, 2712 (1966).
144. K. Gewald and J. Schael, J. Prakt. Chem., <u>315</u>, 39 (1973).
145. P. Schmidt and K. Eichenberger, Swiss Patent 523,282 (1972); C. A., <u>78</u>, 3172 (1973).
146. M. Chaykovsky and M. Lin, J. Med. Chem., <u>16</u>, No. 3, 188 (1973).
147. M. Nakanishi and T. Tahara, J. Med. Chem., <u>16</u>, No. 3, 214 (1973).
148. W. Ried and E. Nyiondi-Bonguen, Liebigs Ann. Chem., 202 (1973).
149. K. Gewald, Z. Chem., <u>3</u>, 26 (1963).
150. K. Gewald, J. Prakt. Chem., <u>31</u>, 214 (1966).
151. K. Gewald, J. Prakt. Chem., <u>32</u>, 26 (1966).
152. G. Purrello, Gazz. Chim. Ital., <u>95</u>, 1089 (1965).
153. A. Bruno and G. Purrello, Gazz. Chim. Ital., <u>96</u>, 986 (1966).
154. S. E. Jolly, US Patent 2,380,531 (1945); C. A., <u>40</u>, 726 (1946).
155. J. Ulbricht and W. Makschin, Faserforsch. Textiltech., <u>22</u>, 381 (1971); C. A., <u>76</u>, 15020 (1972).
156. W. D. McGrath, T. Morrow, and D. N. Dempster, Chem. Commun., 516 (1967).
157. W. Weith, Ber., <u>6</u>, 210 (1873).
158. J. U. Nef, Liebigs Ann. Chem., <u>280</u>, 291 (1894).
159. A. Sabaneeva and M. Prozina, Zh. Russ. Fiz.-Khim. Ova., <u>33</u>, 20 (1901); <u>34</u>, 404 (1902); C. A., <u>24</u>, 834 (1930).
160. J. H. Boyer and V. T. Ramakrishnan, J. Org. Chem., <u>37</u>, 1360 (1972).
161. P. Jacobson and A. Frankenbacher, Ber., <u>24</u>, 1400 (1891).
162. A. Schönberg and E. Frese, Chem. Ber., <u>95</u>, 2810 (1962).
163. N. Latif and I. Fathy, J. Org. Chem., <u>27</u>, 1633 (1962).

164. N. A. Korchevin, V. A. Usov, and M. G. Voronkov, Zh. Org. Khim.,
     12, 2412 (1976).
165. V. A. Usov, N. A. Korchevin, and M. G. Voronkov, Zh. Org. Khim.,
     16, 1337 (1980).
166. M. S. Raasch, J. Org. Chem., 44, 632 (1979).
167. G. Purrello, Ann. Chim. (Rome), 51, 143 (1961); C. A., 55,
     18730f (1961).
168. F. Feher and H. Kulus, Z. Naturforsch., B26, 1071 (1971).
169. N. Ya. Derkach and N. A. Pasmurtseva, Zh. Org. Khim., 9, 1414
     (1973).
170. E. S. Levchenko and A. V. Kirsanov, Zh. Obshch. Khim., 32, 2256
     (1962).
171. E. S. Levchenko, Ya. Bal'on, and A. V. Kirsanov, Zh. Org. Khim.,
     3, 2066 (1967).
172. L. N. Markovskii, G. S. Fedyuk, E. S. Levchenko, and A. V.
     Kirsanov, Zh. Org. Khim., 9, 2502 (1973).
173. L. N. Markovskii and E. S. Levchenko, Zh. Obshch. Khim., 39,
     2786 (1969).
174. E. S. Levchenko and E. M. Dorokhova, Zh. Org. Khim., 8, 1516
     (1969).
175. E. S. Levchenko, G. S. Borovikova, and E. M. Dorokhova, Zh. Org.
     Khim., 13, 103 (1977).
176. J. Geevers and W. P. Trompen, Tetrahedron Lett., 1687 (1974).
177. W. C. Wilson, US Patent 1,732,453 (1929); C. A., 24, 253 (1930).
178. J. A. Elix and G. C. Morris, Tetrahedron Lett., 671 (1969).
179. M. N. Shchukina and G. I. Predvoditeleva, Izv. Akad. Nauk SSSR,
     Ser. Khim., 110, 565 (1956); C. A., 51, 4969b (1957).
180. H. G. Beard, H. H. Hodgson, and R. R. Davies, J. Chem. Soc., 4
     (1944).
181. W. A. Pryor, Mechanisms of Sulfur Reactions, McGraw-Hill,
     New York—San Francisco—Toronto—London (1962), p. 150.
182. I. I. Grandberg, Din Vej pog, and A. N. Kost, Zh. Obshch. Khim.,
     31, 941 (1961).
183. V. Harley, C. R. Acad. Sci. (Paris), C 224, 568 (1947).
184. A. A. Morton and D. Horvitz, J. Am. Chem. Soc., 57, 1860 (1935).
185. F. Asinger, M. Thiel, W. Dathe, O. Hample, E. Mittag,
     E. Plaschil, and Ch. Schröder, Liebigs Ann. Chem., 639, 146
     (1961).
186. F. Asinger, H. Offermanns, and D. Neuray, Liebigs Ann. Chem.,
     739, 32 (1970).
187. B. Oddo and L. Raffa, Gazz. Chim. Ital., 69, 562 (1933).
188. L. Raffa, Gazz. Chim. Ital., 72, 549 (1942).
189. L. Szperl, Rocz. Chem., 18, 804 (1938); Chem. Zentralbl., II,
     2067 (1939).
190. S. Gabriel, W. Gerhard, and R. Walter, Ber., 56, 1024 (1923).
191. W. Carpenter, M. S. Grand, and H. R. Snyder, J. Am. Chem. Soc.,
     82, 2739 (1960).
192. L. Raffa, Gazz. Chim. Ital., 72, 557 (1942).
193. L. Raffa and G. Rossi, Gazz. Chim. Ital., 44, 104 (1914); Chem.
     Zentralbl., I, 1648 (1914).

194. B. Emmert and M. Groll, Chem. Ber., <u>86</u>, 205 (1953).
195. J. Helen, H. I. Thayer, and B. B. Corson, J. Am. Chem. Soc.,
     <u>70</u>, 2330 (1948).
196. US Patent 2,496,319 (1950); C. A., <u>44</u>, 5920a (1950).
197. W. G. Dauben, J. C. Reid, P. E. Yankwich, and M. Calvin, J. Am.
     Chem. Soc., <u>72</u>, 121 (1950).
198. R. C. Mansfield, J. Org. Chem., <u>24</u>, 1111 (1959).
199. H. D. Porter, J. Am. Chem. Soc., <u>76</u>, 127 (1954).
200. H. Najer, P. Chabrier, R. Giudicelli, and E. Joannic-Voisinet,
     C. R. Acad. Sci. (Paris), C <u>224</u>, 2935 (1957).
201. J. Schmidt and M. Suguet, Bull. Soc. Chim. France, 755 (1956).
202. Ger. Patent 964,142 (1947); C. A., <u>53</u>, 13178b (1959).
203. L. Kramberger, P. Lorenĉak, S. Polanc, B. Verĉek, B. Stanovnik,
     and M. Tiŝler, J. Heterocycl. Chem., <u>12</u>, 337 (1975).
204. B. Emmert, Chem. Ber., <u>91</u>, 1388 (1958).
205. K. V. Martin, J. Am. Chem. Soc., <u>80</u>, 233 (1958).
206. F. Lions and K. V. Martin, J. Am. Chem. Soc., <u>80</u>, 1591 (1958).
207. H. Saikashi and T. Hisano, Chem. Pharm. Bull. (Tokyo), <u>7</u>, 349
     (1959); C. A., <u>55</u>, 1613 (1961).
208. B. Emmert and M. Groll, Chem. Ber., <u>86</u>, 208 (1953).
209. P. E. Miller, G. L. Oliver, L. R. Dann, and J. W. Gates, J. Org.
     Chem., <u>22</u>, 664 (1957).
210. H. Saikashi and T. Hisano, Chem. Pharm. Bull. (Tokyo), <u>7</u>, 716
     (1959); C. A., <u>55</u>, 23494 (1961).
211. E. Kuhle and R. Wegler, US Patent 2,774,757 (1956); C. A., <u>51</u>,
     6705 (1957).
212. H. Saikashi, T. Hisano, and S. Jochina, J. Pharm. Soc. Jpn.,
     <u>74</u>, 1318 (1954); C. A., <u>49</u>, 15892d (1955).
213. B. Emmert and A. Holz, Chem. Ber., <u>87</u>, 676 (1954).
214. A. P. Terent'ev, I. G. Mochalina, and G. V. Panova, Vysokomol.
     Soedin., Ser. B, <u>5</u>, 842 (1963); C. A., <u>59</u>, 8882 (1963).
215. H. Schmidt, R. Bejnisch, F. Mietzsch, and G. Domagk, Natur-
     wissenschaften, <u>33</u>, 314 (1946).
215a. A. Morimoto, H. Takasugi, and Y. Nakai, Jpn. Patent 50,635
     (1972); C. A., <u>78</u>, 124460e (1973).
216. T. P. Sycheva and M. N. Shchukina, Biologically Active Com-
     pounds, Izd. Akad. Nauk SSSR, Moscow-Leningrad (1965);
     p. 42; C. A., <u>64</u>, 6607 (1966).
217. T. Hisano and H. Koga, J. Pharm. Soc. Jpn., <u>90</u>, 552 (1970).
218. A. Edinger, Ber., <u>33</u>, 3769 (1900).
219. A. Edinger, J. Prakt. Chem., <u>64</u>, 182, 196 (1901).
220. Ger. Patent 120,586 (1901); Chem. Zentralbl., I, 1254 (1901).
221. I. Baranowska and W. Karminski, Zesz. Nauk. Politech. Slask.,
     Chem., No. 50, 127 (1969); C. A., <u>73</u>, 98772 (1970).
222. H. Saikashi and T. Hisano, Chem. Pharm. Bull. (Tokyo), <u>8</u>, 51
     (1960); C. A., <u>55</u>, 1614 (1961).
223. W. Miller, Ber., <u>21</u>, 1827 (1888).
224. A. Walker, W. E. Baldwin, H. I. Thayer, and B. B. Corson,
     J. Org. Chem., <u>16</u>, 1805 (1951).
225. J. P. Brown, J. Chem. Soc., C 1074 (1968).

226. K. Gleu and R. Schaarschmidt, Ber., 72, 1246 (1939).
227. H. E. Winberg, J. E. Carnahn, D. D. Coffman, and M. Braun, J. Am. Chem. Soc., 87, 2054 (1965).
228. O. N. Chupakhin, V. A. Trofimov, and Z. V. Pushkareva, Dokl. Akad. Nauk SSSR, 188, 376 (1969); Khim. Geterotsikl. Soedin., 1674 (1970).
229. T. P. Sycheva, I. V. Lebedeva, and M. N. Shchukina, Zh. Vses. Khim. Ova., 5, 234 (1960); C. A., 54, 21049 (1960).
230. B. G. Dzantiev and A. V. Shishkov, in: Radiochemical Methods of Determining Microelements, Izd. Akad. Nauk SSSR (1965), p. 185; C. A., 63, 8338 (1965).
231. H. Saikashi and T. Hisano, J. Pharm. Soc. Jpn., 79, 1305 (1959); C. A., 55, 1614 (1961).
232. K. Baker and H. E. Fierz-David, Helv. Chim. Acta, 33, 2011 (1950).
233. A. V. El'tsov, Zh. Org. Khim., 3, 199 (1967).
234. M. Z. Girshovich and A. V. El'tsov, Zh. Obshch. Khim., 39, 941 (1969).
235. A. V. El'tsov and K. M. Krivosheiko, Zh. Org. Khim., 2, 189 (1966).
236. W. Walter and K.-D. Bode, Angew. Chem., 78, 517 (1966).
237. G. Zigeuner, H. Hamberger, E. Pinter, and R. Ecker, Monatsh. Chem., 104, 585 (1973).
238. G. Zigeuner, H. Hamberger, and R. Ecker, Monatsh. Chem., 101, 881 (1970).
239. O. N. Chupakhin, E. O. Sidorov, S. M. Shein, and I. I. Bel'kis, Zh. Org. Khim., 12, 2464 (1974).
240. J. Daunis, M. Guerret-Rigail, and R. Jacquier, Bull. Soc. Chim. France, 3198 (1972).
241. F. H. McMillan and J. A. King, J. Am. Chem. Soc., 70, 4143 (1948).
242. F. Kirchhof, Gummi Ztg., 39, 892 (1925); C. A., 19, 2424 (1925).
243. M. Beytrup, J. Chem. Soc., Perkin Trans., 507 (1975).
244. M. Beytrup, Acta Chim. Scand., B29, 141 (1975).
245. H. Becker, D. Nagel, and H. Timple, J. Prakt. Chem., 315, 97 (1973).
246. R. Walentowski and H. Wanzlick, Z. Naturforsch., 25b, 1421 (1970).
247. H. Schönherr and H. Wanzlick, Liebigs Ann. Chem., 731, 176 (1970).
248. H. Wanzlick, H. Kleiner, I. Lasch, and H. Füldner, Liebigs Ann. Chem., 708, 155 (1967).
249. A. G. Bayer, Ger. Patent 2,422,955 (1975); C. A., 85, 44087x (1976).
250. H. Lecoq, Bull. Soc. R. Sci. (Liège), 11, 260 (1942); C. A., 38, 3658 (1944).
251. O. Lange, Die Schwefelfarbstoffe, 2 Aufl., Leipzig (1925).
252. H. Zollinger, Chemie der Azofarbstoffe, Birkhäuswer Verlag, Basel und Stuttgart (1953).

253. M. A. Chekalin, Chemistry and Technology of Organic Dyes, Goskhimizdat, Moscow (1956).
254. Organic Synthetic Dyes (A Catalog), Khimiya, Moscow (1968).
255. H. E. Fierz-David, Grundlegende Operationen der Farbenchemie, 3. Aufl., Springer-Verlag, Berlin (1924).
256. I. M. Khmel'nitskaya and G. E. Bonvech, Sulfur Dyes, Goskhimizdat, Moscow—Leningrad (1934).
257. I. M. Kogan, Chemistry of Dyes, Goskhimizdat, Moscow (1956).
258. Colour Index, 2nd ed., London, Vols. 1-4 (1956-1957); Vol. 5 (1963).
259. L. Tigler, Am. Dyestuff Rep., 57, P333 (1968).
260. A. A. Myasnikov, Sulfur Water-Soluble Dyes, Moscow (1960).
261. K. Venkataraman (ed.), The Chemistry of Synthetic Dyes, Vol. III, Academic Press, New York—London (1970).
262. N. G. Laptev and B. M. Bogoslovskii, Chemistry of Dyes, Khimiya, Moscow (1970).
263. G. Schultz and H. Beyschlag, Ber., 42, 743, 753 (1909).
264. Ger. Patent 215,547; 215,548 (1909); Chem. Zentralbl., II, 1741 (1909).
265'. B. Jaeckel, Chem. Ber., 83, 578 (1950).
266. A. Żuk, M. Wejchan-Judek, and G. Lewandowicz, Eighth International Symposium on Organic Sulfur Chemistry, Abstracts of papers, Portoroż, Yugoslavia (1978), pp. 260, 286.
267. A. Bernthsen, Ber., 17, 2857 (1884); Liebigs Ann. Chem., 230, 182 (1885).
268. Brit. Patent 6,319 (1888).
269. W. N. Jones, Chem. Rev., 36, 291 (1945).
270. Ch. Lauth, Ber., 9, 1035 (1876).
271. A. Bernthsen, Ber., 17, 611, 2854 (1884).
272. A. Bernthsen, Ber., 17, 512 (1884).
273. A. Koch, Ber., 12, 2069 (1879).
274. H. Caro, Ger. Patent 1,886 (1877).
275. A. Koch, Ber., 12, 592 (1879).
276. A. Bernthsen, Ber., 16, 1025 (1883).
277. H. H. Hodgson, J. Soc. Dyers Colourist, 42, 76 (1926); C. A., 20, 1717 (1926).
278. O. Baither, Ber., 20, 1731 (1887).
279. Ger. Patent 204,958 (1907); Chem. Zentralbl., I, 602 (1909).
280. P. Buchler, US Patent 2,967,752; C. A., 55, 10905 (1961).
281. F. By, Fr. Patent 1,488,822; C. A., 68, 106072 (1968).
282. Ger. Patent 273,340 (1914); Chem. Zentralbl., I, 1793 (1914).
283. V. I. Minaev, The Chemistry of Indigo and Indigoid Dyes, Goskhimizdat, Moscow—Leningrad (1934).
284. M. A. Chekalin, B. V. Passet, and B. A. Ioffe, The Technology of Organic Dyes and Intermediates, Khimiya, Leningrad (1972).

8
Reactions of Sulfur with
Organometallic Compounds

8.1. ORGANIC DERIVATIVES OF THE ALKALI METALS

Organolithium compounds, RLi, react with sulfur under mild conditions with the evolution of heat and formation of lithium thiolates, R-SLi, which on acid hydrolysis give the corresponding thiols, R-SH [1,2]:

$$RLi + S \rightarrow RSLi \xrightarrow{H_3O^+} RSH \qquad (8.1)$$

Organic polysulfides $R-(S)_n-R$ are also formed in this reaction; for example, Boscato and co-workers have investigated the reaction of organolithium compounds (e.g., sec-butyllithium) with excess sulfur and observed the formation of dialkyl polysulfides, $R-(S)_n-R$, where n = 3,4. The reactions at $RLi:S_8$ mole ratio $\geqslant$ 2:1 give dialkyl sulfides with a lower sulfur ratio and lithium alkanethiolates [3,4].

It has been shown that n-dodecyllithium [2], phenyllithium, and 4-dimethylaminophenyllithium [1] on treatment with sulfur and subsequent hydrolysis give the corresponding thiols in 50-60% yield. However, this method is only rarely used for the synthesis of aliphatic and aromatic thiols because more convenient synthetic methods are available [5].

It has been established [6] that diethylphosphonates, $(EtO)_2-P(O)CH(R)Li$, react readily with sulfur in tetrahydrofuran to form the α-phosphoryl thiols, $(EtO)_2P(O)CH(R)SH$, where R = H, Me, or Ph. The preparation of S,S- and O,S-thioacetals of formylphosphonates [6,7] proceeds in a similar manner, e.g.,

$$(EtO)_2P(O)CH_2OCH_3 \xrightarrow[\begin{array}{l}\text{2. } S_8 \\ \text{3. MeI}\end{array}]{\text{1. } \underline{n}\text{-BuLi}/\underline{tert}\text{-BuOK}} (EtO)_2P(O)CH\begin{array}{l}SMe \\ OMe\end{array} \qquad (8.2)$$

Thioamides, $RCH_2\text{-}CS\text{-}NEt_2$, may be prepared in high yields by the addition of powdered sulfur to a solution of lithium alkylacetylides in excess diethylamine [8]. The reaction of lithium alkylacetylides, $R\text{-}C\equiv CLi$, with sulfur in diethyl ether followed by treatment of the reaction mixture with excess alkanethiol, $R'\text{-}SH$, and potassium hydroxide solution leads to the formation of thioesters, $RCH_2\text{-}CS\text{-}SR'$ [9]. It has been assumed that the corresponding thioketene, $R\text{-}CH\text{=}C\text{=}S$, is an intermediate in this process; the alkanethiol then adds to the ketene.

Lithium alkynethiolates, $R\text{-}C\equiv C\text{-}SLi$, prepared by the action of sulfur on the corresponding carbanion, react with allyl bromide and 1-bromo-2-butyne to give $CH_2\text{=}CHCH_2SC\equiv CR$ and $CH_3C\equiv CCH_2SC\equiv CR$, (R = $\underline{tert}$-Bu and Me_3Si), respectively [10]. The reaction of $Me_3SiC\equiv CLi$ with sulfur and chlorotrimethylsilane in diethyl ether at $-45°C$ leads to bis(trimethylsilyl)thioketene in 90% yield. It is interesting to note, however, that the similar reaction of $Me_3SiC\equiv CSLi$ with bromotrimethylsilane leads to an expected isomeric product, $Me_3SiC\equiv CSSiMe_3$, in 80% yield. This silylated alkynethiol isomerizes on heating to bis(trimethylsilyl)thioketene [11].

An exothermic reaction of sulfur with poly-4-lithiostyrene in benzene has been described. This reaction, upon hydrolysis of the mixture, gives a polymer containing 17% sulfur [12]:

$$\begin{bmatrix} -CH-CH_2- \\ \bigcirc \\ Li \end{bmatrix}_n \xrightarrow[\text{2) } H_3O^+]{\text{1) } S} \begin{bmatrix} -CH-CH_2- \\ \bigcirc \\ SH \end{bmatrix}_n \qquad (8.3)$$

Divinylbenzene-styrene copolymer resins containing thiol groups on some of their aromatic rings may be prepared by treating the lithiated copolymer with sulfur followed by reduction of the resulting product [13].

Naphtho[1,8-$\underline{cd}$]-1,2-dithiol can be prepared by treating sulfur with 1,8-dilithionaphthalene. Its selenium and tellurium analogs were similarly prepared and may also be obtained by treating 1,8-dibromonaphthalene with one equivalent of butyllithium and a chalcogen, then with another equivalent of butyllithium and a different chalcogen [14-16]. The reaction of sulfur with 5,6-dilithioacenaphthylene leads to the formation of acenaphtho[5,6-$\underline{cd}$]-1,2-dithiol-($\underline{1}$) in 28% yield [17,18]:

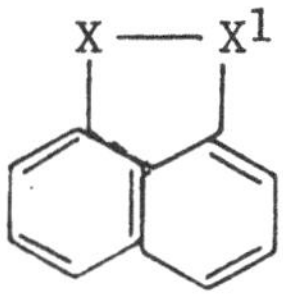

X = X^1 = S, Se, Te; or X = S, X^1 = Se, Te

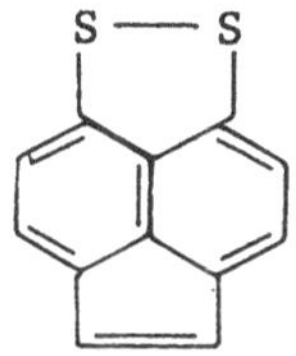

Polyfluorinated organolithium compounds such as 1-lithioundeca-fluorobicyclo[2.2.1]heptane [19] and pentafluorophenyllithium [20, 21], react with sulfur to give the corresponding lithium thiolates:

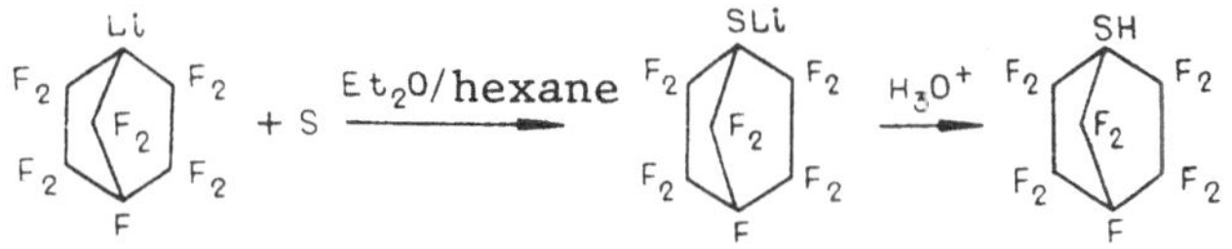

$$(8.4)$$

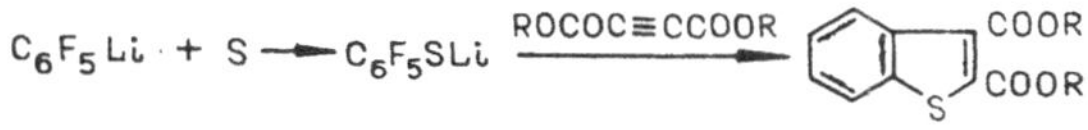

$$C_6F_5Li + S \longrightarrow C_6F_5SLi \xrightarrow{ROCOC\equiv CCOOR}$$

(8.5)

$$R = C_2H_5$$

The above synthesis has found much wider use in thiophene chemistry (cf. Table 8.1). The use of this route has been facilitated by the development of suitable methods giving lithiated derivatives in the thiophene series. The method has been successfully used for the synthesis of unsubstituted 2- and 3-mercapto-thiophenes [22-24], and of a large number of methyl-, methylthio-, chloro-, and bromo-2-thiophenethiols [25-27], e.g.,

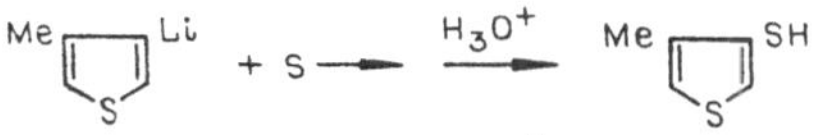

(8.6)

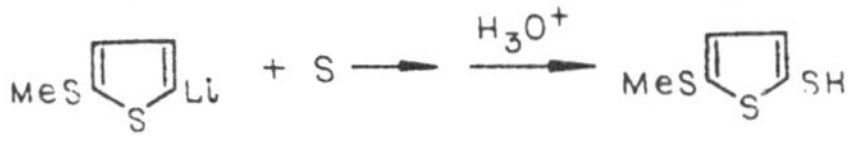

(8.7)

It can be seen from Table 8.1 that, in the thiophene series, the synthetic methods are considerably more numerous because of the possibility of utilizing the lithium thiolates formed in subsequent reactions with alkyl halides, esters of halosubstituted carboxylic acids, and similar compounds. Thus, for example, the synthesis of 4-bromothieno[2,3-b]thiophene-2-carboxylic acid has been published [28]. The synthesis consists of successive reactions of 3,4-di-bromothiophene with one equivalent of butyllithium, dimethylfor-mamide, a second equivalent of butyllithium, elemental sulfur, methyl bromoacetate, and sodium ethoxide without isolation of the inter-mediates:

$$(8.8)$$

The reaction of 6-lithio-2$\underline{H}$-thiapyran with sulfur followed by treating the reaction mixture with propargyl bromide leads to the formation of a product which undergoes cyclization by heating in HMTA to form 2$\underline{H}$,7$\underline{H}$-thiapyrano[2,3-$\underline{b}$]thiapyran [29]:

$$(8.9)$$

An analogous method has been successfully used for the synthesis of furan derivatives [30,32]:

$$(8.10)$$

$$R=H, CH_3, C_2H_5; \quad R'=CH_3, C_2H_5, CH_2=CHCH_2, \text{ etc.}$$

The reaction of sulfur with 2 and 1 equivalents of lithium triethylborohydride in tetrahydrofuran under dry nitrogen afforded solutions of Li_2S and Li_2S_2, respectively. Subsequent addition of alkyl, acyl, arylalkyl or cycloalkyl halides gives sulfides or disulfides in high yields [33]. Dialkyl selenides and dialkyl diselenides have been synthesized in a similar way [34].

The isomeric 1,2-dimercapto-1,2-dicarba-, 1,7-dimercapto-1,7-dicarba-, and 1,12-dimercapto-1,12-dicarba-closo-dodecaboranes were obtained in high yield by the reaction of sulfur with the corresponding lithium derivatives followed by acid hydrolysis of the lithium thiolates formed [35-37]:

$$(8.11)$$

(100%)

Sulfur reacts with a solution of 1,1'-dilithioferrocene in boiling 1,2-dimethoxyethane to give 2,3-trithia-[3]-ferrocenophene (2) which is converted to a dithiol upon reduction with lithium aluminium hydride [38]:

$$(8.12)$$

The structure of compound (2) has been studied [39].

Elemental sulfur reacts under mild conditions with $(H_5C_6)_3SiLi$ [40], $(H_5C_6)_3GeLi$ [42], $(H_5C_6)_3SnLi$ [43,44], $(H_5C_6)_2SnLi_2$ [45], and $(H_5C_6)_3PbLi$ [42] and is inserted into the metal–alkali metal bond, e.g.,

$$(C_6H_5)_3 MLi + S \xrightarrow{THF} (C_6H_5)_3 MSLi$$
$$M = Si,\ Ge,\ Sn,\ Pb$$

$$(8.13)$$

The lithium thiolates formed in this reaction are very reactive and readily react with benzoyl chloride and the organic halides of germanium, tin, and lead [46], e.g.,

$$(C_6H_5)_2Sn(SLi)_2 + 2\,C_6H_5COCl \longrightarrow (C_6H_5)_2Sn(SCOC_6H_5)_2$$

$$(8.14)$$

$$(C_6H_5)_3MSLi + (C_6H_5)_3M'Cl \longrightarrow (C_6H_5)_3MSM'(C_6H_5)_3$$

$$M' = Ge,\ Sn,\ Pb$$

$$(8.15)$$

It may be assumed [19,46,47] that the primary act in reactions (8.4) and (8.13) is nucleophilic attack of the carbanion (i.e., $C_7F_{11}^-$ and $(C_6H_5)_3M^-$, respectively) upon the cyclooctasulfur molecule and that this attack causes ring opening. This step is followed by a gradual degradation of the sulfur chain [19], e.g.,

$$C_7F_{11}^- + S_8 \rightarrow C_7F_{11} - (S)_7 - S^- \xrightarrow{C_7F_{11}^-} C_7F_{11} - (S)_6 - S^- + C_7F_{11}S^-$$
$$\downarrow C_7F_{11}^-$$
$$C_7F_{11} - (S)_5 - S^- + C_7F_{11}S^- \text{ etc.}$$

$$(8.16)$$

The reactions of sulfur with organosodium compounds have been studied to a lesser extent. 1-Alkynic sulfides, R-C≡C-SR', are formed in moderate yields according to the following equation [48, 50]. Under these conditions, 3-thia-1-pentyne does not undergo disproportionation whereas 3-selena-1-pentyne undergoes disproportionation and is converted to bis-ethylselenoacetylene [51]:

$$RC{\equiv}CNa \xrightarrow[\text{liq. } NH_3]{S} RC{\equiv}CSNa \xrightarrow[\text{liq. } NH_3]{R'Hal} RC{\equiv}CSR'$$

$$(8.17)$$

$$R = H,\ CH_3,\ CH_2{=}CH,\ C_2H_5,\ \underline{n}\text{-}C_3H_7,\ \text{etc.}$$

   Sodium phenylacetylide reacts with sulfur in the presence of
carbon disulfide to form 4-phenyl-1,3-dithiol-2-thione. This compound may also be obtained in a stepwise manner, i.e., by converting
sodium phenylacetylide into the corresponding sodium thiolate and
subsequent treatment of the latter with carbon disulfide in a solvent
with pK = 18-25 (ethanol, tert-butyl alcohol, acetone, ethyl acetate,
acetonitrile). The solvent serves as a proton donor. This has been
proved by carrying out the reaction in hexadeuteroacetone, $(D_3C)_2CO$,
[52-55]:

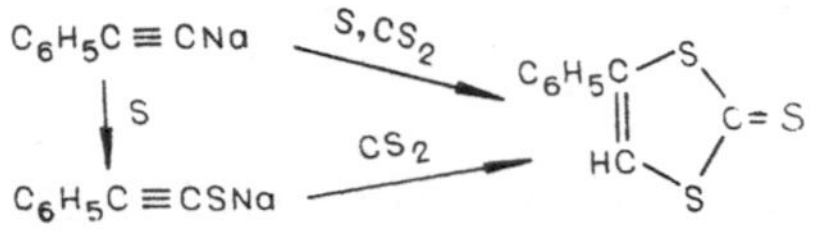

$$C_6H_5C \equiv CNa \xrightarrow{S,CS_2}$$

$$\downarrow S$$

$$C_6H_5C \equiv CSNa \xrightarrow{CS_2}$$

(8.18)

   The reaction of sodium acetylide with sulfur in the presence
of carbon diselenide affords, in addition to the expected 1,3-thia-
selenole-2-selone (3), five other related compounds [56,57]:

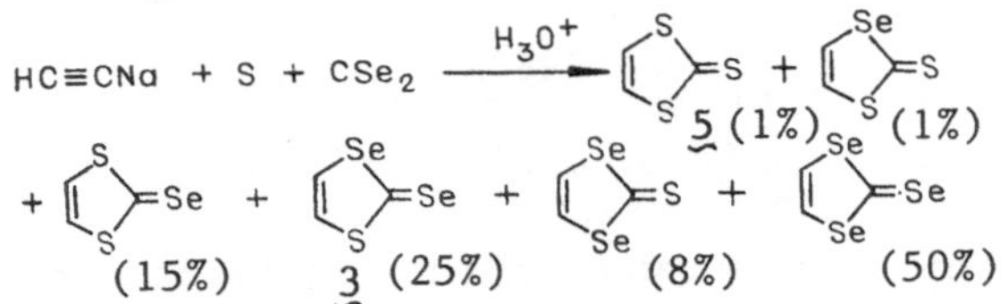

$$HC \equiv CNa + S + CSe_2 \xrightarrow{H_3O^+}$$

(8.19)

   The yield of tetrathiafulvalene (4), obtained by the condensation of sodium acetylide with sulfur and carbon disulfide, desul-
furization of the dithiolthione (5) with peracetic acid, and self-
coupling of (6), was improved by purifying (5) with petroleum ether
and using $NaPF_6 \cdot Et_3N$ as the coupling agent for (6) [58]:

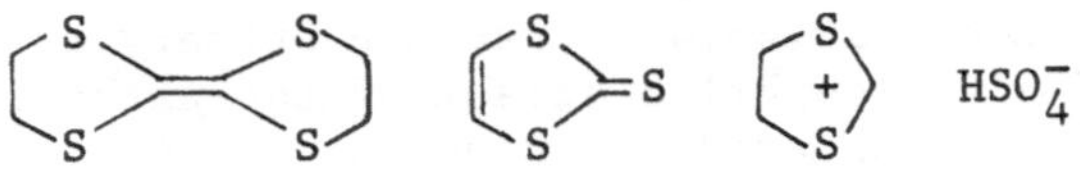

   The reactions of organic compounds of potassium, rubidium, and
cesium with sulfur have not been studied so far. However, the
chemistry of perfluoroalkylcarbanions has made considerable progress
in recent years. These ions are generated in situ by the addition
of the fluoride ion to perfluoro-olefins in polar aprotic solvents.
Thus, the system "fluoro-olefin-alkali metal fluoride" can be viewed
as a perfluoroalkyl derivative of an alkali metal:

$$CF_3CF = CF_2 + MF \rightleftharpoons (CF_3)_2CF^- + M^+$$

(8.20)

$$M = K, Cs$$

   These systems give a number of different reactions. To a considerable extent, they compensate for the impossibility of the use
of organolithium and organomagnesium compounds for synthetic purposes
in the case of perfluoroalkyl compounds. The reactions of the above

systems with sulfur occurs via the intermediate formation of per-
fluoroalkanethiolates. The composition of the final reaction
products depends on the nature of the respective perfluoroalkyl
group and on the reaction conditions [59-61]. Thus, for example,
the interaction of equimolar amounts of hexafluoropropylene and
sulfur in the presence of a catalytic amount of potassium fluoride
gives 2,2,4,4-tetrakis-trifluoromethyl-1,3-dithietane as the principal reaction product. Its formation can be explained as follows:

$$CF_3CF{=}CF_2 \xrightarrow{F^-} (CF_3)_2CF^- \xrightarrow{S} (CF_3)_2CFS^- \xrightarrow[-F^-]{}$$

$$\rightarrow (CF_3)_2C{=}S \rightarrow 1/2\,(CF_3)_2C{\overset{S}{\underset{S}{<}}}C(CF_3)_2 \qquad (8.21)$$

It may be assumed that the primary reaction is the formation of the
hexafluoroisopropyl anion which is then converted to the perfluoro-
2-propanethiolate anion. The latter gives hexafluorothioacetone by
α-elimination of the fluoride ion and the hexafluorothioacetone
dimerizes with the formation of the dithietane [59].

Similarly, perfluoroisobutylene can be converted to bis-per-
fluoro-tert-butyl trisulfide in polar solvents in the presence of
catalytic amounts of potassium fluoride [59,60] or cesium fluoride
[62]:

$$(CF_3)_2C{=}CF_2 \xrightarrow{F^-} (CF_3)_3C^- \xrightarrow{S} (CF_3)_3CS^- \xrightarrow[-S^{2-}]{S} (CF_3)_3CS\cdot \qquad (8.22)$$

$$2(CF_3)_3CS\cdot + S \longrightarrow (CF_3)_3C{-}(S)_3{-}C(CF_3)_3 \qquad (8.23)$$

The reaction products, obtained from the reaction of alkali
metals with sulfur, and their yields are summarized in Table 8.1.

8.2. ORGANIC COMPOUNDS OF COPPER AND SILVER

Alkylcopper complexes, $RCu(PPh_3)_2$ (R = Me, Et, Pr, Me_2CHCH_2),
react smoothly with sulfur in tetrahydrofuran at low temperatures
to produce dialkyl disulfides. This indicates the insertion of
sulfur into the alkyl-copper bond initially takes place followed by
coupling of the two alkylthio-copper bonds [108].

Perfluoro-tert-butylsilver gives an exothermic reaction with
sulfur in acetonitrile [111]. The silver perfluoro-tert-butyl
mercaptide formed is a good source of the perfluoro-tert-butylthio
group:

$$(CF_3)_3CAg \xrightarrow[CH_3CN]{S} (CF_3)_3CSAg \xrightarrow{CH_2{=}CHCH_2I} (CF_3)_3CSCH_2CH{=}CH_2 \quad (8.24)$$

The reaction of sulfur with perfluorocyclobutylsilver proceeds in
an analogous manner [111]:

$$CF_2 \overbrace{\underset{CF_2}{\overset{CF_2}{\diagup}}}^{CF_2} CFAg + S \longrightarrow CF_2 \overbrace{\underset{CF_2}{\overset{CF_2}{\diagup}}}^{CF_2} CFSAg \xrightarrow{C_2H_5I} CF_2 \overbrace{\underset{CF_2}{\overset{CF_2}{\diagup}}}^{CF_2} CFSC_2H_5$$

$$(8.25)$$

Reactions of copper and gold organometallic compounds with sulfur have not so far been studied.

8.3. ORGANOMAGNESIUM COMPOUNDS

Similarly to the case of organolithium compounds, under certain conditions, the reaction of sulfur with Grignard reagents gives thiols (after hydrolysis):

$$RMgX + S \rightarrow RSMgX \xrightarrow{H_3O^+} RSH \qquad (8.26)$$

Because the reaction products readily undergo oxidation, the reaction has to be carried out in an inert atmosphere. Furthermore, finely powdered sulfur has to be added to the Grignard reagent slowly and in a slightly smaller than stoichiometric amount. Otherwise the yield of the final reaction product is decreased because of side reactions leading to the formation of di- and monosulfides:

$$2RSMgX + S \longrightarrow RSSR + S(MgX)_2 \qquad (8.27)$$
$$RSSR + RMgX \longrightarrow RSR + RSMgX \qquad (8.28)$$

Due to these side reactions [112], the above method of synthesis is less important when compared with other methods of preparation of thiols. For the same reason, it is more convenient to obtain CH_3SMgI by the reaction of methylmagnesium iodide with methanethiol [113]. However, under controlled conditions the yield of thiols from the reaction of Grignard reagents with sulfur can be quite high. Thus, for example, benzenethiol can be obtained from phenylmagnesium bromide in a yield better than 80% [114]. The reaction of sulfur with tert-butylmagnesium chloride gives the corresponding thiol in 70-75% yield [115].

It can be seen from Table 8.2 that the above method has been used to synthesize a number of thiols labelled with radioactive sulfur isotopes, such as butanethiol-^{35}S [116,117], sec-butanethiol-^{35}S [118], phenylmethanethiol-^{35}S [119,120], phenylmethanethiol-^{34}S [121], and also deuterated benzenethiol, C_6D_5SH [122]. The products obtained from the reaction of Grignard reagents with sulfur are summarized in Table 8.2.

Table 8.1. Reactions of Organic Alkali Metal Compounds, RM, with Sulfur

RM	Reagents	Reaction Products	Yield (%)	Ref.
NaC≡CNa	1)S; 2)CH₃I	H₃CSC≡CSCH₃	31	63
HC≡CNa	1)S; 2)C₃H₇Br	H₇C₃SC≡CSC₃H₇	30	63
	1)S; 2)H₂NNa; 3)S; 4)C₂H₅Br	H₅C₂SC≡CSC₂H₅	35-38	51
	1)S; 2)CH₃I	HC≡CSCH₃	20	50
	1)S; 2)C₂H₅X (X=halogen)	HC≡CSC₂H₅	35-39	50
	1)S; 2)C₄H₉X (X=halogen)	HC≡CSC₄H₉	20	50
	1)S; 2)CS₂; 3)H₃O⁺	(cyclic dithiole-thione ring)	25	52,54
LiC—CLi (B₁₀H₇Br₃)	S, H₃O⁺	HSC—CSH (B₁₀H₇Br₃)		35
LiC—CLi (B₁₀H₉Br)	S, H₃O⁺	HSC—CSH (B₁₀H₉Br)		35
LiC—CLi (B₁₀H₁₀)	S, H₃O⁺	HSC—CSH (B₁₀H₁₀)	100	35
m-LiCB₁₀H₁₀CLi	S, H₃O⁺	m-HSCB₁₀H₁₀CSH		35
p-LiCB₁₀H₁₀CLi	S, H₃O⁺	p-HSCB₁₀H₁₀CSH	90	36,37

Table 8.1 (continued)

RM	Reagents	Reaction Products	Yield (%)	Ref.
m-HCB$_{10}$H$_{10}$CNa	S, H$_3$O$^+$	m-HCB$_{10}$H$_{10}$CSH	89	64
H$_3$CC≡CLi	S, ClCN	H$_3$CC≡CSCN	48	65
	S, H$_3$CSSO$_2$CH$_3$	H$_3$CC≡CSSCH$_3$	58	41
	S, (H$_3$C)$_2$NSCl	H$_3$CC≡CSSN(CH$_3$)$_2$	62	41
H$_3$CC≡CNa	S, CH$_3$I	H$_3$CC≡CSCH$_3$	39	50
	S, C$_2$H$_5$X (X=halogen)	H$_3$CC≡CSC$_2$H$_5$	40	50
	S, (H$_3$C)$_2$CHX (X=halogen)	H$_3$CC≡CSCH(CH$_3$)$_2$	10	50
MeC—CLi / 3-F-B$_{10}$H$_9$	S, H$_3$O$^+$	MeC—CSH / 3-F-B$_{10}$H$_9$		66
MeC—CM / B$_{10}$H$_{10}$	S, H$_3$O$^+$	MeC—CSH / B$_{10}$H$_{10}$	90–96	67
(M=Li,Na,K)	S, CH$_3$I	MeC—CSMe / B$_{10}$H$_{10}$	80–90	67
	S, I$_2$	(MeC—CS- / B$_{10}$H$_{10}$)$_2$	95	67
m-H$_3$CCB$_{10}$H$_{10}$CNa	S, H$_3$O$^+$	m-H$_3$CCB$_{10}$H$_{10}$CSH	81–94	64

RM	Reagents	Reaction Products	Yield (%)	Ref.
Cl-thiophene-Li	S, H_3O^+	Cl-thiophene-SH	59	25
Li-thiophene-Li	S, CH_3I	Cl-thiophene-SMe	50.8	68
LiS-thiophene-Li	S, CH_3I	MeS-thiophene-SMe	27	71
	S, H_3O^+	HS-thiophene-SH	49	80
$H_2C=CHC\equiv CNa$	S, CH_3I	$H_2C=CHC\equiv CSCH_3$	51–60	50,69
	S, C_2H_5X (X=halogen)	$H_2C=CHC\equiv CSC_2H_5$	49	69
	S, C_3H_7X	$H_2C=CHC\equiv CSC_3H_7$	52	69
furan-Li	S, H_3O^+	furan-SH	traces	32
	S, CH_3I	furan-SMe	35	32
	S, C_2H_5I	furan-SEt	53	30
	S, $H_2C=CHCH_2Br$	furan-$SCH_2CH=CH_2$	70	32
	S, $C_6H_5CH_2Cl$	furan-SCH_2Ph	60	32

Table 8.1 (continued)

RM	Reagents	Reaction Products	Yield (%)	Ref.
2-thienyllithium	S, H_3O^+	thiophene–SH	70–82	22,23
	1) S; 2) H_3O^+; 3) NaOH; 4) I_2	thiophene–SS–thiophene	87.3	22
	S, $K_3Fe(CN)_6$	thiophene–SS–thiophene	67.5	70
	S, CH_3I	thiophene–SMe	81	71
	S, C_2H_5I	thiophene–SEt	72	71
	S, $BrCH_2CH_2Br$	thiophene–SCH_2CH_2S–thiophene	58	70
3-thienyllithium	1) S; 2) $ClCH_2COOK$; 3) H_3O^+	thiophene–SCH_2COOH	80	72
	S, $ClCH_2COCH_3$	thiophene–SCH_2COMe	56.6	73
	S, $ClCH_2CO_2CH_3$	thiophene–SCH_2CO_2Me	84.5	74
	1) S; 2) $H_3CCHBrCOOK$; 3) H_3O^+	thiophene–$SCH(Me)CO_2H$	70	76
	S, $BrCH_2CO_2Et$	thiophene–SCH_2CO_2Et	65.6	75

RM	Reagents	Reaction Products	Yield (%)	Ref.
2-thienyllithium	S, $(H_3C)_3CBr$	thienyl-$SC(CH_3)_3$	10	77
	p-H_3CO-C_6H_4-N_2^+	thienyl-$SC_6H_4OCH_3$-p	48.4	78
	S, H_3O^+	thienyl-SH	63,80	23,24
	S, CH_3I	thienyl-SMe	impure	81
	1) S; 2) $ClCH_2CO_2K$; 3) H_3O^+	thienyl-SCH_2CO_2H	62	76
	S, $BrCH_2COCH_3$	thienyl-SCH_2COMe	40	79
	S, $ClCH_2CO_2CH_3$	thienyl-SCH_2CO_2Me	63.5	79
	1) S; 2) $H_3CCHBrCO_2K$; 3) H_3O^+	thienyl-$SCH(Me)CO_2H$	54	76
N-methylthiazolyllithium	1) S; 2) $BrCH_2COOEt$; 3) hydrolysis	Me-thiazolyl-SCH_2COOH	36	31
	1) S; 2) $Hg(OCOCH_3)_2$	(Me-thiazolyl-S—Hg)$_2$	96	31

Table 8.1 (continued)

RM	Reagents	Reaction Products	Yield (%)	Ref.
(5-Me-thiazol-2-yl)Li	1) S; 2) BrCH$_2$COOC$_2$H$_5$	(5-Me-thiazol-2-yl)SCH$_2$COOEt	80	31
(5-Me-thiazol-2-yl)Li	1) S; 2) Hg(OCOCH$_3$)$_2$	(5-Me-thiazol-2-yl)SHgOCOCH$_3$	100	31
EtC≡CLi	S, C$_2$H$_5$SH	H$_5$C$_2$CH$_2$CSSC$_2$H$_5$	50	9
EtC≡CNa	S, CH$_3$I	H$_5$C$_2$C≡CSCH$_3$	48–50	50
	S, C$_2$H$_5$Br	H$_5$C$_2$C≡CSC$_2$H$_5$	45–52	48,50
(tetrachloropyridyl)Li	S, H$_3$O$^+$	tetrachloropyridine-SH	28	49
MeC≡CC≡CNa	S, CH$_3$I	MeC≡CC≡CSCH$_3$	65	51
(pyridyl)Li	S, H$_3$O$^+$	pyridine-2-thione-N-OH	8	86,87
(pyridyl)Na	S	2-SNa-pyridine-N-oxide		84,85
(5-Me-furyl)Li	S, ClCH$_2$CO$_2$CH$_3$	(5-Me-furyl)SCH$_2$CO$_2$Me	50.7	82
(5-Me-furyl)Li	S, C$_4$H$_9$Br	(5-Me-furyl)SC$_4$H$_9$	66	30

RM	Reagents	Reaction Products	Yield (%)	Ref.
2-lithio-5-methylfuran	S, $(H_3C)_2CHCH_2X$ (X = halogen)	2-($SCH_2CH(CH_3)_2$)-5-methylfuran	59	30
	1)S; 2)$ClCH_2CO_2K$ 3)H_3O^+	Me-thiophene-SCH_2CO_2H	73	76
	S, $BrCH_2CO_2C_2H_5$	Me-thiophene-SCH_2CO_2Et		27
3-lithio-5-methylthiophene	1)S; 2)$ClCH_2CO_2K$ 3)H_3O^+	Me-thiophene-SCH_2CO_2H	62	76
2-lithio-5-methylthiophene	S, H_3O^+	SH Me-thiophene	62	26
3-lithio-5-methylthiophene	S, H_3O^+	Me-thiophene-SH	50	26
3-lithio-5-MeS-thiophene	S, H_3O^+	MeS-thiophene-SH	64	26
3-lithio-5-MeS-thiophene	S, CH_3I	MeS-thiophene-SMe	57	27
3-lithio-5-MeS-thiophene	S, H_3O^+	MeS-thiophene-SH	55	26
3-lithio-5-SMe-thiophene	S, H_3O^+	SH-thiophene-SMe	43	26

Table 8.1 (continued)

RM	Reagents	Reaction Products	Yield (%)	Ref.
MeS–[thiophene]–Li (2-lithio-5-methylthiothiophene)	S, H_3O^+	MeS–[thiophene]–SH	63	26
$(H_3C)_2C(ONa)C{\equiv}CNa$	S, C_2H_5I	$(H_3C)_2C(OH)C{\equiv}CSC_2H_5$	22.8	83
$H_7C_3C{\equiv}CLi$	S, C_2H_5SH	$H_7C_3CH_2CSSC_2H_5$	46	9
$(H_3C)_2CHC{\equiv}CLi$	S, C_2H_5SH	$(H_3C)_2CHCH_2CSSC_2H_5$	28	9
$H_7C_3C{\equiv}CNa$	S, C_2H_5X (X = halogen)	$H_7C_3C{\equiv}CSC_2H_5$	44–51	50
$p\text{-}LiC_6F_4Li$	S, H_3O^+	$\underline{p}\text{-}HSC_6F_4SH$	24	88,89
C_6F_5Li	S, H_3O^+	C_6F_5SH	46	20,21
	S, EtOCOC≡COOEt	[benzothiophene, F, with two COOEt groups]		21
[dichloro-trifluoro-phenyl lithium: ring with Cl, Cl, F, F, F, Li]	S, H_3O^+	[ring with Cl, F, Cl, F, SH, F]	85	91
$\underline{o}\text{-}HC_6F_4Li$	S, H_3O^+	$\underline{o}\text{-}HC_6F_4SH$	67	20,21, 89
		$(\underline{o}\text{-}HC_6F_4S)_2$		90
	S, CH_3I	$\underline{o}\text{-}HC_6F_4SCH_3$		89
$\underline{m}\text{-}HC_6F_4Li$	S, H_3O^+	$\underline{m}\text{-}HC_6F_4SH$		89
$\underline{p}\text{-}HC_6F_4Li$	S, H_3O^+	$\underline{p}\text{-}HC_6F_4SH$	69	20,21, 89
C_6H_5Li	S, H_3O^+	C_6H_5SH	62	1

RM	Reagents	Reaction Products	Yield (%)	Ref.
C_6H_5Li	S, $(C_2H_5)_2NCH_2CH_2$—CH_2 (epoxide O)	$C_6H_5CH_2CH(OH)CH_2N(C_2H_5)_2$	56	1
$C_2H_5C{\equiv}CC{\equiv}CNa$	S, CH_3I	$C_2H_5C{\equiv}CC{\equiv}CSCH_3$	67	50
(Cl, Me pyridine-Li with N→O)	S, H_3O^+	(pyridine-2-thione; Cl, Me, N–OH substituents)	11.5	86,87
$C_2H_5CH{=}CHC{\equiv}CNa$	S, CH_3I	$C_2H_5CH{=}CHC{\equiv}CSCH_3$	40	50
(Me pyridine-Li, N→O)	S, H_3O^+	(Me pyridine-2-thione, N→O)	39	86
(Et furan-Li)	S, C_2H_5I	(Et thiophene-SEt)	76	30
(Et furan-Li)	S, H_3O^+	(Et thiophene-SH)	72.5	22
	1)S; 2)H_3O^+; 3)NaOH; 4)I_2	(Et thiophene S–S disulfide)	68.8	22
	S, CH_3I	(Et thiophene-SMe)	84.3	75
	S, C_2H_5I	(Et thiophene-SEt)	74	71
(Et thiophene-Li)	S, $ClCH_2COCH_3$	(Et thiophene-SCH_2COMe)	62.2	73

Table 8.1 (continued)

RM	Reagents	Reaction Products	Yield (%)	Ref.
Et-thienyl-Li	S, BrCH₂CO₂C₂H₅	Et-thienyl-SCH₂CO₂Et	53.2	75
	S, BrCH₂COCH₃	Et-thienyl-SCH₂COMe	45.5	79
	S, ClCH₂CO₂CH₃	Et-thienyl-SCH₂CO₂Me	61.5	92
Li-pyrimidine(OMe, MeO)	1)S; 2)ClCH₂CO₂Na; 3)H₃O⁺	HOOCCH₂S-pyrimidine(OMe, MeO)	4.5	93
$H_9C_4C{\equiv}CLi$	S, C_2H_5SH	$H_9C_4CH_2CSSC_2H_5$	48	9
$(H_3C)_3CC{\equiv}CLi$	S, C_2H_5SH	$(H_3C)_3CCH_2CSSC_2H_5$	30	9
	S, $H_3CSSO_2CH_3$	$(H_3C)_3CC{\equiv}CSSCH_3$	75	41
	S, $(H_3C)_3CSSO_2C(CH_3)_3$	$(H_3C)_3CC{\equiv}CSSC(CH_3)_3$	85	41
	S, $(CH_3)_2NSCl$	$(H_3C)_3CC{\equiv}CSSN(CH_3)_2$	75	41
$H_9C_4C{\equiv}CNa$	S, C_2H_5X (X = halogen)	$H_9C_4C{\equiv}CSC_2H_5$	59-63	50
$(H_3C)_3CC{\equiv}CNa$	S, C_2H_5X	$(H_3C)_3CC{\equiv}CSC_2H_5$	48-57	50
$[(CH_3)_3Si]_2NNa$ (a large excess)	S	$[(CH_3)_3Si]_2NSNa$		94
$[(CH_3)_3Si]_2NNa$	S (1:1)	$\{[(CH_3)_3Si]_2NS\text{-}\}_2$	20	94
		$\{[(CH_3)_3Si]_2NS\text{-}\}_2S$	40	94

RM	Reagents	Reaction Products	Yield (%)	Ref.
perfluoro-C$_6$F$_{11}$-Li (F, CF$_2$, F$_2$ ring)	S, H$_3$O$^+$	perfluoro ring–SH (F, CF$_2$, F$_2$)		19
p-LiC$_6$F$_4$CO$_2$Li	S, H$_3$O$^+$	p-HSC$_6$F$_4$CO$_2$H	68.7	95
H–C$_6$F$_{10}$–Li (F$_2$, CF$_2$, F$_2$ ring)	S, H$_3$O$^+$	H–ring–SH (CF$_2$, F$_2$)		19
p-H$_3$COC$_6$H$_4$Li	S, H$_3$O$^+$	p-H$_3$COC$_6$H$_4$SH	66	96
3,4-dilithio(OCH$_2$O)benzene	2S, H$_3$O$^+$	benzene(OCH$_2$O) with 2 SH	35	97
3-lithio(OCH$_2$O)benzene	S, H$_3$O$^+$	benzene(OCH$_2$O)–SH	61	97
thiophene, CH(OCH$_3$)$_2$, Li	1)S; 2)ClCH$_2$COOCH$_3$; 3)H$_3$O$^+$; 4)C$_2$H$_5$ONa	thieno[...]thiophene–COOH	94	98
thiophene, Br, CH(OLi)NMe$_2$, Li	1)S; 2)BrCH$_2$CO$_2$CH$_3$; 3)H$_3$O$^+$; 4)C$_2$H$_5$ONa	Br–thienothiophene–COOH	25	28
selenophene, Br, CH(OLi)NMe$_2$, Li	1)S; 2)BrCH$_2$CO$_2$CH$_3$; 3)H$_3$O$^+$; 4)C$_2$H$_5$ONa	Br–selenothiophene(Se/S)–COOH	30	28
pyrimidine, OMe, Li, OMe, MeO	1)S; 2)ClCH$_2$CO$_2$Na; 3)H$_3$O$^+$	pyrimidine OMe, SCH$_2$CO$_2$H, OMe, MeO	33	93

Table 8.1 (continued)

RM	Reagents	Reaction Products	Yield (%)	Ref.
pyrimidine: OMe, Li / N, MeO, N (4-lithio-2,6-dimethoxypyrimidine)	1)S; 2)CH₃CH(Cl)CO₂Na; 3)H₃O⁺	pyrimidine: OMe, SCH(Me)CO₂H / MeO, N, OMe, N	33	93
$H_{11}C_5C{\equiv}CLi$	S, C_2H_5SH	$H_{11}C_5CH_2CSSC_2H_5$	40	9
$C_6H_5C{\equiv}CNa$	S, H_3O^+	$C_6H_5C{\equiv}CSH$		99
	S, CS_2, H_3O^+	PhC–S, HC–S ring, C=S	53–60	52,53
	S, CH_3I	$C_6H_5C{\equiv}CSCH_3$		99
	S, C_2H_5X (X = halogen)	$C_6H_5C{\equiv}CSC_2H_5$	58–60	50
dithienyl–Li (thiophene rings)	S, $(H_3C)_2CHCH_2I$	thienyl–S–S–$CH_2CH(CH_3)_2$	70	
benzothiophene–Li	S, H_3O^+	benzothiophene–SH		100
$NaC{\equiv}C(CH_2)_4C{\equiv}CNa$	2S, 2CH_3I	$H_3CSC{\equiv}C(CH_2)_4C{\equiv}CSCH_3$	40	50
$(H_3C)_2N$–C₆H₄–Li	S, H_3O^+	$(H_3C)_2N$–C₆H₄–SH	50	1
tert-BuS–thiophene–Li	S, H_3O^+	tert-BuS–thiophene–SH	54	77
$n{-}C_6H_{13}C{\equiv}CLi$	S, EtSH	$n{-}C_6H_{13}CH_2C(S)SC_2H_5$	44	9

RM	Reagents	Reaction Products	Yield (%)	Ref.
PhC—CM carborane ($B_{10}H_{10}$) (M=Li,Na,K)	S, H_3O^+	PhC—CSH carborane ($B_{10}H_{10}$)	90–96	67
	S, CH_3I	PhC—CSMe carborane ($B_{10}H_{10}$)	80–90	67
	S, I_2	(PhC—CS—)$_2$ carborane ($B_{10}H_{10}$)	95	67
thiophene—$CH(OEt)_2$, Li	S, $ClCH_2CO_2Me$	thiophene with $CH(OEt)_2$ and SCH_2CO_2Me	80	101
	1) S; 2) $ClCH_2CO_2Me$; 3) H_3O^+	thiophene with CHO and SCH_2CO_2Me	80	101
$(EtO)_2CH$—thiophene—Li	S, $ClCH_2CO_2Me$	$(EtO)_2CH$—thiophene—SCH_2CO_2Me	64	102
$(H_3C)_2CH(CH_2)_2$S—thiophene—Li	S, H_3O^+	$(H_3C)_2CH(CH_2)_2$S—thiophene—SH	41.5	77
ferrocene (Li, Li)	S	ferrocene S—S—S bridge	52	38,39

Table 8.1 (continued)

RM	Reagents	Reaction Products	Yield (%)	Ref.
(dioxolane–dithiophene structure)	S	(three fused bicyclic thiophene structures)	6	103
$(H_7C_3O)_2PCH_2(OCH_2)_2C\equiv CNa$ (P=O)	S, EtBr	$(H_7C_3O)_2PCH_2(OCH_2)_2C\equiv CSC_2H_5$ (P=O)		104
(difluorophenyl dilithium biphenyl structure, Li···Li with F substituents)	2S, H$_3$O$^+$	(bis-mercapto perfluorobiphenyl structure, HS···SH)		89
Ph_2SnLi_2	2PhCOCl	$Ph_2Sn(SCOPh)_2$	49	45
$C_{12}H_{25}Li$	S, H$_3$O$^+$	$C_{12}H_{25}SH$	50	2
$H_5C_6CHNaCHNaC_6H_5$	S	$PhCH=CHPh$	75	105
$H_5C_6CH=CHPC_6H_5$ (K)	S	$H_5C_6CH=CHP=S$ (C_6H_5, SK)	72.5	106

RM	Reagents	Reaction Products	Yield (%)	Ref.
(o-$AsPh_2$-C$_6$H$_4$)Li	S, H_3O^+	(o-$AsPh_2$-C$_6$H$_4$)SH		107
Ph_3SiLi	S, MeI	Ph_3SiSCH_3	52	40
	S, $PhCH_2Cl$	$Ph_3SiSCH_2C_6H_5$	36	40
	S, PhCOCl	$Ph_3SiSCOC_6H_5$	36	40
Ph_3GeLi	S, PhCOCl	$Ph_3GeSCOPh$	59	42
	S, Ph_3GeBr	$(Ph_3Ge)_2S$		42
	S, Ph_3SnCl	$Ph_3GeSSnPh_3$		42
	S, Ph_3PbCl	$Ph_3GeSPbPh_3$		42
Ph_3SnLi	S	$(Ph_3SnLi)_2$	95	43
	S, PhCOCl	$Ph_3SnSCOPh$	46.5	43
	S, Ph_3GeBr	$Ph_3SnSGePh_3$	49	43
	S, Ph_3PbCl	$Ph_3SnSPbPh_3$	50	43
Ph_3PbLi	S, PhCOCl	$Ph_3PbSCOPh$		42
	S, Ph_3SnCl	$Ph_3PbSSnPh_3$		42
	S, Ph_3PbCl	$(Ph_3Pb)_2S$		42

Table 8.2. Reactions of Sulfur with Organomagnesium Compounds

RMgX	Reagents	Reaction Products	Yield (%)	Ref.
CH$_2$=CHMgBr	S	CH$_2$=CHSMgBr		162
EtMgBr	S, H$_3$O$^+$	EtSH		123
Cl–(thienyl)–MgI	S, H$_3$O$^+$	Cl–(thienyl)–SH	20	25
(thienyl)–MgI	S, H$_3$O$^+$	(thienyl)–SH	67–85	124–126
(thienyl)–MgI	S, MeI	(thienyl)–SMe	53–60	127–129
(thienyl)–MgI	S, H$_3$O$^+$	(thienyl)–SH		125
BuMgBr	35S, H$_3$O$^+$	Bu–35SH		116,117
sec–BuMgBr	35S, H$_3$O$^+$	sec–Bu–35SH	35	118
tert–BuMgCl	S, H$_3$O$^+$	tert–BuSH	70–75	115
Me–(thienyl)–MgBr	S, H$_3$O$^+$	Me–(thienyl)–SH	65	130
Me–(thienyl)–MgBr	S, H$_3$O$^+$	(thienyl)–SH	52	26
cyclopentyl–MgBr	S, H$_3$O$^+$	cyclopentyl–SH and cyclopentyl–S–S–cyclopentyl	35.5 5	131

RMgX	Reagents	Reaction Products	Yield (%)	Ref.
C_6Br_5MgBr	S, H_3O^+	C_6Br_5SH	58	134
C_6F_5MgBr	S, H_3O^+	C_6F_5SH	67	20,21
$\underline{p}-FC_6H_4MgBr$	S, H_3O^+	$\underline{p}-FC_6H_4SH$	26	135
C_6H_5MgX ($X = Br, Cl$)	S, H_3O^+	C_6H_5SH		110,114, 136,137
C_6D_5MgBr	S, H_3O^+	C_6D_5SH	50	122
	S, H_3O^+			123
	S, H_3O^+			123,132
	S, H_3O^+			123
cyclo-$C_6H_{11}MgBr$	S, H_3O^+	cyclo-C_6H_{11}-SH (cyclo-C_6H_{11}-S-)$_2$	60 3,4	131,133
	S, H_3O^+		49	140
$PhCH_2MgCl$	$^{35}S, H_3O^+$	$PhCH_2-^{35}SH$	80-88	120,139
	$^{35}S,$ ClCH$_2$CHCOOH$\underset{NH_2 \cdot HCl}{\mid}$	$PhCH_2-^{35}SCH_2CHCOOH\underset{NH_2}{\mid}$	38-44	138

Table 8.2 (continued)

RMgX	Reagents	Reaction Products	Yield (%)	Ref.
PhCH$_2$MgBr	34S, H$_3$O$^+$	PhCH$_2$-34SH	88	121
	35S, H$_3$O$^+$	PhCH$_2$-35SH	78	116,117, 119
p-MeOC$_6$H$_4$MgBr	S, H$_3$O$^+$	p-MeOC$_6$H$_4$SH		141
[cyclohexyl-Me MgCl]	S, H$_3$O$^+$	[cyclohexyl-Me SH]		123,132
[benzothienyl MgI]	S, H$_3$O$^+$	[benzothienyl SH]		109
[indole MgBr]	S	[indolyl-S]$_2$		144,145
[o-vinylphenyl MgBr], CH=CH$_2$	S, H$_3$O$^+$	[o-vinylphenyl SH], CH=CH$_2$		142
[F-mesityl MgBr], Me	S, H$_3$O$^+$	[F-mesityl SH], Me	48	143
2,6-Me$_2$C$_6$H$_3$MgBr	S, H$_3$O$^+$	2,6-Me$_2$C$_6$H$_3$SH	78	146
[cyclohexyl-Me CH$_2$MgX]	S, H$_3$O$^+$	[cyclohexyl-Me CH$_2$SH]		123
[indolyl-Me MgBr]	S, CH$_3$COCl	[indolyl-Me SCOMe]		148

RMgX	Reagents	Reaction Products	Yield (%)	Ref.
2-methylindole-MgBr (Me, MgBr, N–H)	S, PhCOCl	2-methylindole SCOPh derivative (Me, SCOPh, N–H)		148
	S, PhCOCl	{indole $(S)_3$ trisulfide (Me Me, N–H) and SCOPh derivative (Me, SCOPh, N–H)}	main product	148
o-(CH$_2$CH=CH$_2$)C$_6$H$_4$MgCl	S, H$_3$O$^+$	o-(CH$_2$CH=CH$_2$)C$_6$H$_4$SH		142
p-Me$_3$SiC$_6$H$_4$MgBr	S, H$_3$O$^+$	p-Me$_3$SiC$_6$H$_4$SH	main product	147
m-Me$_3$SiC$_6$H$_4$MgBr	S, H$_3$O$^+$	(p-Me$_3$SiC$_6$H$_4$S)$_2$, m-Me$_3$SiC$_6$H$_4$SH		147
4-bromonaphthalene-1-MgBr		4-bromonaphthalene-1-SH, [4-bromonaphthalene-1-S–]$_2$		150
4-chloronaphthalene-1-MgBr		4-chloronaphthalene-1-SH, [4-chloronaphthalene-1-S–]$_2$		150
1-C$_{10}$H$_7$MgCl	S, H$_3$O$^+$	1-C$_{10}$H$_7$SH		136

Table 8.2 (continued)

RMgX	Reagents	Reaction Products	Yield (%)	Ref.
1-$C_{10}H_7$MgBr	S, H_3O^+	1-$C_{10}H_7$SH	55	149
2-$C_{10}H_7$MgBr	S, H_3O^+	2-$C_{10}H_7$SH	65	149
(cyclohexyl)MgCl	S, H_3O^+	(cyclohexyl)SH		151,153
(decahydronaphthyl)MgX	S, H_3O^+	(decahydronaphthyl)SH; [(decahydronaphthyl)S—]$_2$		123,132
(decahydronaphthyl)MgX	S, MeI	(decahydronaphthyl)SMe		123
(phenanthryl)MgX	S, H_3O^+	(phenanthryl)SH		
Ph_3CMgBr	S, H_2O/NH_4Cl	Ph_3CSH	70	154
$Ph_2C=C(Ph)$MgBr	S, H_3O^+	$Ph_2C=C(SH)Ph$		152

8.4. ORGANIC DERIVATIVES OF ZINC, CADMIUM, AND MERCURY

The reaction of sulfur with perfluoroisopropylzinc iodide in a sealed tube at 243°C (13 hours) gives a mixture of sulfides [155]. It is assumed that the reaction follows a free-radical mechanism:

$$(CF_3)_2CFZnI + S \longrightarrow [(CF_3)_2CF]_2S + [(CF_3)_2CF]_2S_2$$
$$+ [(CF_3)_2CF]_2S_3 \qquad (8.29)$$

The reaction of dimesitylzinc with an equimolar amount of sulfur in refluxing tetrahydrofuran leads to the formation of zinc sulfide and a mixture of products including R_2S, R_2S_2 (main product), R_2S_3, and R_2S_4, where R = mesityl [155a].

Organic derivatives of mercury are dealkylated (or dearylated) when treated with sulfur. In general, the reaction products are sulfides, R_2S, or disulfides, R_2S_2. Thus, for example, when diphenylmercury is heated with sulfur in a sealed tube at 220-230°C, diphenyl sulfide is formed in a yield up to 66% [161]. Bis-pentafluorophenylmercury is dearylated by sulfur under more drastic conditions (250°C) to give an 82% yield of bis-pentafluorophenyl sulfide [156]. However, the reaction of dibenzylmercury with sulfur gives a quantitative yield of bibenzyl, instead of the expected sulfide [124]:

$$R_2Hg + 2S \longrightarrow R_2S + HgS \qquad (8.30)$$
$$R = C_6H_5,\ C_6F_5,\ o\text{-}CH_3C_6H_4\ ,\ etc.$$

The reaction of sulfur with (bromodichloromethyl)phenylmercury takes a very interesting course [158,159]. The reaction products are bromophenylmercury and tetrachlorothiirane, obtained in 92% and 36% yield, respectively. The reaction involves conversion of sulfur into thiophosgene by interaction with dichlorocarbene or $C_6H_5HgCCl_2Br$ followed by the addition of the carbene to the C=S bond:

$$RHgCCl_2Br + S \xrightarrow[-RHgBr]{} Cl_2C{=}S \xrightarrow{RHgCCl_2Br} Cl_2C{-}CCl_2 + RHgBr \qquad (8.31)$$
$$\underset{S}{\diagdown\diagup}$$
$$R = C_6H_5$$

The reaction of arylmercury chlorides and chlorododecylmercury with sulfur in sulfolane gives disulfides in good yields [160]. Under these conditions, 3-pyridylmercury chloride is not converted to the corresponding disulfide:

$$2RHgCl + 2S \xrightarrow{140-180°C} RSSR + Hg_2Cl_2 \qquad (8.32)$$

A general method for the preparation of perfluorothioketones is based on the reaction of perfluoro-sec-alkyl derivatives of mercury [163-165]. Thus, for example, hexafluorothioacetone is obtained in 60% yield by the addition of bis-perfluoroisopropyl-mercury to boiling sulfur:

$$[(CF_3)_2CF]_2Hg + 2S \xrightarrow{440°C} (CF_3)_2C=S \qquad (8.33)$$

Under milder conditions (70-100°C), sulfur reacts with bis-perfluoroalkyl derivatives of mercury in dimethylformamide in the presence of catalytic amounts of potassium fluoride [59,166] or cesium fluoride [167]. The corresponding bis-perfluoroalkanethio-lates of mercury are thus obtained in high yield, e.g.,

$$(R)_2Hg + S \xrightarrow[70-100°C]{KF \ in \ DMF} (R\,S)_2Hg \qquad (8.34)$$

$$R = (CF_3)_2CF, \ (CF_3)_3C$$

It is assumed that the fluoride ion is coordinated to the mercury atom and this enhances the carbanionic nature of the carbon atom bonded to the mercury atom. The complex formed then reacts with sulfur [59]:

$$(R)_2Hg \underset{-F^-}{\overset{F^-}{\rightleftharpoons}} (R)_2HgF^- \underset{-S}{\overset{S}{\rightleftharpoons}} R \overset{\cdots}{\underset{\cdots\ S}{\underset{S}{\cdots}}} Hg-R \longrightarrow R\,S-Hg-R \qquad (8.35)$$

Sulfur reacts exothermally with bis-triethylsilylmercury at room temperature and is inserted into the Si-Hg bond. When the ratio of the reagents is 1:1 and the reaction is carried out without any solvent, an almost quantitative yield of the corresponding mercury-containing sulfide is obtained [168]:

$$(R_3Si)_2Hg + S \longrightarrow R_3SiSHgSiR_3 \qquad (8.36)$$

$$R = C_2H_5$$

Under comparable conditions, the reaction between sulfur and bis-triethylgermylmercury is accompanied by demercuration. Mercury and bis-triethylgermyl sulfide are formed as reaction products.

The undesirable demercuration can be suppressed when the reaction between sulfur and silyl or germyl derivatives of mercury is carried out in pentane or hexane under controlled temperature conditions (0-20°C) [169,170]:

$$(R_3M)_2Hg + S \xrightarrow{0-20°C} R_3MS-Hg-MR_3 \qquad (8.37)$$

$$M = Si, \ Ge; \ R = C_2H_5, \ CH(CH_3)_2$$

The reaction of sulfur with bis-(triperfluorophenylgermyl)-mercury carried out in tetrahydrofuran at 50°C takes place in an analogous manner [171]. It is assumed that the primary process is the ring opening of the cyclic S_8 molecule. This process takes place in a four-center activated complex whose formation is facili-tated by the coordination of the sulfur atoms with the atoms in the Si-Hg (or Ge-Hg) bond, e.g.,

$$R_3Si \cdots HgSiR_3$$
$$\begin{array}{c} | \quad\quad | \\ S \cdots S \\ \searrow S_6 \nearrow \end{array} \rightarrow R_3SiS-(S)_6-SHgSiR_3 \qquad (8.38)$$

This is followed by degradation of the sulfur chain as a result of the interaction of the polysulfide formed with the starting silyl or germyl derivative of mercury.

In agreement with the above, the reaction of sulfur with the asymmetric ethyl(triethylsilyl)mercury and ethyl(triethylgermyl)-mercury occurs with selective insertion of the chalcogen atom into the Si–Hg and Ge–Hg bond, respectively [169,170]. An analogous mechanism has been suggested for the reaction of sulfur with bis-triethylgermylmercury [186]:

$$RHg-MR_3+S \longrightarrow RHg-SMR_3 \qquad (8.39)$$
$$M = Si, \; Ge; \; R = C_2H_5$$

Reaction (8.40) takes place at room temperature in toluene. When an equimolar ratio of the reagents is used, the reaction products are triethylgermyl(triethylgermylthio)cadmium (76% yield) and bis-(triethylgermylthio)cadmium (12%). The former of the above two compounds is a dimer. The molecular weight of the latter compound was not determined because of its poor solubility [186]:

$$3\,(R_3Ge)_2\,Cd + 4S \xrightarrow[\text{toluene}]{\sim 20°C} [R_3GeS-CdGeR_3]_2 + (R_3GeS)_2\,Cd \qquad (8.40)$$
$$R = C_2H_5$$

The reactions of organic derivatives of mercury with elemental sulfur are summarized in Table 8.3.

8.5. ORGANIC DERIVATIVES OF THE BORON SUBGROUP ELEMENTS

The reaction of cobalt vapor with pentaborane, cyclopentadiene and elemental sulfur yielded the first known example of a dithia-metalloborane cluster, $6,8,7,9-(\eta^5-C_5H_5)CO_2S_2B_5H_5$ [187].

In contrast to the reactions of sulfur with organolithium and organomagnesium compounds, the reactions of trialkylboranes with sulfur are endothermic. When trialkylboranes are heated with sulfur, small yields of esters of dialkylthioborinic acids are obtained [188–190]:

$$R_3B+S \longrightarrow R_2BSR \qquad (8.41)$$
$$R = C_3H_7, \; C_4H_9$$

Table 8.3. Reactions of Organic Compounds of Mercury with Sulfur

R_2Hg or RHgX	Reagents and reaction conditions	Reaction products	Yield (%)	Ref.
$(F_3CCCl_2)_2Hg$	S, 440°C (reflux)		37	166,172
$(F_3CCClF)_2Hg$	S, 440°C (reflux)		80	163,164
$(F_3C)_3CHgCl$	S and KF in DMF, 70°C, 15 hr	$[(F_3C)_3CS]_2Hg \cdot 0.5$ DMF	68	166
$(F_5C_2)_2Hg$	S, 250°C	$(F_3CCF_2S)_2$		173
$[(F_3C)_2N]_2Hg$	S, 165°C, 15 hr	$[(F_3C)_2N]_2S$ $[(F_3C)_2NS]_2$		174
F_5C_6HgCl	S, 250°C, 6 days	$(F_5C_6)_2S$	10	156
	S, 230°C	$(F_5C_6)_2S$ $(F_5C_6S)_2$	5	175
$[(F_3C)_2CF]_2Hg$	S, 445°C (reflux)	$(F_3C)_2C{=}S$	60	163–165
	S, 220°C, stainless steel bomb	$[(F_3C)_2CF]_2S_2$ $[(F_3C)_2CF]_2S_3$ $[(F_3C)_2CF]_2S_4$		130
	S and KF in DMF, 90–100°C, 15 hr	$[(F_3C)_2CFS]_2Hg$	100	59,166
	S in sulfolane, 180–190°C, 24 hr		60–75	160

R_2Hg or RHgX	Reagents and reaction conditions	Reaction products	Yield (%)	Ref.
$\underline{o}$-FC_6H_4HgCl	S in sulfolane, 140-180°C, 24 hr	$(\underline{o}\text{-}FC_6H_4S\text{-})_2$	60-75	160
$PhHgCl$	S in sulfolane, 140-180°C, 24 hr	$PhSSPh$	60-75	160
$\underline{o}$-$MeOC_6H_4HgCl$	S in sulfolane, 140-150°C, 24 hr	$(\underline{o}\text{-}MeOC_6H_4S\text{-})_2$	60-75	160
$\left(ClCF_2CF_2\!\!\diagdown\!\!\underset{CF_3}{CF}\right)_2 Hg$	S, 440°C (reflux)	$ClCF_2\text{—}CF_2\diagdown\underset{CF_3}{\overset{}{C}}{=}S$		165
$[(CF_3)_3Cl]_2Hg$	S and KF in DMF, 70°C, 20 min	$[(CF_3)_3CS]_2Hg\cdot 0.5$ DMF	79	59,166
$Et_3SiHgEt$	S in hexane, 20°C, 4 hr	$Et_3SiSHgEt$	85	170
$Et_3GeHgEt$	S in hexane, 20°C, 4 hr	$Et_3GeSHgEt$	98	170
$2,4,6\text{-}Me_3C_6H_2HgCl$	S in sulfolane, 140-180°C, 24 hr	$(2,4,6\text{-}Me_3C_6H_2S\text{-})_2$	60-75	160
$2\text{-}C_{10}H_7HgCl$	S in sulfolane, 140-180°C, 24 hr	$(2\text{-}C_{10}H_7S\text{-})_2$	60-75	160
$\left(\underset{F}{\overset{F}{\diagdown}}\!\!\overset{F}{\underset{F}{\diagup}}\!\!\!\!\overset{-}{Br}\right)_2 Hg$	S, 230°C, 22 hr	(dibenzothiophene-type structure)		182
$(C_6F_5)_2Hg$	S, 250°C, 7 days	$(C_6F_5)_2S$	82	156

Table 8.3 (continued)

R_2Hg or $RHgX$	Reagents and reaction conditions	Reaction products	Yield (%)	Ref.
$(o\text{-}HC_6F_4)_2Hg$	S, 230°C	$(o\text{-}HC_6F_4)_2S$	94	178
Ph_2Hg	S, sealed tube 220–230°C, 6 hr	Ph_2S $PhSSPh$	66	161,177
	S, fused	Ph_2S $(PhS)_2Hg$ $PhSH$		176
$(Et_3Si)_2Hg$	S in pentane, 20°C, 4 hr	$Et_3SiSHgSiEt_3$	54.7	168–170
$(Et_3Ge)_2Hg$	S in pentane, 20°C, 4 hr	$Et_3GeSHgGeEt_3$	68.1	169–170
$Et_3SiHgGeEt_3$	S in pentane, 20°C, 10 hr	$Et_3SiSHgGeEt_3$	69.0	169–170
$n\text{-}C_{12}H_{25}HgCl$	S in sulfolane, 140–150°C, 24 hr	$(n\text{-}C_{12}H_{25}S\text{-})_2$	60–75	160
	S and CsF in DMF			167
	S and CsF in DMF			167

R_2Hg or $RHgX$	Reagents and reaction conditions	Reaction products	Yield (%)	Ref.
(o-MeC$_6$H$_4$)$_2$Hg	S, sealed tube, 225–230°C, 12 hr and 285°C, 4 hr	(o-MeC$_6$H$_4$)$_2$S		179
(PhCH$_2$)$_2$Hg	S, sealed tube, 100°C, 24 hr	PhCH$_2$CH$_2$Ph	100	157
(i-Pr$_3$Ge)$_2$Hg	S in hexane, 20°C, 15 hr	i-Pr$_3$GeSHgGe(i-Pr)$_3$	74.5 85	169,170
diferrocenyl mercury	S, sealed tube, 160–180°C, 18 hr	diferrocenyl disulfide	small amount	180
	S, 300°C			183
[(C$_6$F$_5$)$_3$Ge]$_2$Hg	S in THF, 50°C, 1 hr	(C$_6$F$_5$)$_3$GeSHgGe(C$_6$F$_3$)$_3$	63.5	171
{[(CH$_3$)$_3$CCH$_2$]$_3$Sn}$_2$Hg	S in hexane, 20°C	{[(CH$_3$)$_3$CCH$_2$]$_3$Sn}$_2$S	77	184
[(C$_6$H$_5$CH$_2$)$_3$Si]$_2$Hg	S in xylene, 100°C, 24 hr	[(C$_6$H$_5$CH$_2$)$_3$Si]$_2$S	36	185

It is assumed that the primary process is the formation of a donor-
acceptor complex between the reacting molecules, $R_3B \leftarrow S_8$ [189].
Reaction (8.41) can be utilized for the formation of dialkyl disulfides when the reaction is distilled in the presence of an organic
base (e.g., piperidine) [191]. Diphenyl disulfide can be similarly
obtained:

$$R_3B + S \xrightarrow[\text{10 hr}]{130°C} R_2BSR \xrightarrow{\text{hydrolysis}} RSH + R_2BOH$$

$$\downarrow$$

$$RSSR \qquad (8.42)$$

$$R = \text{cyclo-}C_6H_{11} \qquad (33-48\%)$$

Because the B–C bond is less polar than the Mg–C bond, reactions
(8.41) and (8.42) cannot include a step which would be analogous to
the nucleophilic attack of the Grignard reagent upon the cyclic
sulfur molecule, S_8. It is more likely that these reactions take
place with the intermediate formation of a four-center cyclic complex [191]:

$$
\begin{array}{ccc}
R_2B\text{------}R & \quad & R_2B \quad R \\
\vdots \quad\quad \vdots & \rightarrow & |\quad\quad| \\
S\text{------}S & & S \quad S \\
\diagdown S_6 \diagup & & \diagdown S_6 \diagup
\end{array}
\qquad (8.43)
$$

The polysulfide $R_2B\text{-}S\text{-}(S)_6SR$ formed in this reaction can react with
the starting trialkylborane and form an ester of a dialkylthioborinic acid (or the corresponding acid with two other organic
groups) (cf. reaction schemes (8.41) and (8.42)), or it can undergo
hydrolysis in the presence of bases such as piperidine or sodium
hydroxide and form a thiol which is quickly converted into the
corresponding disulfide in the presence of a base:

$$R_2BS\text{-}(S)_6\text{-}SR \xrightarrow{BR_3} R_2BS\text{------}S\text{-}(S)_6 R \rightarrow R_2BSR + R_2BS_7R \qquad (8.44)$$

$$\vdots \quad\quad\quad \vdots$$

$$R\text{------}BR_2$$

$$R_2B\text{-}(S)_7\text{-}R \xrightarrow{BR_3} R_2BSR + R_2B\text{-}(S)_6\text{-}R \text{ , etc.} \qquad (8.45)$$

The reaction of sulfur with dialkyliodoboranes is quite excep-
tional in that it leads to the formation of bis-dialkylboryl disulfides [192] regardless of the ratio of the starting reactants:

$$2R_2BI + 2S \rightarrow R_2BSSBR_2 + I_2 \qquad (8.46)$$

$$R = C_2H_5, \ CH(CH_3)_2, \ C_4H_9$$

It has been shown in the case of iododiphenylborane that, even under
drastic conditions, there is no further cleavage of the S–S bonds
according to reaction scheme (8.47):

$$R_2B-S-S-BR_2 + 2R_2BI \nrightarrow 2R_2BSBR_2 + I_2 \qquad (8.47)$$

However, iododiphenylborane reacts with sulfur (even when sulfur is present in excess) giving only B,B'-thiobis-diphenylborane. Reaction (8.48) takes place at room temperature. Under more drastic conditions, the excess sulfur is incorporated into the B-C bond. The difference between reaction schemes (8.46) and (8.48) can be explained by the inductive effect of the alkyl groups stabilizing the
$-\overset{|}{B}-S-S-\overset{|}{B}-$ structure [192]:

$$2R_2BI+S \longrightarrow R_2BSBR_2+I_2 \qquad (8.48)$$
$$R=C_6H_5$$

The reaction of sulfur with compounds of the type RBI_2 (where R = I, C_6H_5, p-$CH_3C_6H_4$, or CH_3) leads to the formation of five-membered heterocyclic compounds according to the following scheme [193-195]:

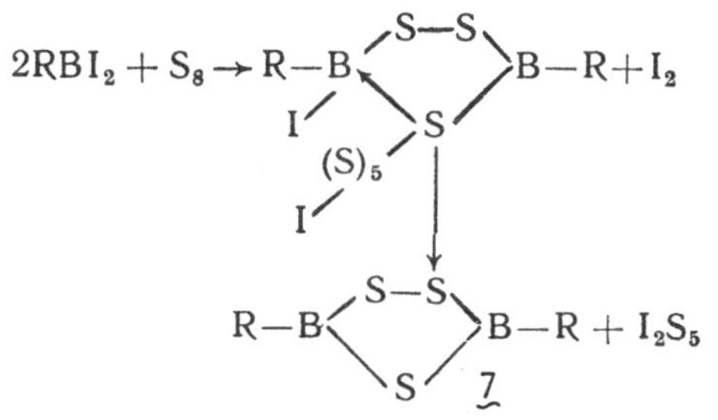

$$R=I, \; CH_3, \; C_2H_5, \; \text{p-}CH_3C_6H_4$$

The molecular structure of heterocycle ($\underline{7}$, R = CH_3) has been studied [196]. Iodomesitylborane does not react according to scheme (8.49), even upon prolonged heating at 90°C. Under more drastic conditions, polymeric products of ill-defined structure are formed [195].

The interaction of equimolar amounts of sulfur and trialkyl-aluminium compounds gives products of the type $RSAlR_2$ where the sulfur atom is incorporated into the C-Al bond [197-198]. These compounds can be distilled under reduced pressure without decomposition and their hydrolysis yields the corresponding thiol:

$$R_3Al + S \longrightarrow RSAlR_2 \xrightarrow{H_2O} RSH \qquad (8.50)$$
$$R = C_2H_5, \; C_3H_7, \; CH_2CH(CH_3)_2$$

The reaction of trialkylaluminium with two equivalents of sulfur gives a mixture of dialkyl(alkylthio)aluminium, dialkyl disulfide, as well as other reaction products:

$$R_3Al+S \longrightarrow RSAlR_2+R_2S+R_2S_2 \qquad (8.51)$$

A similar complex is also obtained in the reaction of triisobutyl-aluminium with three equivalents of sulfur [197]. This reaction affords (upon hydrolysis) a sulfide, R_2S, disulfide, R-S-S-R, tri-sulfide, $R-(S)_3-R$ (where $R = CH_2CH(CH_3)_2$), and other reaction products.

According to two patents [199,200], alkanethiols, dialkyl disulfides, and dialkyl polysulfides containing from 2 to 30 carbon atoms in the alkyl groups can be obtained by the reaction of tri-alkylaluminium derivatives with sulfur followed by decomposition of the aluminium thiolates formed.

Treatment of alkylaluminiums, $(Me_2CHCH_2)_2AlCH=CHR$ where R = Et, Pr, Bu, or n-hexyl, with elemental sulfur affords $(Me_2CHCH_2)_2AlSCH=$ $=CH-R$ which were treated without separation with acetic anhydride to give AcSCH=CHR with an overall yield of 52-64% and 93-96% selec-tivity to the _trans_ isomer [201].

Bis-pentafluorophenylthallium bromide reacts with sulfur to give bis-pentafluorophenyl sulfide [202,203]:

$$(C_6F_5)_2TlBr + S \longrightarrow 2TlBr + (C_6F_5)_2S \qquad (8.52)$$

8.6. ORGANIC DERIVATIVES OF THE SILICON SUBGROUP ELEMENTS

Sulfur reacts with tetrabutylsilane under drastic conditions (380°C) with the formation of hexabutyldisilathiane [204]. It is assumed that the reaction takes place via the intermediate formation of tributyl(butylthio)silane. Under comparable conditions, the reaction with tetraphenylsilane gives diphenyl sulfide and tarry products.

$$R_4Si + S \xrightarrow{380°C} R_3SiSR \rightarrow (R_3Si)_2S \qquad (8.53)$$

$$R = C_4H_9$$

The corresponding reaction with tetrabutylgermane at 230°C takes place with the formation of dibutyl sulfide and hexabutyl-2,4,6-trigerma-1,3,5-trithiane $[(C_4H_9)_2GeS]_3$, whose yield can reach 45%. Triphenylgermane reacts with sulfur above 270°C with the intermediate formation of hexaphenyl-2,4,6-trigerma-1,3,5-trithiane $[(C_6H_5)_2GeS]_3$, which undergoes further decomposition with the for-mation of metallic germanium and diphenyl sulfide [204].

It is known that above 160°C the cyclic molecules of sulfur, S_8, are in equilibrium with linear polar fragments, $^+\underline{S}-(\underline{S})_n-\underline{S}^-$. It is assumed [204] that the linear sulfur fragments cause the polar-ization of the M-C bonds in compounds of the type R_4M (M = Si, Ge, Sn). In the limiting case, this effect may lead to the formation of carbanions R^-. The nucleophilic attack of the latter species upon the polar sulfur fragments leads to the cleavage of the S-S bonds. An analogous mechanism has been postulated for the reaction of sulfur with 1,1-dialkylgermacyclobutanes and similar compounds [205]. These reactions are complete within several minutes at 250°C and are accompanied by the incorporation of a sulfur atom into the endocyclic Ge-C bond [205,206]:

$$R_2Ge\overbrace{\qquad}R' \; + \; {}^+S-(S)_n-S^- \; \xrightarrow[\text{5 min}]{250°C} \; R_2Ge\overbrace{\qquad}^{R'}_S \; + \; {}^+S-(S)_{n-1}-S^- \qquad (8.54)$$

$$R = C_2H_5, \; C_4H_9; \; R' = H \; \text{or} \; CH_3$$

The reaction of sulfur with dialkylsilacyclobutanes takes place in an analogous manner, but under more drastic conditions (270°C) [207]. However, the insertion of sulfur into 1,1-di-tert-butyl-1-silacyclobutane yields 2,4-di-tert-butyl-2,4-dipropyl-1,3,-2,4-dithiadisiletane instead of the expected 1,1-di-tert-butyl-1-sila-2-thiacyclopentane [208]. 1,1,3,3-Tetramethyl-1,3-disila-cyclobutane fails to react with sulfur in refluxing decalin [209]. In contrast, the exothermic reaction of 2,2-dimethyl-1-phenyl-2,1-silaphosphetane with sulfur in tetrahydrofuran involves ring expansion via the insertion of sulfur into the silicon—phosphorus bond [210]:

$$Me_2Si\text{---}PPh \; + \; S \; \xrightarrow[\text{r.t.}]{\text{THF}} \; Me_2Si\diagup^S\diagdown P(S)Ph \qquad (8.55)$$

Siliranes (silacyclopropanes) are extremely reactive compounds, much more so than the silacyclobutanes. Unlike the results given in the preliminary communication [211] the slightly exothermic reaction of dispirosilirane (8) with sulfur in tetrahydrofuran at room temperature gave a mixture of four isomeric products (9, 10, 11, and 12) with a ratio 18:9:3:1. This reaction is discussed in terms of a free-radical mechanism [212]:

The free-radical reaction of hexamethylsilirane with elemental
sulfur affords a high yield of 3,3,4,4,5,5-hexamethyl-1,2-dithia-
3-silacyclopentane [212]:

$$Me_2C-CMe_2 \underset{Me}{\overset{Me}{\underset{}{Si}}} + S_8 \xrightarrow[\text{1 hr}]{\text{THF, r.t.}} Me_2Si \overset{Me_2 \quad Me_2}{\underset{S}{\diagup \diagdown}} S \qquad (8.56)$$

In the case of germa-4-spiro[3.4]octane, sulfur is only inserted
into the four-membered ring [205]:

$$\boxed{\diagup}Ge\diamondsuit + S \xrightarrow[\text{40 min}]{250°C} \boxed{\diagup}Ge\diagdown S \qquad (8.57)$$

Compound (13) and spiro-compound (14) show a considerable
difference in their reactivity [182]. When compound (13) is heated
with sulfur in a vacuum at 230°C, it undergoes cleavage with the
formation of octafluorothianthrene and diphenylgermanium sulfide.
Compound (14) does not react with sulfur at 360°C under reduced
pressure:

13   14

In the reaction of sulfur with tetraphenylstannane, the composition of the reaction products depends on the ratio of the reagents,
the temperature, and the duration of the reaction. Depending on the
above mentioned factors, diphenyl sulfide, diphenyl disulfide, and
thianthrene can be formed [213]. When tetraphenylstannane is heated
with sulfur in a sealed tube at 225°C, tin(II) sulfide and diphenyl
disulfide are obtained [157]. Under milder conditions, at around
200°C, the reaction takes place with stepwise insertion of sulfur
into the Sn-C bonds of tetraphenylstannane. The intermediates in
this reaction are triphenyl(phenylthio)stannane and diphenyl(diphenylthio)stannane. The latter compound is not stable at this
temperature and decomposes with the formation of diphenyl sulfide
and hexaphenyl-2,4,6-tristanna-1,3,5-trithiane [214] whose structure
is discussed in [215]:

$$3R_4Sn \xrightarrow[\sim200°C]{3S} 3R_3SnSR \xrightarrow{3S} 3R_2Sn(SR)_2 \qquad (8.58)$$
$$\downarrow$$
$$3R_2S + cyclo\text{-}(R_2SnS)_3$$
$$R = C_6H_5$$

The following facts are in agreement with the proposal that a
stepwise insertion of sulfur into the Sn-C bonds takes place. The

stable final product, cyclo-[$(C_6H_5)_2SnS]_3$, can be obtained by
heating sulfur with triphenyl(phenylthio)stannane, as well as by the
thermal decomposition of diphenyl(diphenylthio)stannane [47,214].
Hexaphenyl-2,4,6-tristanna-1,3,5-trithiane is obtained in a 98%
yield by heating hexaphenyldistannathiane with two equivalents of
sulfur [216]:

$$3(R_3Sn)_2S + 6S \longrightarrow 3R_2S + 2 \text{ cyclo-} (R_2SnS)_3$$
$$R = C_6H_5$$
(8.59)

Finally, when tetraphenylstannane is heated with more than two
equivalents of sulfur at 190-200°C, diphenyl sulfide and polymeric
phenylated tin sulfides are obtained as the reaction products. The
more sulfur that is introduced into the reaction, the smaller the number
of phenyl groups attached to the tin atoms [216]. When tetra-
(4-methylphenyl)stannane is heated with sulfur, di-(4-methylphenyl)
sulfide is formed [217].

The reaction of sulfur with tetrabutylstannane at 150-190°C
gives dibutyl sulfide and hexabutyl-2,4,6-tristanna-1,3,5-trithiane
(45% yield) [158].

Triethylstannylacetone, $(C_2H_5)_3Sn - CH_2COCH_3$, reacts with sulfur
at room temperature with the formation of hexaethyldistannathiane
and resinous products [218].

In contrast to the organotin analogs, tetrabutylplumbane and
tetraphenylplumbane do not form lead-containing organic sulfides on
heating with sulfur. Thus, for example, tetrabutylplumbane does not
react with sulfur below 125°C [204]. A gradual increase of temperature leads to an almost explosive reaction with the formation of
dibutyl sulfide and lead(II) sulfide. Tetraphenylplumbane behaves
similarly at 150°C. It has been proposed that these reactions
follow a free-radical mechanism [157,204]:

$$R_4Pb + 3S \longrightarrow PbS + 2R_2S$$
$$R = C_4H_9, C_6H_5$$
(8.60)

The reaction of organic halosilanes with sulfur have not been
studied so far. The only information available is the statement that,
when triethyliodosilane is refluxed with sulfur for twenty-four
hours, unchanged starting material can be regenerated [219]. When
tributylchlorostannane is heated with two equivalents of sulfur
(190°C, 14 hours), dibutyl sulfide as well as a mixture of dibutyl-
(butylthio)chlorostannane and hexabutyldistannathiane are obtained
[220]. On the basis of the above mentioned mechanism for the
reaction of sulfur with compounds of the type R_4M (M = Si, Ge, Sn),
the conclusion has been drawn that the more polar Sn-Cl bond is more
stable with respect to the polarizing effect of sulfur than the
Sn-C bond. Because of this, it is assumed that only carbanions
participate in the degradation of sulfur atom chains [9,220]:

$$ClR_2\overset{\delta+}{Sn} - \overset{\delta-}{R+} {}^+S - (S)_6 - S^- \longrightarrow ClR_2Sn - S - (S)_6 - SR \qquad (8.61)$$

$$7ClR_2SnR + ClR_2Sn - (S)_8 - R \longrightarrow 8ClR_2SnSR \qquad (8.62)$$

$$2\,ClR_2SnSR \longrightarrow R_2S + (ClR_2Sn)_2S \qquad (8.63)$$

$$R = \underline{n} - C_4H_9$$

The reaction of sulfur with chlorotriphenylstannane is more complex. At 190–200°C, with two equivalents of sulfur, diphenyl sulfide, dichlorodiphenylstannane, and an organotin compound of the composition $(C_6H_5)_5Sn_3S_5Cl$ are obtained as the reaction products [220].

Dibutyldichlorostannane and butyltrichlorostannane react with sulfur above 220°C. It is likely that cleavage of the C–C bonds takes place. Tin(II) sulfide, hydrogen sulfide, and cracking products were isolated from the reaction mixture [220].

The reaction of dichlorodiphenylstannane with one equivalent of sulfur is accompanied by a disproportionation:

$$3\,Ph_2SnCl_2 + 3\,S \longrightarrow 2\,C_6H_6 + 2\,PhSnCl_3 + SnS + \text{[thianthrene structure]} \qquad (8.64)$$

Trichlorophenylstannane does not react with sulfur below 240°C. The reaction mixture affords only disproportionation products (i.e., tin(IV) chloride and dichlorodiphenylstannane) [220].

Organometallic cyanides of the group IV elements react with sulfur with the formation of isothiocyanates [221–225]. The mechanism of these reactions is discussed in [226]:

$$(C_6H_5)_3SiCN + S \longrightarrow (C_6H_5)_3SiNCS \qquad (8.65)$$
$$(CH_3)_3MCN + S \longrightarrow (CH_3)_3MNCS \qquad (8.66)$$
$$M = Si,\ Ge,\ Sn$$

Reactions of sulfur with compounds of the type R_3MH (where M = Si, Ge, Sn) take place according to the general equation:

$$R_3M - H + S \longrightarrow R_3M - SH \qquad (8.67)$$

Thus, for example, triphenylsilanethiol, $(H_5C_6)_3SiSH$, is formed in yields up to 80% by the reaction of triphenylsilane with sulfur in boiling decalin or in the absence of a solvent [227–231]. Tri-4-methylphenylsilane [233], triethylsilane [234], tris(pentafluorophenyl)germane [171], and triethylgermane [235–237] react in a similar way.

In the case of triethylsilane, reaction (8.67) is complicated by the formation of bis-triethylsilyl trisulfide, $[(C_2H_5)_3Si]_2S_3$ [234]. This can be explained by the following side processes:

$$2R_3SiSH \xrightarrow[-H_2S]{240°C} (R_3Si)_2S \xrightarrow{2S} (R_3Si)_2S_3 \qquad (8.68)$$

$$R = C_2H_5$$

According to scheme (8.68), triethylgermane reacts with sulfur at 140°C with the formation of triethylgermanethiol, hexaethylgerma-thiane, and hydrogen sulfide [236]. Also, it is known that hexa-ethyldisilathiane and hexamethyldisilathiane react with sulfur at 200°C with the formation of a mixture of polysulfides, $(R_3Si)_2S_n$ where n = 2-9 [238,239].

In the above-mentioned reactions, the stability of the hydride bond with respect to the action of sulfur decreases in the order Si-H > Ge-H $\geqslant$ Sn-H. Whereas triethylsilane reacts with sulfur at a noticeable rate at 240°C [234], the reaction with triethylstannane occurs at room temperature [236]. The reaction products in this reaction are hexaethyldistannathiane, hydrogen sulfide, and hydrogen. The formation of these products can be explained in terms of the following processes:

$$R_3SnH + S \longrightarrow R_3SnSH \qquad (8.69)$$

$$2R_3SnSH \longrightarrow H_2S + (R_3Sn)_2S \qquad (8.70)$$

$$R_3SnSH + R_3SnH \longrightarrow H_2 + (R_3Sn)_2S \qquad (8.71)$$

$$R = C_2H_5$$

The reactions of tris(pentafluorophenyl)stannane, $(C_6F_5)_3SnH$, and tributylstannane, $(C_4H_9)_3SnH$, with sulfur are similar [236,241].

The reaction of diphenylsilane, $(C_6H_5)_2SiH_2$, with sulfur leads to the formation of hexaphenyl-2,4,6-trisila-1,3,5-trithiane (15) [240]. On the other hand, the reaction of $(C_6F_5)_2GeH_2$ with sulfur gives 2,4-bis(pentafluorophenyl)-2,4-digerma-1,3-dithietane (16) in a high yield [171]. The reaction of sulfur with phenylsilane, $C_6H_5SiH_3$ [240], and pentafluorophenylgermane, $C_6F_5GeH_3$ [171], takes place with the formation of hydrogen sulfide and compounds (17) with adamantane-like structures:

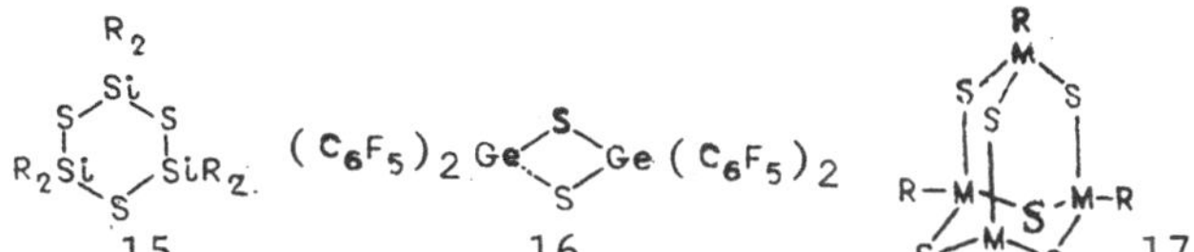

The synthesis of tetraalkyl-1,3,2,4-dithiadisiletanes via copyrolysis of the disilanes $R_3Si-SiR_3$ (R = CH_3, i-C_3H_7) with sulfur at 320°C is described. The same compounds are obtained by the reaction of disilanes $R_2HSi-SiHR_2$ (R = CH_3, i-C_3H_7, tert-C_4H_9,

cyclo-C_6H_{11}) with sulfur [208]. It is interesting to note that
hexakis(pentafluorophenyl)digermane [242,243] and 1,1,1-triethyl-
2,2,2-tris(pentafluorophenyl)digermane [244] can react with sulfur
under mild conditions. These compounds react in tetrahydrofuran
with one equivalent of sulfur at 60°C and 100°C, respectively, to
form digermathianes $(C_6F_5)_3GeSGeR_3$ where R = C_6F_5 or C_2H_5. In
contrast, attempts to obtain hexaethyldigermathiane [236] and hexa-
phenyldigermathiane [242] by the reaction of sulfur with hexaethyl-
digermane in the absence of a solvent at 200-215°C and hexaphenyl-
digermane in tetrahydrofuran at 100°C (a sealed tube), respectively,
have been unsuccessful.

The insertion of sulfur into the Ge-Ge bond was also observed
in the case of a sterically strained heterocyclic compound such as
1,1,2,2-tetraethyl-1,2-digermolane [245]:

$$R_2Ge{-}GeR_2 \overset{\bigsqcup}{} + S \xrightarrow[\text{15 min}]{160°C} R_2Ge \overset{S}{\diagdown\diagup} GeR_2 \tag{8.72}$$

$$R = C_2H_5$$

The corresponding heterocyclic analog with a six-membered ring,
i.e., 1,1,2,2-tetraethyl-1,2-digermacyclohexane, does not react with
sulfur under comparable conditions.

Hexamethyldistannane [246] and hexaethyldistannane [235,236]
react with sulfur at 20-85°C with the formation of the corresponding
hexa-alkyldistannathianes. In contrast, hexabenzyl- and hexaphenyl-
distannane do not react with excess sulfur in boiling chloroform
[247].

The reaction of hexaphenyldiplumbane with sulfur in benzene at
∿ 20°C leads to the formation of hexaphenyldiplumbathiane (59%
yield) [250,251]. Tetraphenylplumbane and polydiphenylthioplumbane,
$(-R_2PbS-)_n$, are formed as by-products.

Decamethylcyclopentasilane reacts with sulfur or selenium in
decalin at 190°C to give six-membered heterocycles $(Me_2Si)_5S$ and
$(Me_2Si)_5Se$. Dodecamethylcyclohexasilane is unreactive towards
sulfur or selenium under the same conditions [252].

Polydiarylstannanes, $(R_2Sn)_n$, react with sulfur with cleavage
of the Sn-Sn bonds and are converted into polydiarylstannathianes,
$(-R_2SnS-)_n$. Thus, for example, polydiphenylstannathiane may be
obtained by refluxing a mixture of polydiphenylstannane, $[(C_6H_5)_2{-}Sn]_n$, and sulfur in toluene, or by keeping the same reaction mixture
at room temperature for several days [248].

The heterocyclic sulfur compound (18) is obtained from the
corresponding organotin compound and sulfur in benzene [249]:

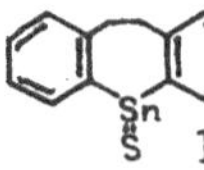

18

1,3-Di-$\underline{tert}$-butyl-2,2-dimethyl-1,3,2,4-λ^2-diazasilastannetidine
is prepared as a monomer [253]. Its further oxidation with sulfur
proceeds according to the scheme:

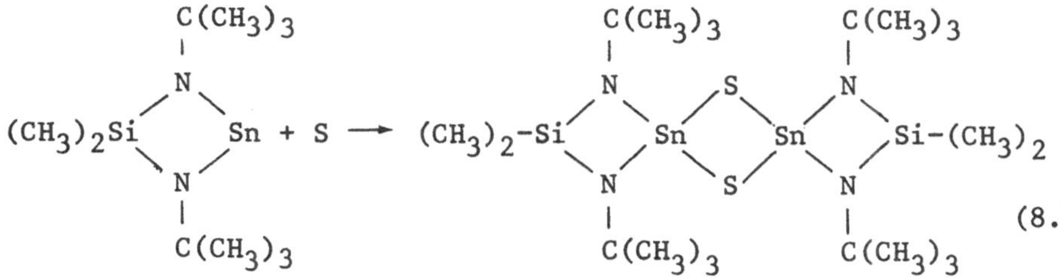

(8.73)

Reactions of sulfur with compounds of type $(C_6H_5)_3MLi$ (M = Si,
Ge, Sn, Pb) have been discussed in Section 8.1. Reactions of sulfur
with bis-triethylgermylcadmium and compounds of type $(R_3M)_2Hg$ (where
M = Si, Ge) are described in Section 8.4.

The kinetics of the reaction of the radicals $(\underline{n}-C_4H_9)_3Sn\cdot$ with
sulfur have been investigated [254].

8.7. ORGANIC DERIVATIVES OF THE PHOSPHORUS SUBGROUP ELEMENTS

The reactivity of sulfur with organophosphorus compounds has
been studied in some detail. The reactions of butylphosphine [255],
phenylphosphine [256,257], and benzylphosphine [258] with sulfur
are quite complex. Thus, for example, the reaction of phenylphosphine with one equivalent of sulfur gives an unstable product,
$C_6H_5P(S)H_2$, which decomposes under the reaction conditions employed
according to the equation:

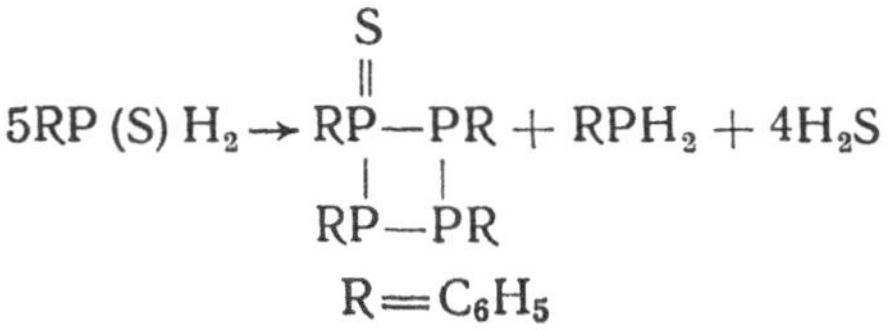

$$5RP(S)H_2 \rightarrow \underset{\substack{| \\ RP-PR}}{\overset{\substack{S \\ \parallel \\ RP-PR \\ | }}{}} + RPH_2 + 4H_2S$$

(8.74)

$$R = C_6H_5$$

When the ratio of the reagents is 1:2 or 1:3, compounds (19) and
(20) are obtained, respectively [257]:

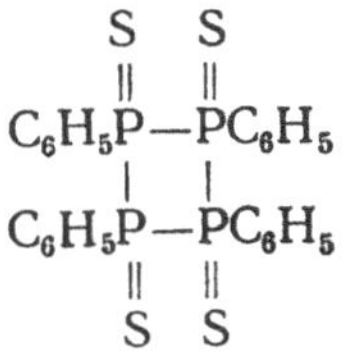

19 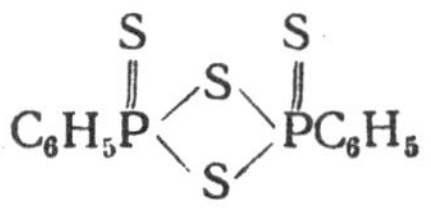   20

On the other hand, when sulfur is allowed to react with alkylphos-
phines, RPH_2, only tetralkylcyclotetraphosphine tetrasulfides are
formed, i.e., analogs of compound (19) [255].

The treatment of secondary phosphines with sulfur under controlled conditions gives the corresponding phosphine sulfides [259-
262]:

$$R_2PH + S \longrightarrow R_2P(S)H \qquad (8.75)$$

Their further oxidation with sulfur leads to the formation of dithiophosphinic acids in high yield [263]:

$$R_2P(S)H + S \longrightarrow R_2P(S)SH \qquad (8.76)$$

Dithiophosphinic acids, $R_2PS\text{-}SH$, are usually obtained by treating
two equivalents of sulfur with a secondary phosphine (cf. Table 8.4).
These reactions occur via the intermediate formation of polysulfides,
$R_2P(S)\text{-}(S)_3\text{-}P(S)R_2$ [264] or $R_2P(S)\text{-}S\text{-}P(S)R_2$ [265]. An exception is
the reaction of bis-trifluoromethylphosphine with sulfur which takes
place under drastic conditions and leads to a complex mixture of
sulfur-containing products [266]:

$$(CF_3)_2PH + S \xrightarrow[\text{4 days}]{160°C} (CF_3)_2P(S)SH + [(CF_3)_2P]_2S$$
$$+ (CF_3)_2PSH + H_2S \qquad (8.77)$$

The corresponding arsine, $(F_3C)_2AsH$, reacts with sulfur at 100°C
with the formation of a single product, $[(CF_3)_2As]_2S$ [267].

The reaction of dimethylarsine with sulfur depends on the ratio
of the reagents and gives either $[(CH_3)_2As]_2S$ or $(CH_3)_2As(S)\text{-}SAs\text{-}$
$(CH_3)_2$ [268,269]. In solution the latter compound exists in equilibrium with tetramethyldiarsine disulfide [270]:

$$R_2As(S)\text{-}SAsR_2 \rightleftharpoons (R_2As)_2S_2 \qquad (8.78)$$
$$R = CH_3$$

The usual synthetic procedure for the preparation of salts of
dialkyldithiophosphinic acids is based on the reaction of dialkyl-
phosphines with sulfur in dilute aqueous ammonium hydroxide [260]:

$$R_2PH + 2S + NH_4OH \longrightarrow R_2P(S)SNH_4 + H_2O \qquad (8.79)$$

Bis-2-cyanoethylphosphine is an exception. Under these conditions,
it is converted into bis-2-aminothiocarbonylethylphosphinic acid
[273]:

$$(NCCH_2CH_2)_2PH + 2S \xrightarrow{NH_4OH} \xrightarrow{H_2O} (H_2N\overset{\overset{\displaystyle S}{\|}}{C}CH_2CH_2)_2P(O)OH \qquad (8.80)$$

A salt of bis-2-cyanoethyldithiophosphinic acid is obtained when the reaction is carried out in acetonitrile and gaseous ammonia is used as the base [263]:

$$(NCCH_2CH_2)_2PH + 2S \longrightarrow (NCCH_2CH_2)_2P(S)SNH_4 \qquad (8.81)$$

Most tertiary phosphines are oxidized by sulfur according to scheme (8.82):

$$R_3P + S \longrightarrow R_3P = S \qquad (8.82)$$

The analogous reactions in the arsenic and antimony series have been much less studied. Furthermore, in contrast to other ditertiary stibines, which react according to the above scheme, the compound $R_2Sb(CH_2)_4SbR_2$ (where R = tert-C_4H_9) reacts with sulfur with the formation of Sb_2S_5 [289]. Also, it has been found [290] that carboranylstibines (21) are stable toward sulfur. Reactions analogous to scheme (8.82) do not apply to derivatives of trivalent bismuth:

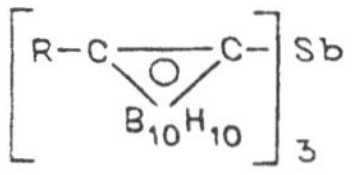

$$\left[R-C \underset{B_{10}H_{10}}{\overset{O}{\diagdown\diagup}} C- \right]_3 Sb \qquad \underset{\sim}{21}$$

$$R = CH_3, C_6H_5$$

The reaction between trialkylphosphines and sulfur is exothermic [291,292] and very fast and this makes kinetic studies difficult. Kinetic data for the formation of triarylphosphine sulfides has been obtained for several cases [293-295]. In the case of the reaction of triphenylphosphine with sulfur, the reaction is strictly second order. The rate of the process is strongly dependent on the polarity of the solvent. The rate-determining step is the ring opening of S_8 which occurs as the result of the nucleophilic attack of triphenylphosphine [293].

Thus, the reaction mechanism involves an ionic ring opening of the S_8 molecule with the formation of a dipolar ion and subsequent intramolecular nucleophilic attack of the terminal sulfur anion upon the sulfur atom which is in the β-position with respect to the phosphorus atom:

$$R_3P + S_8 \longrightarrow R_3\overset{+}{P} - S - S - (S)_5 - S^- \longrightarrow R_3PS + S_7 \qquad (8.83)$$

$$R_3P + S \longrightarrow R_3\overset{+}{P} - S - S - (S)_4 - S^- \longrightarrow R_3PS + S_6 , \text{ etc.} \qquad (8.84)$$

$$R = C_6H_5$$

The reaction with S_6 is 25,000 times faster than with S_8 [296]. This finding supports the above mechanism.

Table 8.4. Reactions of Primary and Secondary Phosphines and Arsines with Sulfur

Starting compounds	Reaction products	Yield (%)	Ref.
$(F_3C)_2PH$	$(F_3C)_2PSSH$ $[(F_3C)_2P]_2S$ $(F_3C)_2PSH$		266
$BuPH_2$	$cyclo\text{-}(BuPS)_4$	82.6	255
Et_2PH	Et_2PSSH		265
$EtPHCH_2CH_2NH_2$	$EtPS(SH)CH_2CH_2NH_2$	76	181
$EtPHCH_2\text{-}PH\text{-}Et$	$EtPS(H)CH_2PS(H)Et$	45	261
Cl–C$_6$H$_4$–PF$_3$ (H)	Cl–C$_6$H$_4$–PF$_2$=S	90	274
$PhPH_2 + S$ (1:1)	PhP–P(S)Ph PhP–PPh		257
(1:2)	PhP——PPh PhP(S)–P(S)Ph	68	257
(1:3)	PhP(S)–S S–P(S)PS		257
$(NCCH_2CH_2)_2PH$	$(NCCH_2CH_2)_2P\text{-}SNH_4$ (1)	47	263

354   CHAPTER 8

Starting compounds	Reaction products		Yield (%)	Ref.
	$(NCCH_2CH_2)P-CMe_2$ (2)	$\overset{\text{S}}{\underset{\parallel}{}}$ $\overset{}{\text{OH}}$	70	263
EtPHCH₂CH₂PHEt	$EtPH(CH_2)_2PHEt$ $\underset{\text{SH}}{\overset{\text{S}}{\parallel}}$		63	275
PhCH₂PH₂	$(PhCH_2P-)_2S$ $\underset{\text{S}}{\parallel}$			258
EtPh(CH₂)₃PHEt	$EtPH(CH_2)_3PHEt$ $\underset{\text{S}}{\overset{\text{S}}{\parallel}}$		72	261
PhPHCH₂CH₂SH	$PhPCH_2CH_2SH$ $\underset{\text{SH}}{\parallel}$		82.3	271
	$PhPCH_2CH_2SH$ (1) $\underset{\text{SNH}_4}{\parallel}$		78	271

Table 8.4 (continued)

Starting compounds	Reaction products	Yield (%)	Ref.
PhPCH$_2$CH$_2$NH$_2$	PhPCH$_2$CH$_2$NH$_2$ with $\overset{S}{\overset{\|}{}}$ and SH	95	277
(NH — PH — (CH$_2$)$_5$ cyclic)	NH$_2^+$ P(S)S$^-$ (CH$_2$)$_5$ cyclic	71	276
C$_8$H$_{17}$PH$_2$	cyclo-(C$_8$H$_{17}$P-S)$_4$	80	255
Bu$_2$PH	Bu$_2$P$\overset{S}{\underset{H}{=}}$	36	259
	Bu$_2$P$\overset{S}{\underset{SNH_4}{=}}$ (1)	55	260
	Bu$_2$P-CHCCl$_3$ with $\overset{S}{\overset{\|}{}}$ and OH (3)	85	263
i-Bu$_2$PH	i-Bu$_2$P$\overset{S}{\underset{H}{=}}$	65	259
tert-Bu$_2$PH	tert-Bu$_2$P$\overset{S}{\underset{SNa}{=}}$ (4)		279
(BuO)$_2$PH	(BuO)$_2$P$\overset{S}{\underset{H}{=}}$	100	262
EtPH(CH$_2$)$_4$PHEt	EtPH(CH$_2$)$_4$PHEt with $\overset{S}{\overset{\|}{}}$ $\overset{S}{\overset{\|}{}}$	81	261

Starting compounds	Reaction products	Yield (%)	Ref.
$C_6H_{13}PH(CH_2)_2NH_2$	$C_6H_{13}P(CH_2)_2NH_2$ (with $\overset{S}{\underset{SH}{\parallel}}$ on P)	76	181
$PhP(H)CH_2CH(OH)CH_3$	$PhP(S)(SH)CH_2CH(OH)CH_3$	80	280
$PhP(H)CH_2CH_2CH_2NH_2$	$PhP(S)(S^-)CH_2CH_2CH_2NH_3^+$	75	278
$PhCH_2P(H)CH_2CH_2NH_2$	$PhCH_2P(S)(SH)CH_2CH_2NH_2$	82	181
$C_6H_{11}P(H)CH_2CH_2CN$	$C_6H_{11}P(S)(H)CH_2CH_2CN$	56	259
$Et(H)P(CH_2)_5P(H)Et$	$Et(H)P(S)(CH_2)_5P(S)(H)Et$	23	261
	$EtP(S)(SH)(CH_2)_5P(S)(SH)Et$		261
$\underline{tert\text{-}Bu(Ph)PH}$	$\underline{tert\text{-}Bu(Ph)P(S)SH}$	60	288
$Ph(H)P(CH_2)_4NH_2$	$PhP(S)(SH)(CH_2)_4NH_2$		300
$(EtOCOCH_2CH_2)_2PH$	$(EtOCOCH_2CH_2)_2P(S)(SH)$	82	259
$Ph(H)PCH_2CH_2NHEt$	$PhP(S)(SH)CH_2CH_2NHEt$	76	181
$Et(H)P(CH_2)_6P(H)Et$	$Et(S)(SH)(CH_2)_6P(S)(SH)Et$		261
(cyclic structure: EtN, PH, Ph, H)	(cyclic structure: $EtNH^+$, $P(S)S^-$, Ph, H)	81	276
Ph_2PH	$Ph_2P(S)SH$	91	259,263
	$Ph_2P(S)C(OH)Me_2$ (2)	75	263
	$[Ph_2P(S)S]_2$ (5)		263

Table 8.4 (continued)

Starting compounds	Reaction products	Yield (%)	Ref.
$Ph(H)PC_6H_4SH\text{-}\underline{o}$	$PhP(S)(SNH_4)C_6H_4SH\text{-}\underline{o}$ (1)	84	271
$Ph(H)PCH_2CH_2NEt_2$	$PhP(S)(SH)CH_2CH_2NEt_2$	77	181
$(C_6H_{11})_2PH$	$(C_6H_{11})_2P(S)H$	90	260
	$(C_6H_{11})_2P(S)SNH_4$ (1)	85	260
(i-Pr, Pr-i substituted 1,3-dioxa-phosphorinane ring with PH)	(i-Pr, Pr-i ring) $P(S)SNH_4$ (1)		260
	(i-Pr, Pr-i ring) $P(S)SNa$ (4)	46	260
$Ph(H)P(CH_2)_4NH(Pr\text{-}i)$	$PhP(S)(SH)(CH_2)_4NH(Pr\text{-}i)$		281
$Ph(H)P(CH_2)_2P(H)Ph$	$PhP(S)(SH)(CH_2)_2P(S)(SH)Ph$	70	287
$\underline{n}\text{-}Pr_2P(CH_2)_2P(H)Ph$	$\underline{n}\text{-}Pr_2P(S)(CH_2)_2P(S)(SH)Ph$		284
$(C_6H_{11})(H)P(CH_2)_2P(S)(C_6H_{11})$	$C_6H_{11}(H)P(S)(CH_2)_2P(S)(H)C_6H_{11}$		275
$Ph(H)P(CH_2)_3P(H)Ph$	$PhP(S)(SH)(CH_2)_3P(S)(SH)Ph$		285
$Ph(H)PCH(Ph)CH_2COMe$	$PhP(S)(SH)CH(Ph)CH_2COMe$	85	282
$(PhCH_2CH_2)_2PH$	$(PhCH_2CH_2)_2P(S)SNH_4$ (1)	55	260
$Ph(H)P(CH_2)_4P(H)Ph$	$PhP(S)(SH)(CH_2)_4P(S)(SH)Ph$	61–84	285,286
$(\underline{n}\text{-}C_8H_{17})_2PH$	$(\underline{n}\text{-}C_8H_{17})_2P(S)SNH_4$ (1)	25	260

Starting compounds	Reaction products	Yield (%)	Ref.
$n\text{-}Bu_2P(CH_2)_2P(H)(\underline{n}\text{-}C_6H_{13})$	$\underline{n}\text{-}Bu_2P(S)(CH_2)_2P(S)(SH)(\underline{n}\text{-}C_6H_{13})$		284
$C_6H_{11}(H)P(CH_2)_4P(H)C_6H_{11}$	$C_6H_{11}P(S)(SH)(CH_2)_4P(S)(SH)C_6H_{11}$	57	283
$C_6H_{11}(H)P(CH_2)_6P(H)C_6H_{11}$	$C_6H_{11}P(S)(SH)(CH_2)_6P(S)(SH)C_6H_{11}$	86	283
$Ph(H)PCH(Ph)CH_2COBu\text{-}\underline{tert}$	$PhP(S)(SH)CH(Ph)CH_2COBu\text{-}\underline{tert}$	71	282
$Ph_2P(CH_2)_2P(H)Ph$	$Ph_2P(S)(CH_2)_2P(S)(SH)Ph$	70	284
$Ph(H)PCH(Ph)CH_2COPh$	$PhP(S)(SH)CH(Ph)CH_2COPh$	86	282
$(\underline{n}\text{-}C_{12}H_{25})_2PH$	$(\underline{n}\text{-}C_{12}H_{25})_2P(S)SNH_4$ (1)	68	260
Compounds with As–H bonds:			
$(CF_3)_2AsH$	$[(CF_3)_2As]_2S$		267
Me_2AsH	$\begin{cases}(Me_2As)_2S\\(Me_2As)_2S_2\end{cases}$		268

<u>Note</u>. (1) The reaction is carried out in the presence of NH_3 or NH_4OH. (2) The reaction is carried out in the presence of acetone. (3) The reaction is carried out with chloral added. (4) The reaction is carried out in the presence of aqueous NaOH.

The reaction rate is strongly affected by small amounts of phenol, methanol, and acetonitrile whose catalytic activity is due to the formation of hydrogen bonds in the transition state. Electron-donating substituents in the p-position of the aryl ring of triaryl-phosphines facilitate the addition of sulfur. The value of Hammett's constant is $\rho = -2.5$. This finding, together with other observations, suggests that the transition state is strongly polar and possesses a strong P-S bond [293].

Reactions of sulfur with optically active phosphines [297-300] and arsines [301,302] take place with retention of configuration:

$$(+)-(\underline{S})-Me(\underline{n}-Pr)PhP + S \xrightarrow{\text{benzene}} (+)-(\underline{R})-Me(\underline{n}-Pr)PhPS \qquad (8.85)$$

$$(+)-Me(\underline{n}-Bu)PhCH_2P + S \longrightarrow (+)-Me(\underline{n}-Bu)PhCH_2PS \qquad (8.86)$$

$$(+)-Me(Et)PhAs + S \xrightarrow{\text{benzene}} (+)-Me(Et)PhAsS \qquad (8.87)$$

Similar sulfurations are observed for other chiral tertiary phosphines [303], e.g., $(-)-(\underline{R})-Me(\underline{n}-Pr)PhP$. Absence of racemization indicates that the intermediate dipolar ions (cf. schemes (8.83) and (8.84)) are not in equilibrium with the cyclic tautomers [297]:

$$(8.88)$$

The high stereospecificity of these additions of sulfur has been utilized to prove the presence of two asymmetric phosphorus atoms and the existence of the d, ℓ- and meso-diastereomers in the case of the following linear tetraphosphine [304]:

$$(8.89)$$

$$R = C_6H_5$$

Another example of a stereospecific addition of sulfur is the formation of cis-2-hydroxy-4-methyl-1,3,2-dioxaphosphorinane-2-thione [306] (cf. also [300,305]):

$$\text{(structure)} + S \longrightarrow \text{(structure)} \qquad (8.90)$$

$$R = CH_3$$

Replacement of alkyl or aryl groups in R_3P with electron-accepting substituents leads to a decrease of the phosphorus nucleophilicity and a sharp drop in the reaction rate of the interaction with sulfur, e.g., in the case of pentafluorophenyl derivatives of trivalent phosphorus, the ease of sulfur addition decreases in the following order (R = CH_3 or C_2H_5) [307]:

$$R_2(C_6F_5)P > (C_6H_5)_2(C_6F_5)P > R(C_6F_5)_2P > (C_6F_5)_3P -$$
$$(RO)_2(C_6F_5)P > (RO)(C_6F_5)_2P > (C_6H_5O)_2(C_6F_5)P$$
$$> (C_6H_5O)(C_6F_5)_2P > (C_6F_5)_3P$$

The first members of these two series add sulfur in boiling benzene [307], whereas the reaction with $(F_5C_6)_3P$ occurs only at 160°C after several days [308]. Tris-trifluoromethylphosphine does not react with sulfur even at 180°C [309]. Formation of the phosphine sulfide (50% yield) is observed only after heating at 200°C for four days [266].

In contrast to this, the addition of sulfur to the bis-trifluoromethylphosphino group in π-$C_5H_5Fe(CO)_2P(CF_3)_2$ occurs, quite unexpectedly, under mild conditions (60°C) without evolution of carbon dioxide [310,311]. It is assumed that the nucleophilicity of phosphorus in this compound is increased because of the considerable polarization of the iron—phosphorus σ-bond due to the electropositive character of the iron atom and, furthermore, because of π-bonding in the metal—phosphorus group [310]:

$$\pi\text{-}C_5H_5Fe(CO)_2P(CF_3)_2 + S \longrightarrow \pi\text{-}C_5H_5Fe(CO)_2P(S)(CF_3)_2 \qquad (8.91)$$

Transition metal arsines of the type η^5-$C_5H_5(CO)_2LMAs(CH_3)_2$, where M = Mo, W, L = CO, $P(CH_3)_3$, react with sulfur in carbon disulfide to give the corresponding transition-metal-substituted arsine sulfides [313].

Reactions of tetraalkyl and tetraaryldiphosphines with sulfur take place under controlled conditions and yield tetraalkyl- and tetraaryldiphosphine disulfides [311,312]. The disulfides formed react with sulfur with cleavage of the P-P bond [255]:

$$(C_2H_5)_2P—P(C_2H_5)_2 + 2S \longrightarrow (C_2H_5)_2P(S)—P(S)(C_2H_5)_2 \qquad (8.92)$$

$$(C_6H_5)_2P—P(C_6H_5)_2 + 2S \longrightarrow (C_6H_5)_2P(S)—P(S)(C_6H_5)_2 \qquad (8.93)$$

$$(C_6H_5)_2P(S)—P(S)(C_6H_5)_2 \xrightarrow{S} (C_6H_5)_2P(S)—S—P(S)(C_6H_5)_2 \qquad (8.94)$$

When reaction (8.94) is carried out in the presence of sodium sulfide [279,314,315], salts of dialkyldithiophosphinic acids are formed:

$$R_2P(S)-P(S)R_2 + Na_2S \cdot 9H_2O + S \xrightarrow{100\cdot C} 2R_2P(S)SNa \qquad (8.95)$$
$$(27-90\%)$$
$$R=CH_3,\ C_2H_5,\ C_3H_7,\ C_4H_9$$

When these reactions are carried out in the presence of heavy metals or their sulfides, complex salts of the above acids are obtained [316,317]:

$$1/2R_2P(S)-P(S)R_2 + M^{n} + S \xrightarrow{150-200\ C} R_2P \overset{S}{\underset{S}{\diagdown}} M/n \qquad (8.96)$$

$$R=C_2H_5,\ C_3H_7;\ C_4H_9,\ C_5H_{11};\ M=Zn,\ Cd,\ Hg,\ As,\ Sb,\ \mathbf{Bi}$$

$$R_2P(S)-P(S)R_2 + S + MS \xrightarrow{150-200°C} R_2P \overset{S}{\underset{S}{\diagdown}} M/n \qquad (8.97)$$

$$R=C_2H_5,\ C_3H_7;\ M=Zn,\ Hg,\ Pb^{11},\ Tl^{1},\ In$$

The interaction of diphosphines with four organic groups with excess sulfur leads to cleavage of the P–P bond even under mild conditions [318-322]:

$$R_2P-PR_2 + S \longrightarrow [R_2P(S)]_2S + [R_2P(S)S]_2 \qquad (8.98)$$
$$R = CF_3,\ \underline{tert}-C_4H_9$$

$$(8.99)$$

$$R=C_6H_5$$

The reaction of sulfur with cyclo-$(CF_3P)_4$ is quite unusual [323,324]. It leads to the formation of tetrakis-trifluoromethyl-1,2,3,4,5-thiatetraphospholane, $(F_3C-P)_4S$, in which the phosphorus atoms are trivalent.

The reaction of tetraaryldiarsines with sulfur starts by attack upon the As–As bond [325,326]:

$$R_2AsAsR_2 + S \longrightarrow (R_2As)_2S \qquad (8.100)$$
$$R=C_6H_5,\ \underline{p}-NO_2C_6H_4$$

Replacement of the alkyl groups in trialkylphosphines by halogen atoms decreases the reactivity of these compounds towards sulfur. Thus, for example, chlorodiethylphosphine reacts with sulfur only upon heating [327]. However, in the presence of catalytic amounts of aluminium chloride or, more rarely used, iron(III) chloride or zinc chloride, alkyldihalophosphines react with sulfur at room temperature to give compounds of the type $RP(S)X_2$ where X = halogen [328-330]. Compounds $(F_3C)_2P(S)Cl$ [331], $F_5C_6P(S)X_2$, $(F_5C_6)_2P(S)X$, and $R(C_6F_5)P(S)X$ [332] can be obtained in a similar manner. In the case of $R(C_6F_5)P(S)X$, the reactivity of the phosphines decreases in the following order:

$$Me(C_6F_5)PX > Et(C_6F_5)PX > \underline{tert}\text{-}Bu(C_6F_5)PX > Ph(C_6F_5)PX > (C_6F_5)_2PX$$

The reaction of sulfur with diiodophenylphosphine in boiling benzene is anomalous [333]:

$$C_6H_5PI_2 + 2S \longrightarrow I_2 + (C_6H_5PS_2)_n \qquad (8.101)$$

Chlorides of dialkylthiophosphinic acids and dichlorides of alkylthiophosphinic acids may also be obtained by treating sulfur with the complexes $R_2PCl_3 \cdot AlCl_3$ and $RPCl_4 \cdot AlCl_3$, respectively, in the presence of potassium chloride [334,335]:

$$R_2PCl_3 \cdot AlCl_3 + 2S + KCl \xrightarrow{200°C} R_2P(S)Cl + SCl_2 + KAlCl_4 \qquad (8.102)$$
$$R = C_2H_5$$

Esters and thioesters of phosphonous, phosphonic, and phosphorous acids react with sulfur and the trivalent phosphorus is converted to tetra-coordinated phosphorus (hexyldiphenylphosphine sulfide) [336], e.g.,

$$Ph_2P\text{-}S\text{-}C_6H_{13} + S \longrightarrow Ph_2P \lessgtr^{S}_{SC_6H_{13}} \qquad (8.103)$$

On the other hand, an analogous reaction between sulfur and the thioester $C_6H_5(C_4H_9)AsSC_4H_9$ does not take place even upon prolonged heating at 160-180°C [337,338].

When (ethylthio)dipropylarsine is heated with sulfur in a sealed tube (120°C, 6 hours) dipropyl(dipropylarsinothio)arsine sulfide is formed [339]:

$$2(C_3H_7)_2AsSC_2H_5 \xrightarrow{S} (C_3H_7)_2\overset{\text{S}}{\underset{\parallel}{As}}SAs(C_3H_7)_2 \qquad (8.104)$$

Monoesters of thiophosphonic and phosphorothionic acids are obtained by treating sulfur with monoalkyl phosphonates and dialkyl phosphonates in the form of their alkali metal salts [340-346] or their salts with ammonia [347] or amines [348-350]. Subsequent acid hydrolysis gives free thioacids:

$$[R(R'O)PO]^- Na^+ + S \rightarrow [R(R'O)PSO]^- Na^+ \xrightarrow{H_3O^+} R(R'O)P(S)OH \qquad (8.105)$$

$$[(RO)_2 PO]^- Na^+ + S \rightarrow [(RO)_2 PSO]^- Na^+ \xrightarrow{H_3O^+} (RO)_2 P(S)OH \qquad (8.106)$$

$$\langle\!\!\!\!\bigcirc\!\!\!\!\langle N CH_2-CH_2 P(O)H + S \longrightarrow \langle\!\!\!\!\bigcirc\!\!\!\!\langle \overset{+}{N}H CH_2 CH_2 P(S)O^- \qquad (8.107)$$
$$\qquad\qquad OC_4H_9 \qquad\qquad\qquad OC_4H_9$$

The same method has been used to obtain the ammonium salts of tetraalkylphosphorodiamidothiolic acids and their alkyl derivatives [351]:

$$(R_2N)_2 P(O)H + S + B \rightarrow [(R_2N)_2 POS]^- [BH]^+$$
$$\downarrow R'X$$
$$(R_2N)_2 P(O)SR' \qquad (8.108)$$

R = CH_3, C_2H_5; R' = C_2H_5, C_4H_9; B = an amine; X = Br, I

^{35}S-Labelled ester, $(CH_3O)_2 P(S)SCH_2 CONHCH_3$, is prepared in 57% yield by an isotopic exchange reaction between the unlabelled ester and elemental sulfur ^{35}S in benzene or toluene at 130°C (sealed ampoule) [352]. Synthesis of the zinc salt of O,O-di-isobutyl ester phosphorodithioic-^{35}S is carried out using an exchange reaction between the ammonium salt of O,O-di-isobutyl dithiophosphate and elemental sulfur ^{35}S [353].

Amidophosphinites, diamidophosphinites, and similar compounds with P–N bonds react with sulfur according to scheme (8.82) and the phosphorus atom becomes tetra-coordinated. Thus, for example, $RN(PCl_2)_2$ with R = alkyl or aryl react with sulfur in the presence of aluminium chloride at 150°C to give mono- or disulfides [354]. In the absence of a catalyst the reaction does not occur, even at 200°C:

$$RN(PCl_2)_2 \xrightarrow[AlCl_3]{S} Cl_2 P NR P(S)Cl_2 \xrightarrow[AlCl_3]{S} RN[P(S)Cl_2]_2 \qquad (8.109)$$

The reaction of bis(N-diphenyl-N-methylphosphinoamino)phenyl-phosphine with sulfur in benzene carried out under normal conditions gives only the disulfide $(H_5C_6)_2 P(S)N(CH_3)P(C_6H_5)N(CH_3)P(S)(C_6H_5)_2$. The addition of the third sulfur atom to the central phosphorus atom takes place only upon prolonged refluxing of the reaction mixture containing excess sulfur [355]. It is assumed that the disulfide is formed as a result of the cleavage of the S_8 molecules by one of the terminal phosphorus atoms followed by nucleophilic attack of a second terminal phosphorus atom upon the S* atom of intermediate (22):

$$(C_6H_5)_2\overset{+}{P}-N(CH_3)-P(C_6H_5)-N(CH_3)-P(C_6H_5)_2$$
$$| \qquad\qquad\qquad\qquad\qquad\qquad\qquad\qquad$$
$$S-S*-(S)_5-S^-$$
$$22$$

The reaction of sulfur with (dimethylarsino)methylamino(di-_tert_-butyl)phosphine in ether is exothermic, the sulfur atom adds exclusively to the phosphorus atom [356]:

$$(R_2As) \, N(Me)P(CR_3)_2 + S \longrightarrow R_2As \, N(Me) \, P(S) \, (CR_3)_2 \qquad (8.110)$$
$$R = CH_3$$

1,3-Diaza-2,4-diphosphetidine-2-thiones which contain a trivalent phosphorus atom as well as a tetra-coordinated phosphorus atom were obtained for the first time by partial oxidation of 1,3-dialkyl-2,4-dichloro-1,3-diaza-2,4-diphosphetidines with sulfur [357]:

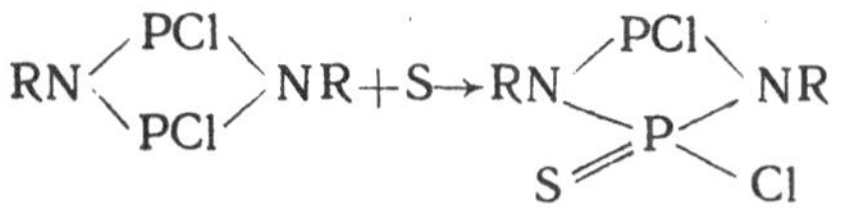

$$(8.111)$$

$$R = CH(CH_3)_2, \; C(CH_3)_3$$

The reaction of 1,3,2-dioxaphosphole with sulfur in ether takes an interesting course [358]:

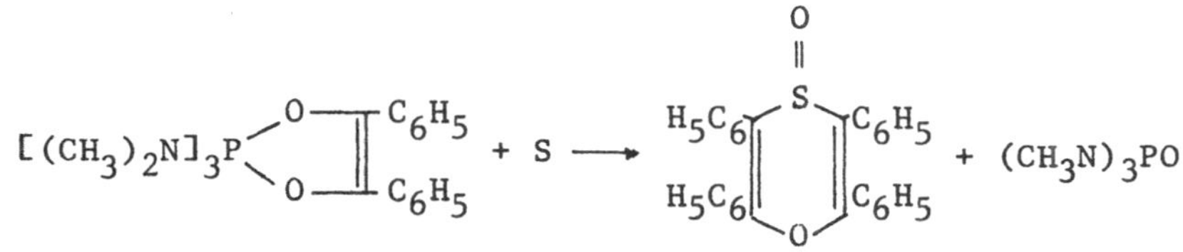

$$(8.112)$$

A five-membered arsenic-containing heterocycle is obtained in low yield when compound (23) is treated with sulfur [359]:

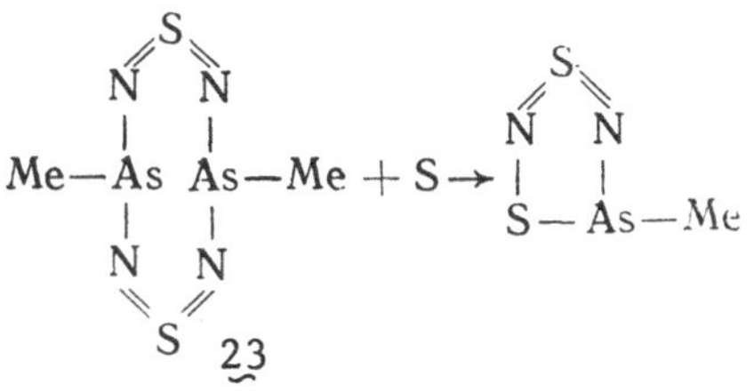

$$(8.113)$$

The expected sulfide is not formed when $[(H_3C)_2N]_2B-P(C_2H_5)_2$ is treated with sulfur [360]. The reaction proceeds according to the following scheme:

$$2(Me_2N)_2BPEt_2 + 5S \longrightarrow$$

$$\longrightarrow B_2(NMe_2)_4 + Et_2P(S)(S)_3P(S)Et_2 \qquad (8.114)$$

The reaction of diphenyltriphenylstannylphosphine with sulfur in boiling benzene gives a quantitative yield of triphenylstannyldiphenylthiothiophosphinate [361]:

$$R_3SnPR_2 + 2S \longrightarrow R_3SnSP(S)R_2 \qquad (8.115)$$
$$R = C_6H_5$$

On the other hand, the interaction of sulfur with bis(triphenyl-stannyl)phenylphosphine or tris(triphenylstannyl)phosphine does not give the expected esters of thiophosphoric acid, but rather their decomposition products [361]:

$$(R_3Sn)_2PR + S \longrightarrow (R_3Sn)_2S + (RPS_3)_2 \qquad (8.116)$$
$$(R_3Sn)_2P + S \longrightarrow (R_3Sn)_2S + P_2S_5 \qquad (8.117)$$
$$R = C_6H_5$$

Phosphonium ylides react with sulfur with the formation of the corresponding phosphine sulfide and thioketone. Thus, cleavage of diphenylmethylenetriphenylphosphorane with sulfur occurs according to the following scheme [362]:

$$(C_6H_5)_2\overset{..}{\underset{\ominus}{C}}-\overset{\oplus}{P}(C_6H_5)_3 + 2S \rightarrow (C_6H_5)_3PS + (C_6H_5)_2C=S \qquad (8.118)$$

Fluorene-9-thione and triphenylphosphine sulfide are the products of the oxidation of fluorenylidenetriphenylphosphorane with sulfur [363]. Reactions of this type probably start by nucleophilic attack of the ylide carbon atom upon sulfur with intermediate formation of the betaine (24). This is followed by transfer of the sulfur to the phosphorus atom (via a four-center transition state), giving a thioketone and a dipolar phosphonium polysulfide ion. The latter is then attacked by a second molecule of the ylide on the β-sulfur atom and the whole sequence repeats itself:

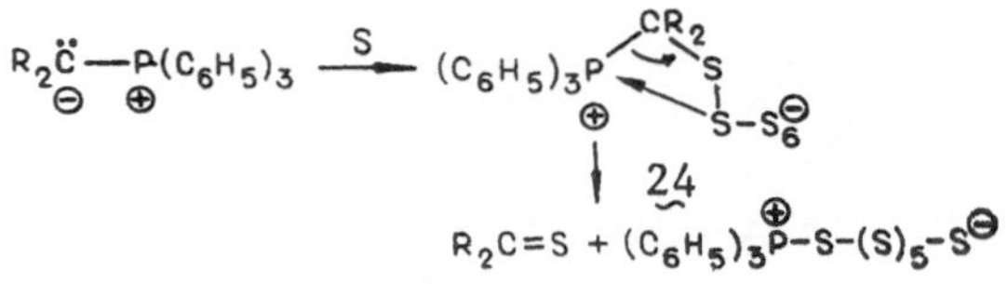

$$(8.119)$$

The reaction of tris-pentafluorophenylbismuth with sulfur leads to substitution of the bismuth atom by the chalcogen [364]:

$$2(C_6F_5)_3Bi + 3S \xrightarrow[\text{7 days}]{184°C} 3(C_6F_5)_2S + 2Bi \qquad (8.120)$$

8.8.   ORGANIC DERIVATIVES OF SELENIUM AND TELLURIUM

The reaction of selenocarbonyl compounds, RR'CSe, with sulfur for 20-180 minutes at 120-140°C in the absence of a solvent leads to the formation of thiocarbonyl compounds RR'S (where R = R' = tert-Bu; R = Ph, R' = OEt, NMe$_2$) in 80-95% yield [365].

   Diphenyl selenide reacts with sulfur at 300°C to give diphenyl sulfide and elemental selenium in almost quantitative yield:

$$(C_6H_5)_2Se + S \xrightarrow{300°C} (C_6H_5)_2S + Se \qquad (8.121)$$

In an analogous reaction between diphenyltellurium and sulfur at 220°C, diphenyl sulfide, diphenyl disulfide, and elemental tellurium were obtained [366].

   The above-mentioned processes in which selenium and tellurium are replaced by sulfur are general in nature. Thus, for example, treating bis-pentafluorophenyl selenide and telluride with sulfur at 230°C gives bis-pentafluorophenyl sulfide in quantitative yield [175]. Octafluorodibenzothiophene and octafluorothianthrene can be obtained in a similar way [178]:

$$(8.122)$$

$$M = Se, Te$$

$$(8.123)$$

Tellurium is substituted by sulfur in bis-triethylstannyl telluride at room temperature [367,368]. Similar reactions of sulfur with bis-triethylsilyl telluride and bis-triethylgermyl selenide require more drastic conditions (170°C, 8-12 hours) [367,368]:

$$[(C_2H_5)_3Sn]_2Te + S \xrightarrow[\substack{90 \\ hrs}]{20°C} [(C_2H_5)_3Sn]_2S + Te \qquad (8.124)$$

   Substitution of selenium by sulfur in bis(diperfluoromethyl)-phosphino selenide occurs at 100°C. Under more drastic conditions, this process is complicated by the addition of sulfur to one of the phosphorus atoms [320]:

$$[(CF_3)_2P]_2Se + S \longrightarrow \begin{array}{l} \xrightarrow[\text{2 days}]{100°C} [(CF_3)_2P]_2S \\ \xrightarrow[\text{6 days}]{150°C} (CF_3)_2P(S)-SP(CF_3)_2 \end{array} \qquad (8.125)$$

8.9. ORGANIC DERIVATIVES OF THE TRANSITION METALS

   No systematic studies of reactions of sulfur with organic transition metal complexes have been carried out so far.

   The interaction of bis-dicyclohexylphosphinetitanium with sulfur in hexane gives tetracyclohexyldiphosphine disulfide, $(H_{11}C_6)_2P(S)-P(S)(C_6H_{11})_2$, and the compound $[(H_{11}C_6)_2P(S)-STiS]_2S$ [369].

Pentasulfides of di-η^5-cyclopentadienyltitanium(IV), -zirconium (IV), and -hafnium(IV) are obtained in good yield by the reaction of $(\eta^5-C_5H_5)_2MCl_2$ (M = Ti, Zr and Hf) with Li_2S_2 and sulfur [370]. Titanocene pentasulfide is also prepared by the photolysis of dialkyl-titanocene compounds, $(\eta^5-C_5H_5)_2TiR_2$ (R = CH_3, $C_6H_5CH_2$) in the presence of sulfur [371], by the reaction of sulfur with $(\eta^5-C_5H_5)_2$-$Ti(CO)_2$ [372], and by the interaction of sulfur with $(\eta^5-C_5H_5)Ti(SH)_2$ in chloroform at room temperature [373], e.g.,

$$(\eta^5-C_5H_5)_2Ti(SH)_2 + S \xrightarrow[\text{r.t.}]{\text{CHCl}_3} (\eta^5-C_5H_5)_2Ti\begin{smallmatrix} S-S \\ | \quad \searrow \\ S-S \quad S \end{smallmatrix} \qquad (8.126)$$

The crystal and molecular structure of this compound has been determined [374,375].

The reaction of sulfur with tetrakis(triphenylsiloxy)vanadium occurs according to the scheme:

$$2\,[(C_6H_5)_3SiO]_2V + S \longrightarrow [(C_6H_5)_3SiO]_2VS$$
$$\underset{25}{}$$
$$+\,[(C_6H_5)_3SiO]_2VO + [(C_6H_5)_3Si]_2O \qquad (8.127)$$

Compound (25) is stable upon heating at 200°C and is not hydrolyzed with cold water [376].

When $(\pi-C_5H_5)V(CO)_4$ is heated with sulfur in boiling toluene, a diamagnetic complex, $[(\pi-C_5H_5)_2V_2S_5]_n$, is formed, whose structure has not been elucidated [377].

The reaction of chromium pentacarbonyl(methoxyarylcarbenes) with sulfur in refluxing ether gave O-methyl arenethiocarboxylates [378]:

$$(OC)_5Cr-C\begin{smallmatrix} \diagup OCH_3 \\ \diagdown C_6H_4R\text{-}\underline{p} \end{smallmatrix} + S \longrightarrow \underline{p}\text{-}RC_6H_4C(S)OCH_3 \qquad (8.128)$$
$$R = H,\ Cl,\ CH_3,\ CH_3O$$

Pentacarbonyl-bis(phenylmercapto)methylenetriphenylphosphorane chromium on treatment with the reactive cyclohexasulfur gave an orange-red compound which analyzed as pentacarbonyl-bis(phenyl-mercapto)carbenechromium [379]:

$$(OC)_5Cr-\overset{\displaystyle SR}{\underset{\displaystyle PR_3}{\overset{|}{\underset{|}{C}}}}-SR - S_6 \xrightarrow[20°C]{\text{benzene}} (OC)_5Cr-C\begin{smallmatrix} \diagup SR \\ \diagdown SR \end{smallmatrix} + R_3PS \qquad (8.129)$$
$$R = C_6H_5$$

An orange carbene complex of another type may be isolated from the reaction of ylide complex (26) with cyclohexasulfur [379,380]:

$$(OC)_5 W-C \underset{P(C_6H_5)_3}{\overset{C \nwarrow O}{\diagdown}} + S_6 \xrightarrow[\text{toluene}]{110°C} (OC)_5 W-C \diagdown + (C_6H_5)_3 PS \tag{8.130}$$

26

New complexes of the type $(\eta^5-C_5H_5)_2MS_4$ (M = Mo, W) have been prepared in 58-63% yields by treating metallocene hydrides $(\eta^5-C_5H_5)_2MH_2$ with sulfur in toluene at room temperature. The tetrathio chain formed a bidentate chelating ligand [381].

Complex (27) reacts with sulfur to give a polymer of type (28) [382]:

$$+S \longrightarrow \tag{8.131}$$

27   28

Iron pentacarbonyl does not react with sulfur at room temperature in carbon tetrachloride and xylene [383]; however, it does react with sulfur at 80°C to give low yields of $Fe_2(CO)_6S_2$ and $Fe_3(CO)_9S_2$. The main product of this reaction is FeS [384,385]. Sulfur substitutes carbonyl ligands in iron carbonyls such as $Fe_2(CO)_9$ [384,385] and $Fe_3(CO)_2$ [384-386] in benzene at 50-60°C to give $Fe_2(CO)_6S_2$ and $Fe_3(CO)_9S_2$, respectively. However, an x-ray diffraction study of $Fe_3(CO)_9S_2$ has revealed that the reaction gives a crystalline 1:1 mixture of two sulfur-containing iron carbonyls, i.e., $S_2Fe(CO)_9 \cdot S_2Fe_2(CO)_6$, whose structure has been investigated [387].

Sulfur reacts with excess cyclohexene in the presence of triiron dodecacarbonyl to give a mixture of products including $Fe_2-(CO)_6S_2$, $Fe_3(CO)_9S_2$, 1,2,3,4-tetrathiadecalin (29), 2,3,4,5,6-pentathiabicyclo[5.4.0]undecane (30) and complex (31) [388]:

29   30   31

A diamagnetic complex with a cubane-like structure, $Fe_4(NO)_4$-$(\mu_3-S)_4$, is obtained from the reaction of sulfur with $[Fe(CO)_3NO]_2Hg$ in boiling toluene [389].

Prolonged heating $[(\pi-C_5H_5)Fe(CO)_2]_2$ with sulfur in boiling toluene gives the complex $Fe_4(\pi-C_5H_5)_4(\mu_3-S)_4$ [377]. The crystal structure of this compound has been studied [389,390].

The photochemical reaction of the iron complex (32) with sulfur in benzene at room temperature gives tetrakis-pentafluorophenyl-thiophene [391]:

$$\text{(8.132)}$$

$$R = C_6F_5$$

The reaction of sulfur with dicobalt octacarbonyl in hexane at 30-40°C gives a mixture of polynuclear complexes. When the molar ratio is 1:1, the principal reaction product is $[Co_2S(CO)_5]_n$ (33% yield) [392]. Furthermore, the complex $Co_3S(CO)_9$ is also formed [393], the structure of which has been determined [394]. With two moles of sulfur, a 2.5% yield of a complex with the empirical formula $Co_3S_2(CO)_7$ is obtained [395]. This compound, according to the results obtained from structural analytical data, is a dimer, i.e., $[SCo_3(CO)_7]_2S_2$, in which two $SCo_3(CO)_7$ fragments are connected via a disulfide bridge [396].

The compound $Co_2Fe(CO)_9$ is formed upon addition of sulfur to a mixture of $Co_2(CO)_8$ and $Fe(CO)_5$ under hydroformylation conditions (200°C, 185 atm) [397]. It is assumed that this complex, similarly to the complex $Co_3S(CO)_9$, is a trigonal pyramid with the sulfur atom at the apex and three metal atoms at the base.

The reaction of elemental sulfur with π-cyclopentadienylcobalt dicarbonyl in hexane at room temperature gives the complex $Co_4(\eta^5-C_5H_5)S_6$ [398]. A 65% yield of this complex has been obtained using a modified procedure [399].

The binuclear complex $\eta-C_5H_5Co(PMe_3)(\mu-CO)_2Mn(CO)\eta-C_5H_4Me$ reacts with stoichiometric amounts of S_8 to form $\eta-C_5H_5(PMe_3)CoS_5$ in quantitative yields. The same product results when the complex $\eta-C_5H_5(PMe_3)Co(h-CS_2)$ is treated with sulfur in benzene at room temperature [399a]. The chain form of the six-membered CoS_5 ring in this cobalt pentathia heterocycle corresponds to that of the compound $(C_5H_5)_2TiS_5$ [374,375]:

The reaction of sulfur with complex (33) in benzene at 70°C gives 4,5-diphenyl-1,2-dithia-4-cyclopentene-3-thione. The yield is almost quantitative [400]:

$$\text{(8.133)}$$

Substituted thiophenes may be synthesized by the reaction of sulfur with a cobaltocyclopentadiene complex [401,402]. The latter reacts with sulfur in the same manner as the ferrocyclopentadienes [403]:

$$\text{(8.134)}$$

$R^1 = C_6H_5$, CO_2CH_3; $R^2 = CH_3$, C_6H_5, CO_2CH_3; $R^3 = H$, CH_3, C_6H_5

CO_2CH_3; $R^4 = CH_3$, C_6H_5, $p\text{-}CH_3C_6H_4$, CO_2CH_3

Irradiation of the cobalt-containing oximes (34) in methanol saturated with S_8 (visible light) gives the complexes (35) and dimeric products [404]:

$$\text{(8.135)}$$

Py = pyridine; R = C_2H_5, C_3H_7, C_5H_{11}, cyclo-C_6H_{11}, etc.

Bis-pentafluorophenylcobalt (obtained from the reaction of pentafluorophenylmagnesium bromide with cobalt bromide in tetra-hydrofuran and existing as a coordination complex with the solvent and magnesium bromide) reacts with sulfur with the formation (after hydrolysis) of pentafluorobenzenethiol, bis-pentafluorophenyl disul-fide, and decafluorobiphenyl [405]. It is assumed that process (8.136) is accompanied by free-radical reactions leading to the formation of decafluorobiphenyl:

$$(C_6F_5)_2 Co + S \xrightarrow{THF} (C_6F_5S)_2 Co \rightarrow (C_6F_5S)_2 \qquad (8.136)$$
$$\downarrow H_3O^+ \qquad (35\%)$$
$$C_6H_5SH$$
$$(30\%)$$

Paramagnetic tris-cyclopentadienylnickel disulfide, $(\eta^5-C_5H_5)_3-Ni_3S_2$, was obtained by treating the dimeric π-cyclopentadienylnickel carbonyl, $(\eta^5-C_5H_5NiCO)_2$, with sulfur in an organic solvent [406].

Sulfur reacts with the complexes $[Rh(CH_3)_2PCH_2CH_2P(CH_3)_2]_2Cl$ and $[Ir(diphos)_2]Cl$, $[diphos = (C_6H_5)_2PCH_2CH_2(C_6H_5)_2]$, with the formation of the complexes $[Rh(S_2)(CH_3)_2PCH_2CH_2P(CH_3)_2]_2Cl$ and $[Ir(S_2)(diphos)_2]Cl$ in which the S_2-ligand is stablilized by co-ordination with the transition metal [407]. The structure of the sulfur-containing iridium complex, $[Ir(S_2)(diphos)_2]Cl \cdot CH_3CN$, has been studied [408].

The complex RhLCl containing the tridentate ligand, $L = H_5C_6P-[CH_2CH_2CH_2P(C_6H_5)_2]_2$, reacts with sulfur in benzene with the for-mation of the complex $Rh(S_2)LCl$ [409,410]. This complex contains a symmetrically π-bonded S_2-ligand similarly to the complex $[Ir(S_2)-(diphos)_2]Cl \cdot CH_3CN$ [408].

The interaction of $(\pi-C_5H_5)Rh[P(C_6H_5)_3]_2$ with sulfur in benzene gives a complex with the composition $(\pi-C_5H_5)Rh[P(C_6H_5)_3]S_5$ in 22% yield [411]. By analogy with the complex $(\pi-C_5H_5)_2TiS_5$ [374,375] the latter has been assigned the following structure:

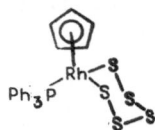

Rigby and co-workers [412] found that triphenylphosphine could be conveniently removed from the complexes (36) simply by treatment with elemental sulfur:

$$2[Rh(\pi-C_5H_5)X_2(PPh_3)] + S \rightarrow [Rh_2(\pi-C_5Me_5)X_4] + Ph_3PS \qquad (8.137)$$
$$\underset{\sim}{36}$$

$$X = N_3 \text{ or } SCN$$

The reaction of rhodium complex (37) with sulfur in refluxing xylene leads to substitution of the fragment PhCl(PPh$_3$)$_2$ by a sulfur atom [413]:

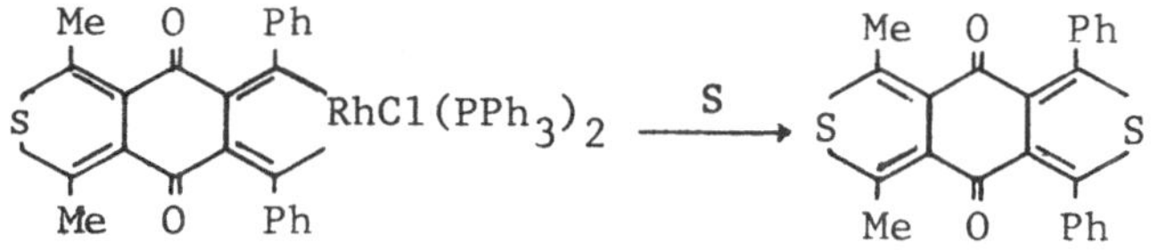

$$(8.138)$$

The reaction of sulfur with complexes (38) to (42) proceeds in an analogous manner [413-418]:

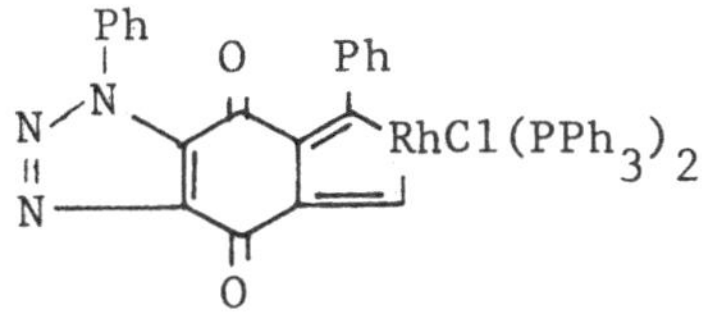

38

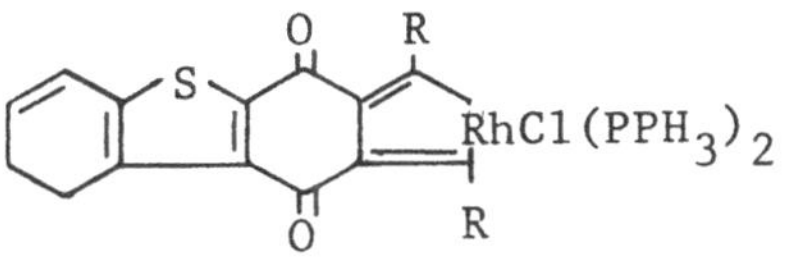

R = Et, Ph   39

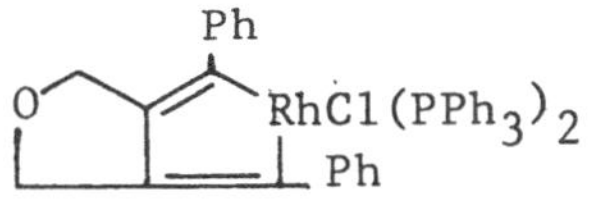

41

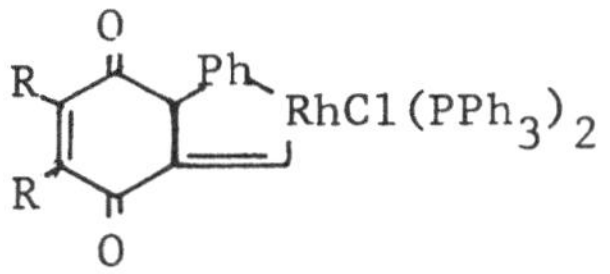

40   R = Me, Et, Ph

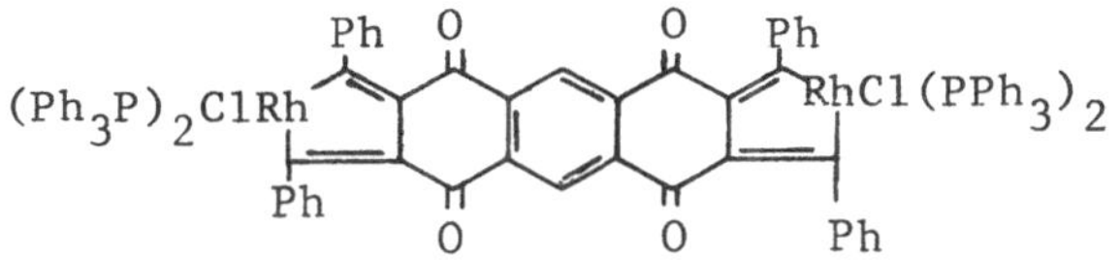

42

The complex Pd$_2$(dpm)$_2$Cl$_2$ [dpm = bis(diphenylphosphino)methane] reacts with sulfur to give a sulfide-bridged species Pd$_2$(dpm)$_2$-(μ-S)Cl$_2$. An X-ray structure determination reveals that the complex thus obtained has a geometry nearly identical with that of Pd$_2$(dpm)$_2$-(μ-SO$_2$)Cl$_2$ [419]. The related diplatinum complex Pt$_2$(dpm)$_2$(μ-S)Cl$_2$ is formed in good yield (79%) merely by stirring a dichloromethane solution of Pt$_2$(dpm)$_2$Cl$_2$ with a suspension of sulfur [420,421].

REFERENCES

1.   H. Gilman and L. Fullhart, J. Am. Chem. Soc., 71, 1478 (1949).
2.   T. V. Talalaeva and K. A. Kocheshkov, Methods of Heteroorganic Chemistry. The Organic Compounds of Lithium, Sodium, Potassium, Rubidium, and Cesium, Nauka, Moscow (1971).

3. J. F. Boscato, J. M. Catala, E. Franta, and J. Brossas, Makromol. Chem., 180, 1571 (1979).
4. J. F. Boscato, J. M. Catala, E. Franta, and J. Brossas, Tetrahedron Lett., 1519 (1980).
5. Weygand-Hilgetag, Organisch-chemische Experimentierkunst. 3. Auflage, J. A. Barth Verlag, Leipzig (1968).
6. M. Mikolajczyk, S. Grzejszak, A. Chetczynska, and A. Zatorski, J. Org. Chem., 44, 2967 (1979).
7. M. Mikolajczyk, B. Costisella, S. Grzejszak, and A. Zatorski, Tetrahedron Lett., 447 (1976).
8. R. S. Sukhai, R. de Jong, and L. Brandsma, Synthesis, 888 (1977).
9. P. J. Schuijl, L. Brandsma, and J. F. Arens, Recl. Trav. Chim. Pays-Bas, 85, 889 (1966).
10. R. S. Sukhai and L. Brandsma, Recl. Trav. Chim. Pays-Bas, 98, 55 (1979).
11. S. J. Harris and D. R. M. Walton, J. Chem. Soc. Chem. Commun., 1008 (1976).
12. D. Braunn, Chimia, 14, 24 (1960).
13. J. M. J. Frechet, M. D. DeSmet, and M. J. Favall, Polymer, 20, 675 (1979).
14. D. Dauplaise, J. Meinwald, J. C. Scott, H. Temkin, and J. Clardy, Ann. N.Y. Acad. Sci., 313, 382 (1978).
15. J. Meinwald, D. Dauplaise, and J. Clardy, J. Am. Chem. Soc., 99, 7743 (1977).
16. J. Meinwald, D. Dauplaise, F. Wudl, and J. J. Hauser, J. Am. Chem. Soc., 99, 255 (1977).
17. R. L. Y. Chiang, Ph.D. Thesis, Cornell University (1980); Diss. Abstr. Int. B, 41, 1367 (1980).
18. L. Y. Chiang and J. Meinwald, Tetrahedron Lett., 21, 4565 (1980).
19. F. Hardwick, R. Stephens, J. C. Tatlow, and J. R. Taylor, J. Fluorine Chem., 3, 151 (1973).
20. G. M. Brooke, B. S. Furniss, W. K. R. Musgrave, and A. Quasem, Tetrahedron Lett., 2991 (1965).
21. G. M. Brooke and A. Quasem, J. Chem. Soc., C, 865 (1967).
22. Ya. L. Gol'dfarb, M. A. Kalik, and M. L. Kirmalova, Zh. Obshch. Khim., 32, 222 (1962).
23. L. Brandsma and H. J. T. Bos, Recl. Trav. Chim. Pays-Bas, 88, 732 (1969).
24. S. Gronowitz and R. Hakansson, Arkiv Kemi, 16, 309 (1960).
25. E. Jones and I. M. Moodie, Tetrahedron, 21, 1333 (1965).
26. S. Gronowitz, P. Moses, and A. B. Hörnfeldt, Arkiv Kemi, 17, 237 (1961).
27. Ya. L. Gol'dfarb, M. A. Kalik, and M. L. Kirmalova, Zh. Obshch. Khim., 29, 3631 (1959).
28. Ya. L. Gol'dfarb, I. P. Konyaeva, and V. P. Litvinov, Izv. Akad. Nauk SSSR, Ser. Khim., 1570 (1974).
29. R. Grafing and L. Brandsma, Synthesis, 578 (1978).
30. Ya. L. Gol'dfarb, Ya. L. Danyushevskii, and M. A. Vinogradova, Dokl. Akad. Nauk SSSR, 151, 332 (1963); C. A., 59, 8681 (1963).

31. D. E. Horning and J. M. Muchowski, Can. J. Chem., 52, 2950 (1974).

32. E. Niwa, H. Aoki, H. Tanaka, K. Munakata, and M. Namiki, Chem. Ber., 99, 3215 (1966).

33. J. A. Gladysz, V. K. Wong, and B. S. Jick, J. Chem. Soc. Chem. Commun., 838 (1978).

34. J. A. Gladysz, J. L. Hornby, and J. E. Garbe, J. Org. Chem., 43, 1204 (1978).

35. H. D. Smith, C. O. Obenland, and S. Papetti, Inorg. Chem., 5, 1013 (1966).

36. N. S. Semenuk, S. Papetti, and H. Schroeder, Inorg. Chem., 8, 2441 (1969).

37. L. I. Zakharkin, V. N. Kalinin, and G. G. Zhigareva, Izv. Akad. Nauk SSSR, Ser. Khim., 912 (1970).

38. J. J. Bishop, A. Davison, M. L. Katcher, D. W. Lichtenberg, R. E. Merrill, and J. C. Smart, J. Organomet. Chem., 27, 241 (1971).

39. A. Davison and J. C. Smart, J. Organomet. Chem., 19, P7 (1969).

40. H. Gilman and G. D. Lichtenwalter, J. Org. Chem., 25, 1064 (1960).

41. J. Meijer, H. E. Wijers, and L. Brandsma, Recl. Trav. Chim. Pays-Bas, 91, 1423 (1972).

42. H. Schumann, K. F. Thom, and M. Schmidt, J. Organomet. Chem., 4, 28 (1965).

43. H. Schumann, K. F. Thom, and M. Schmidt, J. Organomet. Chem., 1, 167 (1963).

44. H. Schumann, K. F. Thom, and M. Schmidt, Angew. Chem., 75, 138 (1963).

45. H. Schumann, K. F. Thom, and M. Schmidt, J. Organomet. Chem., 2, 97 (1964).

46. H. Schumann and M. Schmidt, Angew. Chem., 77, 1049 (1965); Angew. Chem., Int. Ed. Engl., 4, 1007 (1965).

47. H. Schumann, in: Sulphur in Organic and Inorganic Chemistry, Vol. 3, A. Senning (ed.), Dekker, New York (1972), Chapter 21.

48. L. Brandsma, H. E. Wijers, and J. F. Arens, Recl. Trav. Chim. Pays-Bas, 81, 583 (1962).

49. B. Iddon, H. Suschitzky, and A. W. Thompson, J. Chem. Soc., Perkin Trans. I, 2300 (1974).

50. L. Brandsma, H. E. Wijers, and C. Jonker, Recl. Trav. Chim. Pays-Bas, 83, 208 (1964).

51. L. Brandsma, Recl. Trav. Chim. Pays-Bas, 83, 307 (1964).

52. R. Mayer, B. Gebhardt, J. Fabian, and A. K. Müller, Angew. Chem., 76, 143 (1964); Angew. Chem., Int. Ed. Engl., 3, 134 (1964).

53. R. Mayer, B. Hunger, R. Prousa, and A. K. Müller, J. Prakt. Chem., 35, 294 (1967).

54. R. Mayer and B. Gebhardt, Chem. Ber., 97, 1298 (1964).

55. R. Mayer and K. Gewald, Angew. Chem., 79, 298 (1967); Angew. Chem., Int. Ed. Engl., 6, 294 (1967).

56. E. M. Engler and V. V. Patel, J. Chem. Soc. Chem. Commun.,
     671 (1975).
57. E. M. Engler and V. V. Patel, J. Org. Chem., 40, 387 (1975).
58. Hsue-Fen Li, Su-Zheng Li, Eu-Shing Yoa, and Chang-Ye, Tzu-Jan
     Tsa Chih, 2, 136 (1979); C. A., 91, 39393 (1980).
59. B. L. Dyatkin, S. R. Sterlin, L. G. Zhuravkova, B. I. Martynov,
     E. I. Mysov, and I. L. Knunyants, Tetrahedron, 29, 2759 (1973).
60. B. L. Dyatkin, S. R. Sterlin, L. G. Zhuravkova, and I. L.
     Knunyants, Dokl. Akad. Nauk SSSR, 183, 598 (1968).
61. S. R. Sterlin, B. L. Dyatkin, and I. L. Knunyants, Izv. Akad.
     Nauk SSSR, Ser. Khim., 2583 (1967).
62. C. G. Krespan and D. C. England, J. Org. Chem., 33, 1850 (1868).
63. S. I. Radchenko and A. A. Popov, Zh. Org. Khim., 13, 40 (1977).
64. L. I. Zakharkin and G. G. Zhigareva, Zh. Obshch. Khim., 45,
     789 (1975).
65. J. Meier and L. Brandsma, Recl. Trav. Chim. Pays-Bas, 90, 1098
     (1971).
66. L. I. Zakharkin, V. N. Kalinin, and V. V. Gedymin, J. Organomet.
     Chem., 16, 371 (1969).
67. L. I. Zakharkin and G. G. Zhigarev, Izv. Akad. Nauk SSSR, Ser.
     Khim., 1358 (1967).
68. Ya. L. Gol'dfarb, V. P. Litvinov, and A. N. Sukiasyan, Dokl.
     Akad. Nauk SSSR, 182, 340 (1968).
69. A. A. Petrov, S. I. Radchenko, K. S. Mingaleva, I. G. Sarich,
     and V. B. Lebedev, Zh. Obshch. Khim., 34, 1899 (1964).
70. B. P. Fedorov and F. M. Stoyanovich, Zh. Obshch. Khim., 33, 2251
     (1963).
71. Ya. L. Gol'dfarb, M. A. Kalik, and M. L. Kirmalova, Zh. Obshch.
     Khim., 29, 2034 (1963).
72. S. Gronowitz and P. Moses, Acta Chem. Scand., 16, 155 (1962).
73. Ya. L. Gol'dfarb and V. P. Litvinov, Izv. Akad. Nauk SSSR,
     Ser. Khim., 1621 (1963).
74. Ya. L. Gol'dfarb, V. P. Litvinov, and S. A. Ozolin', Izv. Akad.
     Nauk SSSR, Ser. Khim., 1432 (1966).
75. Ya. L. Gol'dfarb and V. P. Litvinov, Izv. Akad. Nauk SSSR, Ser.
     Khim., 343 (1963).
76. S. Gronowitz, U. Rudén, and B. Gestblom, Arkiv Kemi, 20, 297
     (1963).
77. B. P. Fedorov and F. M. Stoyanovich, Zh. Org. Khim., 1, 194
     (1965).
78. B. P. Fedorov and F. M. Stoyanovich, Zh. Obshch. Khim., 31,
     238 (1961).
79. Ya. L. Gol'dfarb, V. P. Litvinov, and S. A. Ozolin', Izv. Akad.
     Nauk SSSR, Ser. Khim., 510 (1965).
80. S. Gronowitz and P. Moses, Acta Chem. Scand., 16, 105 (1962).
81. S. Gronowitz, Arkiv Kemi, 13, 269 (1958).
82. Ya. L. Gol'dfarb and V. P. Litvinov, Izv. Akad. Nauk SSSR, Ser.
     Khim., 2088 (1964).
83. L. A. Krichevskii and A. V. Shchelkunov, Zh. Org. Khim., 7,
     1888 (1971).

84. R. A. Damico, US Patent 3,700,676 (1972); C. A., <u>78</u>, 43278z (1973).
85. R. A. Damico, US Patent 3,773,770 (1973); C. A., <u>80</u>, 47851v (1974).
86. R. A. Abramovitch and E. E. Knaus, J. Heterocycl. Chem., <u>6</u>, 989 (1969).
87. R. A. Abramovitch and E. E. Knaus, J. Heterocycl. Chem., <u>12</u>, 683 (1975).
88. R. H. Mobbs, Chem. Ind., 1562 (1965).
89. R. H. Mobbs, Brit. Patent 1,135,773 (1968); C. A., <u>70</u>, P57415 (1969).
90. P. G. Eller and D. W. Meek, J. Organomet. Chem., <u>22</u>, 631 (1970).
91. S. Hayashi and N. Ishikawa, Yuki Gosei Kagaku Kyokai Shi, <u>28</u>, 533 (1970); C. A., <u>73</u>, 45241c (1970).
92. Ya. L. Gol'dfarb and V. P. Litvinov, Izv. Akad. Nauk SSSR, Ser. Khim., 352 (1963).
93. M. P. L. Caton, M. S. Grant, D. L. Pain, and R. Slack, J. Chem. Soc., 5467 (1965).
94. M. Schmidt and O. Scherer, Naturwissenschaften, <u>50</u>, 302 (1963).
95. C. Tamborski and E. J. Soloski, J. Org. Chem., <u>31</u>, 743 (1966).
96. R. D. Chambers and D. J. Spring, Tetrahedron, <u>27</u>, 669 (1971).
97. F. Dallacker and H. Zegers, Liebigs Ann. Chem., <u>689</u>, 156 (1965).
98. S. Gronowitz and B. Persson, Acta Chem. Scand., <u>21</u>, 812 (1967).
99. M. Schmidt and V. Potschka, Naturwissenschaften, <u>50</u>, 302 (1963).
100. R. B. Mitra, L. J. Pandya, and B. D. Tilak, J. Sci. Ind. Res. (India), B, <u>16</u>, 348 (1957); C. A., <u>52</u>, 5371h (1958).
101. Ya. L. Gol'dfarb, S. A. Ozolin', and V. P. Litvinov, Khim. Geterotsikl. Soedin., 935 (1967).
102. V. P. Litvinov and G. Fraenkel, Izv. Akad. Nauk SSSR, Ser. Khim., 1828 (1968).
103. G. J. Grol, Tetrahedron, <u>30</u>, 3621 (1974).
104. R. K. Zaripov, I. N. Azerbaev, K. Sh. Shamgunov, and U. A. Aimakov, Papers of the IVth All-Union Conference on Acetylene, Alma-Ata (1972), p. 22; C. A., <u>79</u>, 78894 (1973).
105. G. S. Meyers, H. H. Richmond, and G. W. Wright, J. Am. Chem. Soc., <u>312</u>, 571 (1970).
106. K. Issbeib, H. Böhme, and C. Rockstroh, J. Prakt. Chem., <u>312</u>, 571 (1970).
107. R. D. Cannon, B. Chiswell, and L. M. Venanzi, J. Chem. Soc., A, 1277 (1967).
108. M. Kubota and A. Yamamoto, Bull. Chem. Soc. Jpn., <u>51</u>, 3622 (1978).
109. V. V. Gaisas and B. D. Tilak, J. Sci. Ind. Res. (India), B, <u>16</u>, 345 (1957); C. A., <u>52</u>, 5370i (1958).
110. H. E. Ramsden, US Patent 2,921,964 (1960).
111. B. L. Dyatkin, B. I. Martynov, L. G. Martynova, N. G. Kizim, S. R. Sterlin, A. A. Stumbrevichute, and L. A. Fedorov, J. Organomet. Chem., <u>57</u>, 423 (1973).
112. M. S. Kharasch and O. Reinmuth, Grignard Reactions of Nonmetallic Substances, Prentice-Hall, New York (1954), p. 1274.

113. K. A. Hooton and A. L. Allred, Inorg. Chem., 4, 671 (1965).
114. H. Wuyts, Bull. Soc. Chim. France, /4/, 5, 405 (1909).
115. H. Rheinboldt, F. Mott, and E. Motzkus, J. Prakt. Chem., 134, 257 (1932).
116. J. L. Wood, J. R. Rachele, C. M. Stevens, F. H. Carpenter, and V. du Vigneaud, J. Am. Chem. Soc., 70, 2547 (1948).
117. A. Murray, III, and D. L. Williams, Organic Syntheses with Isotopes, Part II, Wiley-Interscience, New York—London (1958).
118. G. Ayrey and J. Label, Compounds, 2, 51 (1966).
119. R. C. Thomas and L. J. Reed, J. Am. Chem. Soc., 77, 5446 (1955).
120. P. T. Adams, J. Am. Chem. Soc., 77, 5357 (1955).
121. G. W. Kilmer and V. du Vigneaud, J. Biol. Chem., 154, 247 (1944).
122. E. N. Prilezhaeva, I. L. Mikhelashvili, and V. S. Bogdanov, Zh. Org. Chem., 8, 1505 (1948).
123. M. Mousseron, H. Bousquet, and G. Marret, Bull. Soc. Chim. France, /5/, 15, 84 (1948).
124. W. H. Houff and R. D. Schuetz, J. Am. Chem. Soc., 75, 6316 (1953).
125. P. D. Caesar and P. D. Branton, Ind. Eng. Chem., 44, 122 (1952).
126. E. Profft, Chem. Ztg., 82, 295 (1958).
127. J. Cymerman-Graig and J. W. Loder, J. Chem. Soc., 237 (1954).
128. J. Cymerman-Graig and J. W. Loder, Org. Syntheses, 35, 85 (1955).
129. F. Challenger and S. A. Miller, J. Chem. Soc., 894 (1938).
130. S. Gronowitz and R. A. Hoffman, Arkiv Kemi, 15, 499 (1960).
131. I. N. Tits-Skvortsova, A. I. Leonova, and R. Ya. Levina, Essays in General Chemistry, Vol. 2, Izd. Akad. Nauk SSSR, Moscow—Leningrad (1953), p. 1135.
132. M. Mousseron, C. R. Acad. Sci., Ser. C, 215, 201 (1942).
133. A. Mailhe and M. Murat, Bull. Soc. Chim. France, /4/, 7, 288 (1910).
134. C. F. Smith, G. J. Moore, and C. Tamborski, J. Organomet. Chem., 42, 257 (1972).
135. M. Seyhan, Ber., 72, 594 (1939).
136. Fr. Patent 1,209,860 (1956); Chem. Zentralbl., 133, 7691 (1962).
137. H. Wuyts and G. Cosyns, Bull. Soc. Chim. France, /3/, 29, 689 (1903).
138. J. L. Wood and L. Van Middleworth, J. Biol. Chem., 179, 529 (1949).
139. A. M. Seligman, A. M. Rutenburg, and H. Banks, J. Clin. Invest., 22, 275 (1949).
140. R. G. Karpenko, F. M. Stoyanovich, and Ya. L. Gol'dfarb, Zh. Org. Khim., 5, 2000 (1969).
141. J. Szmuskovicz, Org. Prep. Proced. Int., 1, 43 (1969).

142. M. Aresta and R. S. Nyholm, J. Organomet. Chem., 56, 395 (1973).
143. D. N. Kravtsov, B. A. Kvasov, L. S. Golovchenko, E. M. Rokhlina, and E. I. Fedin, J. Organomet. Chem., 39, 107 (1972).
144. R. V. Jardine and R. K. Brown, Can. J. Chem., 42, 2626 (1964).
145. W. Madelung and M. Tencer, Ber., 48, 949 (1915).
146. D. N. Kravtsov, A. S. Peregudov, E. M. Rokhlina, and L. A. Fedorov, J. Organomet. Chem., 77, 199 (1974).
147. Y. Sakata and T. Hashimoto, Yakugaku Zasshi, 79, 872 (1959); C. A., 54, 357 (1960).
148. B. Oddo and L. Raffa, Gazz. Chim. Ital., 71, 242 (1941); C. A., 36, 2854 (1942).
149. R. Wilputte and R. H. Martin, Bull. Soc. Chim. Belg., 65, 874 (1956).
150. F. Taboury, C. R. Acad. Sci., Ser. C., 138, 982 (1904); Chem. Zentralbl., I, 1413 (1904).
151. W. Borsche and W. Lange, Ber., 39, 2346 (1906); Chem. Zentralbl., II, 518 (1906).
152. C. F. Koelsch and G. Ullyot, J. Am. Chem. Soc., 55, 3883 (1933).
153. J. Haraszti, J. Prakt. Chem., 149, 301 (1937); C. A., 32, 5285 (1938).
154. W. E. Bachmann and R. F. Cockerill, J. Am. Chem. Soc., 55, 2932 (1933).
155. R. D. Chambers, W. K. R. Musgrave, and J. Savory, J. Chem. Soc., 1993 (1962).
155a. W. Seidel and I. Burger, Z. Anorg. Allgem. Chem., 473, 166 (1981).
156. J. Burdon, P. L. Coe, and M. Fulton, J. Chem. Soc., 2094 (1965).
157. G. A. Razuvaev and M. M. Koton, Zh. Obshch. Khim., 5, 361 (1935).
158. D. Seyferth and W. Tronich, J. Am. Chem. Soc., 91, 2138 (1969).
159. D. Seyferth, W. Tronich, R. S. Marmor, and W. E. Smith, J. Org. Chem., 37, 1537 (1972).
160. G. M. la Roy and E. C. Kooyman, J. Organomet. Chem., 7, 357 (1967).
161. F. Krafft and R. E. Lyons, Ber., 27, 1768 (1894).
162. S. Nozakura, Y. Yamamoto, and S. Murahashi, Polym. J., 5, 55 (1973); C. A., 80, 48432 (1974).
163. W. J. Middleton, E. G. Howard, and W. H. Sharkey, J. Am. Chem. Soc., 83, 2589 (1961).
164. W. J. Middleton, E. G. Howard, and W. H. Sharkey, J. Org. Chem., 30, 1375 (1965).
165. E. G. Howard, Jr., and W. J. Middleton, US Patent 2,970,173 (1961); C. A., 55, 14312a (1961).
166. B. L. Dyatkin, S. R. Sterlin, B. I. Martynov, and I. L. Knunyants, Tetrahedron Lett., 345 (1971).
167. F. Hardwick, A. E. Pedler, R. Stephens, and J. C. Tatlow, J. Fluorine Chem., 4, 9 (1974).
168. E. N. Gladyshev, V. S. Andreevichev, N. S. Vyazankin, and G. A. Razuvaev, Zh. Obshch. Khim., 40, 939 (1970).

169. E. N. Gladyshev, N. S. Vyazankin, V. S. Andreevichev, A. A. Klimov, and G. A. Razuvaev, J. Organomet. Chem., 28, C42 (1971).

170. E. N. Gladyshev, V. S. Andreevichev, A. A. Klimov, N. S. Vyazankin, and G. A. Razuvaev, Zh. Obshch. Khim., 42, 1077 (1972).

171. M. N. Bochkarev, L. P. Maiorova, N. S. Vyazankin, and G. A. Razuvaev, J. Organomet. Chem., 82, 65 (1974).

172. W. J. Middleton, US Patent 3,113,936 (1963); C. A., 60, P4012h (1964).

173. C. G. Krespan, US Patent 2,844,614 (1958); C. A., 53, 3061e (1959).

174. J. A. Young, S. N. Tsoukalas, and R. D. Dresdner, J. Am. Chem. Soc., 82, 396 (1960).

175. S. C. Cohen, M. L. N. Reddy, and A. G. Massey, J. Organomet. Chem., 11, 563 (1968).

176. E. Dreher and R. Otto, Ann. Chem., 154, 93 (1871).

177. E. Dreher and R. Otto, Ber., 542 (1869).

178. S. C. Cohen and A. G. Massey, Adv. Fluorine Chem., 6, 83 (1970).

179. H. Zeiser, Ber., 28, 1670 (1895).

180. M. D. Rausch, J. Org. Chem., 26, 3579 (1961).

181. K. Issleib, R. Kümmel, H. Oehme, and I. Meissner, Chem. Ber., 101, 3612 (1968).

182. S. C. Cohen and A. G. Massey, J. Organomet. Chem., 12, 341 (1968).

183. C. M. Woodward, G. Hughes, and A. G. Massey, J. Organomet. Chem., 112, 9 (1976).

184. O. A. Kruglaya, B. V. Fedot'ev, I. B. Fedot'eva, and N. S. Vyazankin, Zh. Obshch. Khim., 46, 1517 (1976).

185. O. A. Kruglaya, L. I. Belousova, B. V. Fedot'ev, and N. S. Vyazankin, Izv. Akad. Nauk SSSR, Ser. Khim., 975 (1975).

186. V. T. Bychkov, N. S. Vyazankin, O. V. Linzina, L. V. Aleksandrova, and G. A. Razuvaev, Izv. Akad. Nauk SSSR, Ser. Khim., 2614 (1970).

187. G. I. Zimmerman, Ph.D. Thesis. University of Pennsylvania, 1980; Diss. Abstr. Int. B, 41, 953 (1980).

188. B. M. Mikhailov and Yu. N. Bubnov, Izv. Akad. Nauk SSSR, Otd. Khim. Nauk, 172 (1959).

189. B. M. Mikhailov and Yu. N. Bubnov, Zh. Obshch. Khim., 1648 (1959).

190. B. M. Mikhailov and Yu. N. Bubnov, Izv. Akad. Nauk SSSR, Ser., Khim., 531 (1961).

191. Z. Yoshida, T. Okushi, and O. Manabe, Tetrahedron Lett., 1641 (1970).

192. W. Siebert, E. Gast, and M. Schmidt, J. Organomet. Chem., 23, 329 (1970).

193. M. Schmidt, W. Siebert, and F. Rittig, Chem. Ber., 101, 281 (1968).

194. M. Schmidt and W. Siebert, Chem. Ber., 102, 2752 (1969).

195. W. Siebert, K. J. Schaper, and M. Schmidt, J. Organomet. Chem., 25, 315 (1970).

196. H. M. Seip, R. Seip, and W. Siebert, Acta Chem. Scand., $\underline{27}$, 15 (1973).
197. L. I. Zakharkin and V. V. Gavrilenko, Izv. Akad. Nauk SSSR, Otd. Khim. Nauk, 1391 (1961).
198. H. Jenkner, Ger. Patent 1,031,306 (1958); C. A., $\underline{54}$, 17269g (1960).
199. R. E. Leech and J. P. Knap, US Patent 3,012,077 (1961).
200. R. E. Leech and J. P. Knap, US Patent 2,998,455 (1961).
201. S. Warwel and B. Ahlfaenger, Chem. Ztg., $\underline{101}$, 103 (1977).
202. G. B. Deacon and J. C. Parrott, J. Organomet. Chem., $\underline{17}$, P17 (1969).
203. G. B. Deacon and J. C. Parrott, J. Organomet. Chem., $\underline{22}$, 287 (1970).
204. M. Schmidt and H. Schumann, Z. Anorg. Allgem. Chem., $\underline{325}$, 130 (1963).
205. P. Mazerolles, J. Dubac, and M. Lesbre, J. Organomet. Chem., $\underline{12}$, 143 (1968).
206. P. Mazerolles, J. Dubac, and M. Lesbre, J. Organomet. Chem., $\underline{5}$, 35 (1966).
207. J. Dubac and P. Mazerolles, C. R. Acad. Sci., Ser. C, $\underline{267}$, 411 (1968).
208. M. Weidenbruch, A. Schäfer, and R. Rankers, J. Organomet. Chem., $\underline{195}$, 171 (1980).
209. A. M. Devine, P. A. Griffin, R. N. Haszeldine, M. J. Newlands, and A. R. Tipping, J. Chem. Soc., Dalton Trans., 1434 (1975).
210. C. Couret, J. Escudie, J. Stage, J. D. Andriamizaka, and B. Saint-Roch, J. Organomet. Chem., $\underline{182}$, 9 (1979).
211. D. Seyferth, J. Organomet. Chem., $\underline{100}$, 237 (1975).
212. D. Seyferth, D. P. Dunkan, and C. K. Haas, J. Organomet. Chem., $\underline{164}$, 305 (1979).
213. R. W. Bost and P. Borgstrom, J. Am. Chem. Soc., $\underline{51}$, 1922 (1929).
214. M. Schmidt, H. J. Dersin, and H. Schumann, Chem. Ber., $\underline{95}$, 1428 (1962).
215. H. Schumann, Z. Anorg. Allgem. Chem., $\underline{354}$, 192 (1967).
216. M. Schmidt and H. Schumann, Chem. Ber., $\underline{96}$, 462 (1963).
217. R. W. Bost and H. R. Baker, J. Am. Chem. Soc., $\underline{55}$, 1112 (1933).
218. S. V. Ponomarev and I. F. Lutsenko, Zh. Obshch. Khim., $\underline{34}$, 3450 (1964).
219. M. G. Voronkov and Yu. I. Khudobin, Zh. Obshch. Khim., $\underline{26}$, 584 (1956).
220. H. Schumann and M. Schmidt, Chem. Ber., $\underline{96}$, 3017 (1963).
221. T. A. Bither, W. H. Knoth, R. V. Lindsey, Jr., and W. H. Sharkey, J. Am. Chem. Soc., $\underline{80}$, 4151 (1958).
222. D. Seyferth and N. Kahlen, J. Org. Chem., $\underline{25}$, 809 (1960).
223. J. J. McBride, Jr., and H. C. Beachell, J. Am. Chem. Soc., $\underline{74}$, 5247 (1952).
224. J. J. McBride, Jr., J. Org. Chem., $\underline{24}$, 2029 (1959).
225. R. A. Cummins and P. Dunn, Aust. J. Chem., $\underline{17}$, 411 (1964).
226. J. S. Thayer, Inorg. Chem., $\underline{7}$, 2599 (1968).

227. R. Calas, N. Duffaut, B. Martel, and C. Paris, Bull. Soc. Chim. France, 886 (1961).
228. B. Martel, N. Duffaut, and R. Calas, Bull. Soc. Chim. France, 758 (1967).
229. L. Birkofer, A. Ritter, and H. Goller, Chem. Ber., $\underline{96}$, 3289 (1963).
230. M. N. Bochkarev, L. P. Sanina, and N. S. Vyazankin, Zh. Obshch. Khim., $\underline{39}$, 135 (1969).
231. W. Fink, Helv. Chim. Acta, $\underline{56}$, 1117 (1973).
232. M. Wojnowska and W. Wojnowski, Z. Anorg. Allgem. Chem., $\underline{403}$, 179 (1974).
233. F. Fehér and H. Goller, Z. Naturforsch., $\underline{22b}$, 1223 (1967).
234. N. S. Vyazankin, M. N. Bochkarev, and L. P. Sanina, Zh. Obshch. Khim., $\underline{37}$, 1037 (1967).
235. N. S. Vyazankin, M. N. Bochkarev, and L. P. Sanina, Zh. Obshch. Khim., $\underline{36}$, 166 (1966).
236. N. S. Vyazankin, M. N. Bochkarev, and L. P. Sanina, Zh. Obshch. Khim., $\underline{36}$, 1961 (1966).
237. N. S. Vyazankin, M. N. Bochkarev, and L. P. Sanina, Zh. Obshch. Khim., $\underline{36}$, 1154 (1966).
238. M. Schmeisser and W. Burgemeister, Angew. Chem., $\underline{69}$, 782 (1957).
239. Ger. Patent 1,094,262 (1960); C. A., $\underline{55}$, 22132 (1961).
240. F. Fehér and R. Luepschen, Z. Naturforsch., $\underline{26b}$, 1191 (1971).
241. M. N. Bochkarev, N. S. Vyazankin, L. O. Maiorova, and G. A. Razuvaev, Zh. Obshch. Khim., $\underline{48}$, 2706 (1978).
242. M. N. Bochkarev, G. A. Razuvaev, N. S. Vyazankin, and O. Yu. Semenov, J. Organomet. Chem., $\underline{74}$, C4 (1974).
243. M. N. Bochkarev, G. A. Razuvaev, and N. S. Vyazankin, Izv. Akad. Nauk SSSR, Ser. Khim., 1820 (1975).
244. M. N. Bochkarev, N. S. Vyazankin, L. B. Bochkarev, and G. A. Razuvaev, J. Organomet. Chem., $\underline{110}$, 149 (1976).
245. P. Mazerolles, M. Lesbre, and M. Joanny, J. Organomet. Chem., $\underline{16}$, 227 (1969).
246. C. A. Kraus and W. V. Sessions, J. Am. Chem. Soc., $\underline{47}$, 2361 (1925).
247. J. Otera, T. Kadowaki, and R. Okawara, J. Organomet. Chem., $\underline{19}$, 213 (1969).
248. H. G. Kuivila and E. R. Jakusik, J. Org. Chem., $\underline{26}$, 1430 (1961).
249. H. G. Kuivila and O. F. Beumel, Jr., J. Am. Chem. Soc., $\underline{80}$, 3250 (1958).
250. A. W. Krebs and M. C. Henry, J. Org. Chem., $\underline{28}$, 1911 (1963).
251. L. C. Willemsens and G. J. M. van der Kerk, J. Organomet. Chem., $\underline{15}$, 117 (1968).
252. M. Wojnowska, W. Wojnowski, and R. West, J. Organomet. Chem., $\underline{199}$, C1 (1980).
253. M. Veit, Z. Naturforsch., $\underline{33B}$, 1 (1978); C. A., $\underline{88}$, 121291 (1978).
254. J. Spanswick and K. U. Ingold, Int. J. Chem. Kinet., $\underline{2}$, 157 (1970).
255. L. Maier, Helv. Chim. Acta, $\underline{48}$, 1190 (1965).
256. H. Köhler and A. Michaelis, Ber., $\underline{10}$, 807 (1877).
257. L. Maier, Helv. Chim. Acta, $\underline{46}$, 1812 (1963).

258. E. A. Letts and R. F. Blake, J. Chem. Soc., 58, 767 (1890).

259. G. Peters, J. Am. Chem. Soc., 82, 4751 (1960).

260. M. M. Rauhut, H. A. Currier, and V. P. Wystrach, J. Org. Chem. 26, 5133 (1961).

261. K. Issleib and G. Döll, Z. Anorg. Allgem. Chem., 324, 259 (1963).

262. I. F. Lutsenko, M. V. Proskurnina, and A. A. Borisenko, Organomet. Chem. Syn., 1, 169 (1970/1971).

263. G. Peters, J. Org. Chem., 27, 2198 (1962).

264. A. W. V. Hofmann and F. Mahla, Ber., 25, 2436 (1892).

265. L. Malatesta, Gazz. Chim. Ital., 77, 518 (1948).

266. R. C. Dobbie, D. F. Doty, and R. G. Cavell, J. Am. Chem. Soc., 90, 2015 (1968).

267. W. R. Cullen, Can. J. Chem., 41, 2424 (1963).

268. W. M. Dehn and B. B. Wilcox, Am. Chem. J., 35, 1 (1906).

269. N. Camerman and J. Trotter, J. Chem. Soc., 219 (1964).

270. R. A. Zingaro, K. J. Irgolic, D. H. O'Brien, and L. J. Edmonson, Jr., J. Am. Chem. Soc., 93, 5677 (1971).

271. K. Issleib and K. D. Franze, J. Prakt. Chem., 315, 471 (1973).

272. R. B. King and P. R. Heckley, Phosphorus, 3, 209 (1974).

273. M. M. Rauhut, H. A. Currier, G. A. Peters, F. C. Schaefer, and V. P. Wystrach, J. Org. Chem., 26, 5135 (1961).

274. Zh. M. Ivanova and A. V. Kirsanov, Zh. Obshch. Khim., 31, 3991 (1961).

275. K. Issleib and G. Döll, Chem. Ber., 96, 1544 (1963).

276. K. Issleib, H. Oehme, R. Kümmel, and E. Leissring, Chem. Ber., 101, 3619 (1968).

277. K. Issleib and H. Oehme, Chem. Ber., 100, 2685 (1967).

278. K. Issleib, H. Oehme, and E. Leissring, Chem. Ber., 101, 4032 (1968).

279. W. Kuchen and A. Rohrbeck, Chem. Ber., 105, 132 (1972).

280. K. Issleib and H. R. Roloff, J. Prakt. Chem., 312, 578 (1970).

281. K. Issleib, H. Oehme, and K. Mohr, Z. Chem., 13, 139 (1973).

282. K. Issleib and P. V. Malotki, J. Prakt. Chem., 312, 366 (1970).

283. K. Issleib and G. Döll, Chem. Ber., 94, 2664 (1961).

284. K. Issleib and H. Weichmann, Z. Chem., 11, 188 (1971).

285. K. Issleib and D. Jacob, Chem. Ber., 94, 107 (1961).

286. K. Issleib and H. Oehme, Z. Anorg. Allgem. Chem., 343, 268 (1966).

287. K. Issleib and H. Weichmann, Chem. Ber., 101, 2998 (1968).

288. H. Hoffmann and P. Schellenbeck, Chem. Ber., 99, 1134 (1966).

289. K. Issleib, B. Hamann, and L. Schmidt, Z. Anorg. Allgem. Chem., 339, 298 (1965).

290. V. I. Bregadze, D. N. Sadzhaya, and O. Yu. Okhlobystin, Soobshch. Akad. Nauk Gruz. SSR, 63, 77 (1971); C. A., 75, 98644j (1971).

291. A. Cahours and A. W. Hofmann, Liebigs Ann. Chem., 104, 1 (1857).

292. A. Michaelis, Liebigs Ann. Chem., 181, 355 (1876).

293. P. D. Bartlett and G. Meguerian, J. Am. Chem. Soc., 78, 3710 (1956).

294. R. D. Davis, J. Phys. Chem., 63, 307 (1959).
295. H. Goetz, F. Nerdel, and E. Busch, Liebigs Ann. Chem., 665, 14 (1963).
296. P. D. Bartlett, E. F. Cox, and R. E. Davis, J. Am. Chem. Soc., 83, 103 (1961).
297. D. P. Young, W. E. McEwen, D. C. Velex, J. W. Johnson, and C. A. Van der Werf, Tetrahedron Lett., 359 (1964).
298. L. Horner and H. Winkler, Tetrahedron Lett., 175 (1964).
299. G. Zon, K. E. De Bruin, K. Naumann, and K. Mislow, J. Am. Chem. Soc., 91, 7023 (1969).
300. C. L. Bodkin and P. Simpson, J. Chem. Soc., B, 1136 (1971).
301. L. Horner and H. Fuchs, Tetrahedron Lett., 203 (1962).
302. Yu. F. Gatilov and F. D. Yambushev, Zh. Obshch. Khim., 43, 1132 (1973).
303. J. Omelanczuk, W. Perlikowska, and M. Mikolajczyk, J. Chem. Soc., Chem. Commun., 24 (1980).
304. R. B. King, P. R. Heckley, and J. C. Cloyd, Jr., Z. Naturforsch., 29b, 574 (1974).
305. D. B. Cooper, J. M. Harrison, T. D. Inch, and G. J. Lewis, J. Chem. Soc., Perkin Trans. I, 1049 (1974).
306. M. Mikolajczyk, Angew. Chem., 81, 495 (1969); Angew. Chem., Int. Ed. Engl., 8, 511 (1969).
307. M. Fild, Z. Anorg. Allgem. Chem., 358, 257 (1968).
308. H. J. Eméleus and J. M. Miller, J. Inorg. Nucl. Chem., 28, 662 (1966).
309. F. W. Bennett, H. J. Eméleus, and R. N. Haszeldine, J. Chem. Soc., 1565 (1953).
310. R. C. Dobbie and P. R. Mason, J. Chem. Soc., Dalton Trans., 1124 (1973).
311. R. C. Dobbie, P. R. Mason, and R. J. Porter, Chem. Commun., 612 (1972).
312. W. Kuchen and W. Grünewald, Chem. Ber., 98, 480 (1965).
313. W. Malisch, R. Janta, and G. Künzel, Z. Naturforsch., 34b, 599 (1979).
314. W. Kuchen, K. Strolenberg, and J. Metten, Chem. Ber., 96, 1733 (1963).
315. W. Kuchen and H. Mayatepek, Chem. Ber., 101, 3454 (1968).
316. W. Kuchen and J. Metten, Angew. Chem., 72, 584 (1960).
317. W. Kuchen, J. Metten, and A. Judat, Chem. Ber., 97, 2306 (1964).
318. K. Issleib and M. Hoffmann, Chem. Ber., 99, 1320 (1966).
319. G. Hägele, W. Kuchen, and H. Steinberger, Z. Naturforsch., 29b, 349 (1974).
320. R. C. Dobbie and M. J. Hopkinson, J. Fluorine Chem., 3, 367 (1974).
321. A. Ecker and U. Schmidt, Chem. Ber., 106, 1453 (1973).
322. A. A. Pinkerton and R. G. Cavell, J. Am. Chem. Soc., 93, 2384 (1971).
323. A. B. Burg and D. M. Parker, J. Am. Chem. Soc., 92, 1898 (1970).
324. A. B. Burg, J. Am. Chem. Soc., 88, 4298 (1966).

325. F. F. Blicke, R. A. Patelski, and L. D. Powers, J. Am. Chem. Soc., 55, 1158 (1933).
326. A. Michaelis, Liebigs Ann. Chem., 321, 141 (1902).
327. B. A. Arbuzov and N. I. Rizpolozhenskii, Izv. Akad. Nauk SSSR, Otd. Khim. Nauk, 854 (1952).
328. B. A. Arbuzov and N. I. Rizpolozhenskii, Dokl. Akad. Nauk SSSR, 89, 291 (1953).
329. A. I. Razumov, O. A. Mukhacheva, and Sim Do-Khen, Izv. Akad. Nauk SSSR, Otd. Khim. Nauk, 894 (1952).
330. W. L. Jensen, US Patent 2,662,917 (1951); C. A., 48, 13711g (1954).
331. K. Gosling and A. B. Burg, J. Am. Chem. Soc., 90, 2011 (1968).
332. M. Fild and T. Stankiewicz, Z. Naturforsch., 29b, 206 (1974).
333. N. G. Feschenko, T. V. Kovaleva, and A. V. Kirsanov, Zh. Obshch. Khim., 42, 287 (1972).
334. S. Z. Ivin and K. V. Karavanov, Zh. Obshch. Khim., 28, 2958 (1958).
335. I. P. Komkov, S. Z. Ivin, and K. V. Karavanov, Zh. Obshch. Khim., 28, 2960 (1958).
336. C. Steubs, W. M. LeSuer, and G. R. Norman, J. Am. Chem. Soc., 77, 3526 (1955).
337. M. P. Osipova, G. Kh. Kamai, and N. A. Chadaeva, Izv. Akad. Nauk, Ser. Khim., 1326 (1969).
338. Yu. F. Gatilov and M. G. Kralichkina, Zh. Obshch. Khim., 38, 1798 (1968).
339. L. S. Sagan, R. A. Zingaro, and K. J. Irgolic, J. Organomet. Chem., 39, 301 (1972).
340. C. Borecki, J. Michalski, and S. Musierowicz, J. Chem. Soc., 4081 (1958).
341. B. A. Arbuzov and D. Kh. Yarmukhametova, Izv. Akad. Nauk SSSR, Otd. Khim. Nauk, 1881 (1960).
342. R. Appel and R. Milker, Chem. Ber., 107, 2658 (1974).
343. O. Foss, Kgl. Norske, Videnskab. Selskabs. Forh., 15, 119 (1942); C. A., 41, 1599f (1947).
344. N. N. Godovikov, M. Kh. Bekanov, M. Kh. Berkhamov, and M. I. Kabachnik, Zh. Obshch. Khim., 44, 1236 (1974).
345. M. I. Kabachnik, S. T. Ioffe, and T. A. Mastryukova, Zh. Obshch. Khim., 25, 684 (1955).
346. M. I. Kabachnik, T. A. Mastryukova, A. E. Shipov, and T. A. Melent'eva, Tetrahedron, 9, 10 (1960).
347. Z. Pelchowicz and H. Leader, J. Chem. Soc., 3320 (1963).
348. S. Bluj, B. Borecka, A. Lopusinski, and J. Michalski, Rocz. Chem., 48, 329 (1974).
349. M. I. Kabachnik, E. N. Tsvetkov, and Yu Chang Chung, Zh. Obshch. Khim., 32, 3340 (1962).
350. F. Fehér and A. Blümcke, Chem. Ber., 90, 1934 (1957).
351. E. E. Nifant'ev and I. V. Shilov, Zh. Obshch. Khim., 43, 2658 (1973).
352. G. E. Smirnova, V. A. Shalydin, and Ya. D. Zel'venskii, Deposited Doc., 1979, VINITI, p. 471; C. A., 92, 197891 (1980).

353. H. Okabe and Y. Takada, Yukagaku, $\underline{28}$, 106 (1979); C. A., $\underline{91}$, 91104e (1979).
354. R. Keat, J. Chem. Soc., Dalton Trans., 2189 (1972).
355. R. Keat, W. Sim, and D. S. Payne, J. Chem. Soc., A, 2715 (1970)
356. O. J. Scherer and W. M. Janssen, J. Organomet. Chem., $\underline{20}$, 111 (1969).
357. R. Jefferson, J. F. Nixon, T. M. Painter, R. Keat, and L. Stobbs, J. Chem. Soc., Dalton Trans., 1414 (1973).
358. R. Burgada, F. Mathis, and M. Bon, C. R. Acad. Sci., Ser. C, $\underline{264}$, 625 (1967).
359. O. J. Scherer and R. Wies, Angew. Chem., $\underline{84}$, 585 (1972); Angew. Chem., Int. Ed. Engl., $\underline{11}$, 529 (1972).
360. H. Nöth and W. Schrägle, Chem. Ber., $\underline{97}$, 2218 (1964).
361. H. Schumann, P. Jutzi, A. Roth, P. Schwabe, and E. Schauer, J. Organomet. Chem., $\underline{10}$, 71 (1967).
362. H. Staudinger and J. Meyer, Helv. Chim. Acta, $\underline{2}$, 635 (1919).
363. A. Schönberg, K. H. Brosowski, and E. Singer, Chem. Ber., $\underline{95}$, 2144 (1962).
364. G. B. Deacon and I. K. Johnson, Inorg. Nucl. Chem. Lett., $\underline{8}$, 271 (1972).
365. C. P. Klages and J. Voss, Angew. Chem., $\underline{89}$, 743 (1977); Angew. Chem., Int. Ed. Engl., $\underline{16}$, 725 (1977).
366. F. Krafft and O. Steiner, Ber., $\underline{34}$, 560 (1901).
367. N. S. Vyazankin, M. N. Bochkarev, and L. P. Sanina, Zh. Obshch. Khim., $\underline{38}$, 414 (1968).
368. M. N. Bochkarev, L. P. Sanina, and N. S. Vyazankin, Zh. Obshch. Khim., $\underline{39}$, 135 (1969).
369. K. Issleib, E. Wenschuh, and B. Fritzsche, Z. Chem., $\underline{5}$, 143 (1965).
370. J. M. McCall and A. Shaver, J. Organomet. Chem., $\underline{193}$, C37 (1980).
371. E. Samuel and C. Giannotti, J. Organomet. Chem., $\underline{113}$, C17 (1976).
372. E. G. Muller, J. L. Peterson, and L. F. Dahl, J. Organomet. Chem., $\underline{111}$, 91 (1976).
373. H. Köpf, B. Block, and M. Schmidt, Chem. Ber., $\underline{101}$, 272 (1968).
374. E. F. Epstein, I. Bernal, and H. Köpf, J. Organomet. Chem., $\underline{26}$, 229 (1971).
375. E. F. Epstein and I. Bernal, J. Chem. Soc., Chem. Commun., 410 (1970).
376. M. M. Chamberlain, G. A. Jabs, and B. B. Wayland, J. Org. Chem. $\underline{27}$, 3321 (1962).
377. R. A. Schunn, C. J. Fritchie, Jr., and C. T. Prewitt, Inorg. Chem., $\underline{5}$, 892 (1966).
378. E. Fischer and S. Riedmüller, Chem. Ber., $\underline{107}$, 915 (1974).
379. E. Lindner, J. Organomet. Chem., $\underline{94}$, 229 (1975).
380. E. Lindner and H. Berke, Chem. Ber., $\underline{107}$, 1360 (1974).
381. H. Köpf and S. K. S. Hazari, Z. Anorg. Allgem. Chem., $\underline{426}$, 49 (1976).

382. T. S. Piper and G. Wilkinson, J. Am. Chem. Soc., 78, 900 (1956).

383. I. Wender and P. Pino (eds.), Organic Syntheses via Metal Carbonyls, Wiley-Interscience, New York—London—Sydney (1968).

384. N. S. Nametkin, V. D. Tyurin, A. I. Nekhaev, and M. A. Kukina, Izv. Akad. Nauk SSSR, Ser. Khim., 2846 (1975).

385. N. S. Nametkin, V. D. Tyurin, and M. A. Kukina, J. Organomet. Chem., 149, 355 (1978).

386. W. Hieber and J. Gruber, Z. Anorg. Allgem. Chem., 296, 91 (1958).

387. C. H. Wei and L. F. Dahl, Inorg. Chem., 4, 493 (1965).

388. N. S. Nametkin, V. D. Tyurin, I. V. Petrosyan, A. V. Popov, B. I. Kolobkov, and A. M. Krapivin, Izv. Akad. Nauk SSSR, Ser. Khim., 2841 (1980).

389. R. S. Gall, C. Ting-Wah Chu, and L. F. Dahl, J. Am. Chem. Soc., 96, 4019 (1974).

390. C. H. Wei, G. R. Wilkes, P. M. Treichel, and L. F. Dahl, Inorg. Chem., 5, 900 (1966).

391. J. M. Birchall, F. L. Bowden, R. N. Haszeldine, and A. B. P. Lever, J. Chem. Soc., A 747 (1967).

392. L. Markó, G. Bör, and G. Almásy, Chem. Ber., 94, 847 (1961).

393. L. Markó, G. Bör, and E. Klumpp, Chem. Ind., 1491 (1961).

394. C. H. Wei and L. F. Dahl, Inorg. Chem., 6, 1229 (1967).

395. L. Markó, G. Bör, E. Klumpp, B. Markó, and G. Almásy, Chem. Ber., 96, 955 (1963).

396. D. L. Stevenson, V. R. Magnuson, and L. F. Dahl, J. Am. Chem. Soc., 89, 3727 (1967).

397. S. A. Khattab, L. Markó, G. Bör, and B. Markó, J. Organomet. Chem., 1, 373 (1964).

398. V. A. Uchtman and L. F. Dahl, J. Am. Chem. Soc., 91, 3756 (1969).

399. G. L. Simon and L. F. Dahl, J. Am. Chem. Soc., 95, 2164 (1973).

399a. Ch. Buschka, K. Leonhard, and H. Werner, Z. Anorg. Allgem. Chem., 464, 30 (1980).

400. Y. Wakatsuki, H. Yamazaki, and H. Iwasaki, J. Am. Chem. Soc., 95, 5781 (1973).

401. Y. Wakatsuki, T. Kuramitsu, and H. Yamazaki, Tetrahedron Lett., 4549 (1974).

402. Y. Wakatsuki and H. Yamazaki, Jpn. Patent 74,100,073 (1973); C. A., 82, 86419 (1975).

403. K. W. Hübel and E. H. Braye, US Patent 3,280,017 (1966); C. A., 66, 2463 (1967).

404. C. Giannotti, C. Fontaine, B. Septe, and D. Doue, J. Organomet. Chem., 39, C74 (1972).

405. C. F. Smith and C. Tamborski, J. Organomet. Chem., 32, 257 (1971).

406. H. Varenkamp, V. A. Uchtman, and L. F. Dahl, J. Am. Chem. Soc., 90, 3272 (1968).

407. A. P. Ginsberg and W. E. Lindsell, Chem. Commun., 232 (1971).

408. W. D. Bonds and J. A. Ibers, J. Am. Chem. Soc., 94, 3413
      (1972).
409. T. E. Nappier, Jr., D. M. Meek, R. M. Kirchner, and J. A.
      Ibers, J. Am. Chem. Soc., 95, 4194 (1973).
410. T. E. Nappier, Jr., and D. M. Meek, J. Am. Chem. Soc., 94,
      306 (1972).
411. Y. Wakatsuki and H. Yamazaki, J. Organomet. Chem., 64, 393
      (1974).
412. W. Rigby, P. M. Bailey, J. O. McCleverty, and P. Maitlis,
      J. Chem. Soc., Dalton Trans., 371 (1979).
413. E. Müller and W. Winter, Liebigs Ann. Chem., 608 (1975).
414. E. Müller, E. Luppold, and W. Winter, Chem. Ber., 108, 237
      (1975).
415. E. Müller and W. Winter, Liebigs Ann. Chem., 1876 (1975).
416. J. Hambrecht and E. Müller, Liebigs Ann. Chem., 387 (1977).
417. A. Scheller, W. Winter, and E. Müller, Liebigs Ann. Chem.,
      1448 (1976).
418. J. Hambrecht and E. Müller, Z. Naturforsch., 32b, 68 (1977).
419. A. L. Balch, L. S. Benner, and M. M. Olmstead, Inorg. Chem.,
      18, 2996 (1979).
420. M. P. Brown, J. R. Fisher, R. J. Puddephatt, and K. R. Seddon,
      Inorg. Chem., 18, 2808 (1979).
421. M. P. Brown, J. R. Fisher, R. J. Puddephatt, and K. R. Seddon,
      J. Chem. Soc., Chem. Commun., 749 (1978).

Subject Index